Bacteriophages in Health and Disease

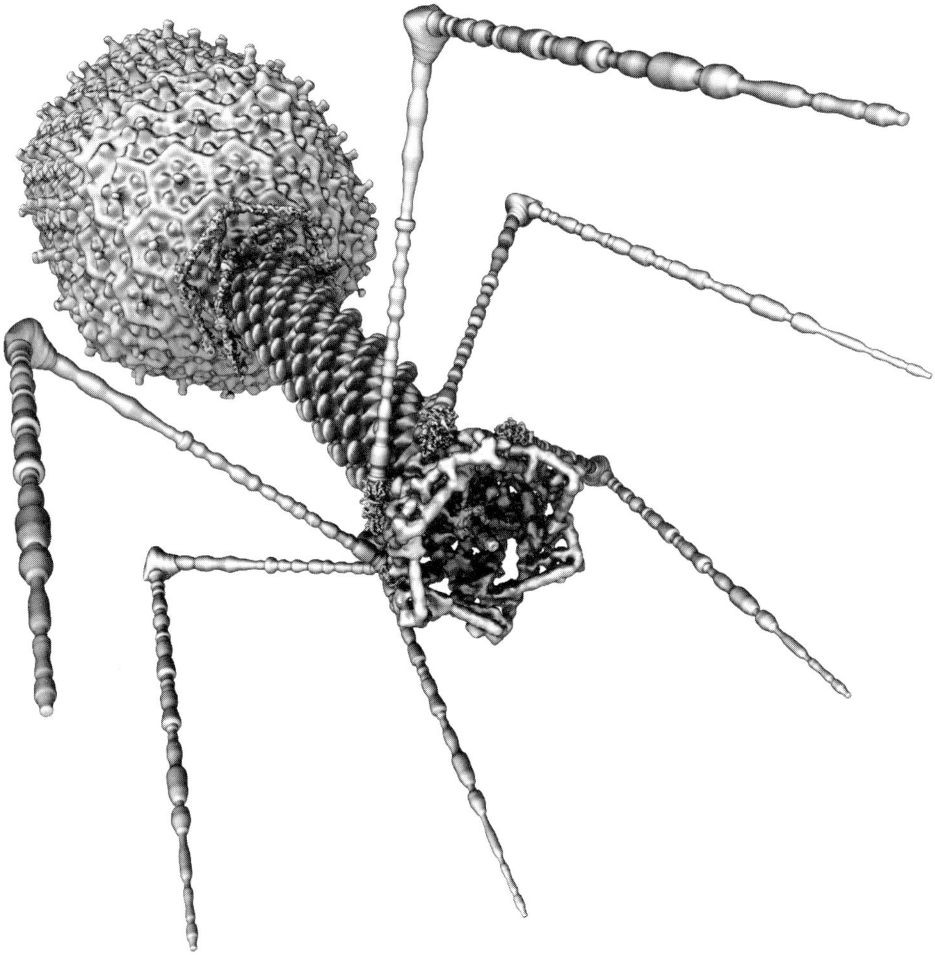

Bacteriophage T4 seen obliquely from the base plate end

Picture created by Steven McQuinn using protein structure data from Rossmann Lab, furnished by the RCSB Protein Databank. The various software utilized includes UCSF Chimera, Accelrys DS Visualizer, MeshLab and DAZ Carrara.

Advances in Molecular and Cellular Microbiology 24

Bacteriophages in Health and Disease

Edited by

Paul Hyman

Department of Biology/Toxicology, Ashland University

and

Stephen T. Abedon

Department of Microbiology, The Ohio State University

www.cabi.org

amcm **Advances in Molecular and Cellular Microbiology**

Through the application of molecular and cellular microbiology we now recognise the diversity and dominance of microbial life forms on our planet, that exist in all environments. These microbes have many important planetary roles, but for we humans a major problem is their ability to colonise our tissues and cause disease. The same techniques of molecular and cellular microbiology have been applied to the problems of human and animal infection during the past two decades and have proved to be immensely powerful tools in elucidating how microorganisms cause human pathology. This series has the aim of providing information on the advances that have been made in the application of molecular and cellular microbiology to specific organisms and the diseases that they cause. The series is edited by researchers active in the application of molecular and cellular microbiology to human disease states. Each volume focuses on a particular aspect of infectious disease and will enable graduate students and researchers to keep up with the rapidly diversifying literature in current microbiological research.

Series Editor

Titles Available from CABI

17. *Helicobacter pylori* in the 21st Century
Edited by Philip Sutton and Hazel M. Mitchell

18. Antimicrobial Peptides: Discovery, Design and Novel Therapeutic Strategies
Edited by Guangshun Wang

19. Stress Response in Pathogenic Bacteria
Edited by Stephen P. Kidd

20. Lyme Disease: an Evidence-based Approach
Edited by John J. Halperin

22. Antimicrobial Drug Discovery: Emerging Strategies
Edited by George Tegos and Eleftherios Mylonakis

24. Bacteriophages in Health and Disease
Edited by Paul Hyman and Stephen T. Abedon

Titles Forthcoming from CABI

Tuberculosis: Diagnosis and Treatment
Edited by Timothy McHugh

Microbial Metabolomics
Edited by Silas Villas-Bôas and Katya Ruggiero

The Human Microbiota and Microbiome
Edited by Julian Marchesi

Meningitis: Cellular and Molecular Basis
Edited by Myron Christodoulides

Earlier titles in the series are available from Cambridge University Press (www.cup.cam.ac.uk).

CABI is a trading name of CAB International

CABI
Nosworthy Way
Wallingford
Oxfordshire, OX10 8DE
UK

Tel: +44 (0)1491 832111
Fax: +44 (0)1491 833508
E-mail: info@cabi.org
Website: www.cabi.org

CABI
875 Massachusetts Avenue
7th Floor
Cambridge, MA 02139
USA

T: +1 800 552 3083 (toll free)
T: +1 (0)617 395 4051
E-mail: cabi-nao@cabi.org

Cover image. The electron micrograph on the cover is of Salmonella phage bound to cell wall residues forming a 'phage bouquet'. The phage were negatively stained with uranyl acetate. This image was prepared by Jochen Klumpp at the Institute of Food, Nutrition and Health, Zurich, Switzerland.

A catalogue record for this book is available from the British Library, London, UK.

Library of Congress Cataloging-in-Publication Data

Bacteriophages in health and disease / [edited by] Paul Hyman, Stephen T. Abedon.
p. ; cm. -- (Advances in molecular and cellular microbiology ; 24)
Includes bibliographical references and index.
ISBN 978-1-84593-984-7 (alk. paper)
I. Hyman, Paul (Paul Lawrence) II. Abedon, Stephen T. III. C.A.B. International.
IV. Series: Advances in molecular and cellular microbiology ; 24.
[DNLM: 1. Bacteriophages. 2. Biological Therapy--methods. QW 161]

579.2′6--dc23

2012007325

ISBN-13: 978 1 84593 984 7

Commissioning editor: Rachel Cutts
Editorial assistant: Alexandra Lainsbury
Production editor: Simon Hill

Typeset by Columns Design XML, Reading.
Printed and bound in the UK by MPG Books Ltd.

Contents

* Corresponding author

Contributors

Stephen T. Abedon, Ph.D., Department of Microbiology, The Ohio State University, 1680 University Drive, Mansfield, OH 44906, USA; abedon.1@osu.edu

Heather E. Allison, Ph.D., Department of Functional and Comparative Genomics, Institute of Integrative Biology, University of Liverpool, Liverpool L69 7ZB, UK.

James Peter Beck, M.D., Department of Orthopaedics, The University of Utah, 590 Wakara Way, Salt Lake City, UT 84108, USA; George E. Wahlen Department of Veterans Affairs Medical Center, VA Salt Lake City Health Care System, Salt Lake City, UT 84148, USA.

Jan Borysowski, M.D., Department of Clinical Immunology, Institute of Transplantology, Medical University of Warsaw, ul. Nowogrodzka 59, 02-006 Warsaw, Poland.

Ben Burrowes, Ph.D., Ampliphi Biosciences Corporation, Colworth Science Park, Sharnbrook, Bedfordshire, MK44 1LQ, UK.

Todd R. Callaway, Ph.D., Food and Feed Safety Research Unit Agricultural Research Service/ USDA, TX, USA; Todd.Callaway@ars.usda.gov

Gail E. Christie, Ph.D., Molecular Biology and Genetics, School of Medicine, Virginia Commonwealth University, Richmond, VA 23298-0678, USA.

Jason Clark, Ph.D., Novolytics Ltd, ITAC-BIO Daresbury Laboratory, Daresbury Science and Innovation Campus, Warrington, WA4 4AD, UK; BigDNA Ltd, Wallace Building, Roslin BioCentre, Roslin, Midlothian EH25 9PP, UK.

Christopher R. Cox, Ph.D., Colorado School of Mines, Department of Chemistry and Geochemistry, Golden, CO 80401, USA; crcox@mines.edu

Krystyna Dąbrowska, Ph.D., Phage Laboratory and Therapy Unit, Institute of Immunology and Experimental Therapy, Polish Academy of Sciences, ul. Rudolfa Weigla 12, 53-114 Wroclaw, Poland.

David M. Donovan, Ph.D., Animal Biosciences and Biotechnology Lab, ANRI, ARS, USDA, Bldg 230, Room 104, BARC-East 10300 Baltimore Ave, Beltsville, MD 20705, USA.

Lawrence D. Goodridge, Ph.D., Center for Meat Safety and Quality, Department of Animal Sciences, Colorado State University, Fort Collins, CO 80523, USA; Lawrence.Goodridge@ColoState.edu

Andrzej Górski, M.D., Ph.D., Phage Laboratory and Therapy Unit, Institute of Immunology and Experimental Therapy, Polish Academy of Sciences, ul. Rudolfa Weigla 12, 53-114 Wroclaw, Poland; Department of Clinical Immunology, Institute of Transplantology, Medical University of Warsaw, ul. Nowogrodzka 59, 02-006 Warsaw, Poland.

David R. Harper, Ph.D., Ampliphi Biosciences Corporation, Colworth Science Park, Sharnbrook, Bedfordshire, MK44 1LQ, UK; drh@ampliphibio.com

Heather Hendrickson, Ph.D., New Zealand Institute for Advanced Study, Massey University, Private Bag 102 904, North Shore Mail Centre, Auckland, New Zealand; former address: Microbiology Unit, Department of Biochemistry, University of Oxford, South Parks Rd, Oxford, OX1 3QU, UK; H.Hendrickson@massey.ac.nz

Paul Hyman, Ph.D., Department of Biology/Toxicology, Ashland University, 401 College Avenue, Ashland, OH 44805, USA; phyman@ashlands.edu

Andrew M. Kropinski, Ph.D., Department of Molecular and Cellular Biology, University of Guelph, Guelph, ON N1G 2W1, Canada; Public Health Agency of Canada, Laboratory for Foodborne Zoonoses, Guelph, ON N1G 3W4, Canada; kropinsk@queensu.ca

Sarah Kuhl, M.D., Ph.D., VA Northern California Health Care System, Martinez, CA 94553, USA; Contra Costa Regional Medical Center, Martinez, CA 94553, USA; sarah.kuhl@va.gov

John Kuzio, Ph.D., Department of Microbiology and Immunology, Queen's University, Kingston, ON K7L 3N6, Canada.

Rob Lavigne, Ph.D., Division of Gene Technology, Department of Biosystems, Katholieke Universiteit Leuven, Kasteelpark Arenberg 21 Box 2462, B-3001 Leuven, Belgium; rob.lavigne@biw.kuleuvenibe.be

Jeffrey T. LeJeune, D.V.M., Ph.D., Food Animal Health Research Program, The Ohio Agricultural Research and Development Center, Wooster, OH 44691, USA; lejeune.3@osu.edu

Andrey V. Letarov, Ph.D., Winogradsky Institute of Microbiology, Russian Academy of Sciences, Moscow, Russia; letarov@gmail.com

Catherine Loc-Carrillo, Ph.D., Department of Orthopaedics, The University of Utah, 590 Wakara Way, Salt Lake City, UT 84108, USA; George E. Wahlen Department of Veterans Affairs Medical Center, VA Salt Lake City Health Care System, Salt Lake City, UT 84148, USA; C.Loc.Carrillo@hsc.utah.edu

Tim A. McAllister, Ph.D., Agriculture and Agri-Food Canada, Lethbridge, AB, Canada.

W. Michael McShan, Ph.D., Department of Pharmaceutical Sciences, College of Pharmacy, The University of Oklahoma, Oklahoma City, OK 73126-0901, USA.

Carl R. Merril, M.D., 6840 Capri Place, Bethesda, MD 20817, USA.

Michael S. Mitchell, Ph.D., Institute for Bioscience and Biotechnology Research, University of Maryland, Rockville, MD 20850, USA.

Daniel C. Nelson, Ph.D., Institute for Bioscience and Biotechnology Research, University of Maryland, Rockville, MD 20850, USA; Department of Veterinary Medicine, University of Maryland, and Virginia-Maryland Regional College of Veterinary Medicine, College Park, MD 20742, USA; nelsond@umd.edu

Yan D. Niu, Ph.D., Agriculture and Agri-Food Canada, Lethbridge, AB, Canada.

Natasza Olszowska-Zaremba, M.S., Department of Clinical Immunology, Institute of Transplantology, Medical University of Warsaw, ul. Nowogrodzka 59, 02-006 Warsaw, Poland; nataszaolszowska@wp.pl

Yang Shen, Ph.D., Institute for Bioscience and Biotechnology Research, University of Maryland, Rockville, MD 20850, USA.

Don L. Siegel, Ph.D., M.D., Division of Transfusion Medicine, Department of Pathology and Laboratory Medicine, Perelman School of Medicine, University of Pennsylvania, 510 Stellar-Chance Laboratories, 422 Curie Blvd, Philadelphia, PA 19104, USA; siegeld@mail.med.upenn.edu

Kim Stanford, Ph.D., Alberta Agriculture and Rural Development, Lethbridge, AB, Canada.

Travis Steiner, B.S., Center for Meat Safety and Quality, Department of Animal Sciences, Colorado State University, Fort Collins CO 80523, USA.

Jeroen Wagemans, M.S., Division of Gene Technology, Department of Biosystems, Katholieke Universiteit Leuven, Kasteelpark Arenberg 21 box 2462, B-3001 Leuven, Belgium.

Matthew K. Waldor, Ph.D., Department of Medicine, Brigham and Women's Hospital, Boston, MA 02115, USA; and, Channing Laboratory and HHMI, Harvard University, Boston, MA 02115, USA.

Michele L. Williams, D.V.M., Ph.D., Food Animal Health Research Program, The Ohio Agricultural Research and Development Center, Wooster, OH 44691, USA.

Sijia Wu, Department of Orthopaedics, The University of Utah, 590 Wakara Way, Salt Lake City, UT 84108. USA; George E. Wahlen Department of Veterans Affairs Medical Center, VA Salt Lake City Health Care System, Salt Lake City, UT 84148, USA.

Foreword

It has been personally gratifying to examine the sections and chapters that make up this book, 'Bacteriophages in Health and Disease'. The book represents real progress since the now almost half a century when I was first introduced to the bacteriophages at a Cold Spring Harbor Phage Course in 1965. At that time, the course concentrated on just a few phage strains, as these viruses and their bacterial hosts were seen as a potential gateway for the understanding of basic life processes. As I had recently graduated from medical school, I raised the question as to why these bacterial viruses were not employed in antibacterial therapy. One of the course mentors answered that question by suggesting that I read Sinclair Lewis', 1925, book 'Arrowsmith'. On reading the book, it soon became apparent that the fictionalized clinical trials described in the book failed not because of deficiencies in the antibacterial phage strains but because the 'hero'/'antihero' chose to break the double-blind code, so that he could treat all the inhabitants on the fictional island who were in the midst of a major plague epidemic. In other words, the actions of the fictionalized character, 'Arrowsmith', destroyed the capacity of the clinical trial to determine the efficacy of phage therapy, so the book really could not address whether the phage were useful as an antibacterial therapy or not.

At the time of the 1965 course, mouth pipetting represented the state of the art of laboratory technology, so I asked whether the bacteriophage could have an adverse effect on the students in the course. The reply, by one of the instructors, was that they could not because: 'they are bacteriophage, which means bacteria eaters'. I found this reply unsatisfactory, as it occurred to me that the viruses might not be restricted in their actions by a name (bacteriophage) that we had given them. Following the phage course, I returned to my laboratory at the NIH, where, with the help of colleagues, I was able to screen for effects of phage strains on mammalian cells in culture. We found no clear-cut gross effects with the phage strains we tested, such as cell killing or alterations in cellular growth patterns. When transducing phages carrying the bacterial galactose genes were used with human cells defective in certain galactose genes, however, we found evidence for restoration of the defective galactose pathway. In addition, in control experiments we discovered that most of our stocks of fetal calf serum used as an ingredient in mammalian cell cultures were contaminated by bacteriophage. At first we thought that this contamination was limited to our lab, but we soon found similar contamination in all of the fetal calf serum samples that we were able to sample from our colleagues at the NIH.

As we realized that many human vaccines are produced by mammalian cell cultures grown in fetal calf serum, we tested some vaccines that we purchased at a local drugstore and found that they were also contaminated by phages. The FDA confirmed our results. To permit the continued use of the vaccines they obtained a Presidential Executive Order permitting the sale and use of bacteriophage-contaminated vaccines, 'since bacteriophage were known to be specific for bacteria'. In a subsequent presentation, at the NIH, I noted that phage are not entirely innocuous, such as in the case of diphtheria, a disease which is caused by a toxin gene that is carried into and expressed in the infecting bacteria by a phage. Lest we forget, this toxin has enzymatic activity that results in the inactivation of a critical mammalian protein synthesis factor and which has no known effects in bacteria.

We also initiated experiments to study phages in germ-free animals, to simplify the system and to look for interactions between phage and animals. It was in these experiments that we recognized the immediate ability of animals to respond to phages even before antibody production could be initiated. Most phages are destroyed in the liver while a small number remained intact in the spleen (they could still infect bacteria when isolated from the spleens of these animals). It was these experiments that gave us ideas that we used in subsequent experiments to select phage strains that could remain in the circulation to optimize their capacity to serve as antibacterial therapeutic agents.

While our carefully controlled studies using phages to treat animals with experimental systemic infections were very encouraging they did not generate support from our administrators, who often cited the book, 'Arrowsmith', which they clearly hadn't read, as their evidence against phage therapy. In addition, an argument that was raised repeatedly was that 'anyone who worked with phages in a lab should know that phage therapy would never be useful because the development of phage-resistant bacterial strains is inevitable'. I guess they didn't know that this is true for most antibacterial agents including most antibiotics, but a number of lives can be saved in the meantime. Besides, with our capacity to isolate new phage strains and engineer existing ones it should be possible to overcome most phage-resistant bacteria. Of course, we should always be aware of possible phage-mediated detrimental effects such as the ability of phage to carry toxin genes, as in the case of diphtheria. In addition, as noted in the last chapter in this book, on 'Phage-therapy Best Practices' (see Abedon, Chapter 17, this volume), careful considerations also need to be used in developing phage protocols and formulations.

For the above reasons and experiences I am personally pleased to see that despite the naysayers, efforts to use phages as therapeutic agents, both as antibacterial agents and as therapeutic 'delivery vehicles' (see Clark *et al.*, Chapter 7, this volume), have substantially progressed. Studies of phages in the human microbiome may facilitate the development of new therapeutic phage strains and therapeutic strategies. Most importantly, careful clinical trials are being designed and conducted as described in several chapters to help us avoid falling into the trap that undid 'Arrowsmith'. In time, truth will overcome the shadow of that fiction.

Carl R. Merril

Preface

If all things have to have a beginning, then the beginning of this monograph, arguably, can be traced to the 1980s in the laboratory of Harris Bernstein (Department of Microbiology and Immunology, University of Arizona). That was the lab that both of us joined to do our Ph.D. research and where we met. Though we worked on separate projects, both involved phages and this led to many discussions. Thus the team began. After graduating we went our separate ways. While S.T.A. dabbled in what eventually would be all things phage ecological, P.H. followed a much more molecular, then medical, then applied route, but ultimately returned to phage biology and what would be a series of collaboration with S.T.A., starting in 2001. The story of how this monograph came to be is slightly more complicated, however.

S.T.A. – at the time at the beginning of his career at the Ohio State University, in 1995 – founded what would become the Bacteriophage Ecology Group. His management of the associated web site (phage.org, though see also archaealviruses.org) began to lead to chapter invitations in the early 2000s, resulting in part in his contributing to the editing of Rich Calendar's *The Bacteriophages* 2/e, which was published in 2006. Very soon after S.T.A.'s contribution to that monograph ended he was invited to edit a monograph on phage ecology, which would become the 2008 *Bacteriophage Ecology* (Cambridge University Press), included in which was a chapter we coauthored.

Meanwhile, speculators brought down the world's economy, hitting academic publishers quite hard. The series that *Bacteriophage Ecology* had been a part of, *Advances in Molecular and Cellular Microbiology*, was sold to CABI Press (cabi.org), 'a not-for-profit international organization that improves people's lives by providing information and applying scientific expertise to solve problems in agriculture and the environment.' It was 100% their idea to do the current monograph and, based on his experience with the *Bacteriophage Ecology* volume, S.T.A. was recruited to edit it. S.T.A., though, was busy at the time, pulling together an edited volume on phage therapy (see *Current Pharmaceutical Biotechnology*, volume 11, issue 1), so recruited P.H. to join in on yet another collaboration. We developed a formal proposal including recruiting the authors of the many chapters found herein and the result is this monograph.

Bacteriophages in Health and Disease, an early working title that stuck, is an effort to provide an introduction to the breadth of roles that phages play or can play in our everyday lives. They can serve as causes of disease, treatments of disease, indicators of the potential for disease, preventers of disease, and even contribute in various ways to the evolution of bacterial pathogens. To capture this variety of phage roles in human conditions, both natural and applied, we have divided the text into three parts. We've also provided a brief introduction to various concepts and terminology associated with phages (chapter 1); for a glossary covering these basics of phage biology, and much more, see phage.org/terms.

Part I considers the role of phages in the natural state. That is, where phages are, how they contribute directly to disease, the underlying mechanism by which phages do this, and then, especially, how they can contribute *indirectly* to disease, that is, to pathogen evolution. The three basic themes are ones of phage presence, phage genes, and phage-mediated transduction of bacterial DNA between bacteria, though a common thread is that they touch upon, in various ways, issues of lysogeny, which is the long-term incorporation of phages as prophages into the genomes of their bacterial hosts. These issues are covered by chapters 2 through 6.

Part II, chapters 7 through 11, considers various phage-based technologies other than the use of whole phages to combat bacterial infections (i.e., besides phage therapy). This includes in particular the use of both modified and 'disembodied' phage parts. Phages thus can serve as carriers and delivery vehicles of especially DNA but also of other chemicals, including serving as vectors for either gene therapy or DNA vaccines. The potential of phages to serve in these functions stems greatly from their relative safety as well as their genetic malleability, such as one sees for example in phage display technologies. Phage antibacterial properties too can be used as a means of discovering novel targets for action by traditional small-molecule antibacterial chemotherapeutics. In addition, phages can be employed in bacterial detection and identification including as indicators of faecal contamination.

Phages – typically in a relatively unaltered state – can be employed directly as antibacterials, that is, in phage therapy, a theme covered by Part III and chapters 12 through 17. Here we provide a chapter introducing various phage properties as medicinals, including the relative safety associated with phage application to bodies. Phage therapy of humans is then considered in two chapters, which we have divided up roughly into treatment of wounds and non-wound infections. Certain phage parts can be used, on their own, as antibacterial drugs, most notably phage lysins which are covered in a subsequent chapter. The potential phage role in food safety is considered, and we then end the monograph with a chapter targeted to would-be phage therapy experimentalists, one that considers, in light of phage properties, especially how phage therapy protocols may be developed in terms of the use of animal models of bacterial disease.

If your primary interest is in learning about the myriad roles that phages can and do play in human health and disease, read all of the chapters perhaps up to this last one. If you already know all there is to know about phages such that much of this book is superfluous, you may still want to at least read this last chapter! In any case, we see this monograph as appealing to the needs of two not-so-disparate groups, those with a primary interest in phage biology, though from an especially applied perspective, and those who are curious as to how phage biology can impact the practice of medicine. In particular, we provide a snap shot of the current state of the field of phage biology, applied as well as basic, concentrating on the roles that phages can play in both health and disease.

We thank our editors at CAB International, Rachel Cutts and Alex Lainsbury, for their support and help bringing this volume to fruition and dedicate this volume to our common but nonetheless not-so-common mentor, Harris Bernstein.

1 Phages

Stephen T. Abedon[1]
[1]*Department of Microbiology, The Ohio State University*

In a phrase such as 'Bacteriophages in Health and Disease', the least familiar term to anyone will be that of bacteriophages. It is the purpose of this introductory chapter to make sure that this term is less unfamiliar – even to those working in the field – so that greater emphasis may be made, in this monograph, on issues of health and disease. In this chapter, I thus will provide a brief and also not terribly molecular overview of the viruses of bacteria. Greater detail as applicable to particular topics will be found in appropriate chapters. For additional discussion of concepts, see www.phage.org/terms/, as well as the resources list presented by www.ISVM.org. For general references, see www.phage.org/terms/phages.html and www.phage.org/terms/phage_history.html.

Bacteriophages, or phages for short, are viruses that infect members of what is known as the domain Bacteria, which are the common prokaryotic organisms associated with, for example, the human body. To help orient readers towards understanding the role that phages can play as part of the human microbiome (see Letarov, Chapter 2, this volume), in health as well as disease (see Kuhl *et al.*, Chapter 3, this volume), in this chapter I provide an overview of basic phage properties. These include, in particular, what phage virions 'look' like, the basic characteristics of phage infections and, because it can play such a large role in bacterial evolution, the potential for phages to move DNA between bacteria, a process known as transduction. I begin, however, with a brief glimpse at the history of phage research.

History

Although a number of publications in the late 19th and early 20th century may hint at the observation of phage-like phenomena, the discovery of phages is unambiguously traced to the work of Twort and, independently, that of d'Hérelle (see the first volume of the journal, *Bacteriophage*, 2011, particularly issues 1 and 3, for discussion and references). Early phage work by necessity addressed issues of basic phage biology, although these efforts were based on both primitive techniques and minimal understanding of just what the phage phenomenon actually entailed. The promise of phage therapy, with the use of phages to combat bacterial infections in particular, was an important driver of early phage research (see Burrowes and Harper, Chapter 14, this volume, for additional historical consideration of this issue).

It was in the late 1930s and early 1940s that phage research began to make what today we view as its most significant impact on basic biological research. These efforts are associated in large part with the physicist-by-training Max Delbrück and associated members of what became known as the Phage Group. It was within this context that the field of molecular genetics and the related discipline of molecular biology were forged. James Watson, for example, was a member of the Phage Group, and Francis Crick too became an important contributor to phage research following their co-discovery of the structure of DNA.

By contributing to the development of techniques that ultimately would greatly simplify the study of the biology of organisms far more complicated than either phages or their bacterial hosts, in a sense phage research set the stage for its own eclipse, perhaps particularly in terms of funding opportunities. Changes in this situation involved an increased focus on areas other than purely the molecular aspects of phages: consideration of aquatic phage ecology, growing appreciation of the role of phages in horizontal gene transfer and increasing emphasis on the application of phage-based biotechnologies. The latter includes in particular techniques known as phage display (see Siegel, Chapter 8, this volume), the use of phages in bacterial detection as well as identification (see Williams and LeJeune, Cox, and Goodridge and Steiner, Chapters 6, 10 and 11, this volume) and, of course, phage therapy (see Olszowska-Zaremba *et al.*, Loc-Carrillo *et al.*, Burrowes and Harper, Shen *et al.*, Niu *et al.* and Abedon, Chapters 12–17, this volume). All of these issues are considered in this monograph, particularly in the last two parts. The association of phages with bacterial disease is a primary emphasis of Part I (see Kuhl *et al.*, Christie *et al.* and Hendrickson, Chapters 3–5, this volume).

Types of phages

Virus particles consist essentially of two components, what is on the inside and what is on the outside. What is on the inside primarily is nucleic acid and what is on the outside is responsible for transporting that nucleic acid between cells. The nucleic acid component varies substantially between phage types, both in terms of size and structure, and can consist of either RNA or DNA, but not both, with DNA being more common. It can be single-stranded or double-stranded, although for most phages it is double-stranded. It also can be multi-segmented, particularly tripartite (e.g. *Pseudomonas* phage ϕ6), although the vast majority of phage genomes are monopartite.

The outside portion of virions consists of a combination of the capsid, which surrounds the nucleic acid, and various appendages, most of which are involved in virion adsorption to bacteria. While some phages have capsids that contain lipids (again such as phage ϕ6), for most phages capsids consist solely of multiple units of proteins known as capsomeres. For simple phage virions, capsids are either helical/filamentous or icosahedral, which in both cases surround single-stranded nucleic acids (usually DNA, but not always). These phages fall into the phage families *Leviviridae*, *Microviridae* and *Inoviridae*. More specifically, these are single-stranded RNA icosahedral, single-stranded DNA icosahedral and ssDNA filamentous phages (e.g. phages MS2, ϕX174 and M13, respectively). A very small number of phages have dsRNA genomes (such as ϕ6).

The vast majority of phages by contrast possess complex rather than simple virion morphologies; dsDNA, monopartite genomes; and protein capsids lacking lipids. The defining feature of these phages – all members of phage order *Caudovirales* – are their tails, however. Tails are multi-protein appendages involved in virion adsorption (see the cover and prelims for illustration). Members of the phage family *Podoviridae*, such as coliphage T7, possess short, non-contractile tails. The tails of members of the family *Siphoviridae*, including coliphage λ, can be quite long and are also non-contractile. Finally, the tails of members of the family *Myoviridae*, coliphage T4 being the most familiar, are approximately intermediate in length while possessing an ability to contract in the course of virion adsorption to bacteria. All tailed phages are lytic and some are also temperate. Discussion of basic phage morphologies in greater detail can be found in a number of reviews that have been published by Hans-Wolfgang Ackermann.

Virus-like particles

The concept of 'virus-like particle', or VLP, has at least two distinct meanings as used in this monograph. First are entities that appear to be virus-like as viewed in an electron micrograph (see Letarov, Chapter 2, this volume). Secondly, the term is used to describe phages that lack genomes and which therefore are 'virus-like' rather than actually viruses (see Clark *et al.*, Chapter 7, this volume). The term 'ghost' has also been employed in the phage literature to describe these latter entities, including those generated from particles that initially possess genomes.

Types of infection

A phage infection begins with virion attachment to a potential host cell, the culmination of a multi-step process known as adsorption. It then proceeds through the translocation of phage nucleic acid into the cell – variously described as both ejection and injection, as well as uptake – but begins in earnest only once the phage genome has made its way into the bacterial cytoplasm. The results of phage adsorption can vary depending on the characteristics of the phage, the bacterium and circumstances. At its most basic level, the phage may either live or die (that is, produce or not produce replicative products after infecting a host cell), and the same is true for the infected bacterium. Furthermore, and depending on infection type, all four combinations of living versus dying are possible. See Cox (Chapter 10, this volume) for additional discussion of phage infections.

In an abortive infection, both the infecting phage and infected bacterium die. In a lytic infection, the infecting phage lives, producing phage virions, while the infected bacterium both dies and is lysed. In various other circumstances, the infecting phage can die but not the infected bacterium, which is seen particularly when bacteria carry restriction endonucleases, although also when they carry the CRISPR/*cas* systems. Lastly, under certain circumstances both phage and bacterium can live. While one means by which this co-survival can occur is seen with the chronic infections of members of the phage family *Inoviridae*, most commonly simultaneous survival of both infecting phage and infected bacterium is seen with phage exhibition of lysogeny (below). Productive infections, those quickly leading to the release of infectious viral progeny, normally occur after infection by both lytic and chronic phages. Overall, the majority of phages are lytic phages.

Once the virion genome of a lytic virus gets into the bacterial cytoplasm, the phage genes are expressed. This gene expression has the effect of taking over the bacterial metabolism so that phage virion particles are produced, the proteins of which are encoded by additional phage genes (see Wagemans and Lavigne, Chapter 9, this volume, for greater consideration of what goes on, at a molecular level, within a phage-infected bacterium). The products of yet more phage genes contribute, at the end of what is known as the phage latent period, to the destruction of the bacterial cell envelope so that phage progeny can leak into the extracellular environment. For tailed phages, this lysis process involves coordinated action by at least two phage gene products, the holin and endolysin or lysin proteins. In particular, the holin protein is responsible for controlling the timing of host lysis, stimulating the action of the lysin, and, as a by-product, shutting down infection and therefore host metabolism. Lysin is the enzyme that is responsible for degradation of the bacterial cell wall. See Shen *et al.* (Chapter 15, this volume) for additional discussion of phage lysins.

Lysogeny

The other major state that involves both successful phage infection *and* bacterial survival is known as the lysogenic cycle, a phenomenon, as noted in the preface, that serves to unite the chapters found in Part I of this volume. All phages that can display lysogeny are described as temperate, and the majority of temperate phages are tailed (although a small number are instead filamentous, i.e. members of the phage family *Inoviridae*; see Christie *et al.*, Chapter 4, this volume, for discussion of a prominent member of the latter, *Vibrio cholerae* phage CTXϕ). Lysogenic cycles are characterized by two features. First, the phage genome, now called a prophage, is replicated sufficiently rapidly within infected bacteria that daughter bacteria, following binary fission, each inherit at least one prophage copy. Secondly, the infections are not productive, that is, no virions are produced.

The prophage can exist either integrated into the bacterial chromosome or as a plasmid. Integration is typically accomplished by the action of a phage protein termed an integrase. Integrases generally bind to a specific site on the bacterial genome and a corresponding, partially homologous site in the phage genome. The result is site-specific recombination to integrate the phage genome into the bacterial genome.

A bacterium that can undergo a lysogenic cycle is described as a lysogen, the process of lysogen formation is called lysogenization and the conversion of a lysogenic infection into a productive one (typically a lytic one) is called induction. In some lysogens, the prophage only expresses genes whose proteins are needed to prevent induction or, instead, trigger induction upon receiving an appropriate signal. Expression of the repressor proteins in particular prevents such induction but also has the effect of blocking the infection of lysogens by similar (known as homoimmune) phages, a process known as superinfection immunity. Other prophages also express genes that can alter the phenotype of the lysogenic bacterium, a process called lysogenic conversion (see Kuhl *et al.* and Christie *et al.*, Chapters 3 and 4, this volume).

Common terms

To avoid ambiguity when considering different types of phage infection, I provide explicit definitions of the following terms:

- Temperate – description of a phage that is capable of displaying a lysogenic cycle; all temperate phages and indeed all phages also display productive cycles at some point in their life cycles. Note that 'lysogenic phage' is not a synonym for 'temperate phage' but instead is a misnomer (bacteria can be lysogenic, while phages are temperate).
- Lytic – description of a phage that lyses its host in the course of productive infection; note that most temperate as well as most non-temperate phages are lytic phages.
- Chronic – description of a phage that does not lyse its host in the course of productive infection; these phages are released by crossing relatively intact bacterial cell envelopes.
- Obligately lytic – description of a functional phage that is not capable of displaying either lysogenic or chronic infections.
- Professionally lytic – description of a phage that is both obligately lytic and not recently descended from temperate phages.
- Virulent – a common synonym of obligately lytic, although it can also describe the potential for a phage isolate to bring a population of target bacteria under control, particularly through infection that is followed by bacterial lysis and associated phage population growth.
- Phage titre (or just titre) – a measure of the number of phages per millilitre in a liquid stock, and typically a measure of viable phages as determined via plaque counts rather than of virion particles as determined by various forms of microscopy.
- Plaque – a visible clearing on a bacterial lawn, growing in or on agar found in a Petri dish, which is the result of localized inhibition of bacterial growth such as can be mediated by phages.

To this list one might add additional terms such as PFU standing for plaque-forming unit (just as CFU stands for colony-forming unit in bacteria). Note that the filamentous coliphage M13 is a prominent *chronically* infecting phage. See Clark *et al.* and Siegel (Chapters 7 and 8, this volume) for discussion of the utility especially of this phage to biotechnology and also Goodridge and Steiner (Chapter 11, this volume) for further discussion of chronically infecting phages (i.e. members of the family *Inoviridae*).

Transduction

Although covered in greater detail by Kuhl *et al.*, Christie *et al.* and Hendrickson (Chapters 3–5, this volume), I provide here a brief introduction to the idea of phage-mediated movement of bacterial DNA. Any time that phages are identified within an environment, this means that a potential exists for phage-mediated horizontal gene transfer of bacterial DNA between bacteria. Generally, we can differentiate phage-mediated horizontal gene transfer into four categories: generalized transduction, specialized transduction, phage morons and a category of gene movement that technically is not transduction at all but instead is the movement of *phage* genes by temperate phages. I will discuss these briefly.

Generalized transduction is the movement of bacterial DNA that has been packed into phage capsids without accompanying phage DNA. Generalized transduction has the property of being able to transfer large segments of DNA, that is, many tens of thousands of base pairs, such as those associated with bacterial pathogenicity islands (see Hendrickson, Chapter 5, this volume). Specialized transduction, by contrast and as narrowly defined, is the incorporation of bacterial genes that are found adjacent to prophage integration sites into the bacterial chromosome. Typically, only relatively few genes are transferred. As phage genomes tend to be limited in their size by constraints on their packaging into phage capsids (heads), this transduction of even relatively few bacterial genes can result in impairment of phage functioning. This type of specialized transduction is by definition limited to temperate phages that integrate their genomes into the host chromosome in the course of infection.

Specialized transduction, as more broadly defined, involves simply the integration of bacterial genes into phage genomes. An aspect of this form of transduction has become associated with the term moron, meaning 'more DNA'. Morons are typically considered to be bacterial genes that have become incorporated into phage genomes via processes of illegitimate recombination and which do not encode a mechanism for their removal. Lastly are seemingly legitimate phage genes, ones associated with temperate phages that modify the phenotypes of lysogens. Included under this heading are a number of virulence factor genes. The difference between a moron and these *lysogenic converting* genes is one of degree of integration into the phage's genetic structure, with morons more evidently newly acquired by the transferring phage. See Christie *et al.* (Chapter 4, this volume) for consideration of lysogenic conversion in general as it applies to the encoding of bacterial virulence factors by phages.

Conclusion

It is important to keep in mind that phages are highly diverse. This diversity is seen in terms of genotype, phenotype, the proteins produced and interactions with hosts. Phages also vary in terms of their host range (which bacteria they infect), their transducing ability, their virion morphology, and also with respect to their general infection characteristics. Thus, whenever mention is made of a specific phage, it should be kept in mind that substantial effort may be necessary to elucidate the specific properties, especially phenotypic, that are associated with that phage or phage–host combination. On the other hand, various generalizations may also be made. This monograph will present a mix of both generalizations and specifics in considering phage presence in bodies without disease, their role in both bacterial disease and pathogen evolution, and how phages can be employed to combat infectious diseases, in particular of humans.

2 Bacteriophages as a Part of the Human Microbiome

Andrey V. Letarov[1]
[1] *Winogradsky Institute of Microbiology, Russian Academy of Sciences*

Human beings, our problems, our triumphs and our everyday lives, have through the ages been central to philosophy, literature, arts and religion. This anthropocentric comprehension of the Universe has led to numerous, dramatic and sometimes quite controversial considerations in the natural sciences as well. These include, for example, resistance to acceptation of the heliocentric model of the Universe (16th century), the difficulties that the theory of evolution met before gaining wide acceptance (19th century) and, more recently, the numerous problems – ranging from underestimation to overestimation – in interpreting the mental capacities of different animal species, ranging from bees to dogs, dolphins, or even to ourselves (Alcock, 2001).

The anthropocentric view has led to a perception by many modern biologists, whose work is not directly aimed at species-specific aspects of animal biology, to regard at first glance the human animal as a valid model for mammals in general or vice versa. Indeed, almost everyone, scientist as well as non-scientist, is biased towards the idea that humans – or indeed any animal or plant – consists exclusively of cells, which can be described unambiguously as animal or plant cells. Such a perspective, however, is not even close to accurate, as each macroscopic organism hosts many types of microorganisms as part of its normal state: the microbiome of the organism. In this chapter, I emphasize essentially this aspect of the human body that traditionally one wouldn't regard as the human body. Furthermore, my emphasis will not even be on all of the microbiome but instead on the viruses that infect most of those non-human cells, the bacteriophages or phages. See Loc-Carrillo *et al.* (Chapter 13, this volume) for a discussion focusing in part instead on the bacterial aspects of the human normal flora.

This discussion serves as the beginning of an exploration of copious connections between humanity, the most powerful of organisms, and phages, which are the most numerous. Indeed, in this monograph the numerous roles that phages can play – in healthy humans, in contributing to human disease (see Kuhl *et al.*, Christie *et al.* and Hendrickson, Chapters 3–5, this volume), in the treatment of human disease (see Siegel, Wagemans and Lavigne, Olszowska-Zaremba *et al.*, Loc-Carrillo *et al.*, Burrowes and Harper, Shen *et al.* and Abedon, Chapters 8, 9, 12–15 and 17, this volume) and in the prevention of human disease (see Clark *et al.*, Cox, Goodridge and Steiner, and Niu *et al.*,

Chapters 7, 10, 11 and 16; see also Williams and LeJeune, Chapter 6, this volume) – all will be discussed. First, and the emphasis of this chapter, will be the normal state of affairs, which is the phage contribution to the collection of cells and microorganisms that together make up the human body; that is, ourselves and our microbiome, especially as seen during the normal, healthy state.

Phages and the Human Microbiome: an Overview

From the point of view of the microorganism, the human or animal body is merely a system of connected colonizable ecotopes, that is, ecologically distinct features of environments. Each ecotope varies in terms of host factors (the colon versus the lung, for example), in terms of its history (dictating in part what organisms can be present), in terms of interactions between the microorganisms that are there and as a consequence of feedback mechanisms between microorganisms and host. The result is a high potential for variation in microorganism types, including virus types, going from organ to organ, tissue to tissue, individual to individual, and also over time both within individuals and through the generations (both humans and our ancestors). Adding further to these complications is the potential for at least some microorganisms to move between species. For phages, we can also add an ability to modify their bacterial hosts (see Christie *et al.* and Hendrickson (Chapters 4 and 5, this volume), including in terms of phage susceptibility (see Williams and LeJeune, Chapter 6, this volume).

The result of all of these factors is the potential for a high degree of individual variation in the composition of, for example, intestinal bacterial populations in humans, as was reported by Costello *et al.* (2009). The individuality of the associated phage populations can be even higher than that of bacteria (Reyes *et al.*, 2010; Caporaso *et al.*, 2011). Consistently, divergence of both bacterial and phage communities at the same body sites but in different individuals were found to be higher than in the same individuals over time (Costello *et al.*, 2009; Reyes *et al.*, 2010; Caporaso *et al.*, 2011). On the other hand, communities of gut bacteria were much more related in closely connected people such as monozygotic twins and their mothers, although this consistency applies less so to the phage component, which appears to be highly individual but none the less quite stable over time (Reyes *et al.*, 2010).

These data highlight the effect of amplification of slight differences in physiology (or conditions) by the complex events of microbial interactions as well as exposure history, resulting in significantly different states of microbial systems in different subjects, and indicate that phage communities may be more sensitive to colonization history than bacterial microflora (Costello *et al.*, 2009; see also Nemergut *et al.*, 2011). The data on the phage prevalence and activity in different sites of the *human* body, analysed in this chapter, thus may not represent any paradigmal model for phages in animal-associated systems but could instead reflect particular features of our species, and maybe even of individuals or subpopulations included in the studies cited. By comparison, we can consider the current understanding of bacteriophage ecology in other animal-associated systems, as recently reviewed elsewhere (Letarov and Kulikov, 2009).

Despite early work with phages where there was a strong emphasis on the impact of phages on the human antibacterial immunity, along with a strong medical orientation of basic research in microbiology and virology in recent decades, the phage ecology of the human body was substantially neglected until recently. Even now it is poorly understood if compared, for example, with phage ecology in aquatic systems. The first observations of human-associated bacteriophages, however, were published by one of the discoverers of these viruses, Felix d'Hérelle, who demonstrated phages lysing enterobacteria in faeces (d'Hérelle, 1921). Nevertheless, and despite two periods of high interest in phage therapy – in the 1920s to 1930s and from the 1990s until now (Abedon, 2011) these 'endogenous' phages have been subject to systemic research only over the past few years. Thus, our knowledge of bacteriophage impact on microbial ecology and on macro-host homeostasis in humans (and, more widely, in animals) is highly mosaic, with many important parts of this puzzle still missing.

Here, I focus exclusively on the available data on phages in human symbiotic microbiota, presenting a limited subset of data for other species only for comparison. I consider the phage component following standard biogeographical principles, that is, based on body site. As has been evident since early electron-microscopy based studies of non-cultured viral communities from human faeces and other body sites, and later confirmed by metagenomic analysis (reviewed below), phages in fact appear to dominate the human body virome, being much more prevalent than any eukaryotic viruses.

Skin

The skin is the largest organ of the human body and is a reservoir for multiple and highly diverse habitats for symbiotic bacterial communities. The diversity of bacteria on the skin is comparable to that of the gut, although the total microbial biomass is much lower. The variation in species composition of the skin microflora of individual people appears to be highest among all the body sites that have been analysed (Costello *et al.*, 2009). The data on skin-associated bacteriophages, however, is almost non-existent for humans, as well as for other animals.

Alternatively, phages potentially associated with skin microflora have been isolated from downstream habitats such as *Staphylococcus aureus* phages isolated from sewage. Generally, such occurrence of phages infecting human-associated (and animal-associated) hosts from sewage and other 'downstream' sources is a well-known phenomenon. Moreover, this kind of sampling is widely used in many works on the isolation of phages of pathogenic bacteria for phage therapy or diagnostic applications (see Gill and Hyman, 2010, for review). There is always a question, however, of whether these phages come from human or animal microbiomes or instead are indigenous for specific systems (waste-water drains, for example). In numerous cases, it has been difficult to isolate phages directly from animals, but they are easily found in farm waste water. Furthermore, T-even-related bacteriophages are seldom isolated from healthy humans or animals (see below), but they can frequently be isolated from sewage samples (where T4 and T6 were originally found; Abedon, 2000). The details of the life strategy that make a phage better adapted to macro-host-associated or 'downstream' habitats are not yet clear.

Respiratory tract

The respiratory tract of healthy subjects is believed to be poorly colonized by any microflora, thus suggesting that no stable bacteriophage population should be present. In good agreement with this conclusion are the results of a recent metagenomic study of the viruses contained in the sputum samples from five cystic fibrosis (CF) patients and five non-CF subjects (Willner *et al.*, 2009). The diversity of both viral communities was at the level of about 175 viral genotypes in both metagenomes. The metagenomes of CF patients shared significant similarity and it was possible therefore to evaluate the core of the CF-specific phage community (Willner *et al.*, 2009; Willner and Furlan, 2010). These phages were proposed to play significant roles in the pathological microbial ecology in CF patients' lungs, encoding numerous bacterial virulence factors including adhesins, biofilm-formation genes and quorum-sensing genes (Willner and Furlan, 2010; see Christie *et al.*, Chapter 4, this volume, for a general overview of phage encoding of bacterial virulence factors). At the same time, the phages present in the lungs of non-CF persons were found to vary considerably and most likely represented the random samples of environmental viruses. Two non-CF sputum metagenomes collected from people sharing the environment with CF patients were more similar to the CF profile than the other three non-CF samples, and this observation was in good agreement with the assumption that the viral community of the respiratory tract in healthy people is represented by transitionally captured particles originating from the air.

Gastrointestinal tract

The human gastrointestinal tract includes the mouth cavity, throat, oesophagus, stomach and gut, the latter comprising the small intestine, large intestine and rectum. Due to the bactericidal activity of high acidity along with the proteolytic enzymes found in gastric juices, the stomach is poorly colonized by microorganisms and contains only about 10^3 bacterial cells ml^{-1}. In the small intestine, the bacterial population is also limited by the rapid peristalsis combined with the action of the bile and pancreatic secretions and reaches about 10^5 cells ml^{-1} (up to 10^8 ml^{-1} in the ileum; Baranovsky and Kondrashina, 2008). The main reservoir of the gut microbial biomass is the large intestine, harbouring up to a total of 10^{14} bacterial cells per individual (Savage, 1977) and corresponding to hundreds of grams of bacterial biomass. Below, I consider the phage populations in the oral cavity and in the large intestine that represent the major ecotopes of the human gastrointestinal tract colonized by bacteria.

Oral cavity and pharynx

Studies assessing phage presence in the human oral cavity and pharynx are not extensive. Direct electron microscopic observations indicate the presence of a large number of VLPs in some but not in all samples of the dental plaque material (Brady *et al.*, 1977). The presence of VLPs in the matrix of complex multi-species biofilms seems logical, as the data suggest that in some bacterial species, such as *Pseudomonas aeruginosa*, induction of the prophage and production of viral particles may be a normal programmed stage of the biofilm development (Rice *et al.*, 2009).

These observations of high concentrations of VLPs in dental plaque to my knowledge have not been corroborated. Nevertheless, the existence of bacteriophages in dental plaque material was recently confirmed by a small viral metagenomic survey of the dental plaque uncultured viral community collected from a single individual (Al-Jarbou, 2012). Of the 80 sequences obtained, only a total of 21 phage-related sequences were discovered. Despite the paucity of the data set obtained in this study, it indicates clearly the low complexity of the dental plaque viral community.

Another recently published metagenomic study of viral communities of pooled oropharyngeal swabs from 19 healthy individuals (Willner *et al.*, 2011) also indicated that they are formed almost exclusively by phages – a mix of virulent and temperate phages. The sole exception was the eukaryotic Epstein–Barr virus. Of interest, the complete genomes of three phages were assembled and among them was *Escherichia coli* phage T3. The natural reservoir for this virus was never identified, although the data of Willner *et al.* (2011) indicated that it may be a normal inhabitant of the human pharynx. The authors also detected a number of phages of *Propionobacterium*, *Lactobacillus* and other lactic acid bacteria, streptococci and other hosts as well as non-host-attributed bacteriophage-related sequences. The SM1 bacteriophages encoding the platelet-binding factors of *Streptococcus mitis* were also found, suggesting possible involvement of the oropharyngeal microbiota in development of endocarditis. The total abundance of the bacteriophages was not directly determined in this study but would appear not to be very high, as the authors applied an amplification procedure prior to sequencing. The diversity was estimated to be about 236 different viral genomes from the 19 pooled samples.

A much more profound metagenomic study of human saliva viromes has recently been published by Pride *et al.* (2011). The authors collected saliva samples from several healthy subjects at different time points. The viruses were quantified by epifluorescence microscopy and the morphology of VLPs was investigated by electron microscopy. Profound sequencing of the virome metagenomes was performed accompanied by sequencing of bacterial 16S rRNA gene libraries from the same samples. This study confirmed the presence of a robust indigenous phage population dominating the saliva microbiome, comprising up to 10^8 VLPs ml^{-1} of saliva. The composition of the viromes in different subjects was highly individual, but two viromes collected from members of the same household showed much greater relatedness than the others, indicating the impact of externally acquired phages on this system (which makes a striking contrast to the data of Reyes *et al.*, 2010, which demonstrated that faecal viromes are markedly distinct in closely connected subjects such as monozygotic twins and their mothers – see below). The composition of the viromes within subjects at different time points was highly related but nevertheless exhibited significant changes over time, especially at 60–90-day time intervals. The saliva virus population thus appears to be quite dense and at the same time dynamic. The presence of phage integrases in 10% of all contigs and the identification of virus contigs matching distinct regions in sequenced bacterial genomes indicated a high prevalence of lysogeny. At the present time, however, the data are insufficient to estimate the relative significance of virulent and temperate phages (or of phage multiplication in the lytic cycle versus lysogen induction in this system). Overall, the substantial numbers of phage particles detected by Brady *et al.* (1977), Al-Jarbou (2012) and Pride *et al.* (2011) in oral cavity-derived samples is inconsistent with published negative results of attempts to isolate phages from the human oral cavity active against normal indigenous bacteria present in that site (Hitch *et al.*, 2004); these authors were, however, able to isolate a phage of a non-oral pathogen, *Proteus mirabilis*.

A very limited number of successful phage isolations from the oral cavity have been described. For example, Tylenda *et al.* (1985) reported the isolation of actinobacterial phages from ten out of 336 samples of human dental plaque material. The isolation of *Veillonella* bacteriophage from the oral samples was also published by Hiroki *et al.* (1976). More recently, phages for *Enterococcus faecalis* were cultured from human saliva (Bachrach *et al.*, 2003), but it should be noted that in this latter study the authors failed to detect bacteriophages for a number of other species of bacteria of the normal oral microbiota. Temperate *E. faecalis* phages were also successfully induced from root canal isolates of this bacterium (Stevens *et al.*, 2009).

At the moment, it is difficult to build a comprehensive concept of phage ecology in the oral cavity. Summarizing the data, one could conclude that the bacterial community of the human oral cavity and probably also of the pharynx are not substantially impacted by phages. The metagenomic data of Pride *et al.* (2011), however, suggests that the viral community in this site is dynamic and is able to incorporate externally acquired phage strains and to maintain quite an elevated density. No coherent explanation of these contradictions was suggested. Given the limited amount of data on oral bacteriophages from non-human species that has been published, it is difficult to speculate on how these human traits compare with the phage ecology of the oral cavity in mammals.

The gut

The gut, especially the lower intestine, is believed to be the main habitat of the human-associated microbiota including bacteriophages. Being the natural habitat of the world's best-studied bacterium – *E. coli* – the intestinal microbial system of humans and animals has served as a subject of multiple studies of coliphage ecology (see below), starting from d'Hérelle's pioneering work in 1921.

Total viral counts

Transmission electron microscopy-based studies have repeatedly demonstrated a high abundance of VLPs in the faeces and intestinal contents of humans (Flewett *et al.*, 1974), as well as in other species. The latter include cattle and sheep (Paynter *et al.*, 1969; Hoogenraad *et al.*, 1967), the rumen of reindeer (Tarakanov, 1971), the forestomachs of Australian marsupials (Hoogenraad and Hird, 1970; Klieve, 1991) and the large intestine contents and faeces of horses (Alexander *et al.*, 1970, Kulikov *et al.*, 2007). In all of these cases, the vast majority of observed VLPs belonged to tailed bacteriophages. No quantitative data characterizing the morphological diversity of the human gut bacteriophages, however, has been published (in contrast to some animal-associated communities, as recently reviewed by Letarov and Kulikov, 2009).

Metagenomic studies

Currently available data on the diversity of the non-cultured viral community of the human gut has been based mainly on metagenomic data. The first metagenomic analysis of the virome of a single specimen of human faeces, collected from a 30-year-old male subject, was published by Breitbart *et al.* (2003). In their study, about two-thirds of the sequences obtained were database orphans; among the rest, known viral sequences constituted 27%. The predominant viral group, judged by database hits, in human faeces were siphoviruses (bacteriophages with long, non-contractile tails), which are probably most prevalent in the majority of natural habitats (Weinbauer, 2004). The estimated diversity of bacteriophages was about 1200 viral genotypes present in the sample.

Sequences related to eukaryotic viruses comprised only a minor fraction. This is consistent with the fact that such particles are rarely seen in transmission electron microscopy images of faecal viral communities. It is interesting that, in metagenome analyses of RNA-containing VLPs extracted from human faeces, the sequences of plant viruses were highly predominant (Zhang *et al.*, 2006), indicating that ingestion of these particles with food is much higher than the internal production of RNA-containing phages. The later metagenomic analysis of multiple samples of human faeces used high-throughput sequencing technology (pyrosequencing) and revealed very interesting features of these communities.

In order to determine the impact of the genetic background of the macro host on the composition of the human intestinal virome and the stability of the individual phage populations, Reyes *et al.* (2010) collected samples of faeces from four pairs of twins and their mothers over a 1-year period. VLPs were extracted from these samples (32 viromes in total were characterized) and the associated viral metagenomes were sequenced. About 85% of the sequences from the VLP metagenomes did not correspond to any known viruses, while most of the rest matched various known prophages and temperate phages.

The bacterial diversity of the same set of samples was analysed by 16S rRNA gene library sequencing. The diversity of bacterial communities was estimated to be about 800 species-level bacterial phylotypes, while the complexity of the phage community was measured by two different approaches as 52–2773 (median 346) or 10–984 (median 35) predicted virotypes. Both the abundance of phages and the number of phage species per bacterial species was therefore quite low in comparison with known, free-living bacterial communities. The idea of a temperate nature of predominant phage types in the analysed viromes was strongly supported by the identification of a significant number of sequences related to known bacteriophage integrases, the phage-encoded site-specific recombinases responsible for integration of the temperate phage genome into the host chromosome.

The similarities seen in the bacterial communities studied by Reyes *et al.* (2010) strongly correlated with family links between the subjects. In contrast, the VLP metagenome composition was highly individual and varied considerably between the subjects, regardless of family relationships. At the same time, the variability was low within individuals: over the time of the study, the sequences of the VLP metagenomes were almost stable. Moreover, the authors were able to detect the dominant phage that persisted at high levels in one of the individuals for an extended period of time but showed no significant divergence or mutations in its genome.

The proposed low dependence of the phage populations in this environment on their success in competition for host bacteria might facilitate long-term persistence of those phage populations that happen to colonize a particular niche first. Alternatively, if the assumption of high prevalence of temperate phages made in the above-cited studies is correct, then those temperate phages that colonized the *genomes* of bacteria that have established their populations in an individual gut ecosystem may predominate. The high individuality of the phage community in different subjects therefore may reflect the history of colonization of the infant gut by bacteria, phages and phage lysogens (Breitbart *et al.*, 2008).

What does not seem to be present is substantial selective pressure acting on either bacteria or phages, at least over the intervals analysed by metagenomic studies. The data of Reyes *et al.* (2010), in particular, provided no indication of 'Red Queen' dynamics (continuing change and adaptation of host and parasite; Weitz *et al.*, 2005) in human faecal microbial communities, although perhaps the resolution of metagenomic studies could be unable to provide such a signal. Thus, the bacteriophages in this environment do not seem to exert a sufficient influence on the dynamics of bacterial populations in the gut to result in bacterial resistance. Abedon (2011) suggested that this sort of low phage pressure despite ongoing phage presence may be the norm, given bacterial persistence in environments predominantly as biofilms (see also MacFarlane *et al.*, 2011).

This concept of human intestinal phage ecology is in good agreement with the data of Caporaso *et al.* (2011) who used a somewhat different approach for analysis of the metagenomic data. These researchers studied 26 viral metagenomes of human faeces collected from 12 individuals (one to four samples per individual). They compared the composition of these viromes with the viromes obtained from a variety of free-living communities. The authors found that the distances between the individual viromes of human faeces were higher than between the samples of related free-living communities. They also confirmed the predominance of temperate phages and the absence of observable phage/bacteria co-evolution in the human gut.

In agreement with these observations, Minot *et al.* (2011) reported that, following analysis of metagenomic sequences of human faecal viromes collected from five people over a 1-week period, they did not detect any signs of 'Red Queen' dynamics in these systems. The viromes studied also showed high individual variability; however, 1-week diet interventions (as low fat/high fibre or high fat/low fibre diets) led to an increase in similarity of the viromes between subjects fed the same diet. This may indicate that, in humans, alterations in virome composition follow the composition of the microbiome. This may also be interpreted as an argument in favour of the hypothesis of a high impact of lysogen induction in generations of phage VLPs in the human gut. Minot *et al.* (2011) analysed the viral (phage) contigs assembled from their data for genetic relatedness (using various criteria, described in detail in the paper) for temperate phages or prophages and found about 14% of them to be potentially temperate. However, this value was only the lowest estimate, as not all fragments of the temperate phage genomes would fit the criteria applied.

Culture-based analyses

When considering the culture-based data dealing with phage indication and quantification in natural environments, one must always remember that the majority of these viruses tend to be specific for only a subset of the strains found within bacterial species (Hyman and Abedon, 2010). Thus, the phage titre obtained with any given indicator bacterium reflects only a fraction of the phage particles, that is, those able to infect that particular bacterial strain (the phenomenon is known in phage ecology as the 'great plaque count anomaly'; see Weinbauer, 2004). Moreover, some of the phages that are able to infect the bacterial strain used nevertheless may display reduced plating efficiency, at least during the first passage on a host strain. This effect may be due to the action of restriction-modification systems, as well as of many other host resistance mechanisms (Labrie *et al.*, 2010), and can lead to serious – several orders of magnitude – under-estimations of the real abundance of a particular phage strain in a sample. In each case, one therefore has to consider the choice of the bacterial host used for phage quantification, especially if comparison of phage titres in samples collected from different subjects is involved.

Notwithstanding these caveats, the data of culture-based studies are in general agreement with the hypothesis of a low phage impact on bacterial populations in human intestinal ecology. Furuse *et al.* (1983), for example, found that faecal coliphage titres in healthy humans are low, and the pools of free virions in faeces are represented mainly by temperate phages. In contrast to healthy people, the phage populations in some patients with internal and leukaemic diseases contained a substantial fraction of virulent phages (including T-even-related phages), as well as an increased faecal coliphage background (Furuse *et al.*, 1983). In several patients, phage titres increased when the severity of the clinical symptoms increased.

Consistent results of phage isolation from Bangladesh paediatric patients with diarrhoea were reported by Chibani-Chennoufi *et al.* (2004b). About 19% of acute diarrhoeal stools yielded quite divergent T4-related phages infecting the *E. coli* K803 laboratory indicator strain. The detection of phages in the stools from convalescent patients was less frequent. It is interesting that other *E. coli* strains used for phage isolation from the same set of samples yielded completely different *Siphoviridae* coliphages. The occasional presence of both temperate and virulent coliphages as well as some culturable phages infecting other hosts has also been reported in the literature. In most cases, these viruses were present in low titres in healthy subjects (Dhillon *et al.*, 1976; Havelaar *et al.*, 1986; Cornax *et al.*, 1994; Grabow *et al.*, 1995; Calci *et al.*, 1998; Gantzer *et al.*, 2002; Schaper *et al.*, 2002; Cole *et al.*, 2003; Lusiak-Szelachowska *et al.*, 2006; see also Letarov and Kulikov, 2009, for review).

Temperate phages in the gut

The results of metagenomic analysis performed by Reyes *et al.* (2010) suggested that the main source of free phage particles in the human intestine are not productive cycles immediately following phage adsorption, as seen in the majority of other natural communities (see Weinbauer, 2004 for review), but instead are the induction of bacterial lysogens that can occur due to starvation, such as of bacteria in faeces. This conclusion is similar to that of Furuse *et al.* (1983) who first suggested that, in healthy people, most of the released phage particles are produced by induced lysogenic cells and therefore that phage multiplication may have a limited impact on the intestinal coliform microflora.

Interestingly, the vast majority of temperate coliphages isolated by Dhillon *et al.* (1980) from free phage particles of human and animal faeces belonged to the lambdoid group, while phages obtained from cultured bacterial lysogens were all immunologically P2-related. These data may be explained by a higher frequency of induction of the lambdoid prophages present in the studied microbiomes than that of P2-like prophages. Alternatively, conditions favourable for phage multiplication in the lytic cycle may occur in the healthy human gut, perhaps in spatially limited sites or over short periods of time, and this multiplication could allow the reproduction of some phages independent of the induction of lysogenic bacteria. The above-cited results could thus reflect the preferential success of the lambdoid phages within hypothetical windows for lytic cycle multiplication.

F-specific phages

RNA-containing F-pilus-specific (F-RNA) coliphages (members of the family *Leviviridae*) have been found in human as well as in some animal wastes and show certain species specificities. These phages can be subdivided into multiple genetic groups that can be also distinguished serologically. The incidence of these serotypes varies significantly among species: horse faeces, for example, rarely contain them, while the faeces of more than 70% of chickens contain high titres of these phages (10^5–10^7 plaque-forming units (PFU) g^{-1}). Only about 10–20% of human faeces contains F-RNA coliphages, but the occurrence of group II and III F-RNA phages in these samples is much higher (>80% of all isolates) than in animals, where groups I and IV are prevalent at the same level (Furuse *et al.*, 1978; Havelaar *et al.*, 1986; Schaper *et al.*, 2002; Cole *et al.*, 2003). No coherent explanation for this group specificity has yet been suggested. See Goodridge and Steiner (Chapter 11, this volume) for further discussion.

Possible limitations on phage replication in the gut

Among the environmental factors that may contribute to the protection of bacteria from phage attack in the human gut are chemicals that can inhibit phage infection as well as the specific physiological state of bacterial populations. Bile salts and carbohydrates have been shown to inhibit the adsorption of a variety of coliphages (Gabig *et al.*, 2002). This effect is suppressed if Ag43 protein, mediating cell aggregation and attachment, is present on the surface of *E. coli* cells. The expression of Ag43 is regulated in a phase-variation manner. Thus, phages may, under some circumstances, select against increased biofilm formation. For *Bacteroides* phages, the addition of bile salts to the medium had an opposite effect, that is, its addition improved phage plating efficiency (Araujo *et al.*, 2001). Some food-derived phage inhibitors may also be present in the intestine, at least occasionally. For example, Swain *et al.* (1996) demonstrated that tannic acid at physiological concentrations may inhibit bacteriophage replication in the rumen of ruminants. This is consistent with the observations of de Siqueira *et al.* (2006) that tea infusions can inactivate phages. Similar compounds are also ingested by humans with certain types of food and could therefore contribute to low phage lytic activity in their gastrointestinal tract.

Growth in biofilms and on the surfaces of mucosa and food particles may also contribute to bacterial anti-phage protection, although this potential has at best only been demonstrated inconclusively (Abedon, 2011). There is also some evidence that the *E. coli* population in the mouse gut lumen may be starving and the actively replicating population in fact may be limited to microcolonies found on the mucosal surface (see Chibani-Chennoufi *et al.*, 2004a, and references therein). It is not clear, however, if this model of *E. coli* ecology in the gut can be extended to humans and other large animals (Letarov and Kulikov, 2009).

Phage propagation predominantly via lysogenic cycles or associated prophage induction, resulting in a reduced impact of phage infections on the microbial ecology of the human gut, may not be unique among mammalian species (reviewed in detail by Letarov and Kulikov, 2009; Clokie *et al.*, 2011). For example, attempts to isolate bacteriophages from dog faeces using indigenous coliform strains (Ricca and Cooney, 2000) were largely unsuccessful. Over 500 indigenous coliform strains isolated from six specimens of dog kept in private homes did not detect phages in the same samples, and only one of these samples yielded phages on the laboratory *E. coli* C strain. In 16 dogs from a kennel, however, coliphages were detected at variable titres from 0 to 10^7 PFU g^{-1}. The authors suggested that a low abundance of coliphages in home-kept dogs may be due to isolation from other dogs and 'too clean' living conditions. Recontamination by faecal microbes is also limited in humans, which could suggest that phages able to overcome existing barriers for replication in the gut are seldom acquired by humans. Perhaps the occurrence and possible impact of phages in humans would be higher if we lived in a less civilized manner. It would be interesting to compare the phage prevalence in human subpopulations living in the same area but differing in their quotidian hygienic practice (for example, in Bangladeshi state employees and peasants).

The opposite situation to that of highly domesticated animals seems to exist in the horse intestinal ecosystem. In our studies (Golomidova *et al.*, 2007; Kulikov *et al.*, 2007; reviewed by Letarov and Kulikov, 2009), the structure of the indigenous coliform community was compatible with high phage pressure. Of the dozens of horse coliphage isolates that we were able to identify, all belonged to known groups of virulent phages (T-even-related, T5-related, rv5-related, N-related, similar to *Salmonella* SETP3 phage, Felix 01-related, similar to *Caulobacter* phage Cd1, coliphage K1F-related and others; Golomidova *et al.*, 2007, and our unpublished data). The investigation of morphological diversity of uncultured horse faecal VLPs (Kulikov *et al.*, 2007) has demonstrated the high abundances of a diversity of large bacteriophages that are very likely to be virulent. The ecology of intestinal bacteriophages thus appears to be highly dependent on host-species-specific features of digestive tract physiology along with the general environment.

The vagina

The vaginal microbiota of healthy, fertile women is normally dominated by vaginal lactobacilli (Hillier, 2008), usually comprising more than 70% of the total bacterial count in this environment. These bacteria are believed to play an essential role in colonization resistance against pathogenic bacteria and fungi by contributing to the establishment of a low vaginal fluid pH (Servin, 2008; Linhares *et al.*, 2011), along with frequently producing hydrogen peroxide (Martin and Suarez, 2010). Lactobacilli also possess more-specific antagonistic activities against pathogenic bacteria (Servin, 2008). Partial loss of the indigenous vaginal lactobacilli in combination with polymicrobial anaerobic overgrowth on the vaginal mucosa are characteristic features of bacterial vaginosis (BV) – one of the most common reasons women seek medical help (Sobel, 2000). The density of *Lactobacillus* colonization of the vaginal mucosa is quite high at 10^6–10^7 CFU per vaginal swab. Thus, one can assume that episodes of mass killing of lactobacilli by phages may occur, although attempts to detect free bacteriophages in vaginal swabs have been unsuccessful (Kiliç *et al.*, 2001). At the same time, however, lysogenic strains of lactobacilli were shown to be prevalent in this environment (Kiliç *et al.*, 2001).

Although the vagina is subject to relatively low levels of exchange of bacteria and viruses with the external environment, except as mediated by sexual activity, a sudden breakdown of *Lactobacillus* populations is frequently observed in examinations of vaginal swabs from clinically healthy women. Some individuals also develop anaerobic bacterial vaginosis syndrome, when lactobacilli are replaced by anaerobic bacteria such as *Gardnerella vaginalis*, *Prevotella*, *Porphyromonas* and *Mobiluncus* species. This condition has the epidemiology of a sexually transmitted disease (Verstraelen *et al.*, 2010). It was proposed by Blackwell (1999) that the causative agent triggering the breakdown of the normal vaginal microbiota might be a bacteriophage attack. I speculate that such events may take place if a lysogenic strain able to produce a phage infectious for the major resident strain(s) is acquired (Letarov and Kulikov, 2009). A similar scenario was modelled on *in vitro E. coli* populations both experimentally and mathematically by Brown *et al.* (2006).

The results of our recent study of the diversity of individual populations of vaginal lactobacilli at the strain level indicate that individual populations of these bacteria are normally dominated by a single strain, which could be resolved by repetitive element PCR fingerprinting (A. Isaeva, E. Ilina, A. Bordovskaya, A. Ankirskaya, V. Muraviouva and A. Letarov, unpublished data). This is compatible with the proposed mechanism of the rapid drop of lactobacilli counts due to mass phage-mediated lysis, as almost all of the cells in individual populations would (presumably) represent the same phage-susceptibility type. In such a case, a single phage liberated by an invading lysogenic strain potentially could reduce the resident population density and make enough room for a newly acquired lysogenic strain (and secondary lysogens formed in some cells of the resident strain). In contrast to the results of Kiliç *et al.* (2001) described above, however, the search for inducible lysogens in our collection of vaginal isolates yielded no strains producing a phage viable on any other isolate tested (A. Isaeva and A. Letarov, unpublished data). An observation similar to ours was recently reported for a Spanish subpopulation of women (Martin *et al.*, 2009). These authors suggested that high concentrations of H_2O_2 produced by vaginal lactobacilli select for prophage-cured lineages that do not suffer from the mortality induced by elevated phage induction due to permanent oxidative stress characteristic of this environment. Thus, the vaginal ecosystem may be considered as one in which phages exert, ecologically, a low impact on the resident bacterial populations. The possible exceptions, however, require further investigation.

Direct Interaction of the Virome with the Macro Host

The phage ecology of human-associated microbial systems generally should be considered as a tripartite interplay between the bacteriophage, the host bacteria and the macroorganism (i.e. ourselves). In this interplay, the macroorganism influences both bacteria and phages. For example, phage destruction can occur due to the action of digestive enzymes as well as that of macrophages, phage transportation by blood and phage transmission between individuals that are facilitated or restricted due to specific macroorganism behaviour. The most striking phenomenon may be the influence on phage infection by some compounds secreted by the macro host such as bile salts (see above).

The macroorganism seems to be directly and indirectly influenced by both phages and microbes. Direct interactions, once phage particles have reached systemic circulation, include immunomodulatory activity (see Olszowska-Zaremba *et al.*, Chapter 12, this volume). Lepage *et al.* (2008) observed increased concentrations of VLPs on ulcerated intestinal mucosa and speculated that phage-mediated immune effects may play some role in the pathology of inflammatory bowel disease. Phages occurring *in vivo* instead may contribute to a downregulation of the inflammatory response in the gut, as well as regulation of the complex interactions of the gut-associated lymphoid tissue with massive amounts of antigens provided by intestinal microorganisms (Górski *et al.*, 2006). Experimental data on direct interactions of naturally occurring bacteriophages with the immune system, however, are currently almost non-existent.

The direct influence of the macro host on phages within the normal microflora has not yet been studied in any detail. It has been reported that phages administrated orally or rectally in the course of phage therapy penetrated to the bloodstream from the gastrointestinal tract (reviewed in detail in Letarov *et al.*, 2010; Górski *et al.*, 2006; Dabrowska *et al.*, 2005). In recent rigorously controlled studies on animal models (Olivera *et al.*, 2009), however, the penetration of externally administered phages was very low (if at all). The proposed hypothesis of constant translocation of intestinal phages into the bloodstream – so-called 'physiological phagaemia' (Górski *et al.*, 2006) – was based on reports of phage contamination of commercial blood sera from the 1970s (see Górski *et al.*, 2006, for a detailed review and Olszowska-Zaremba *et al.*, Chapter 12, this volume). The presence of natural phages in the blood of humans or animals, by contrast, has not been demonstrated. In our recent work (Letarova *et al.*, 2012), we were unable to detect coliphages in the blood of horses that harboured these viruses at levels of 10^4–10^7 PFU g^{-1} of faeces on the day of blood sampling. The rate of phage translocation from the gastrointestinal tract to the bloodstream (similar to bacteria), however, can vary significantly depending on the physiological conditions and perhaps also on phage characteristics.

Phages delivered into the bloodstream can be rapidly sequestered by the lymphoid organs, spleen and liver and gradually inactivated by the reticuloendothelial system (see Sulakvelidze and Kutter, 2005; Uchiyama *et al.*, 2009) and in some cases by phage-specific antibodies, creating a barrier for phage travel within the blood and, at least within the context of the blood, selecting for viruses with altered surface proteins (Vitiello *et al.*, 2005; Capparelli *et al.*, 2006, 2007). The role of these processes in the ecology of natural phages in humans, however, is not clear, as the concept of penetration of phages into the bloodstream remains elusive.

Conclusions

The permanent presence of a variety of bacteriophages in the normal human microbiome, especially in the intestinal microbial complex, is now well established. Nevertheless, it remains to be seen whether the ecological relationships between intestinal phage and host populations in other species may be similar to humans or dogs, or instead completely different (as in horses), thus highlighting the need for comparative studies of the phage component in the microflora of different animal species. Furthermore, composition of the viromes of different subjects appears to be highly individual in comparison with the bacterial diversity in the same habitats. This makes it difficult to define any profile of the 'normal' composition of the virome of any particular site of the human body, as can be done for bacterial flora. This may be distinct from the 'pathological' composition, as some links between phage diversity and pathology can be established; for example, the presence of T-even-related phages in high titres in the gut is frequently associated with diarrhoea and some other disturbances of the gut physiology. The data available on this subject, however, are scarce and much work is required to create a comprehensive description of the human phage community in health and disease.

Acknowledgements

The work of my laboratory is supported by RFBR grant no. 06-04-48651-a, by a grant of the program of Presidium of RAS 'Basic science to medicine' and by contracts with the Russian Ministry for Education and Science.

References

Abedon, S.T. (2000) The murky origin of Snow White and her T-even dwarfs. *Genetics* 155, 481–486.

Abedon, S.T. (2011) Bacteriophages and biofilms. In: Bailey, W.C. (ed.) *Biofilms: Formation, Development and Properties.* Nova Science Publishers, Hauppauge, New York.

Alcock, J. (2001) *Animal Behavior, an Evolutionary Approach*, 7th edn. Sinauer Associates Inc., Sunderland, MA.

Alexander, F., Davies, M.E. and Muir, A.R. (1970) Bacteriophage-like particles in the large intestine of the horse. *Research in Veterinary Science* 11, 592–593.

Al-Jarbou, A. (2012) Genomic library screening for viruses from the human dental plaque revealed pathogen-specific lytic phage sequences. *Current Microbiology* 64, 1–6.

Araujo, R., Muniesa, M., Méndez, J., Puig, A., Queralt, N., Lucena, F. and Jofre, J. (2001) Optimisation and standardisation of a method for detecting and enumerating bacteriophages infecting *Bacteroides fragilis. Journal of Virological Methods* 93, 127–136.

Bachrach, G., Leizerovici-Zigmond, M., Zlotkin, A., Naor, R. and Steinberg, D. (2003) Bacteriophage isolation from human saliva. *Letters in Applied Microbiology* 36, 50–53.

Baranovsky, A.Y. and Kondrashina, E.A. (2008) [*Gut Disbacteriosis*]. Piter, Moscow, St Petersburg, Russia, pp. 44–45 (in Russian).

Blackwell, A.L. (1999) Vaginal bacterial phaginosis? *Sexually Transmitted Infections* 75, 352–353.

Brady, J.M., Gray, W.A. and Caldwell, M.A. (1977) The electron microscopy of bacteriophage-like particles in dental plaque. *Journal of Dental Research* 56, 991–993.

Breitbart, M., Hewson, B., Felts, J., Mahaffy, M., Nulton, J., Salamon, P. and Rohwer, F. (2003) Metagenomic analyses of an uncultured viral community from human feces. *Journal of Bacteriology* 185, 6220–6223.

Breitbart, M., Haynes, M., Kelley, S., Angly, F., Edwards, R.A., Felts, B., Mahaffy, J.M., Mueller, J., Nulton, J., Rayhawk, S., Rodriguez-Brito, B., Salamon, P. and Rohwer, F. (2008) Viral diversity and dynamics in an infant gut. *Research in Microbiology* 159, 367–373.

Brown, S.P., Le Chat, L., de Paepe, M. and Taddei, F. (2006) Ecology of microbial invasions: amplification allows virus carriers to invade more rapidly when rare. *Current Biology* 16, 2048–2052.

Calci, K.R., Burkhardt, W. III, Watkins, W.D. and Scott. R.R. (1998) Occurrence of male-specific bacteriophage in feral and domestic animals wastes, human feces, and human-associated wastewaters. *Applied and Environmental Microbiology* 64, 5027–5029.

Caporaso, J.G., Knight, R. and Kelley, S.T. (2011) Host-associated and free-living phage communities differ profoundly in phylogenetic composition. *PLoS One* 6, e16900.

Capparelli, R., Ventimiglia, S., Roperto, S., Fenizia,

D. and Iannelli, D. (2006) Selection of an *Escherichia coli* O157:H7 bacteriophage for persistence in the circulatory system of mice infected experimentally. *Clinical Microbiology and Infection* 12, 248–253.

Capparelli, R., Parlato, M., Borrielo, G., Salvatore, P. and Iannelli, D. (2007) Experimental phage therapy against *Staphylococcus aureus* in mice. *Antimicrobial Agents and Chemotherapy* 51, 2765–2773.

Chibani-Chennoufi, S., Bruttin, A., Dillmann, M.L. and Brüssow, H. (2004a) Phage–host interaction: an ecological perspective. *Journal of Bacteriology* 186, 3677–3686.

Chibani-Chennoufi, S., Sidoti, J., Bruttin, A., Dillmann, M.L., Kutter, E., Qadri, F., Sarker, S.A. and Brüssow, H. (2004b) Isolation of *Escherichia coli* bacteriophages from the stool of pediatric diarrhea patients in Bangladesh. *Journal of Bacteriology* 186, 8287–8294.

Clokie, M.R.J., Millard, A.D., Letarov, A.V. and Heaphy, S. (2011) Phages in nature. *Bacteriophage* 1, 31–45.

Cole, D., Long, S.C. and Sobsey, M. (2003) Evaluation of F^+ RNA and DNA coliphages as source-specific indicators of fecal contamination in surface waters. *Applied and Environmental Microbiology* 69, 6507–6514.

Cornax, R., Morinigo, M.A., Gonzalez-Jaen F., Alonso, M.C. and Borrego, J.J. (1994) Bacteriophages presence in human faeces of healthy subjects and patients with gastrointestinal disturbances. *Zentralblatt für Bakteriologie* 281, 214–224.

Costello, E.K., Lauber, C.L., Hamady, M., Fierer, N., Gordon, J.I. and Knight, R. (2009) Bacterial community variation in human body habitats across space and time. *Science* 18, 1694–1697.

Dabrowska, K., Switala-Jelen, K., Opolski, A., Weber-Dabrowska, B. and Górski, A. (2005) Bacteriophage penetration in vertebrates. *Journal of Applied Microbiology* 98, 7–13.

d'Hérelle, F. (1921) *La bactériophage. Son rôle dans l'immunité.* Masson et Cie, Paris.

de Siqueira, R.S., Dodd, C.E. and Rees, C.E. (2006) Evaluation of the natural virucidal activity of teas for use in the phage amplification assay. *International Journal of Food Microbiology* 111, 259–62.

Dhillon, E.K. Dhillon, T.S., Lam, Y.Y. and Tsang, A.H.C. (1980) Temperate coliphages: classification and correlation with habitats. *Applied and Environmental Microbiology* 39, 1046–1053.

Dhillon, T.S., Dhillon, E.K., Chau, H.C., Li, W.K. and Tsang, A.H. (1976) Studies on bacteriophage distribution, virulent and temperate bacteriophage content of mammalian feces. *Applied and Environmental Microbiology* 32, 68–74.

Flewett, T.H., Bryden, A.S. and Davies, H. (1974) Diagnostic electron microscopy of faeces. *Journal of Clinical Pathology* 27, 603–614.

Furuse, K., Sakurai, T., Hirashima, A., Katsuki, M., Ando, A. and Watanbee, I. (1978) Distribution of ribonucleic acid coliphages in South and East Asia. *Applied and Environmental Microbiology* 35, 995–1002.

Furuse, K., Osawa, S., Kawashiro, J., Tanaka, R., Ozawa, A., Sawamura, S., Yanagawa, Y., Nagao, T. and Watanabe, I. (1983) Bacteriophage distribution in human faeces: continuous survey of healthy subjects and patients with internal and leukaemic diseases. *Journal of General Virology* 64, 2039–2043.

Gabig, M., Herman-Antosiewicz, A., Kwiatkowska, M., Los, M., Thomas, M.S. and Wegrzyn, G. (2002) The cell surface protein Ag43 facilitates phage infection of *Escherichia coli* in the presence of bile salts and carbohydrates. *Microbiology* 148, 1533–1542.

Gantzer, C., Henny, J. and Schwartzbrod, L. (2002) *Bacteroides fragilis* and *Escherichia coli* bacteriophages in human faeces. *International Journal of Hygiene and Environmental Health* 205, 325–328.

Gill, J.J. and Hyman, P. (2010) Phage choice, isolation, and preparation for phage therapy. *Current Pharmaceutical Biotechnology* 11, 2–14.

Golomidova, A., Kulikov, E., Isaeva, A., Manykin, A. and Letarov, A. (2007) The diversity of coliphages and coliforms in horse feces reveals a complex pattern of ecological interactions. *Applied and Environmental Microbiology* 73, 5975–5981.

Górski, A., Wazna, E., Weber-Dabrowska, B., Dabrowska, K., Switala-jelen, K. and Miedzybrodzki, R. (2006) Bacteriophage translocation. *FEMS Immunology and Medical Microbiology* 46, 313–319.

Grabow, W.O.K., Neubrech, T.E., Holrzhausen C.S. and Jofre, J. (1995) *Bacteroides fragilis* and *Escherichia coli* bacteriophages: excretion by humans and animals. *Water Science and Technology* 31, 223–230.

Havelaar, A.H., Furuse, K. and Hogeboom, W.M. (1986) Bacteriophages and indicator bacteria in human and animal feces. *Journal of Applied Bacteriology* 60, 55–262.

Hillier, S.L. (2008) Normal vaginal flora. In: Holmes, K.K., Sparling, P.F., Stamm, W.E., Piot, P., Wasserheit, J., Corey, L. and Cohen, M. (eds) *Sexually Transmitted Diseases.* 4th edn. McGraw-Hill, New York, pp. 289–307.

Hiroki, H., Shiiki, J., Handa, A., Totsuka, M. and Nakamura, O. (1976) Isolation of bacteriophages specific for the genus *Veillonella*. *Archives of Oral Biology* 21, 215–217.

Hitch, G., Pratten, J. and Taylor, P.W. (2004) Isolation of bacteriophages from the oral cavity. *Letters in Applied Microbiology* 39, 215–219.

Hoogenraad, N.J. and Hird, F.J.R. (1970) Electron-microscopic investigation of the flora in sheep alimentary tract. *Australian Journal of Biologial Sciences* 23, 793–808.

Hoogenraad, N.J., Hirk, F.J., Holmes, I. and Millis, N.F. (1967) Bacteriophages in rumen contents of sheep. *Journal of General Virology* 1, 575–576.

Hyman, P. and Abedon, S.T. (2010) Bacteriophage host range and bacterial resistance. *Advances in Applied Microbiology* 70, 217–248.

Kiliç, A.O., Pavlova, S. I., Alpay, S., Kiliç, S.S. and Tao, L. (2001) Comparative study of vaginal *Lactobacillus* phages isolated from women in the United States and Turkey: prevalence, morphology, host range, and DNA homology. *Clinical Diagnostic and Laboratory Immunology* 8, 31–39.

Klieve, A.V. (1991) Bacteriophages from the forestomachs of Australian marsupials. *Applied and Environmental Microbiology* 57, 3660–3663.

Kulikov, E.E., Isaeva, A.S., Rotkina, A.S., Manykin, A.A. and Letarov, A.V. (2007) Diversity and dynamics of bacteriophages in horse feces. *Mikrobiologiia* 76, 271–278.

Labrie, S.J., Samson, J.E. and Moineau, S. (2010) Bacteriophage resistance mechanisms. *Nature Reviews Microbiology* 8, 317–327.

Lepage, P., Colombet, J., Marteau, P., Sime-Ngando, T., Doré, J. and Leclerc, M. (2008) Dysbiosis in inflammatory bowel disease: a role for bacteriophages? *Gut* 57, 424–425.

Letarov, A. and Kulikov, E. (2009) The bacteriophages in human- and animal body-associated microbial communities. *Journal of Applied Microbiology* 107, 1–13.

Letarov, A., Golomidova, A. and Tarasyan, K. (2010) Ecological basis for rational phage therapy. *Acta Naturae* 2, 66–79.

Letarova, M., Strelkova, D., Nevolina, S. and Letarov, A. (2012) A test for the "physiological phagemia" hypothesis – natural intestinal coliphages do not penetrate to the blood in horses. *Folia Microbiol (Praha)* [Epub ahead of print]

Linhares, I.M., Summers, P.R., Larsen, B., Giraldo, P.C. and Witkin, S.S. (2011) Contemporary perspectives on vaginal pH and lactobacilli. *American Journal of Obstetrics and Gynecology* 204, 120.e1–5.

Lusiak-Szelachowska, M., Weber-Dabrowska, B. and Górski, A. (2006) The presence of bacteriophages in human feces and their potential importance. *Polski Merkuriusz Lekarski* 124, 381–383.

MacFarlane, S., Bahrami, B. and Macfarlane, G.T. (2011) Mucosal biofilm communities in the human intestinal tract. *Advances in Applied Microbiology* 75, 111–143.

Martin, R. and Suarez, J.E. (2010) Biosynthesis and degradation of H_2O_2 by vaginal lactobacilli. *Applied and Environmental Microbiology* 76, 400–405.

Martin, R., Soberyn, N., Escobedo, S. and Suarez, J. (2009) Bacteriophage induction versus vaginal homeostasis: role of H_2O_2 in the selection of *Lactobacillus* defective prophages. *International Microbiology* 12, 131–136.

Minot, S., Sinha, R., Chen, J., Li, H., Keilbaugh, S.A., Wu, G.D., Lewis, J.D. and Bushman, F.D. (2011) The human gut virome: inter-individual variation and dynamic response to diet. *Genome Research* 21, 1616–1625.

Nemergut, D.R., Costello, E.K., Hamady, M., Lozupone, C., Jiang, L., Schmidt, S.K., Fierer, N., Townsend, A.R., Cleveland, C.C., Stanish, L. and Knight, R. (2011) Global patterns in the biogeography of bacterial taxa. *Environmental Microbiology* 13, 135–144.

Olivera, A., Sereno, R., Nicolau, A. and Azeredo, J. (2009) The influence of the mode of administration in the dissemination of three coliphages in chickens. *Poultry Science*, 88, 728–733.

Paynter, M.J.B., Ewert, D.L. and Chalupa, W. (1969) Some morphological types of bacteriophages in bovine rumen contents. *Applied Microbiology* 18, 942–943.

Pride, D.T., Salzman, J., Haynes, M., Rohwer, F., Davis-Long, C., White, R.A. III, Loomer, P., Armitage, G.C. and Relman, D.A. (2011) Evidence of a robust resident bacteriophage population revealed through analysis of the human salivary virome. *ISME Journal* doi: 10.1038/ismej.2011.169 (Epub ahead of print).

Reyes, A., Haynes, M., Hanson, N., Angly, F.E., Heath, A.C., Rohwer, F. and Gordon, J.I. (2010) Viruses in the faecal microbiota of monozygotic twins and their mothers. *Nature* 466, 334–338.

Ricca, D.M. and Cooney, J.J. (2000) Screening environmental samples for source-specific bacteriophage hosts using a method for simultaneous pouring of 12 Petri plates. *Journal of Industrial Microbiology and Biotechnology* 24, 124–126.

Rice, S.A., Tan, C.H., Mikkelsen, P.J., Kung, V., Woo, J., Tay, M., Hauser, A., McDougald, D., Webb, J.S. and Kjelleberg, S. (2009) The biofilm life cycle and virulence of *Pseudomonas aeruginosa* are dependent on a filamentous prophage. *ISME Journal* 3, 271–282.

Savage, D.C. (1977) Microbial ecology of the gastrointestinal tract. *Annual Review of Microbiology* 31, 107–133.

Schaper, M., Jofre, J., Uys, M. and Grabow, W.O.K. (2002) Distribution of genotypes of F-specific RNA bacteriophages in human and non-human sources of faecal pollution in South Africa and Spain. *Journal of Applied Microbiology* 92, 657–667.

Servin, A.L. (2008) Antagonistic activities of lactobacilli and bifidobacteria against microbial pathogens. *FEMS Microbiology Reviews* 28, 405–440.

Sobel, J.D. (2000) Bacterial vaginosis. *Annual Review of Medicine* 51, 349–356.

Stevens, R.H., Porras, O.D. and Delisle, A.L. (2009) Bacteriophages induced from lysogenic root canal isolates of *Enterococcus faecalis*. *Oral Microbiology and Immunology* 24, 278–284.

Sulakvelidze, A. and Kutter, E. (2005) Bacteriophage therapy in humans. In: Kutter, E. and Sulakvelidze, A. (eds) *Bacteriophages: Biology and Applications.* CRC Press, Boca Raton, FL, pp. 381–436.

Swain, R.A., Nolan, J.V. and Klieve, A.V. (1996) Effect of tannic acid on the bacteriophage population of the rumen. *Microbiology Australia* 17, A87(GWP.27).

Tarakanov, B.V. (1971) The electron-microscopy examination of the microflora of reindeer rumen. *Mikrobiologia* 41, 862–870.

Tylenda, C.A., Calvert, C., Kolenbrander, P.E and and Tylenda, A. (1985) Isolation of *Actinomyces* bacteriophage from human dental plaque. *Infection and Immunity* 49, 1–6.

Uchiyama, J., Maeda, Y., Takemura, I., Chess-Williams, R., Wakiguchi, H. and Matsuzaki, S. (2009) Blood kinetics of four intraperitoneally administered therapeutic candidate bacteriophages in healthy and neutropenic mice. *Microbiology and Immunology* 53, 301–304.

Verstraelen, H., Verhelst, R., Vaneechoutte, M. and Temmerman, M. (2010) The epidemiology of bacterial vaginosis in relation to sexual behaviour. *BMC Infectious Disease* 10, 81.

Vitiello, C.L., Merril, C.R. and Adhya, S. (2005) An amino acid substitution in a capsid protein enhances phage survival in mouse circulatory system more than a 1000-fold. *Virus Research* 114, 101–103.

Weinbauer, M. (2004) Ecology of procaryotic viruses. *FEMS Microbiology Reviews* 28, 127–181.

Weitz, J.S., Hatman, H. and Levin, S.A. (2005) Coevolution arms races between bacteria and bacteriophage. *Proceedings of the National Academy of Sciences USA* 102, 9535–9540.

Willner, D. and Furlan, M. (2010) Deciphering the role of phage in the cystic fibrosis airway. *Virulence* 1, 309–313.

Willner, D., Furlan, M., Haynes, M., Schmieder, R., Angly, F.E., Silva, J., Tammadoni, S., Nosrat, B., Conrad, D. and Rohwer, F. (2009) Metagenomic analysis of respiratory tract DNA viral communities in cystic fibrosis and non-cystic fibrosis individuals. *PLoS One* 4, e7370.

Willner, D., Furlan, M., Schmieder, R., Grasis, J.A., Pride, D.T., Relman, D.A., Angly, F.E., McDole, T., Mariella, R.P. Jr, Rohwer. F. and Haynes, M. (2011) Metagenomic detection of phage-encoded platelet-binding factors in the human oral cavity. *Proceedings of the National Academy of Sciences USA* 108, 4547–4553.

Zhang, T., Breitbart, M., Lee, W.H., Run, J.Q., Wei, C.L., Soh, S.W., Hibberd, M.L., Liu, E.T., Rohwer, F. and Ruan, Y. (2006) RNA viral community in human feces: prevalence of plant pathogenic viruses. *PLoS Biology* 4, e3.

3 Diseases Caused by Phages

Sarah Kuhl[1], Stephen T. Abedon[2] and Paul Hyman[3]
[1]*VA Northern California Health Care System, Martinez;* [2]*Department of Microbiology, The Ohio State University;* [3]*Department of Biology/Toxicology, Ashland University*

Genomic sequencing of bacterial pathogens has demonstrated that many of the virulence factor molecules (also called pathogenicity factors) that allow bacteria to be more effective pathogens are encoded by prophages; in other words, these otherwise seemingly bacterial genes are found within the genomes of temperate phages. Temperate phage genomes can, in turn, reside semi-permanently within bacteria as prophages, forming what are commonly described as bacterial lysogens, or simply lysogens. In some cases, these prophages have only recently been acquired by the bacteria, while in other cases they have resided within the bacterial lineage for many generations. Prophages can also be lost from bacteria, wholly or in part. The acquisition step itself represents a form of horizontal gene transfer (see Hendrickson, Chapter 5, this volume), one that can loosely be described as a kind of transduction (see Abedon, Chapter 1, this volume). Overall, the number of virulence factors that are prophage associated is both large and diverse (see Christie *et al.*, Chapter 3, this volume). The evolution of many bacterial pathogens thus appears to be as greatly mediated by phages as by plasmid-mediated horizontal gene transfer (Relman, 2011). Many of the resulting lysogen-associated diseases are in fact explicitly caused by phage-encoded exotoxins, for example diphtheria, cholera and Shiga toxin-induced diarrhoeas (see below). In a very real sense, then, there exist diseases that can be attributed, to a large extent, to the action of phages.

These various phage-associated bacterial diseases historically have contributed to substantial mortality and morbidity. With clean water, adequate nutrition, vaccinations and effective antibiotics, however, in many cases we have all but forgotten what many of these diseases are like, although that is not necessarily true in countries or regions that lack these health-promoting services. This relative lack of rampant infectious disease indeed is only relatively recently the case, even in what we today describe as the developed world. The grandmother of one of us (S.K.), for example, had six siblings die before the age of five in rural Pennsylvania, several apparently from the 'deadly scourge of childhood', diphtheria, a lysogen-mediated disease that had incited fear for decades. By contrast, few American physicians of more recent generations have even seen diphtheria, although the same cannot be said for Russian physicians practising in the aftermath of the break up of the Soviet Union (Vitek *et al.*, 1999). In Haiti, a cholera epidemic, also a lysogen-mediated disease, followed the breakdown in public heath after the 2010 earthquake. Cholera, which had not been seen in Haiti for 100 years, incited a panic similar to that seen during diphtheria or cholera outbreaks in the past. The dreaded 'bad' *Escherichia coli, E. coli* O157:H7, too is 'bad' particularly because of the prophage-encoded Shiga toxin. Lysogen-mediated diseases thus exist as more than just historical oddities. Rather, they can have both real and recent impacts, resulting in substantial morbidity and mortality, even in our otherwise modern world.

What do these diseases look like? To orient the reader, in this chapter we provide an overview of various diseases that are associated with virulence factors that are prophage encoded. Table 3.1 contains a list of diseases in which prophage-encoded virulence factors play a role. This table is not exhaustive but provides an overview to show the scope of illnesses caused by these virulence factors. The goal is to bring discussions within this monograph into the realm of the purely medical before returning to emphases that are more abstract or molecular. This chapter additionally serves as a counterbalance to the preceding chapter by Letarov (Chapter 2, this volume), which places an emphasis on virions rather than prophages and the normal, healthy state rather than diseases. It also provides a prologue of sorts to the chapter by Christie *et al.* (Chapter 4, this volume), where the variety of toxins at the molecular level is deeply explored. This then is an introduction to the dark side of phages: diseases that are aided and abetted especially by temperate phages and the lysogenic-converting genes that they carry.

Diphtheria

The classical example of a phage-mediated as well as exotoxin-mediated disease is diphtheria. Indeed, this was the first disease for which a linkage between bacterial virulence and the presence of a prophage was firmly established (Freeman, 1951). Diphtheria is caused by infection with *Corynebacterium diphtheriae,* a Gram-positive, facultative anaerobic bacterium. A clinically similar disease can be caused by *Corynebacterium ulcerans* strains that also produce diphtheria toxin, presumably after lysogenic conversion by a toxin-encoding phage (CDC, 2011a). Prior to the introduction of the first effective vaccines in the 1920s, diphtheria was endemic worldwide. Today, it is still seen in parts of Africa, South and Central America, Eastern Europe, Asia and the Middle East (CDC, 2011b). See Christie *et al.* (Chapter 4, this

Table 3.1. Bacteria containing toxigenic prophages and associated diseases.

Organisms with toxigenic prophage(s)	Disease/syndrome
Clostridium botulinum	Botulism
Corynebacterium diphtheriae and *Corynebacterium ulcerans*	Diphtheria, cutaneous infections
Escherichia coli, especially Shiga-toxigenic *E. coli* (STEC) including *E. coli* O157:H7	Diarrhoeal disease of varying severity including enterohaemorrhagic diarrhoeas, haemolytic uraemic syndromes
Pasteurella multocida	Cellulitis and other wound infections
Pseudomonas aeruginosa	Opportunistic pathogen causing wound infections, pneumonias and chronic lung infections of cystic fibrosis patients
Salmonella enterica serovar Typhimurium and related *S. enterica* strains	Food-associated diarrhoeal diseases
Staphylococcus aureus (including MRSA and multidrug-resistant *S. aureus*)	Cellulitis, impetigo, osteomyelitis, pneumonia, endocarditis, septicaemia, toxic shock syndrome (TSS), staphylococcal scalded skin syndrome, food poisoning
Streptococcus pyogenes (group A) and some group C and G *Streptococcus* species	Scarlet fever, necrotizing fasciitis, streptococcal TSS, septicaemia, myositis
Vibrio cholerae	Cholera

Not all of the listed diseases are directly associated with a prophage-encoded virulence factor.

volume) for a more detailed discussion of the phages and conversion factors involved.

Diphtheria, as an upper respiratory tract infection, typically starts as a fever and sore throat but progresses to a swollen neck and a grey 'pseudomembrane' that adheres to the tonsils and throat. The pseudomembrane can block the airway and suffocate a child, particularly a small child with a small airway. Diphtheria toxin can also enter the bloodstream and cause damage to the heart and other organs. Occasionally, a cutaneous infection is seen, causing skin lesions that vary in appearance. Transmission is usually via contact with respiratory secretions, although the exudates in cutaneous lesions are also infectious. The mortality rate of the respiratory form of diphtheria is 5–10% (CDC, 2011a), with children most severely affected. Morbidity and mortality are largely a function of the diphtheria exotoxin, with the phage-encoded virulence factor synthesized particularly in response to iron starvation (Tao *et al.*, 1994).

Both treatment and prevention of diphtheria involve the presence of anti-toxin antibodies, which serve to neutralize the toxin. Prevention is effective via the administration of a toxoid vaccine – the D or d in the DT, dT, DTaP and Tdap vaccines combining diphtheria, tetanus and pertussis toxoids in concentrations appropriate to children and adults. These vaccines can be viewed as anti-phage or at least anti-phage gene-product medical treatments, as well as anti-toxin prophylaxes. Historically, passive immunity to diphtheria, and therefore treatment, was achieved via administration of horse antiserum or anti-toxin (Mortimer, 2011), for which von Behring was awarded the Nobel Prize in 1891. Such treatments, however, are associated with serum sickness, as the patient's immune system reacts to the horse serum. Administration of horse-serum anti-toxin none the less continues to be a treatment option (Ciok, 2000).

Botulism

The clostridia are a group of Gram-positive, spore-forming, obligately anaerobic bacteria that are found especially in soil environments. *Clostridium*-associated diseases include tetanus, gas gangrene and botulism – which are caused by *Clostridium tetani*, *Clostridium perfringens* and *Clostridium botulinum*, respectively – along with *Clostridium difficile*-associated diarrhoea and pseudomembranous colitis. Only in the case of some (but not all) *C. botulinum* toxins are any of these species' pathogenicity factors known to be phage encoded. Further discussion of the molecular aspects of the latter is found in Christie *et al.* (Chapter 4, this volume).

The predominant symptom of botulism is paralysis, first affecting the head (eyes and mouth), resulting in vision problems and difficulty in speaking and swallowing. As the disease progresses, the paralysis descends, affecting neck, arm and hand movements, and may also involve the legs. Eventually, the breathing muscles become impaired, leading to death due to respiratory failure (Sobel, 2005). The rate and extent of paralysis is variable, presumably depending on the toxin dose. While the toxin molecules bind irreversibly to neurons, the cells will regenerate over time if the patient survives, and patients will regain movement over weeks or months with proper supportive and rehabilitation treatment.

Botulism typically presents as a food poisoning. Common sources include home-canned foods that have not been successfully heat treated to kill spores (Sobel, 2005). Despite this common association with improperly canned foods, however, botulism is in fact most often seen in the USA in infants who have been fed honey, which is a consequence of both the presence of *C. botulinum* spores in honey and the relatively high pH of an infant's gastric secretions, which allows the spores to survive and germinate in the infant. Wound botulism can also occur when spores from soil contaminate and then grow within a wound, resulting in paralysis. Thus, unlike botulism food poisoning, which is a consequence of ingestion of toxin rather than spores, botulism in infants as well as wound botulism are consequences of direct exposure to spores.

Although the botulinum toxin is considered to be the most potent of the known neurotoxins, 'Botox' treatments involve the local injection of the toxin to cause a localized temporary muscle paralysis, as used for the treatment of deep facial wrinkles associated with facial muscle contraction. This could be considered a medical application of a phage-encoded toxin, although there are both phage-encoded and bacterially encoded botulinum toxins. In 2009, the US Food and Drug Administration issued a 'black box warning' that the effects of the botulinum toxin may spread from the area of injection to other areas of the body, causing symptoms similar to those of botulism, including potentially life-threatening swallowing and breathing difficulties and even death.

Only two of the seven known botulinum toxins are encoded by prophage genes. The other five are produced from genes in the *C. botulinum* genome (see Christie *et al.*, Chapter 4, this volume). This raises the question of the utility that the toxin could provide to phages other than supplying some useful function to bacterial hosts upon lysogen formation (Abedon and LeJeune, 2005). It may be that the botulinum toxin in these phages is an example of a 'moron' – more DNA located in a genome by evolutionarily recent random acquisition rather than a gene maintained by selection and adaptation. See Hyman and Abedon (2008) for additional discussion on the role of pathogenicity factors in phage-bacterial interactions.

As with diphtheria, the treatment of botulism poisoning can involve the administration of anti-toxin derived from horses. In addition, the California Department of Health has developed BabyBIG®, Botulism Immune Globulin Intravenous (Human) (BIG-IV), for infants less than 1 year of age. BIG-IV is a human-derived botulism anti-toxin antibody preparation purified from pooled plasma from adults selected for their high titres of neutralizing antibody against botulinum neurotoxin types A and B following immunization with pentavalent botulinum toxoid. Even with anti-toxin, supportive care, often including weeks on a ventilator for patients with respiratory failure, is the mainstay of care.

Unlike diphtheria, no anti-botulism vaccine is generally available. Botulism instead tends to be avoided through proper food handling. The latter can include proper cooking, as the botulinum toxin is permanently denatured at relatively low temperatures (well below boiling). Killing spores requires pressure cooking or autoclaving at 121°C (250°F) for at least 3 min. Providing conditions that prevent the spores from growing – such as an acidic or osmotically unfavourable environment, e.g. traditional sugared jams and jellies – is also a successful method of prevention, as is avoiding feeding infants honey. Notwithstanding this phage association and indeed cause of some of this food-borne disease, phages can alternatively be employed as protective agents of food supplies, as discussed by Niu *et al.* (Chapter 16, this volume), although not specifically against *C. botulinum*.

Staphylococcus Infection

Gram-positive bacteria are common components of the normal flora in humans, especially on skin (see Loc-Carrillo *et al.*, Chapter 13, this volume, for an additional discussion of human bacterial normal flora). Among Gram-positive pathogens possessing phage-associated pathogenicity, perhaps the most notorious is *Staphylococcus aureus*. *S. aureus* has numerous pathogenicity factors, a substantial number of which can be phage encoded (see Christie *et al.*, Chapter 4, this volume, for additional discussion). Indeed, an array of phage-encoded pathogenicity factors can be found in this pathogen: superantigens, including toxic shock syndrome (TSS) toxin; Panton–Valentine leukocidin (PVL) toxin; staphylococcal enterotoxins (associated with food poisoning and TSS) (Novick *et al.*, 2010) which are discussed below. Other phage-encoded virulence factors include staphylokinase, chemotaxis inhibitory protein, staphylococcal complement inhibitor and factors associated with biofilm formation (Verkaik *et al.*, 2011; see also Christie, *et al.*, Chapter 4, this volume).

Although not a direct consequence of phage-encoded proteins, *S. aureus* strains are becoming increasingly resistant to common antibiotics, with some strains of methicillin-resistant *S. aureus* (MRSA) presenting with very limited antibiotic choices, sometimes limited to intravenous vancomycin, as oral vancomycin is not absorbed (but see Williams and LeJeune, Chapter 6, this volume, for a brief discussion of the potential role of phages in antibiotic resistance horizontal gene transfer, and Hendrickson, Chapter 5, this volume, for additional discussion). The coincidence of the acquisition of methicillin resistance with at least partially phage-mediated evolution of a virulent community-acquired strain has focused much public and medical concern on drug-resistant strains of *S. aureus*, especially MRSA. The repeated use of vancomycin has also resulted in the appearance of increasingly vancomycin-resistant and -intermediate-resistant strains. Drug-sensitive strains of *S. aureus* still cause disease and mortality. Penicillin-sensitive strains, however, are rarely seen except in recrudescence of infections that originally occurred before the antibiotic era, such as in World War II veterans.

S. aureus normally colonizes the skin and nasopharynx of humans and animals and can produce a variety of diseases, from relatively benign skin infections such as folliculitis (hair follicle infections) and small abscesses to life-threatening infections such as deep abscesses, osteomyelitis and bone infections (see Loc-Carrillo *et al.*, Chapter 13, this volume), pneumonia, septicaemia, endocarditis and heart-valve infections (Gordon and Lowy, 2008; Que and Moreillon, 2010). In addition, several diseases are mediated by toxin production including staphylococcal scalded-skin syndrome (SSSS), staphylococcal TSS, and food poisoning or intoxication.

Skin and lung infections

S. aureus is the major pathogenic cause of cellulitis (skin infections with diffuse inflammation) and of abscesses (collections of pus). Most people are familiar with an infected wound, hair follicle or fingernail containing pus, typically caused by *S. aureus*. Recently, the epidemic USA300 strain of *S. aureus* has presented with a localized area of inflammation, often described as a 'spider bite'. It may eventually form a pustule, and may progress to more serious infection including septic shock.

S. aureus can also cause deep abscesses in muscle or soft tissue, osteomyelitis or bone infections, pneumonia, which is often necrotizing, septicaemia, endocarditis and heart-valve infections. Necrotizing fasciitis ('flesh-eating bacteria') is typically caused by *Streptococcus pyogenes* but can be caused by *S. aureus*. This fasciitis typically presents with severe pain in a moderate infection, pain that is typically out of proportion to the physical signs of infection. Emergency surgical debridement is indicated in these cases. These infections may occur due to infected intravenous lines or intravenous drug abuse, surgical site infection or infection of a traumatic wound.

Two forms of disseminated skin infections are associated with the production of exfoliative toxins including prophage-encoded exfoliative toxin A. The first, impetigo, is most commonly seen in children between 2 and 5 years of age. It presents as an area with small sores that may be eroded (non-bullous) or fluid filled (bullous, i.e. consisting of fluid-filled blisters) (Stulberg *et al.*, 2002). Each sore is a site of localized staphylococcal infection. The blisters in bullous impetigo are fluid filled, painless and caused by a localized lifting off of the outer layers of the skin due to the effects of the exfoliative toxins. Patients will often have fever and malaise. In most cases, impetigo is treatable with topical antibiotics, although there is a small risk of systemic infection developing if the bacteria are able to penetrate to deeper tissue. Impetigo is sometimes considered a localized form of SSSS.

SSSS, also known as Ritter's disease, is a serious, sometimes life-threatening disease that presents with widespread erythema (redness) and large bullae covering large portions of the skin (Bukowski *et al.*, 2010). Most commonly seen in children, it is the bullae that give the appearance of scalded skin. Like impetigo, patients with SSSS often have fever and lethargy. In contrast to the bullae seen in impetigo, SSSS bullae often do not contain staphylococci. This can occur because SSSS is primarily mediated by toxin production, often from a localized infection. Further challenges occur as the patient improves with antibiotic treatment, but desquamation of the superficial layer of skin typically occurs. After this, it is important to protect the exposed skin layers from secondary infections as well as from fluid loss.

The role of phage-encoded toxins in *Staphylococcus* infections can be seen explicitly with PVL. PVL-producing *Staphylococcus* strains have been shown to be associated with severe or invasive skin infections and pneumonia since PVL was discovered in the 1930s (Gordon and Lowy, 2008; see Christie *et al.*, Chapter 4, this volume). In the current epidemic of the community-acquired MRSA strain USA300, expression of PVL appears to correlate with more-rapid onset and increased severity of symptoms, as seen within the first 3 days of infection, although it does not increase the virulence of this strain in animal studies (Diep and Otto, 2008; Kobayashi *et al.*, 2011). Frequently, human skin infection with USA300 appears as an area of redness or erythema, followed by localized tissue necrosis. This occurs without much pus, unlike typical purulent staphylococcal infections, which can cause PVL-containing infections to present like 'spider bites'. Staphylococcal pneumonias are often fulminant with high fevers, and necrotizing pneumonias are not uncommon. Pneumonia typically presents with shortness of breath, cough and fever. Necrotizing fasciitis has also been described with this strain.

Gastrointestinal intoxication

Enterotoxins are associated with staphylococcal food poisoning that can occur upon consumption of *S. aureus*-contaminated food or drink. This food poisoning presents 2–6 h after eating with nausea, vomiting, diarrhoea and abdominal cramping. Symptoms can also include prostration, hypotension, headache and muscle cramps, and typically resolve within 6–12 h. Treatment is rehydration, orally if possible, intravenously if severe. Foods may be contaminated during preparation especially those that are typically prepared by hand. Milk and cheese have also been sources of infection. Because *Staphylococcus* is salt tolerant, salted foods such as ham or fish may be a source of infection. Even cooked foods may remain contaminated with staphylococcal toxins, some of which are heat stable, although the bacteria are killed during cooking (CDC, 2006). Staphylococcal enterotoxins A and P are typically phage encoded and are associated with staphylococcal food poisoning.

Staphylococcal enterotoxin A, a prophage-encoded protein, is also a superantigen (Balaban and Rasooly, 2000). This means it binds directly to MHC class II molecules in a non-specific manner and activates T cells, resulting in a massive cytokine response, causing the patient to develop systemic inflammatory response syndrome (SIRS). SIRS can occur in response to sepsis or other methods of cytokine activation such as burns or massive trauma.

Toxic shock syndrome

Another staphylococcal disease that is predominantly toxin-mediated is TSS. In TSS, as with generalized SSSS, the infection may remain localized, but the toxaemia causes systemic effects, such as skin exfoliation and SIRS. TSS is associated with *S. aureus*-produced exotoxins, especially the eponymous toxic shock syndrome toxins. In the 1980s, TSS was associated with superabsorbent tampons, which were inadvertently well suited to supporting *S. aureus* growth (and subsequently were removed from the marketplace). TSS continues to occur in post-surgical and other sporadic infections. Wounds infected with *S. aureus* may not be inflamed, presumably due to the toxin inhibiting the influx of macrophages.

The presentation of TSS is variable and is thought to relate to the specific toxins excreted by particular infecting strains. TSS patients typically present with localized pain, high fever, hypotension (low blood pressure) and sometimes fatigue and confusion (Stevens, 1995). TSS can quickly progress to include an erythematous rash (sometimes described as sunburn-like in appearance, but often with macules, or spots), and often with the involvement of other organ systems resulting in vomiting, diarrhoea, myalgias (muscle pain), low platelet counts, confusion, renal insufficiency and/or hepatic inflammation. Treatment includes removal of the infection focus, antibiotics, supportive care, often including 10–20 l of intravenous fluid, and vasopressors to counter the hypotension due to capillary leakage.

Streptococcus Infection

Members of the genus *Streptococcus* are another Gram-positive pathogen for which phages play important roles in pathogenicity (see Christie *et al.*, Chapter 4, this volume). *S. pyogenes* (group A *Streptococcus* or GAS) is the major pathogen, but other species also have phage-encoded virulence factors. GAS has multiple clinical presentations, including streptococcal pharyngitis or sore throat, cellulitis and occasionally abscesses, necrotizing fasciitis, scarlet fever and TSS (sometimes described as toxic shock-like syndrome to differentiate it from the syndrome caused by *S. aureus*). Other streptococcal skin and soft-tissue infections caused include cellulitis, erysipelas (which presents with raised demarcated bright-red lesions), necrotizing fasciitis, and myositis or myonecrosis. The clinical presentation depends on the combination of infection site and which virulence factors are produced by the particular infecting strain. Additional streptococcal species containing toxigenic prophages include group C and G streptococci and *Streptococcus mitis*. These species typically colonize the respiratory and gastrointestinal tract, but can cause infections similar to those caused by GAS, as well as endocarditis, osteomyelitis and infections at other body sites.

Streptococcal virulence factors include adhesion factors, superantigens, lipase, DNase, streptokinase and hyaluronidase (Hynes *et al.*, 1995; Banks *et al.*, 2005). The streptococcal pyrogenic exotoxins (Spes) are a family of superantigens that cause symptoms of systemic shock, similar to the superantigens in *S. aureus*. With only a few exceptions, these are all encoded by genes found in prophages (see Christie *et al.*, Chapter 4, Table 4.2, this volume, for specific references for each Spe). Spes are believed to be particularly associated with streptococcal TSS, necrotizing fasciitis and other severe streptococcal infections (Bisno and Stevens, 2010).

Streptococcal TSS is toxin mediated, as is staphylococcal TSS, and presents with a streptococcal infection associated with sudden onset of SIRS. *Streptococcus*, like *Staphylococcus*, can also present with bacteraemia and/or superantigen-mediated activation of a cytokine response. This is related to the SIRS, which is often seen in conjunction with sepsis or septicaemia. Typically, these patients may also present with fever, chills, hypotension and tachycardia (rapid heart rate). The septic patient is often delirious and confused, in part because the brain is poorly perfused due to vascular leakage and hypotension.

Additional phage-encoded proteins include hyaluronidase, DNase, streptokinase and SpeB (a potent protease), all of which are thought to facilitate the liquefaction of pus and spread of streptococci through tissue planes. DNases are thought to also help resist neutrophil extracellular trap (NET)-mediated killing by neutrophils. SpeB in addition cleaves IgG bound to GAS, interfering with ingestion and killing by phagocytes.

SpeA and SpeC are known as the scarlatinal toxins due to their association with scarlet fever. Scarlet fever results from infections with strains that secrete these pyrogenic exotoxins. This disease is most often associated with streptococcal pharyngitis or sore throat, but can follow wound or other infections. Scarlet fever usually affects children between the ages of 5 and 18, but is sometimes seen in adults. Typically, scarlet fever presents with fever, 'strawberry tongue' and a fine erythematous rash that spares the face but is worse in skin folds (known as Pastia lines) (CDC, 2011c).

In contrast to *S. aureus*, *S. pyogenes* remains exquisitely sensitive to penicillin, a cell-wall synthesis inhibitor, which as a consequence often remains the drug of choice, although once-daily cephalosporins are sometimes used for ease of delivery with a negligible decrease in efficacy. In cases of severe streptococcal infection, such as necrotizing fasciitis, the protein synthesis inhibitor clindamycin is typically added to inhibit the exotoxin production by the bacteria. The failure of penicillin in cases of high inoculum (the inoculum effect) is thought to be due to the ineffectiveness of penicillin against stationary-phase bacteria.

Pasteurella Infection

Pasteurella are Gram-negative bacteria that are responsible for zoonotic infections of humans, including after bites by pets or by wild animals (Zurlo, 2010). Infections are sometimes seen after licking of the patient's skin by a pet, especially where there is a break in the skin such as after injury or surgery, including after knee or hip replacement (Heydemann *et al.*, 2010). Morbidity is also seen in infected animals.

Infections most commonly present as a severe inflammatory cellulitis or diffuse inflammation of the skin around the bite or wound site, and are caused by the species *Pasteurella multocida*. Some strains of *P. multocida* produce *P. multocida* toxin, a phage-encoded exotoxin that can rapidly cause severe inflammation and swelling at the site of the bite, sometimes necessitating surgical debridement (Pullinger *et al.*, 2004). *P. multocida* remains antibiotic sensitive, usually treated by amoxicillin-clavulanate in mild cases, or in severe cases with an intravenous antibiotic such as a β-lactam/β-lactamase inhibitor or a carbapenem.

Diarrhoeal diseases

Diarrhoeal diseases are the second leading cause of worldwide morbidity and mortality due to infectious diseases, after pneumonia and other lower respiratory tract diseases. According to the World Health Organization (WHO, 2008), there were over 2 million deaths due to diarrhoeal diseases in 2004. These typically present with loose, watery and frequent stools that may be bloody or not depending on the pathogen. In severe disease, such as some cases of cholera, the profuse amounts of watery diarrhoea can cause profound and sometimes fatal dehydration. Treatments range from oral rehydration with fluid and electrolytes to intravenous rehydration, antibiotics and in some cases renal dialysis due to toxin effects on the kidneys.

Cholera

Cholera is considered to be one of the great plagues that has had a tremendous impact on human history (Bray, 1996; Sherman, 2006). In the recent cholera epidemic in Haiti, patients became rapidly dehydrated in as little as 2 h, resulting in death when rehydration was not available. Cholera stools are described as 'rice water' stools because the flecks of mucous, intestinal epithelial cells and bacteria appear similar to rice. Phages play a direct role in the pathology of cholera, as the major cholera toxin, CT, is encoded by a gene in CTXϕ, a temperate filamentous phage that forms a stable lysogen in *Vibrio cholerae* (Waldor and Mekalanos, 1996; see also Christie *et al.*, Chapter 4, this volume).

Cholera is most commonly spread by faecal contamination of water and therefore is best controlled by properly functioning water treatment plants or other water purification systems. As a consequence, the last cholera outbreak in the USA occurred in 1910, killing 11 people in New York City. In parts of the world with seasonal flooding, regular outbreaks are seen when flooding brings a connection between water supplies and contaminated waste water. There also appears to be a role of phages in at least some of these outbreaks. Seasonal epidemics of cholera in the waters of Bangladesh, for example, appear to be controlled by lytic bacteriophages (Faruque *et al.*, 2005). One of the earliest applications of phages to populations, rather than individuals, in fact, took place when d'Hérelle and others treated cholera in India (Morison, 1932; d'Hérelle, 1946).

The mainstay of cholera treatment of individual patients is rehydration with fluid and electrolytes, orally if possible and intravenously if not. Antibiotics such as tetracyclines and fluoroquinolones will shorten the duration of diarrhoea by about 50% and shorten the duration of *Vibrio* shedding by about 1 day. Macrolide antibiotics are typically used in children and pregnant women. Outside the USA, there are two licensed commercial oral vaccines. The killed whole-cell B-subunit vaccine, WC-rBS (Dukoral®), contains killed whole cells of *V. cholerae* O1 plus recombinant cholera toxin B subunit. It is WHO pre-qualified, allowing it to be licensed in over 60 countries. Dukoral also provides some protection against infection with enterotoxigenic *E. coli*. The second type of vaccine is a modified bivalent whole-cell-type vaccine produced in two countries. These mORCVAX and Shanchol vaccines contain killed whole cells of *V. cholerae* O1 and *V. cholerae* O139. These vaccines are licensed in India and Vietnam, respectively, and appear to be safe and effective in endemic populations (Lopez *et al.*, 2008; CDC, 2010).

Shiga toxin-associated diseases

Dysentery is diarrhoea with blood or mucous in the stool. The two bacteria that cause bloody diarrhoea are *Shigella* and *E. coli* strains such as O157:H7 (see Christie *et al.*, Williams and LeJeune, Goodridge and Steiner, and Niu *et al.*, Chapters 4, 6, 11 and 16, this volume, for further discussion, particularly of *E. coli* O157:H7). *Shigella* typically causes fever and acute abdominal pain, and can also cause severe dehydration as well as haemolytic–uraemic syndrome (HUS) due to the production and systemic release of Shiga toxin. Severe disease in children can cause seizures, encephalopathy (brain disease with headache, confusion and lethargy) and stiff neck. HUS leads, in some cases, to a problem with coagulation called disseminated intravascular coagulation (DIC). DIC can result in a layer of fibrin forming in the glomerular capillary bed. Acute renal failure is also seen with HUS and DIC due to a combination of the effects of Shiga and other toxins. HUS, DIC and renal failure are most likely to occur in children, the elderly and pregnant women. Renal failure can become chronic, requiring dialysis or kidney transplantation.

Shiga-like toxins (Stx) in *E. coli* are typically located in the genome of one of several lambdoid prophages. The bacteria lack an export mechanism for the Stx so release of the toxin usually follows induction of the prophage, which then lyses the cell as part of the phage lytic cycle. The recent outbreak of a Stx-producing *E. coli* in Germany appears to be the result of a toxin-producing phage infecting an entero-aggregative *E. coli* O104:H4 strain (Rasko *et al.*, 2011), which otherwise tend to be less virulent, although still causing chronic diarrhoea.

Treatment of shigellosis (*Shigella*-caused dysentery) typically consists of hydration in combination with supportive care along with antibiotic treatment, typically with ampicillin or trimethoprim-sulfamethoxazole. This appears to shorten the duration of diarrhoea and microbial shedding. In contrast, supportive care and monitoring for HUS and DIC is the mainstay of treatment of enterohaemorrhagic *E. coli* (EHEC). Antibiotic treatment of EHEC in particular is not recommended, as it can induce the expression and release of Stx (Wong *et al.*, 2000). Targeting the toxin directly using hyperimmune anti-toxin antisera or oral toxin-binding agents has been proposed but remains experimental (Gupta *et al.*, 2011).

Pseudomonas Infections

Pseudomonas aeruginosa contains many prophages that appear to be involved in serotype conversion (Christie *et al.*, Chapter 4, this volume). *P. aeruginosa* strain 158, for example, contains a cytotoxin (a leukocidin) that is encoded by the temperate phage CTX (Hayashi *et al.*, 1990). More generally, *Pseudomonas* is a soil organism and tends only to be an opportunistic pathogen, rarely causing infection in healthy people except by inoculation, such as with a rusty nail through a tennis shoe. *Pseudomonas*, however, can cause severe wound and post-surgical infections, as well as pneumonias. Pneumonias are seen especially with patients undergoing respiratory therapy or on ventilators. *P. aeruginosa* in addition can infect immunocompromised persons and the elderly. These infections are complicated by the innate antibiotic resistance of *P. aeruginosa* and related species, with resistance to multiple antibiotics commonly seen in samples isolated from patients.

P. aeruginosa is a particularly serious pathogen for persons with the genetic defect causing cystic fibrosis. The respiratory tracts of these patients are often colonized first by *S. aureus* and later, after antibiotic treatment, with *P. aeruginosa* (and sometimes subsequently with *Burkholderia cepacia*). *Pseudomonas* infections tend to develop into chronic infections that are treated intermittently to reduce colony counts. In these cases, *Pseudomonas* usually develops a mucoid phenotype due to increased production of alginate, which causes a chronic lung infection in cystic fibrosis. *P. aeruginosa* and related species possess an innate antibiotic resistance and are becoming increasingly resistant to the few antibiotics that are currently used to treat these infections. Doctors have resorted, as a consequence, to again using more toxic antibiotics such as colistin. See, however, Burrowes and Harper (Chapter 14, this volume) for discussion of the use of phage therapy against various *Pseudomonas* infections.

Conclusion

Phages can be viewed as playing at least four roles in humans and other environments. They are an important component of the normal flora (see Letarov, Chapter 2, this volume), they can strongly influence the evolution of bacteria (see Hendrickson, Chapter 5, this volume), they can be employed in various guises as antibacterial agents (i.e. phage therapy; see the last section of this volume), and, as considered in this chapter, they can contribute directly to bacterial diseases. The last is a consequence of prophages encoding bacterial pathogenicity factors including exotoxins (see Christie *et al.*, Chapter 4, this volume). It is important to note, however, that these bacterial pathogens often possess non-phage-encoded bacterial virulence factors as well, although the acquisition of these too, if a consequence of horizontal transfer, can be due to the phage-mediated process of generalized transduction. Respectively, and metaphorically, phages thus may be viewed, under these different circumstances, as presenting to humans a multitude of qualities including serving as benign neighbours (normal flora; Letarov, Chapter 2, this volume), scheming anarchists (Hendrickson, Chapter 5, this volume), trusted police (phage therapy) and outright criminals (Christie *et al.*, Chapter 4, this volume).

Here, we have considered the consequences especially of phage 'criminal behaviour' in the guise of prophages encoding bacterial virulence factors. In some cases, such as infection by opportunistic soil bacteria such as *C. botulinum*, the phages appear to be accidental or incidental criminals. In more professional pathogens such as *C. diphtheriae*, *S. aureus* and *V. cholerae*, the phages appear to be the major contributors to the 'criminal behaviour' of their bacterial hosts, coding for major pathogenicity factors. Indeed, a common theme throughout this monograph is the idea that bacteria in fact can be very bad indeed, although as recounted particularly in Chapters 6–17, much good can come instead from human harnessing of bacteriophages for the detection, prevention and treatment of disease.

References

Abedon, S.T. and LeJeune, J.T. (2005) Why bacteriophage encode exotoxins and other virulence factors. *Evolutionary Bioinformatics Online* 1, 97–110.

Balaban, N. and Rasooly, A. (2000) Staphylococcal enterotoxins. *International Journal of Food Microbiology* 61, 1–10.

Banks, D.J., Beres, S.B. and Musser, J.M. (2005) Contribution of phages to Group A *Streptococcus* genetic diversity and pathogenesis. In: Waldor M.K., Friedman D.I.

and Adhya S.L. (eds) *Phages: Their Role in Bacterial Pathogenesis and Biotechnology.* ASM Press, Washington, DC, pp. 319–334.

Bisno, A.L. and Stevens, D.L. (2010) *Streptococcus pyogenes.* In: Mandell G.L., Bennett J.E. and Dolin R. (eds) *Mandell, Douglas, and Bennett's Principles and Practice of Infectious Diseases.* Churchill Livingstone/Elsevier, Philadelphia, PA, pp. 2593–2610.

Bray, R. S. (1996) *Armies of Pestilence: The Impact of Disease on History.* James Clark & Co., Cambridge, UK.

Bukowski, M., Wladyka, B. and Dubin, G. (2010) Exfoliative toxins of *Staphylococcus aureus. Toxins (Basel)* 2, 1148–1165.

CDC (2006) Staphylococcal food poisoning. <http://www.cdc.gov/ncidod/dbmd/diseaseinfo/staphylococcus_food_g.htm>.

CDC (2010) Cholera – prevention and control. <http://www.cdc.gov/cholera/prevention.html>.

CDC (2011a) Diphtheria. <http://www.cdc.gov/ncidod/dbmd/diseaseinfo/diptheria_t.htm>.

CDC (2011b) Infectious diseases related to travel – diphtheria <http://wwwnc.cdc.gov/travel/yellowbook/2012/chapter-3-infectious-diseases-related-to-travel/diphtheria.htm>.

CDC (2011c) Scarlet fever: a group A streptococcal infection. <http://www.cdc.gov/Features/ScarletFever/>.

Ciok, A.E. (2000) Horses and the diphtheria antitoxin. *Academic Medicine* 75, 396.

d'Hérelle, F. (1946) *L'étude d'une Maladie, le Choléra: Maladie a Paradoxes.* Rouge, Lausanne, Switzerland.

Diep, B.A. and Otto, M. (2008) The role of virulence determinants in community-associated MRSA pathogenesis. *Trends in Microbiology* 16, 361–369.

Faruque, S.M., Naser, I.B., Islam, M.J., Faruque, A.S.G., Ghosh, A.N., Nair, G.B., Sack, D.A. and Mekalanos, J.J. (2005) Seasonal epidemics of cholera inversely correlate with the prevalence of environmental cholera phages. *Proceedings of the National Academy of Sciences USA* 102, 1702–1707.

Freeman, V.J. (1951) Studies on the virulence of bacteriophage infected strains of *Corynebacterium diphtheriae. Journal of Bacteriology* 61, 675–688.

Gordon, R.J. and Lowy, F.D. (2008) Pathogenesis of methicillin-resistant *Staphylococcus aureus* infection. *Clinical Infectious Diseases* 46 Suppl 5, S350-S359.

Gupta, P., Singh, M.K., Singh, Y., Gautam, V., Kumar, S., Kumar, O. and Dhaked, R.K. (2011) Recombinant Shiga toxin B subunit elicits protection against Shiga toxin via mixed Th type immune response in mice. *Vaccine* 29, 8094–8100.

Hayashi, T., Baba, T., Matsumoto, H. and Terawaki, Y. (1990) Phage-conversion of cytotoxin production in *Pseudomonas aeruginosa. Molecular Microbiology* 4, 1703–1709.

Heydemann, J., Heydemann, J.S. and Antony, S. (2010) Acute infection of a total knee arthroplasty caused by *Pasteurella multocida*: a case report and a comprehensive review of the literature in the last 10 years. *International Journal of Infectious Diseases* 14 (Suppl. 3) e242–e245.

Hyman, P. and Abedon, S.T. (2008) Phage ecology of bacterial pathogenesis. In: Abedon S.T. (ed.) *Bacteriophage Ecology.* Cambridge University Press, Cambridge, UK, pp. 353–385.

Hynes, W.L., Hancock, L. and Ferretti, J.J. (1995) Analysis of a second bacteriophage hyaluronidase gene from *Streptococcus pyogenes*: evidence for a third hyaluronidase involved in extracellular enzymatic activity. *Infection and Immunity* 63, 3015–3020.

Kobayashi, S.D., Malachowa, N., Whitney, A.R., Braughton, K.R., Gardner, D.J., Long, D., Bubeck, W.J., Schneewind, O., Otto, M. and Deleo, F.R. (2011) Comparative analysis of USA300 virulence determinants in a rabbit model of skin and soft tissue infection. *Journal of Infectious Diseases* 204, 937–941.

Lopez, A.L., Clemens, J.D., Deen, J. and Jodar, L. (2008) Cholera vaccines for the developing world. *Human Vaccines* 4, 165–169.

Morison, J. (1932) *Bacteriophage in the Treatment and Prevention of Cholera.* H.K. Lewis, London, UK.

Mortimer, P.P. (2011) The diphtheria vaccine debacle of 1940 that ushered in comprehensive childhood immunization in the United Kingdom. *Epidemiology and Infection* 139, 487–493.

Novick, R.P., Christie, G.E. and Penades, J.R. (2010) The phage-related chromosomal islands of Gram-positive bacteria. *Nature Reviews Microbiology* 8, 541–551.

Pullinger, G.D., Bevir, T. and Lax, A.J. (2004) The *Pasteurella multocida* toxin is encoded within a lysogenic bacteriophage. *Molecular Microbiology* 51, 255–269.

Que, Y.A. and Moreillon, P. (2010) *Staphylococcal aureus* (including staphylococcal toxic shock). In: Mandell G.L., Bennett J.E. and Dolin R. (eds) *Mandell, Douglas, and Bennett's Principles and Practice of Infectious Diseases.* Churchill Livingstone/Elsevier, Philadelphia, PA, pp. 2543–2578.

Rasko, D.A., Webster, D.R., Sahl, J.W., Bashir, A., Boisen, N., Scheutz, F., Paxinos, E.E., Sebra,

R., Chin, C.S., Iliopoulos, D., Klammer, A., Peluso, P., Lee, L., Kislyuk, A.O., Bullard, J., Kasarskis, A., Wang, S., Eid, J., Rank, D., Redman, J.C., Steyert, S.R., Frimodt-Moller, J., Struve, C., Petersen, A.M., Krogfelt, K.A., Nataro, J.P., Schadt, E.E. and Waldor, M.K. (2011) Origins of the *E. coli* strain causing an outbreak of hemolytic-uremic syndrome in Germany. *New England Journal of Medicine* 365, 709–717.

Relman, D.A. (2011) Microbial genomics and infectious diseases. *New England Journal of Medicine* 365, 347–357.

Sherman, I. W. (2006) *The Power of Plagues.* ASM Press, Washington, DC.

Sobel, J. (2005) Botulism. *Clinical Infectious Diseases* 41, 1167–1173.

Stevens, D.L. (1995) Streptococcal toxic-shock syndrome: spectrum of disease, pathogenesis, and new concepts in treatment. *Emerging Infectious Diseases* 1, 69–78.

Stulberg, D.L., Penrod, M.A. and Blatny, R.A. (2002) Common bacterial skin infections. *American Family Physician* 66, 119–124.

Tao, X., Schiering, N., Zeng, H.Y., Ringe, D. and Murphy, J.R. (1994) Iron, DtxR, and the regulation of diphtheria toxin expression. *Molecular Microbiology* 14, 191–197.

Verkaik, N.J., Benard, M., Boelens, H.A., de Vogel, C.P., Nouwen, J.L., Verbrugh, H.A., Melles, D.C., van Belkum, A. and van Wamel, W.J. (2011) Immune evasion cluster-positive bacteriophages are highly prevalent among human *Staphylococcus aureus* strains, but they are not essential in the first stages of nasal colonization. *Clinical Microbiology and Infection* 17, 343–348.

Vitek, C.R., Brisgalov, S.P., Bragina, V.Y., Zhilyakov, A.M., Bisgard, K.M., Brennan, M., Kravtsova, O.N., Lushniak, B.D., Lyerla, R., Markina, S.S. and Strebel, P.M. (1999) Epidemiology of epidemic diphtheria in three regions, Russia, 1994–1996. *European Journal of Epidemiology* 15, 75–83.

Waldor, M.K. and Mekalanos, J.J. (1996) Lysogenic conversion by a filamentous phage encoding cholera toxin. *Science* 272, 1910–1914.

Wong, C.S., Jelacic, S., Habeeb, R.I., Watkins, S.L. and Tarr, P.I. (2000) The risk of the hemolytic-uremic syndrome after antibiotic treatment of *Escherichia coli* O157:H7 infections. *New England Journal of Medicine* 342, 1930–1936.

World Health Organization (2008) The Global Burden of Disease: (2004) Update. <http://www.who.int/healthinfo/global_burden_disease/GBD_report_2004update_full.pdf>.

Zurlo, J. J. (2010) *Pasteurella* species. In: Mandell G.L., Bennett J.E. and Dolin R. (eds) *Mandell, Douglas, and Bennett's Principles and Practice of Infectious Diseases.* Churchill Livingstone/Elsevier, Philadelphia, PA, pp. 2939–2942.

4 Prophage-induced Changes in Cellular Cytochemistry and Virulence

Gail E. Christie[1], Heather E. Allison[2], John Kuzio[3], W. Michael McShan[4], Matthew K. Waldor[5] and Andrew M. Kropinski[6]

[1]*Molecular Biology and Genetics, School of Medicine, Virginia Commonwealth University;* [2]*Department of Functional and Comparative Genomics, University of Liverpool;* [3]*Department of Microbiology and Immunology, Queen's University;* [4]*Department of Pharmaceutical Sciences, The University of Oklahoma;* [5]*Department of Medicine, Brigham and Women's Hospital;* [6]*Department of Molecular and Cellular Biology, University of Guelph; and, Public Health Agency of Canada, Laboratory of foodborne Zoonoses.*

Lysogenic conversion has been recently defined as a 'phage-associated heritable change in the host cell's genotype and phenotype that is independent of the effects expected from repression and integration. In other words, phenomena that are directly associated with lysogenization such as immunity to superinfection (a function of the prophage repressor) and loss of function resulting from insertion into a host gene (a function of the integrase) are expected consequences of lysogenization and should not be defined as lysogenic conversion' (Los *et al.*, 2010). We now realize that this definition is inadequate, as transcriptomic analysis has revealed that the molecular impact of lysogenization can be considerable (Su *et al.*, 2010) and that even virulent phages may induce transient changes in the cell's phenotype (Busch *et al.*, 2011). Therefore, we suggest that the term 'lysogenic conversion' or 'phage conversion' always be used in conjunction with an explicit description of the phenotypic change observed, as in 'phage conversion of serotype' or 'lysogenic conversion to toxigenicity'. What we will discuss in this chapter is the impact of non-essential temperate phage genes, variously described as morons (Juhala *et al.*, 2000) or cargo genes (Brüssow *et al.*, 2004), on host expression that leads to a phenotypic change in virulence or in surface chemistry. The specific emphases will be on the production of secreted toxins by *Corynebacterium*, *Clostridium*, *Escherichia*, *Staphylococcus*, *Streptococcus* and *Vibrio*, the synthesis of membrane proteins, and alterations in lipopolysaccharide (LPS) structure in *Salmonella* and *Pseudomonas*. We will conclude with a brief discussion of the concept of phage-mediated transfer of bacterial genes, or transduction.

Corynebacterium

Working with bacteriophage B, isolated by S. Toshach (School of Hygiene, Toronto, Canada) from a virulent culture of *Corynebacterium diphtheriae*, V.J. Freeman was able to demonstrate that this virus plaqued on avirulent strains of the bacterium, converting them to stable lysogenicity and toxinogeny

 Bacteriophages in Health and Disease (eds P. Hyman and S.T. Abedon)

(Freeman, 1951; Freeman and Morse, 1952). Phage β is directly descended from Toshach's isolate, and is the virus that, along with corynebacteriophage γ, has received the most attention. It is a temperate, UV-inducible member of the family *Siphoviridae* (Holmes and Barksdale, 1970). The *tox* gene has been found to be located adjacent to *attP* (Groman, 1984).

Most strains of *C. diphtheriae* are lysogenic. An extended group of prophages capable, upon induction, of plaque formation have been isolated from *C. diphtheriae* and *Corynebacterium ulcerans* (phage L). One can recognize several subgroups within the five different immunity classes: group A (α^{Tox+}, β^{Tox+}, P^{Tox+}, ω^{Tox+}), group B (γ^{Tox-}, π^{Tox+}), group D (δ^{Tox+}, L^{Tox+}) and two unique Tox^- phages ρ and K (Holmes and Barksdale, 1970). All of these phages showed similar restriction patterns, and, with the exception of phage δ, extensive DNA sequence homology to phage β (Rappuoli *et al.*, 1983; Buck *et al.*, 1985).

The next member of this group was phage γ (gamma) (Groman, 1955). This phage is serologically related to β, displays similar restriction patterns, and heteroduplex analyses indicate that they are 99% similar (Buck *et al.*, 1978), yet they are heteroimmune and γ is Tox^-.

Clostridium

Botulinum neurotoxins (BoNTXs) are associated as a high-molecular-weight complex with haemagglutinin (HA1, HA2, HA3a and HA3b) and a non-toxic non-haemagglutinin protein. BoNTXs are classified into seven antigenic groups, A, B, C1, D, E, F and G (Los *et al.*, 2010).

Lysogenic conversion to toxin production in *Clostridium botulinum* was first shown by K. Inoue and H. Iida (Inoue and Iida, 1970, 1971), and was confirmed by M. Eklund and co-workers (Eklund *et al.*, 1971, 1972). They suggested that botulinum toxin production following phage infection was analogous to the *C. diphtheriae* system with an important difference – botulinum toxin production was a pseudolysogenic conversion: toxigenicity and lysogeny were lost upon serial cultivation, particularly in the presence of antiphage sera (Oguma, 1976). *C. botulinum*-converting phage have been classified by the antigenic nature of the toxin they produce (Oguma *et al.*, 1976; Sugiyama, 1980; Simpson, 1981). Therefore, intraspecies conversions between antigen types is possible by infecting cured toxigenic strains (Tox^-) with the appropriate phage (Eklund and Poysky, 1974). It was also demonstrated that *Clostridium novyi* strains are capable of producing a variety of potent, antigenically distinguishable toxins (Eklund *et al.*, 1976). Again, positive correlation was observed with specific temperate phage infections in different strains of *C. novyi* leading to the production of a characteristic toxin (Eklund *et al.*, 1974; Schallehn *et al.*, 1980). Interspecies conversions between *C. botulinum* and *C. novyi* strains have been reported (Eklund *et al.*, 1974).

We have now the sequences of numerous *C. botulinum* strains (Sebaihia *et al.*, 2007; Smith *et al.*, 2007c; Skarin *et al.*, 2011). Only in the case of *C. botulinum* groups C and D is there a demonstrable relationship between prophages and toxigenicity. *C. botulinum* type C strain Stockholm and type D strain 1873 harbour UV- and mitomycin C-inducible serologically unrelated prophages c-st and D-1873, respectively. Strain C-203 harbours a prophage that is antigenically identical to D-1873, yet carries a C-type neurotoxin (Oguma, 1976). These viruses are unusually large members of the family *Myoviridae* (Inoue and Iida, 1968). In the lysogenic state, these temperate phages exist as plasmid prophages. The 185.7 kb genome of c-st has been completely sequenced (Sakaguchi *et al.*, 2005), while the genome of D-1873 exists in GenBank as eight fragments totalling 142.1 kb (NZ_ACSJ01000012.1– NZ_ACSJ01000019.1). These two phages show only limited DNA sequence similarity. The 203.3 kb plasmid p1BKT015925 (Skarin *et al.*, 2011) also carries the BoNTX operon and shows significantly more regions of sequence similarity to phage c-st than to D-1873. In the other botulinum groups, the BoNTX operons are not associated with phage-like elements.

In addition to the BoNTX operons, both of these phages encode a 25 kDa basic toxin, exoenzyme C3, which possesses N-ADP-ribosyltransferase activity. This protein is also present in a sequence (NCBI protein accession no. A46957) from *Clostridium limosum* (Just *et al.*, 1992), which is listed, without evidence, as being from an unnamed phage and on plasmid p1BKT015925 (Skarin *et al.*, 2011). The gene for C3 exotoxin lies almost 60 kb downstream of the botulinum toxin gene and is flanked by two identical 1.9 kb regions corresponding to insertion sequence (IS) OrfAB-like transposases. Hauser and colleagues suggested that phage c-st contains a large transposable element (Hauser *et al.*, 1993). The complete DNA sequence of c-st, however, indicates that the intra-IS region contains genes encoding virion structural proteins.

Staphylococcus

The pathogenesis of *Staphylococcus aureus* is complex and involves the production of an array of virulence factors that allow the organism to cause infection at a variety of sites and to cause multiple disease symptoms (see Kuhl *et al.*, Chapter 3, this volume). Staphylococcal virulence factors include secreted proteins such as cytotoxins, superantigens and tissue-degrading enzymes, cell surface-bound proteins that promote adhesion to cell surfaces, internalization and resistance to host defences, and components of the cell-wall peptidoglycan and polysaccharide capsule. With the exception of the cell-wall components, which are also essential housekeeping functions, these virulence factors are accessory proteins that are dispensable for bacterial growth. The genes encoding such accessory factors are generally carried on mobile genetic elements, and the evolution of new pathogenic strains of *S. aureus* has been attributed to the accumulation of mobile genetic elements encoding virulence factors and resistance determinants into successful lineages (Novick, 2006; Lindsay, 2010).

Phages play a major role in staphylococcal virulence; they not only provide a vehicle for generalized transduction but have also been implicated in lysogenic conversion and serve as helpers for the high-frequency mobilization of a family of superantigen pathogenicity islands (SaPIs). Sequence analysis of *S. aureus* genomes reveals that temperate bacteriophages are widespread; virtually all sequenced strains carry one to four prophages. *S. aureus* prophages have been shown to carry and/or interrupt virulence genes, so both prophage establishment and prophage loss can result in increased expression of a virulence trait. Resident staphylococcal prophages (and the SaPIs they mobilize) can be induced by a variety of environmental conditions that lead to induction of the SOS response (Wagner and Waldor, 2002; Goerke and Wolz, 2010).

Mobilization of temperate phages has been proposed to play an important role in the evolution of virulent clinical isolates, and a growing body of evidence supports the ongoing occurrence of complex phage dynamics. Not only do closely related methicillin-resistant *S. aureus* (MRSA) isolates exhibit different prophage complements (Goerke *et al.*, 2009; Nübel *et al.*, 2010), but horizontal phage transfer as well as intragenic translocation of prophages to different chromosomal integration sites appears to occur at a higher frequency during host infection by *S. aureus* (Goerke *et al.*, 2006; Goerke and Wolz, 2010). In addition, deletion of resident prophages reduced the virulence of *S. aureus* strain Newman in a mouse model of infection (Bae *et al.*, 2006), consistent with a direct contribution of lysogenic conversion to virulence.

A large-scale genome analysis of staphylococcal phages defined three different phage classes, distinguished by genome size and virion morphology (Kwan *et al.*, 2005). The vast majority were members of a large family of related temperate *Siphoviridae*, while the remainder were either *Podoviridae* or *Myoviridae*. All of the known staphylococcal transducing phages and converting phages belong to the family *Siphoviridae*, and many more examples have been provided by the sequences of prophages in staphylococcal genomes. Despite variations in virion morphology, sequence analysis reveals that these phages share a common modular genetic organization and display a mosaic pattern of relatedness at the genetic level (Kwan *et al.*, 2005; Kahankova *et al.*, 2010). Classification schemes for these siphophages have been proposed based on conservation of morphogenetic genes (Pantucek *et al.*, 2004) and the integrase (Lindsay and Holden, 2004; Goerke *et al.*, 2009). A recent multilocus PCR strategy for classification is based on up to seven discrete genomic modules, which appear to reassort freely among the different members of this group (see Fig. 4.1; Holochova *et al.*, 2010; Kahankova *et al.*, 2010). There is one recently described temperate prophage that appears to define an additional class of staphylococcal phages (Holden *et al.*, 2010). It is similar in sequence to the large RP62a prophage previously identified in *Staphylococcus epidermidis* (Gill *et al.*, 2005) and carries putative virulence genes, but whether it is capable of autonomous replication is not known.

In most cases, the genes for phage-encoded virulence factors lie at the ends of the prophage genomes (Fig. 4.1). While this is

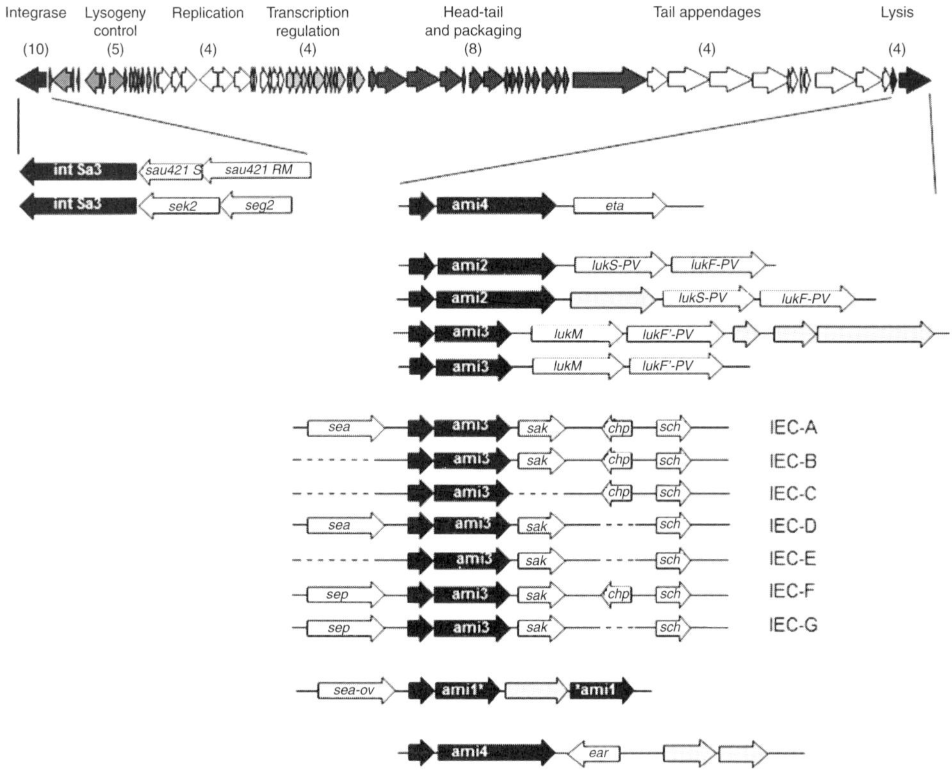

Fig. 4.1. Location and organization of staphylococcal conversion genes. Shown across the top is a prophage map of a typical member of the *S. aureus Siphoviridae*, with genes in the seven different modules defined by Kahankova *et al.* (2010) distinguished by different shading. Numbers in parentheses below each module indicate how many types of each module have been defined thus far. Expanded views of conversion genes at the left and right ends show the integrase or holin and lysin genes, respectively, in black, with the type of module for each specified. Genes in white encode known conversion functions, as indicated; genes in grey encode other (or unknown) functions. The seven different types of immune evasion cluster (IEC; see text) were defined by van Wamel *et al.* (2006). The lysis region in ϕSea-ov contains two open reading frames, one homologous to the N-terminal end of the ami1 type of amidase and the other homologous to the C-terminal end of ami1 (denoted ami1* and *ami1, respectively), separated by an unrelated open reading frame.

suggestive of acquisition by aberrant prophage excision, similar to the generation of specialized transducing phages, it is also possible that these are the locations in the phage genome where insertion of foreign DNA is not affected by and does not interfere with expression of essential phage lytic functions.

Lysogenic conversion

The earliest reports of lysogenic conversion affecting virulence gene expression in *S. aureus* were the loss of the exoproteins β-haemolysin (Winkler *et al.*, 1965) and lipase (Rosendal *et al.*, 1964) as the result of prophage integration into the *hlb* and *geh* genes, respectively (Coleman *et al.*, 1986; Lee and Iandolo, 1986). While this type of negative conversion does not strictly fit the definition of lysogenic conversion as described for this chapter, it is mentioned here because it has been documented extensively in staphylococci and it leads to direct effects on exotoxin expression. Numerous phages have an attachment site in *hlb* (Coleman *et al.*, 1986, 1989; Zabicka *et al.*, 1993; Sumby and Waldor, 2003; Dempsey *et al.*, 2005; Goerke *et al.*, 2006; van Wamel *et al.*, 2006; Kumagai *et al.*, 2007; see Table 4.1). The known *hlb*-converting phages share a conserved integrase and lysis module but are otherwise quite diverse (Kahankova *et al.*, 2010). The precise contribution of the *hlb* gene product to staphylococcal disease has not been clearly established, but this protein, also known as sphingomyelinase C, is produced in large quantities by a number of staphylococcal strains (Dinges *et al.*, 2000). As many characterized *hlb*-converting phages also carry phage-encoded virulence factors, it has been suggested that the integration of phages into this site during infection leads to splitting of the bacterial population into subfractions exhibiting different virulence potentials.

The lipase activity encoded by the *geh* gene has been reported, at least *in vitro*, to affect phagocytosis of *S. aureus* by cells involved in the immune response (Braconier and Rollof, 1991; Izdebska-Szymona *et al.*, 1992). No known virulence factors or other lysogenic conversion genes have yet been identified in this group of phages.

A variety of staphylococcal virulence factors and other functions are carried on prophages and function as typical lysogenic conversion genes. The roles and regulation of these prophage-encoded genes have been reviewed in detail recently (Los *et al.*, 2010) and are therefore only summarized below.

Exfoliative toxin A

Exfoliative toxins (ETs) are exoproteins encoded by *S. aureus* strains that cause blistering skin diseases, including bullous impetigo and staphylococcal scalded skin syndrome. These toxins are glutamate-specific serine proteases that specifically cleave a single peptide bond between cadherin repeats in the extracellular region of desmoglein 1 (Dsg1) (Nishifuji *et al.*, 2008), leading to a loss of cell–cell adhesion between keratinocytes and subsequent epidermal splitting. Of the three isoforms of ET implicated in human disease (ETA, ETB and ETD), only ETA has been shown to be carried on a phage (Ladhani *et al.*, 1999).

Leukotoxins

Leukotoxins are two-component, β-barrel pore-forming cytotoxins that assemble into oligomeric transmembrane complexes in polymorphonuclear leukocytes. This activates calcium channels and leads to an influx of divalent cations and ultimately to cell lysis (Finck-Barbancon *et al.*, 1993; Staali *et al.*, 1998). In addition, sublytic concentrations of leukotoxins activate neutrophils and monocytes, leading to the release of inflammatory mediators (König *et al.*, 1995). Lysogenic conversion of *S. aureus* to leukocidin production was first reported in 1972 (van der Vijver *et al.*, 1972), and two of the leukotoxins, Panton–Valentine leukocidin (PVL) and lukM/F′, are known to be phage encoded. PVL, first described in 1932 (Panton *et al.*, 1932), is comprised of subunits LukS-PV and LukF-PV and targets phagocytic cells. It shows specificity for cells of human and

Table 4.1. Summary of *S. aureus*-converting phages for which the complete phage/prophage sequence is available. The serotype (as defined by the morphogenesis module) and type of integrase and lytic module (Kahankova *et al.*, 2010) is shown for each phage. Phages are grouped by the conversion functions carried (or genes interrupted).

Phage or prophage	GenBank accession no.	Sero-type	Int module	Lysis module	Negative	Positive conversion
ETA	NC_003288 (Yamaguchi *et al.*, 2000)	B	Sa1	ami4		*eta*
ETA2	NC_008798	B	Sa1	ami4		*eta*
ETA3	NC_008799	B	Sa1	ami4		*eta*
SLT	NC_002661 (Narita *et al.*, 2001)	A	Sa2	ami2		*lukS-PV*, *lukF-PV*
Sa2mw	NC_003923 (Baba *et al.*, 2002)	A	Sa2	ami2		*lukS-PV*, *lukF-PV*
2958PVL	NC_011344 (Ma *et al.*, 2008)	A	Sa2	ami2		*lukS-PV*, *lukF-PV*
Sa2USA	NC_007793 (Diep *et al.*, 2006)	A	Sa2	ami2		*lukS-PV*, *lukF-PV*
SLT-USA300	CP000730 (Highlander *et al.*, 2007)	A	Sa2	ami2		*lukS-PV*, *lukF-PV*
tp310-1	NC_009761	F	Sa2	ami2		*lukS-PV*, *lukF-PV*
PVL108	NC_008689 (Ma *et al.*, 2006)	F	Sa2	ami2		*lukS-PV*, *lukF-PV*
PVL-CN125	NC_012784	F	Sa2	ami2		*lukS-PV*, *lukF-PV*
PVL	NC_002321 (Kaneko *et al.*, 1998)	F	Sa2	ami2		*lukS-PV*, *lukF-PV*
PV83	NC_002486 (Zou *et al.*, 2000)	F	Sa5	ami3		*lukM*, *lukF-PV*
Saov3	CP001996 (Guinane *et al.*, 2010)	F	Sa5	ami3		*lukM*, *lukF-PV*
Sa3mr(252)	NC_002952 (Holden *et al.*, 2004)	F	Sa3	ami3	*hlb*	IEC-A
NM3	NC_008617 (Bae *et al.*, 2006)	F	Sa3	ami3	*hlb*	IEC-A
13	NC_004617 (Iandolo *et al.*, 2002)	F	Sa3	ami3	*hlb*	IEC-B
tp310-3	NC_009763	F	Sa3	ami3	*hlb*	IEC-B
Sa3USA	NC_007793 (Diep *et al.*, 2006)	F	Sa3	ami3	*hlb*	IEC-B
βC-USA300	CP000730 (Highlander *et al.*, 2007)	F	Sa3	ami3	*hlb*	IEC-B
SA3(JKD6159)	CP002114 (Chua *et al.*, 2010)	F	Sa3	ami3	*hlb*	IEC-B
Sa3JH1	NC_009632	F	Sa3	ami3	*hlb*	IEC-B
Sa3JH9	NC_009487	F	Sa3	ami3	*hlb*	IEC-B
Sa3(TW20)	FN433596 (Holden *et al.*, 2010)	F	Sa3	ami3	*hlb*	IEC-B
Mu3A	NC_009782 (Neoh *et al.*, 2008)	F	Sa3	ami3	*hlb*	IEC-D
Mu50A	NC_002758 (Kuroda *et al.*, 2001)	F	Sa3	ami3	*hlb*	IEC-D
Sa3ms(476)	NC_002953 (Holden *et al.*, 2004)	F	Sa3	ami3	*hlb*	IEC-D, *sek2*, *seg2*
Sa3mw	NC_003923 (Baba *et al.*, 2002)	F	Sa3	ami3	*hlb*	IEC-D, *sek2*, *seg2*
Sa3(T0131)	CP002643	F	Sa3	ami3	*hlb*	IEC-D
Mu3A	NC_009782 (Neoh *et al.*, 2008)	F	Sa3	ami3	*hlb*	IEC-D
Mu50A	NC_002758 (Kuroda *et al.*, 2001)	F	Sa3	ami3	*hlb*	IEC-D

Phage or prophage	GenBank accession no.	Sero-type	Int module	Lysis module	Negative	Positive conversion
N315	NC_004740 (Kuroda *et al.*, 2001)	F	Sa3	ami3	*hlb*	IEC-F
N315(ECT-R2)	FR714927	F	Sa3	ami3	*hlb*	IEC-F
N315(ST5G)	CP001844 (Nübel *et al.*, 2010)	F	Sa3	ami3	*hlb*	IEC-F
71	NC_007059 (Kwan *et al.*, 2005)	B	Sa1	ami4		*ear*
Sa1JH1	NC_009632	B	Sa1	ami4		*ear*
Sa1JH9	NC_009487	B	Sa1	ami4		*ear*
SaST5K	CP001844 (Nübel *et al.*, 2010)	B	Sa1	ami4		*ear*
Av1	NC_013450 (Lowder *et al.*, 2009)	F	Sa1	ami4		*ear*
L54a (Col)	NC_002951	A	Sa6	ami2	*geh*	
Sa6S0385	AM990992 (Schijffelen *et al.*, 2010)	A	Sa6	ami2	*geh*	
tp310-2	NC_009762	A	Sa6	ami2	*geh*	
Saov1	CP001996 (Guinane *et al.*, 2010)	A	Sa6	ami2	*geh*	
ROSA	NC_007058 (Kwan *et al.*, 2005; Liu *et al.*, 2004)	B	Sa6	ami2	*geh*	
52A	NC_007062 (Kwan *et al.*, 2005)	B	Sa6	ami1	*geh*	
Sa6 JH1	NC_009632	B	Sa6	ami1	*geh*	
Sa6 JH9	NC_009487	B	Sa6	ami1	*geh*	
Sa1 (TW20)	FN433596 (Holden *et al.*, 2010)	B	Sa6	ami1	*geh*	
77	NC_005356 (Liu *et al.*, 2004)	F	Sa6	ami2	*geh*	
Saov2	CP001996 (Guinane *et al.*, 2010)	F	Sa7	ami1		*sea-ov*

rabbit origin but not those of mouse, rat or cattle (Szmigielski *et al.*, 1998). The lukM/F′toxin, comprised of subunits LukM and LukF′-PV, is associated with mastitis in ruminants (Choorit *et al.*, 1995) and shows a binding specificity for bovine leukocytes (Fromageau *et al.*, 2010).

PVL has been associated with strains causing both severe skin infections and necrotizing pneumonia (Panton *et al.*, 1932; Lina *et al.*, 1999; Gillet *et al.*, 2002). The prevalence has been increasing due to the spread of PVL-producing community-associated MRSA (CA-MRSA; Holmes *et al.*, 2005). The five predominant CA-MRSA clonal lineages that are associated with widespread disease outbreaks carry PVL, and multiple reports over the past decade or so have noted a strong epidemiological association between PVL and outbreaks of disease caused by CA-MRSA (for recent reviews, see Diep and Otto, 2008; Diep *et al.*, 2008; Kobayashi and DeLeo, 2009).

The immune evasion cluster

Co-conversion of staphylokinase (SAK) and enterotoxin by *hlb*-integrating phages has been well documented over the past two decades (Coleman *et al.*, 1989; Carroll *et al.*, 1993; Zabicka *et al.*, 1993; Sumby and Waldor, 2003; Dempsey *et al.*, 2005; Kumagai *et al.*, 2007). Two additional prophage-encoded innate immune modulators associated with *hlb*-converting phages, staphylococcal complement inhibitor (SCIN) and chemotaxis inhibitory protein of *S. aureus* (CHIPS), have been described more recently (de Haas *et al.*, 2004; Rooijakkers *et al.*, 2005; van Wamel *et al.*, 2006). All of these modulators are encoded in an 8 kb region at the 3′ end of the prophage genome, which has been named the immune evasion cluster (IEC; van Wamel *et al.*, 2006). These genes are prevalent in clinical isolates and seven different variants, or IEC types, with different combinations of these genes have been defined by PCR and Southern blot analysis (van Wamel *et al.*, 2006; see Fig. 4.1). The IEC phages are quite diverse; the same IEC module is found on different phages, and the four IEC modules present in sequenced phage genomes are associated with at least seven different modular phage variants, comprising five immunity groups (Kahankova *et al.*, 2010).

Conversion to SAK production was the earliest example of a phage-encoded virulence factor in *S. aureus* (Winkler *et al.*, 1965). SAK is a potent plasminogen activator that binds to fibrin-bound plasminogen, leading to the dissolution of fibrin clots (Bokarewa *et al.*, 2006). Disruption of the fibrin net that often forms around an infectious focus is thought to promote invasion of the staphylococci into deeper host tissue (Lahteenmaki *et al.*, 2005). SAK also neutralizes the bactericidal effects of α-defensins, contributing to virulence by evasion of the host immune system (Jin *et al.*, 2004). The gene for SAK (*sak*) is generally found on *hlb*-converting phages, where it is associated with one or more additional genes of the IEC (Fig. 4.1). Several *hlb*$^+$ *sak*$^+$ phages/lysogens have also been described (Kondo and Fujise, 1977; Sako and Tsuchida, 1983; Sako *et al.*, 1983; Goerke *et al.*, 2006). These prophages are integrated at other chromosomal sites. A partial sequence for one phage confirms that *sak* is located at the right end of the prophage genome and is associated with a different integrase module than the more common *hlb*-converting phages carrying *sak* as part of the IEC (Goerke *et al.*, 2006, 2009).

CHIPS is a 14.1 kDa secreted protein that helps modulate immune evasion by blocking an early step in the inflammatory response, the chemotaxis of neutrophils and monocytes towards the site of infection. The gene for CHIPS (*chp*) lies in the opposite orientation from the other genes in the IEC (Fig. 4.1) and was found in over 60% of clinical *S. aureus* isolates surveyed by van Wamel *et al.* (2006).

SCIN is another small secreted protein that acts to interfere with an early step of the innate immune response. The 9.3 kDa SCIN polypeptide binds to C3 convertases on the bacterial cell surface and blocks cleavage of the central complement protein C3 into C3a and C3b (Rooijakkers *et al.*, 2005), which prevents opsonization and protects the bacteria from phagocytosis. In addition, as C3b is required for the formation of C5 convertases, SCIN strongly attenuates C5a-induced neutrophil activation and chemotaxis (Rooijakkers *et al.*, 2006). van Wamel *et al.* (2006) found the gene for SCIN in 90% of the clinical isolates they examined. This gene is annotated in a number of prophage genomes as 'fibrinogen binding protein'. The *scn* gene lies at the end of the IEC and is transcribed divergently from *chp* (Fig. 4.1).

Genes for several enterotoxins are also associated with the IEC. These belong to a large family of related pyrogenic exotoxins that function as superantigens, (reviewed by Fraser and Proft, 2008). The exotoxins fall into two groups, the enterotoxins (SEs) and toxic shock syndrome toxins (TSSTs). Both groups have been implicated in toxic shock (Todd *et al.*, 1978; McCormick *et al.*, 2001), and the enterotoxins also have a well-established role in staphylococcal food poisoning (Tranter, 1990). Over 20 serologically distinct superantigens have been identified in *S. aureus*; these are related proteins that share 15–90% amino acid identity. Enterotoxin A (SEA), the most prevalent serotype of staphylococcal enterotoxin in human clinical isolates (Mathews and Novick, 2005), was shown to be phage encoded over 25 years ago (Betley and Mekalanos, 1985). Enterotoxin P (SEP), which has not been extensively characterized, shares 77% amino acid identity with SEA (Kuroda *et al.*, 2001). The *sep* gene was reported in place of *sea* 21% of the time among human clinical isolates carrying an enterotoxin gene in the IEC (van Wamel *et al.*, 2006; see Fig. 4.1) and in 29% of enterotoxigenic *S. aureus* food isolates (Bania *et al.*, 2006). Unlike the rest of the genes in the IEC, the *sea* (or *sep*) gene is located upstream of the phage lysis genes, which argues against initial acquisition of this gene by aberrant prophage excision.

Two additional enterotoxin genes, *seg*2 and *sek*2, have been described in phage ϕSa3ms (Sumby and Waldor, 2003). These genes lie at the right end of the prophage genome, between the repressor and integrase genes (Fig. 4.1).

The *ear* gene

The *ear* gene was first identified in a pathogenicity island present in *S. aureus* strain COL (Yarwood *et al.*, 2002) and is widespread among clinical isolates. The function of *ear* has not been demonstrated. The *ear* gene lies at the right end of the genome of prophages that carry the same integrase and lysis modules as the ETA-converting phages but are otherwise quite mosaic (Kwan *et al.*, 2005; Lowder *et al.*, 2009; Kahankova *et al.*, 2010; Nübel *et al.*, 2010).

The *sau42I* gene

Staphylococcal phage ϕ42 is a serotype F *hlb*-inactivating phage that encodes *sak* and *entA* (Coleman *et al.*, 1989). Like prophages carrying *seg2* and *sek2*, this phage carries additional genes that map just upstream of the *int* gene at the left end of the prophage genome (Fig. 4.1). These two slightly overlapping open reading frames, *sau42RI* S and *sau42I* RM, encode a *Bcg*I-like restriction modification system (Dempsey *et al.*, 2005) that confers resistance to lysis by all phages of the International Basic Set of *S. aureus* typing phages (Carroll and Francis, 1985).

High-frequency mobilization of SaPIs

Phages of *S. aureus* are not only carriers of virulence genes themselves but also play a major role in the horizontal transmission of genes for enterotoxins and other virulence factors carried by the SaPI family of pathogenicity islands. Twenty-two different SaPIs, which utilize six different phage-like chromosomal integration sites, have been identified thus far in sequenced staphylococcal genomes (Novick and Subedi, 2007; Guinane *et al.*, 2010; Novick *et al.*, 2010; Viana *et al.*, 2010). These highly mobile elements, members of a larger group of elements termed 'phage-related chromosomal islands', are dependent on helper phages for their propagation and are therefore included in this discussion of prophage-induced changes in bacterial phenotype. The SaPI genome, which is stably integrated into the host chromosome, is derepressed following induction of a resident helper prophage or after helper phage infection of a SaPI-containing strain.

SaPIs are widespread among *S. aureus*; most strains carry at least one SaPI and many carry two. As most *S. aureus* strains also carry multiple resident prophages, there is significant potential for SaPI spread following prophage induction. Not all phages can transduce SaPIs at high frequency. Nevertheless, there are a number of reported examples of SaPI mobilization by endogenous prophages (Kwan *et al.*, 2005; Ubeda *et al.*, 2005; Subedi *et al.*, 2007). Horizontal transmission of SaPIs also readily occurs, albeit at a lower frequency, by generalized transduction. The biology of SaPIs is described in detail in a recent review (Novick *et al.*, 2010, and references therein).

Streptococcus

Streptococci, lysogeny and toxigenic conversion

The prophages of *Streptococcus pyogenes* are an essential source of genetic variation and virulence factors in this bacterium. The first evidence that a bacterial toxin (the scarlet fever toxin) could be transferred between strains in sterile-filtered culture media was demonstrated by Frobisher and Brown in the 1930s, perhaps the earliest reported example of toxigenic conversion (Frobisher, 1927, 1934).

Fourteen published *S. pyogenes* genomes have revealed a large number of prophage-encoded toxins, superantigens and streptodornases (DNases) that have proposed or confirmed roles in pathogenesis (Table 4.2). With the exception of SmeZ (and occasionally SpeG and SpeJ), all of the identified streptococcal superantigens are elements of endogenous prophages or prophage remnants (Ferretti *et al.*, 2001).

Table 4.2. Prophage and prophage-like elements in *S. pyogenes* strains. The serotype (M type), number of intact prophages with their encoded virulence factors, and associated disease or site of anatomical isolation is listed for each of the published *S. pyogenes* strains. Alternative gene names of virulence factors are shown in parentheses.

Strain	M type	Number of prophages	Prophage-encoded virulence factors	GenBank accession no.	Associated disease or isolation[a]
SF370	1	4	*speC*, *speH*, *speI*, *spd1* (mf2), *spd3* (mf3)	AE004092	Severe wound
MGAS5005	1	3	*speA2*, *spd3*, *sda*D2	CP000017	Cerebrospinal fluid
MGAS10270	2	5	*speC*, *spd1*, *spd3*, *speK*	CP000260	Pharyngeal swab
MGAS315	3	6	*ssa*, mf4, *speK*, *sla*, *speA3*, *sdn*	AE14074	STSS
SSI-1	3	4	*speA*, *speG*, *speK*, *speL*, *ssa*, mf4.1, *sda*	BA000034	STSS
MGAS10750	4	4	*speJ*, *spd1*, *spd3*, *speA2*,	CP000262	Pharyngitis
Manfredo	5	5	*spd2* (MF4.1), *spd3*, *speH*, *speI*, *speC*	AM295007	ARF
MGAS10394	6	8	*speA*, *speC*, *spd1*, *speI*, *speH*, *spd3*, *speK*, *sla*, *sdn*, *sda*, *mefA*	CP000003	Pharyngitis
MGAS2096	12	2	*speC*, *spd1*, *srtA*, *sda*	CP000261	APSGN
MGAS9429	12	3	*speC*, *spd1*, *speH*, *speI*, *srtA*, *sda*	CP000259	Pharyngeal swab
MGAS8232	18	5	*speA1*, *speC*, *spd1*, *speL*, *speM*, *spd3*, *sda*	AE009949	ARF
MGAS6180	28	4	*srtA*, *speC*, *spd1*, *speK*, *sla*	CP000056	Invasive blood isolate
NZ131	49	3	*speH*, *spd3*	CP000829	APSGN

[a]STSS, streptococcal toxic shock syndrome; ARF, acute rheumatic fever; APSGN, acute post-streptococcal glomerulonephritis.

The ability of many of these toxins to function as superantigens has been confirmed (Proft *et al.*, 1999; Beres *et al.*, 2002; Proft and Fraser, 2003; Proft *et al.*, 2003a,b). Genome analysis showed that pathogenic strains contained multiple prophages or prophage-like elements (ranging from two to eight per genome). Furthermore, it is common for a given prophage to carry more than a single virulence gene (for example, exotoxins SpeH and SpeI are often encoded by the same prophage).

In addition to toxin genes, *S. pyogenes* prophages are vectors for virulence factors to assist the bacterium in avoiding innate immunity, particularly DNases. Neutrophils have been shown to release chromatin DNA and histones along with other intracellular components that combine to form neutrophil extracellular traps (NETs), which can trap and kill pathogenic bacteria, minimizing their spread (Brinkmann *et al.*, 2004). However, many isolates of *S. pyogenes* are resistant to NET-mediated killing, employing prophage-encoded DNases to degrade the NETs (Buchanan *et al.*, 2006).

SpyCI and the mutator phenotype

Another unique group of prophage-related mobile genetic elements are commonly found in *S. pyogenes* (Scott *et al.*, 2008). These *S. pyogenes* chromosomal islands (SpyCIs) lack any discernable virulence factors but regulate host gene expression by integrating into the 5′ end of *mutL*, an essential gene for DNA mismatch repair. The inhibition of mismatch repair has been shown to promote homologous recombination in other bacterial species, so the SpyCI may promote recombination between different prophage chromosomes to create diversity. An in-progress genome project of the *Streptococcus anginosus* strain F0211 genome has uncovered a SpyCI-related element integrated into the identical *mutL* attachment site as in *S. pyogenes*, so the distribution of these mutator phage-like chromosomal islands may extend to even more streptococcal species.

Recombination and the dissemination of virulence genes

Comparison of the genomes of prophages demonstrates how modular evolution has contributed to the diversification and dissemination of toxin genes and DNases to create novel prophages.

A conserved genetic module, containing genes for DNA packaging, head and tail proteins and tail fibre proteins, is found combined with an assortment of virulence genes. Additionally, different integrase genes that target a variety of bacterial attachments sites (*attB*) are also associated with this conserved structural module, leading to additional diversity. A highly conserved open reading frame has been identified, usually positioned between the phage-encoded virulence gene and the attachment site, which may promote homologous recombination and lead to phage diversification (Aziz *et al.*, 2005). This open reading frame, named paratox, was found in 18 out of 24 *S. pyogenes* prophages in genome strains SF370, MGAS8232, MGAS315 and SSI-1, and was always located adjacent to a toxin gene. Homologues of paratox also exist in some *Streptococcus agalactiae* and *Streptococcus thermophilus* phages (Aziz *et al.*, 2005).

Other related streptococcal species contain toxigenic prophages. Neither of the medically important species of *S. agalactiae* (group B *Streptococcus*) and *S. pneumoniae* has been found to be a significant carrier for toxigenic prophages, but the group C and group G *Streptococcus* species have provided many examples. These virulence factors are all close homologues of *S. pyogenes* prophage toxins (SlaA, SpeL, SpeM, SpeH and SpeI), strengthening the argument that these streptococcal species share a genetic pool of prophages. The group G *Streptococcus dysgalactiae* subsp. *equisimilis* GGS_124 contains a prophage with a streptodornase-like DNase (Shimomura *et al.*, 2011). Surveys of group C streptococcal isolates have found evidence for other superantigens, some of which may be prophage associated (Sachse *et al.*, 2002). Among the oral streptococci, a prophage of *Streptococcus mitis* was identified that encoded a platelet-binding protein (Bensing *et al.*, 2001). *S. mitis* is an important cause of bacterial endocarditis, and the binding of platelets by bacteria is a proposed central mechanism in the pathogenesis of this disease.

Escherichia coli

It was not until the early 1980s that temperate coliphages were really associated with the toxigenic conversion of their hosts (O'Brien *et al.*, 1984; Strockbine *et al.*, 1986). Acquisition of these toxigenic converting phages was the end result in the evolution of a pathogen (Feng *et al.*, 1998) that was not formally implicated in disease until a food-borne outbreak in 1982 (Karmali *et al.*, 1983; Riley *et al.*, 1983; O'Brien *et al.*, 1984). The phages responsible for the conversion of pathogenic *E. coli* serovars to the notorious pathotype, Shiga toxin-producing *E. coli* (STEC) and the smaller but more virulent subdivision enterohaemorrhagic *E. coli* (EHEC) (reviewed in more depth by Kaper *et al.*, 2004), are collectively known as Shiga toxin-encoding phages (Stx phages), although historically they have also gone by the nomenclature verotoxigenic phages (VT phages) and Shiga-like toxigenic phages (SLT-phages; Allison, 2007).

All Stx phages characterized to date have been lambdoid phages (Allison, 2007). The only trait shared by all Stx phages is their carriage of an *stx* operon (Allison, 2007). The genes encoding Stx, an AB_5 holotoxin related in structure to both cholera toxin (also phage encoded) and ricin (O'Loughlin and Robins-Browne, 2001), are located in a small, bicistronic operon, which is associated with the late gene regulatory region of Stx phages and prophages (Fig. 4.2). However, all Stx phages do not share the same genes encoding Stx. In fact, the term Shiga toxin describes two families of toxin (Allison, 2007).

Normally, the *stx* operon lies just downstream of the late gene-associated anti-terminator, Q, but there are instances reported in the literature of non-inducible, apparently remnant Stx prophages, where the *Q* gene has been lost while the prophage has retained the toxin operon in the expected location (Teel *et al.*, 2002). The location of the toxin-

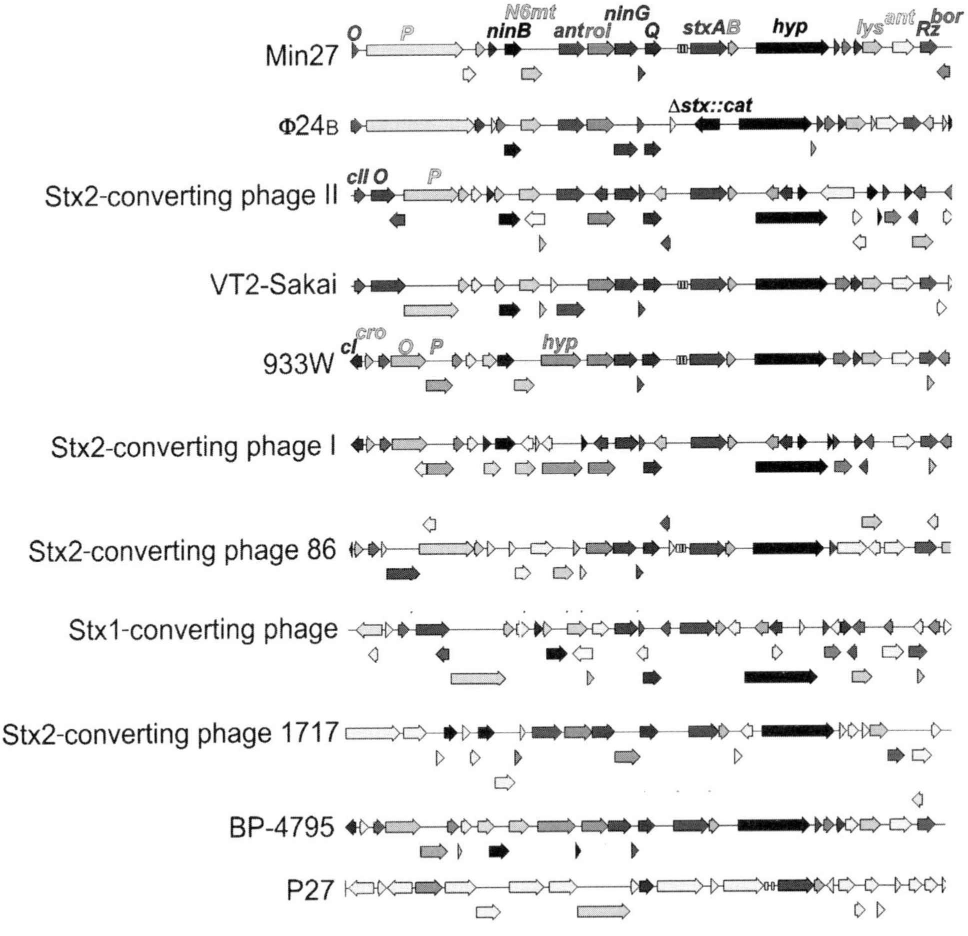

Fig. 4.2. Genetic map of the late gene regulatory region of sequenced Stx phages. Genes possessing significant gene homology are shaded identically. The first shaded variant of a gene is labelled accordingly. Gene abbreviations: *O*, origin-specific replication initiation factor; *P*, replication protein; *ninB* and *ninG*, genes comprising λ phage's *ninR* region involved in both RecF and RecBCD recombination as well as Red-mediated recombination (Tarkowski *et al.*, 2002); *ant*, putative anti-repressor (Fogg *et al.*, 2011); *roi*, similar to the gene encoded on HK2022 involved in controlling lytic propagation (reduced phage lytic propagation on integration host factor-negative cells; Clerget and Boccard, 1996); *Q*, anti-terminator controlling late gene expression; *stxAB*, Shiga toxin operon (or its inactivated derivative, Δstx::Cat; Allison *et al.*, 2003); *hyp*, hypothetical proteins with no predictable function; *lys*, phage lysin (1,4-β-*N*- acetylmuramidase); *Rz*, endopeptidase; *bor*, gene encoding the lipoprotein Bor; *cII*, transcription regulator involved in the establishment of lysogeny; *cI*, phage repressor involved in the maintenance of lysogeny; *cro*, regulator involved in prophage induction.

encoding operon in the prophage is crucial to its regulation. In most instances, Stx production is linked to the phage regulatory mechanism controlling the lytic cycle. *E. coli* lacks a dedicated secretion system for Stx, although limited secretion of Stx2 can occur in EHEC strains (Shimizu *et al.*, 2009), and maximal release of Stx is linked to phage-directed lysis of the cell (Wagner *et al.*, 2001).

Although there is at least one exception, φP27 (Recktenwald and Schmidt, 2002), Stx phages typically have a genome that is at least 10 kb larger than the 48.5 kb genome of λ phage (Table 4.3). The limited amount of DNA, 1.2 kb, linked to the presence of the *stx* operon in the phage genome does not solely account for the genomic size variation of Stx phages from λ phage. Genes such as *lom* and *bor* are carried by λ phage that are not essential to either its replication or its biology. These two genes have been demonstrated to aid the λ lysogen in colonization of the mammalian host (Barondess and Beckwith, 1990). These genes are also associated with many Stx phages (Fig. 4.2), but Stx phages appear to encode other traits that also increase their survival in their environment. Expression of Stx provides the lysogen with the ability to survive protozoan grazing (Lainhart *et al.*, 2009), which is an important attribute in both the ruminant gut and in soil environments, two crucial traits for organisms that are associated with cattle, cow pats and farm environments. The data indicate that Stx phages are not simply toxigenic converting phages, but that, in addition to converting the toxigenic phenotype of their host, these phages are capable of altering the fitness and phenotypic traits of their lysogen.

More than 500 serogroups of *E. coli* as well as other members of the *Enterobacteriaceae* have been associated with Shiga toxin production and human disease (Allison, 2007). Stx phages are driving this emergence. It is interesting to note that no two identical Stx phages have ever been described or sequenced (Muniesa *et al.*, 2003, 2004a,b; Smith *et al.*, 2007b; H.E. Allison, unpublished data). These phages possess genomes that are highly mosaic. Additionally, it has been demonstrated that many members of the *Enterobacteriaceae* can support Stx phage adsorption where the appropriate short tail is present, even where those bacterial species may not support infection (Smith *et al.*, 2007a); it may be that only simple abortive infection systems, restriction and modification systems or other mechanisms (Chopin *et al.*, 2005; Hyman and Abedon, 2010; Labrie *et al.*, 2010) are all that has, up to this time, protected us from the further emergence of new Shiga toxin-encoding pathogens. However, the serogroup range associated with Shiga toxin production continues to expand. As the host range for Stx phages grows, so too will the diversity of these phages as they continue to mix and form new phage mosaics with the endogenous prophages and remnant (defective) phages found in their extended host range (Allison, 2007).

Table 4.3. Characteristics of fully sequenced Stx phages

Stx phage	Genome size (kb)	Toxin variant	GenBank accession no.
BP-4795	57.9	Stx1	NC_004813
Min27	63.4	Stx2	NC_010237
Stx converting phage II	62.7	Stx2	NC_004914
Stx1-converting bacteriophage	59.9	Stx1	NC_004913
Stx2-converting phage I	61.8	Stx2	NC_003525
Stx2-converting phage 1717	62.1	Stx2c	NC_011357
Stx2-converting phage 86	60.2	Stx2	NC_008464
VT2-Sakai	60.9	Stx2	NC_000902
φ24B	58.5	Stx2	HM_208303
φ933W	61.7	Stx2	NC_000924
φP27	42.6	Stx2e	NC_003356

One further lambdoid phage-associated product is cytolethal distending toxin (CDT), first identified by Johnson and Lior in 1987. Induction of CDT-1 cells with mitomycin C results in enhanced toxin production and the release of CDT-1Φ, a member of the *Siphoviridae* (Asakura *et al.*, 2007). This phage possesses a 47 kb genome that is normally integrated into the *prfC* gene, which encodes peptide chain-release factor RF-3. The prophage shows similarity to the serotype-converting *Shigella* phage SfV (Allison *et al.*, 2002). Interestingly, unlike phage SfV and Stx λ phages, the toxin genes are found upstream of the integrase gene.

Vibrio cholerae

The filamentous phage CTXϕ has played a critical role in the evolution of the cholera pathogen, *V. cholerae* (Waldor and Mekalanos, 1996). The genes encoding cholera toxin (*ctxAB*), an AB_5-subunit toxin whose activity accounts for the profuse watery diarrhoea that is characteristic of cholera, are carried by CTXϕ (Fig. 4.3). In contrast to filamentous coliphages such as M13 which replicate as plasmids CTXϕ integrates into the *V. cholerae* genome, creating a stable lysogen bearing *ctxAB* (Davis and Waldor, 2003; McLeod *et al.*, 2005).

CTXϕ virion production does not result in host cell lysis; instead, similar to other filamentous phages, CTXϕ particles are secreted through an outer-membrane pore (secretin) (Davis *et al.*, 2000a). Both CTXϕ particles and cholera toxin exit the cell through the same secretin (Davis *et al.*, 2000a).

Like many other temperate phages, but in contrast to the filamentous coliphages, environmental conditions that lead to DNA damage and induce the host SOS response lead to increased production of CTXϕ virions. However, there are two notable differences in the consequences of SOS induction of the CTXϕ prophage: (i) CTXϕ prophage induction does not kill the host cell; and (ii) induction does not lead to excision of the CTXϕ genome from the chromosome. Instead, prophage DNA is used as a template to generate the single-stranded DNA copies that are packaged into virions via a replication mechanism that depends on the presence of tandem prophages (Davis and Waldor, 2000; Moyer *et al.*, 2001). Some of the biotechnological applications of CTXϕ have been reviewed recently (Rakonjac *et al.*, 2011).

Similar to other temperate phages, CTXϕ lysogeny is maintained by a phage-encoded repressor, RstR. LexA, the host-encoded SOS repressor, acts along with RstR to inhibit transcription from the promoter (P_{rstA}) that controls expression of most CTXϕ genes (Quinones *et al.*, 2005). The majority of toxin transcripts initiate from a distinct promoter found immediately upstream of *ctxA* (Davis *et al.*, 2002; Quinones *et al.*, 2006).

The classical and El Tor biotypes of toxigenic *V. cholerae* O1 harbour distinct forms of CTXϕ. Classical CTXϕ lysogens do not produce virions, whereas most El Tor isolates

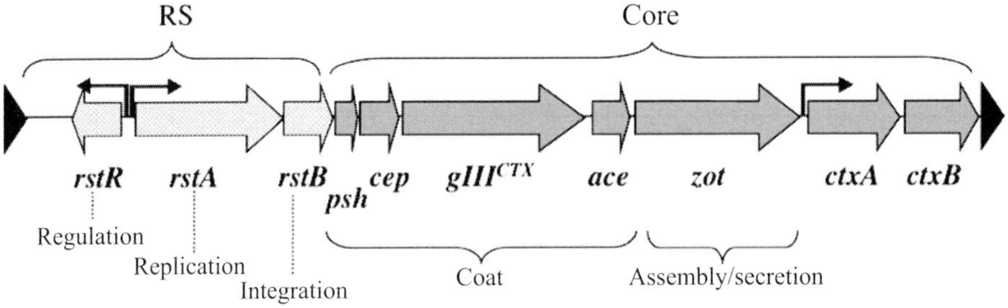

Fig. 4.3. Genetic structure of the CTX prophage. The genome is divided into an RS and core region. The black triangles indicate the direct repeats that flank the integrated CTXϕ. The arrows indicate the direction of transcription of the respective open reading frames and the bent arrows represent promoters.

do; the arrangement of prophages in classical strains explains why they do not produce phage particles (Davis *et al.*, 2000b). The classical and El Tor CTXϕs have distinct repressors as well as polymorphisms throughout the phage genome, including in the cholera toxin genes. In recent years, hybrid CTXϕs have been detected (Lee *et al.*, 2009). Such hybrid phages often contain the classical *ctxB* allele, which may yield a more potent toxin, in an otherwise El Tor strain. The existence of such hybrid phage genomes strongly suggests that there has been (and may still be) recombination between CTXϕs, generating new variants of toxigenic *V. cholerae*.

Salmonella

That *Salmonella enterica* serovar Typhimurium carried 'symbiotic bacteriophages' was first recognized by Boyd in 1950. It is noteworthy that most salmonellae carry the genomes of temperate phage – *Salmonella* Typhimurium strains usually carry four to five prophages, while *S. enterica* serovar Typhi strains Ty2 and CT18 both possess seven prophages (Thomson *et al.*, 2004). These prophages may induce three phenotypic changes to the cells that carry them – serotype conversion, a change in phage type and an impact on virulence. The presence of ST64B, for example, changes the phage type of *Salmonella* Typhimurium from definitive (phage) type (DT) 41 to DT 44 (Tucker and Heuzenroeder, 2004), while lysogenization by ST64T is probably responsible for converting DT 9 to DT 64, DT 135 to DT 16, and DT 41 to DT 29 (Mmolawa *et al.*, 2002).

Expression of the genes associated with lysogenic conversion can also impact on the virulence of the host. This has been recently reviewed (Lemire *et al.*, 2008). For example, Fels-1 prophage carries *nanH* (neuraminidase) and *sodC3* (superoxide dismutase; Figueroa-Bossi *et al.*, 2001; Bossi and Figueroa-Bossi, 2005), while Gifsy-1 carries a number of potential virulence modulating genes including *gipA* in what is equivalent to the λ *b2* region. The encoded protein is involved in colonization of the small intestine, and its deletion results in reduced bacterial virulence.

As phage conversion of O-antigenicity was reviewed recently in depth (Los *et al.*, 2010), here we shall only touch on what is known. Among the members of subgenus I (*S. enterica* subsp. *enterica*), a total of 57 O-antigens can be distinguished serologically (Grimont *et al.*, 2007). Some of these, O1, O5, O6, O14, O15, O20, O27 and O34, are in certain cases known to be phage encoded. The types of change noted in the structure of the LPS O-side-chain polysaccharide include O-acetylations, O-glucosylations and changes in bonding between adjacent O-antigen repeats. In the simplest case, this is manifest by *Salmonella* phage P22, an α-linked glucosyl residue (antigen O:1) to the 6 position of galactose moieties in the LPS O-antigenic branched tetrameric repeat.

This phage carries a conversion module consisting of three genes downstream of the phage attachment site (*attP*). This consists of GtrA (120 amino acids, 13.5 kDa; glucosylated undecaprenyl phosphate translocase or flippase), GtrB (311 amino acids, 35.1 kDa; bactoprenol glucosyl transferase) and GtrC (486 amino acids, 55.4 kDa; glucosyl transferase) (Vander Byl and Kropinski, 2000; Pedulla *et al.*, 2003). The sequence of GtrC shows more diversity than that of GtrA or GtrB, presumably as a result of differences in the chemical structure of the O-antigen (Fig. 4.4).

In certain other phages, rather than a glucosyl residue being added, one of the side-chain sugars gets acetylated. This has been observed with temperate *Salmonella* phage g341c (GenBank accession no. NC_013059), *E. coli* phage ϕV10 (Perry *et al.*, 2009) and *Pseudomonas aeruginosa* phage D3 (see below). In each of these cases, like the GtrC homologues, the O-acetylases are high-molecular-weight proteins with multiple transmembrane domains.

Pseudomonas

Lysogeny and polylysogeny are common in *P. aeruginosa* (Holloway *et al.*, 1971). Numerous early studies have shown O-antigenic conversion followed by infection

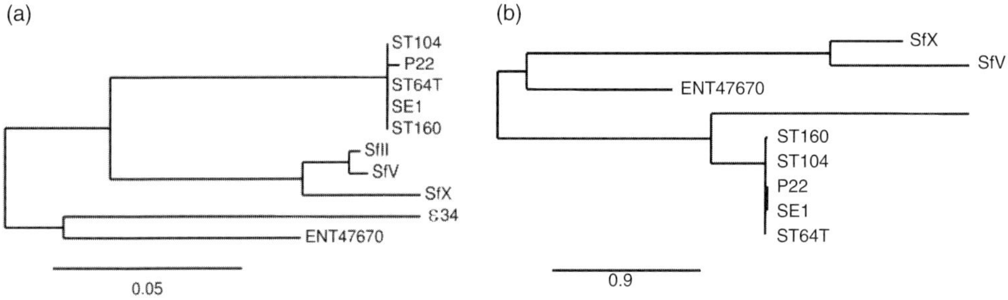

Fig. 4.4. Phylogenetic analysis of GtrB (a) and GtrC (b) homologues. The diagrams were generated using 'one click' analysis at http://www.phylogeny.fr/ (Dereeper *et al.*, 2008).

with a variety of phages (Liu, 1969; Castillo and Bartell, 1974; Bergan and Midtvedt, 1975; Lanyi and Lantos, 1976; Dimitracopoulos and Bartell, 1979). The definitive story of serotype conversion in this bacterium is with temperate siphovirus D3. In 1962, Holloway and Cooper reported that *P. aeruginosa* D3 lysogens lost the ability to adsorb this phage and displayed an antigenic shift. Kuzio and Kropinski (1983) confirmed that the lysogens did not support phage adsorption, and that LPS from the convertant did not neutralize D3 or the serologically and morphologically unrelated phage E79. Nuclear magnetic resonance analysis of the O-side-chain polysaccharides from strains PAO1 and PAO1(D3) revealed that, in a manner analogous to *Salmonella* phage ε15, the bonding of tetrameric O-antigenic repeats changed from α(1→4) to β(1→4) and the D-fucosamine (2-amino-2,6-dideoxy-D-galactose) residues were O-acetylated. Sequencing the genome of this phage (Kropinski, 2000) resulted in the discovery of the fucosamine O-acetylase, but it was not until Newton cloned the conversion region (Newton *et al.*, 2001) that we fully understood the role of each of the components in the conversion. The conversion module consists of the acetyltransferase, a β-LPS polymerase and Iap (inhibitor of alpha polymerase), which is a 31 amino acid (3884 Da) transmembrane peptide which inhibits the α-LPS polymerase. Interestingly, this polypeptide is extremely difficult to express or produce chemically due to its inherent insolubility (J. Lam, personal communication).

The only other phage about which some details are known is FIZ15 (Vaca *et al.*, 1993; Vaca-Pacheco *et al.*, 1999), which unfortunately is now lost (S. Vaca, personal communication). In the lysogenic state, this virus was found to: (i) increase resistance to phagocytosis by mouse peritoneal macrophages; (ii) increase resistance to killing by normal human serum; and (iii) increase adhesion to human buccal epithelial cells, which are considered to be a function of changes to the LPS.

Transduction

The three common methods of genetic exchange within bacterial species are transformation, conjugation and transduction. The last is the transfer of genetic information between bacterial cells mediated by phage particles. Two distinct types have been recognized: (i) specialized transduction; and (ii) generalized transduction (Thierauf *et al.*, 2009). In the former case, which is best exemplified by coliphage λ, genes adjacent to an integrated prophage (*att*λ site) can be co-excised with the phage upon induction. The defining characteristics of the transducing particles are that they carry a strictly limited number of host markers and that they contain host DNA physically associated with phage DNA. Lysogenic conversion could be viewed as a unique example of specialized transduction.

Generalized transduction, in which any gene can be transduced, is best exemplified by coliphage P1 and *Salmonella* phage P22.

Here, the transducing particles are derived through the accidental packaging of host DNA, rather than phage DNA. With renewed interest in the use of bacteriophages in therapy, biosanitation, biocontrol and biopreservation, the ability of the phage preparations to transduce (especially toxin and other virulence-factor genes) is one of the issues that those who characterize phages for practical uses need to consider. While most of the transducing phages are temperate, that is, are capable of entering into a lysogenic state with a host bacterium, lytic phages may also transduce (Table 4.4).

Genetic exchange via all three mechanisms is commonly observed. In contrast, unlike many of the bacterial species mentioned above, transduction is the only mechanism of genetic exchange observed in *S. pyogenes* and is mediated by both lytic and temperate phages. It was first reported by Leonard and co-workers when they identified five phages, two temperate and three lytic, that were able to transduce streptomycin resistance (Leonard *et al.*, 1968). The highest frequency of transduction was observed for lytic phage A25 (1×10^{-6} plaque-forming units). Temperate *S. pyogenes* phages are also able of mediating the transfer of antibiotic resistance by transduction (Ubukata *et al.*, 1975; Hyder and Streitfeld, 1978), presumably generalized transduction.

Transduction between streptococcal groups has been observed and may be a potential means for disseminating virulence genes. Bacteriophages isolated from streptococcal groups A, E and G were isolated that could lyse streptococci from serogroups A, C, G, H and L, and in some cases, these lytic phages were also capable of propagation in one or more of the other serogroups (Cólon *et al.*, 1972; Wannamaker *et al.*, 1973). Horizontal transfer of genetic material in *S. pyogenes* has been shown by numerous recent studies to be important in the dissemination of genes in natural populations (Stevens *et al.*, 1989; Cleary *et al.*, 1992). The role in transduction and bacteriophages in horizontal transfer is assumed, yet no modern studies have directly addressed the question. The majority of studies into streptococcal transduction were carried out before the advent of many of the current techniques in molecular biology, and so the time may have come to re-examine this phenomenon. A better understanding of streptococcal transduction may prove key to understanding the flow of genetic information in natural populations of streptococci. This is

Table 4.4. Lytic phages that can, under controlled conditions, be demonstrated to transduce.

Host	Phage	Reference
Bacillus subtilis	SPP1	Canosi *et al.* (1982); de Lencastre and Archer (1980)
Caulobacter crescentus	φCr30T	Bender (1981)
Citrobacter rodentium	φCR1	Petty *et al.* (2007)
Erwinia carotovora	φM1	Toth *et al.* (1997)
	φKP	Toth *et al.* (1993)
Escherichia coli	T1	Bendig and Drexler (1977); Drexler (1970; 1977)
	T4	Roberts and Drexler (1981)
	RB43	Wilson *et al.* (1979); Young *et al.* (1982)
	RB49	Tianiashin *et al.* (2003)
Pseudomonas spp.	φKZ	Dzhusupova *et al.* (1982)
	φPA3	Monson *et al.* (2011)
	E79	Morgan (1979)
	pf16	Rheinwald *et al.* (1973); Daz *et al.* (1976); Gorbunova *et al.* (1985)
Salmonella typhi	Vil	Cerquetti *et al.* (1993)
Serratia marcescens	φIF3	Petty *et al.* (2006)
Streptococcus pyogenes	A25	Leonard *et al.* (1968)
Xanthomonas campestris	XTP1	Weiss *et al.* (1994)

also true for other pathogenic bacteria as our understanding of the role of horizontal gene transfer in pathogen development is still quite incomplete. Future studies with transducing phages can only add to this understanding.

Acknowledgements

A.M.K. was supported by a Discovery Grant from the Natural Sciences and Engineering Research Council of Canada.

References

Allison, G.E., Angeles, D., Tran-Dinh, N. and Verma, N.K. (2002) Complete genomic sequence of SfV, a serotype-converting temperate bacteriophage of *Shigella flexneri*. *Journal of Bacteriology* 184, 1974–1987.

Allison, G.E., Angeles, D.C., Huan, P. and Verma, N.K. (2003) Morphology of temperate bacteriophage SfV and characterisation of the DNA packaging and capsid genes: the structural genes evolved from two different phage families. *Virology* 308, 114–127.

Allison, H.E. (2007) Stx-phages: drivers and mediators of the evolution of STEC and STEC-like pathogens. *Future Microbiology* 2, 165–174.

Asakura, M., Hinenoya, A., Alam, M.S., Shima, K., Zahid, S.H., Shi, L., Sugimoto, N., Ghosh, A.N., Ramamurthy, T., Faruque, S.M., Nair, G.B. and Yamasaki, S. (2007) An inducible lambdoid prophage encoding cytolethal distending toxin (Cdt-I) and a type III effector protein in enteropathogenic *Escherichia coli*. *Proceedings of the National Academy of Sciences USA* 104, 14483–14488.

Aziz, R.K., Edwards, R.A., Taylor, W.W., Low, D.E., McGeer, A. and Kotb, M. (2005) Mosaic prophages with horizontally acquired genes account for the emergence and diversification of the globally disseminated M1T1 clone of *Streptococcus pyogenes*. *Journal of Bacteriology* 187, 3311–3318.

Baba, T., Takeuchi, F., Kuroda, M., Yuzawa, H., Aoki, K., Oguchi, A., Nagai, Y., Iwama, N., Asano, K., Naimi, T., Kuroda, H., Cui, L., Yamamoto, K. and Hiramatsu, K. (2002) Genome and virulence determinants of high virulence community-acquired MRSA. *Lancet* 359, 1819–1827.

Bae, T., Baba, T., Hiramatsu, K. and Schneewind, O. (2006) Prophages of *Staphylococcus aureus* Newman and their contribution to virulence. *Molecular Microbiology* 62, 1035–1047.

Bania, J., Dabrowska, A., Bystron, J., Korzekwa, K., Chrzanowska, J. and Molenda, J. (2006) Distribution of newly described enterotoxin-like genes in *Staphylococcus aureus* from food. *International Journal of Food Microbiology* 108, 36–41.

Barondess, J.J. and Beckwith, J. (1990) A bacterial virulence determinant encoded by lysogenic coliphage λ. *Nature* 346, (6287) 871–874.

Bender, R.A. (1981) Improved generalized transducing bacteriophage for *Caulobacter crescentus*. *Journal of Bacteriology* 148, 734–735.

Bendig, M.M. and Drexler, H. (1977) Transduction of bacteriophage Mu by bacteriophage T1. *Journal of Virology* 22, 640–645.

Bensing, B.A., Siboo, I.R. and Sullam, P.M. (2001) Proteins PblA and PblB of *Streptococcus mitis*, which promote binding to human platelets, are encoded within a lysogenic bacteriophage. *Infection and Immunity* 69, 6186–6192.

Beres, S.B., Sylva, G.L., Barbian, K.D., Lei, B., Hoff, J.S., Mammarella, N.D., Liu, M.Y., Smoot, J.C., Porcella, S.F., Parkins, L.D., Campbell, D.S., Smith, T.M., McCormick, J.K., Leung, D.Y., Schlievert, P.M. and Musser, J.M. (2002) Genome sequence of a serotype M3 strain of group A *Streptococcus*: phage-encoded toxins, the high-virulence phenotype, and clone emergence. *Proceedings of the National Academy of Sciences USA* 99, 10078–10083.

Bergan, T. and Midtvedt, T. (1975) Epidemiological markers for *Pseudomonas aeruginosa*. *Acta Pathologica Microbiologica Scandanavica*, 83B, 1–9.

Betley, M.J. and Mekalanos, J.J. (1985) Staphylococcal enterotoxin A is encoded by phage. *Science* 229, 185–187.

Bokarewa, M.I., Jin, T. and Tarkowski, A. (2006) *Staphylococcus aureus*: staphylokinase. *International Journal of Biochemistry and Cell Biology* 38, 504–509.

Bossi, L. and Figueroa-Bossi, N. (2005) Prophage arsenal of *Salmonella enterica* serovar Typhimurium, In: Waldor, M.K., Friedman, D.I. and Adhya, S.L. (eds) *Phages: Their Role in Bacterial Pathogenesis and Biotechnology*. ASM Press, Washington, DC, pp. 165–186.

Boyd, J.S. (1950) The symbiotic bacteriophages of *Salmonella typhimurium*. *Journal of Pathology and Bacteriology* 62, 501–517.

Braconier, J.H. and Rollof, J. (1991) Influence by staphylococcal lipase on granulocyte metabolism and killing of bacteria. *Zentralblatt für Bakteriologie* 276, 68–72.

Brinkmann, V., Reichard, U., Goosmann, C., Fauler, B., Uhlemann, Y., Weiss, D.S., Weinrauch, Y. and Zychlinsky, A. (2004) Neutrophil extracellular traps kill bacteria. *Science* 303, 1532–1535.

Brüssow, H., Canchaya, C. and Hardt, W.D.(2004) Phages and the evolution of bacterial pathogens: from genomic rearrangements to lysogenic conversion. *Microbiology and Molecular Biology Reviews* 68, 560–602.

Buchanan, J.T., Simpson, A.J., Aziz, R.K., Liu, G.Y., Kristian, S.A., Kotb, M., Feramisco, J. and Nizet, V. (2006) DNase expression allows the pathogen group A *Streptococcus* to escape killing in neutrophil extracellular traps. *Current Biology* 16, 396–400.

Buck, G., Groman, N. and Falkow, S. (1978) Relationship between β converting and γ non-converting corynebacteriophage DNA. *Nature* 271, 683–685.

Buck, G.A., Cross, R.E., Wong, T.P., Loera, J. and Groman, N. (1985) DNA relationships among some tox-bearing corynebacteriophages. *Infection and Immunity* 49, 679–684.

Busch, A.W., Reijerse, E.J., Lubitz, W., Hofmann, E. and Frankenberg-Dinkel, N. (2011) Radical mechanism of cyanophage phycoerythrobilin synthase (PebS). *Biochemical Journal* 433, 469–476.

Canosi, U., Luder, G. and Trautner, T.A. (1982) SPP1-mediated plasmid transduction. *Journal of Virology* 44, 431–436.

Carroll, J.D., Cafferkey, M.T. and Coleman, D.C. (1993) Serotype F double- and triple-converting phage insertionally inactivate the *Staphylococcus aureus* β-toxin determinant by a common molecular mechanism. *FEMS Microbiology Letters* 106, 147–155.

Carroll, P.J. and Francis, P.G. (1985) The basic phage set for typing bovine staphylococci. *Journal of Hygiene* 95, 665–669.

Castillo, F.J. and Bartell, P.F. (1974) Studies on the bacteriophage 2 receptor of *Pseudomonas aeruginosa*. *Journal of Virology* 14, 902–909.

Cerquetti, M.C., Hooke, A.M., Cerquetti, M.C. and Hooke, A.M. (1993) Vi I typing phage for generalized transduction of *Salmonella typhi*. *Journal of Bacteriology* 175, 5294–5296.

Choorit, W., Kaneko, J., Muramoto, K. and Kamio, Y. (1995) Existence of a new protein component with the same function as the LukF component of leukocidin or γ-hemolysin and its gene in *Staphylococcus aureus* P83. *FEBS Letters* 357, 260–264.

Chopin, M.C., Chopin, A. and Bidnenko, E. (2005) Phage abortive infection in lactococci: variations on a theme. *Current Opinion in Microbiology* 8, 473–479.

Chua, K., Seemann, T., Harrison, P.F., Davies, J.K., Coutts, S.J., Chen, H., Haring, V., Moore, R., Howden, B.P. and Stinear, T.P. (2010) Complete genome sequence of *Staphylococcus aureus* strain JKD6159, a unique Australian clone of ST93-IV community methicillin-resistant *Staphylococcus aureus*. *Journal of Bacteriology* 192, 5556–5557.

Cleary, P.P., Kaplan, E.L., Handley, J.P., Wlazlo, A., Kim, M.H., Hauser, A.R. and Schlievert, P.M. (1992) Clonal basis for resurgence of serious *Streptococcus pyogenes* disease in the 1980s. *Lancet* 339, 518–521.

Clerget, M. and Boccard, F. (1996) Phage HK022 Roi protein inhibits phage lytic growth in *Escherichia coli* integration host factor mutants. *Journal of Bacteriology* 178, 4077–4083.

Coleman, D.C., Arbuthnott, J.P., Pomeroy, H.M. and Birkbeck, T.H. (1986) Cloning and expression in *Escherichia coli* and *Staphylococcus aureus* of the β-lysin determinant from *Staphylococcus aureus*: evidence that bacteriophage conversion of β-lysin activity is caused by insertional inactivation of the β-lysin determinant. *Microbial Pathogenesis* 1, 549–564.

Coleman, D.C., Sullivan, D.J., Russell, R.J., Arbuthnott, J.P., Carey, B.F. and Pomeroy, H.M. (1989) *Staphylococcus aureus* bacteriophages mediating the simultaneous lysogenic conversion of β-lysin, staphylokinase and enterotoxin A: molecular mechanism of triple conversion. *Journal of General Microbiology* 135, 1679–1697.

Cólon, A.E., Cole, R.M. and Leonard, C.G. (1972) Intergroup lysis and transduction by streptococcal bacteriophages. *Journal of Virology* 9, 551–553.

Davis, B.M. and Waldor, M.K. (2000) CTXϕ contains a hybrid genome derived from tandemly integrated elements. *Proceedings of the National Academy of Sciences USA* 97, 8572–8577.

Davis, B.M. and Waldor, M.K. (2003) Filamentous phages linked to virulence of *Vibrio cholerae*. *Current Opinion in Microbiology* 6, 35–42.

Davis, B.M., Lawson, E.H., Sandkvist, M., Ali, A., Sozhamannan, S. and Waldor, M.K. (2000a) Convergence of the secretory pathways for cholera toxin and the filamentous phage, CTXϕ. *Science* 288, 333–335.

Davis, B.M., Moyer, K.E., Boyd, E.F. and Waldor, M.K. (2000b) CTX prophages in classical biotype *Vibrio cholerae*: functional phage genes but dysfunctional phage genomes. *Journal of Bacteriology* 182, 6992–6998.

Davis, B.M., Kimsey, H.H., Kane, A.V. and Waldor, M.K. (2002) A satellite phage-encoded antirepressor induces repressor aggregation and cholera toxin gene transfer. *EMBO Journal* 21, 4240–4249.

Daz, R., de Torrontegui, G. and Canovas, J.L. (1976) [Generalized transduction of *Pseudomonas putida* with a thermosensitive mutant of phage pf16h2]. *Microbiologia Espanola* 29, 33–45 (in Spanish).

de Haas, C.J., Veldkamp, K.E., Peschel, A., Weerkamp, F., van Wamel, W.J., Heezius, E.C., Poppelier, M.J., Van Kessel, K.P. and Vvan Strijp, J.A. (2004) Chemotaxis inhibitory protein of *Staphylococcus aureus*, a bacterial antiinflammatory agent. *Journal of Experimental Medicine* 199, 687–695.

de Lencastre, H. and Archer, L.J. (1980) Characterization of bacteriophage SPP1 transducing particles. *Journal of General Microbiology* 117, 347–355.

Dempsey, R.M., Carroll, D., Kong, H., Higgins, L., Keane, C.T. and Coleman, D.C. (2005) *Sau*42I, a *Bcg*I-like restriction-modification system encoded by the *Staphylococcus aureus* quadruple-converting phage ϕ42. *Microbiology* 151, 1301–1311.

Dereeper, A., Guignon, V., Blanc, G., Audic, S., Buffet, S., Chevenet, F., Dufayard, J.F., Guindon, S., Lefort, V., Lescot, M., Claverie, J.M. and Gascuel, O. (2008) Phylogeny.fr: robust phylogenetic analysis for the non-specialist. *Nucleic Acids Research* 36 (Web Server issue), W465–W469.

Diep, B.A. and Otto, M. (2008) The role of virulence determinants in community-associated MRSA pathogenesis. *Trends in Microbiology* 16, 361–369.

Diep, B.A., Gill, S.R., Chang, R.F., Phan, T.H., Chen, J.H., Davidson, M.G., Lin, F., Lin, J., Carleton, H.A., Mongodin, E.F., Sensabaugh, G.F. and Perdreau-Remington, F. (2006) Complete genome sequence of USA300, an epidemic clone of community-acquired meticillin-resistant *Staphylococcus aureus*. *Lancet* 367, 731–739.

Diep, B.A., Palazzolo-Ballance, A.M., Tattevin, P., Basuino, L., Braughton, K.R., Whitney, A.R., Chen, L., Kreiswirth, B.N., Otto, M., DeLeo, F.R. and Chambers, H.F. (2008) Contribution of Panton–Valentine leukocidin in community-associated methicillin-resistant *Staphylococcus aureus* pathogenesis. *PLoS ONE* 3, e3198.

Dimitracopoulos, G. and Bartell, P.F. (1979) Phage-related surface modification of *Pseudomonas aeruginosa*: effects on the biological activity of viable cells. *Infection and Immunity* 23, 87–93.

Dinges, M.M., Orwin, P.M. and Schlievert, P.M. (2000) Exotoxins of *Staphylococcus aureus*. *Clinical Microbiology Reviews* 13, 16–34.

Drexler, H. (1970) Transduction by bacteriophage T1. *Proceedings of the National Academy of Sciences USA* 66, 1083–1088.

Drexler, H. (1977) Specialized transduction of the biotin region of *Escherichia coli* by phage T1. *Molecular and General Genetics* 152, 59–63.

Dzhusupova, A.B., Plotnikova, T.G., Krylov, V.N., Dzhusupova, A.B., Plotnikova, T.G. and Krylov, V.N. (1982) Detection of transduction by virulent bacteriophage ϕ KZ of *Pseudomonas aeruginosa* chromosomal markers in the presence of plasmid RMS148. *Genetika* 18, 1799–1802.

Eklund, M.W. and Poysky, F.T. (1974) Interconversion of type C and D strains of *Clostridium botulinum* by specific bacteriophages. *Applied Microbiology* 27, 251–258.

Eklund, M.W., Poysky, F.T., Reed, S.M. and Smith, C.A. (1971) Bacteriophage and the toxigenicity of *Clostridium botulinum* type C. *Science* 172, 480–482.

Eklund, M.W., Poysky, F.T. and Reed, S.M. (1972) Bacteriophage and the toxigenicity of *Clostridium botulinum* type D. *Nature New Biology* 235, 16–17.

Eklund, M.W., Poysky, F.T., Meyers, J.A. and Pelroy, G.A. (1974) Interspecies conversion of *Clostridium botulinum* type C to *Clostridium novyi* type A by bacteriophage. *Science* 186, 456–458.

Eklund, M.W., Poysky, F.T., Peterson, M.E. and Meyers, J.A. (1976) Relationship of bacteriophages to alpha toxin production in *Clostridium novyi* types A and B. *Infection and Immunity* 14, 793–803.

Feng, P., Lampel, K.A., Karch, H. and Whittam, T.S. (1998) Genotypic and phenotypic changes in the emergence of *Escherichia coli* O157:H7. *Journal of Infectious Diseases* 177, 1750–1753.

Ferretti, J.J., McShan, W.M., Ajdic, D., Savic, D.J., Savic, G., Lyon, K., Primeaux, C., Sezate, S., Suvorov, A.N., Kenton, S., Lai, H.S., Lin, S.P., Qian, Y., Jia, H.G., Najar, F.Z., Ren, Q., Zhu, H., Song, L., White, J., Yuan, X., Clifton, S.W., Roe, B.A. and McLaughlin, R. (2001) Complete genome sequence of an M1 strain of *Streptococcus pyogenes*. *Proceedings of the National Academy of Sciences USA* 98, 4658–4663.

Figueroa-Bossi, N., Uzzau, S., Maloriol, D. and Bossi, L. (2001) Variable assortment of prophages provides a transferable repertoire of pathogenic determinants in *Salmonella*. *Molecular Microbiology* 39, 260–271.

Finck-Barbancon, V., Duportail, G., Meunier, O. and Colin, D.A. (1993) Pore formation by a two-component leukocidin from *Staphylococcus aureus* within the membrane of human polymorphonuclear leukocytes. *Biochimica et Biophysica Acta* 1182, 275–282.

Fogg, P.C., Rigden, D.J., Saunders, J.R., McCarthy, A.J. and Allison, H.E. (2011) Characterization of the relationship between integrase, excisionase and antirepressor activities associated with a superinfecting Shiga toxin encoding bacteriophage. *Nucleic Acids Research* 39, 2116–2129.

Fraser, J.D. and Proft, T. (2008) The bacterial superantigen and superantigen-like proteins. *Immunology Reviews*, 225, 226–243.

Freeman, V.J. (1951) Studies on the virulence of bacteriophage-infected strains of *Corynebacterium diphtheriae. Journal of Bacteriology* 61, 675–688.

Freeman, V.J. and Morse, I.U. (1952) Further observations on the change to virulence of bacteriophage-infected avirulent strains of *Corynebacterium diphtheriae. Journal of Bacteriology* 63, 407–414.

Frobisher, M. and Brown, J.H. (1927) Transmissible toxicogenicity of Streptococci. Bulletin of the Johns Hopkins Hospital, 41, 107–173.

Fromageau, A., Gilbert, F.B., Prévost, G. and Rainard, P. (2010) Binding of the *Staphylococcus aureus* leucotoxin LukM to its leucocyte targets. *Microbial Pathogenesis* 49, 354–362.

Gill, S.R., Fouts, D.E., Archer, G.L., Mongodin, E.F., DeBoy, R.T., Ravel, J., Paulsen, I.T., Kolonay, J.F., Brinkac, L., Beanan, M., Dodson, R.J., Daugherty, S.C., Madupu, R., Angiuoli, S.V., Durkin, A.S., Haft, D.H., Vamathevan, J., Khouri, H., Utterback, T., Lee, C., Dimitrov, G., Jiang, L., Qin, H., Weidman, J., Tran, K., Kang, K., Hance, I.R., Nelson, K.E. and Fraser, C.M. (2005) Insights on evolution of virulence and resistance from the complete genome analysis of an early methicillin-resistant *Staphylococcus aureus* strain and a biofilm-producing methicillin-resistant *Staphylococcus epidermidis* strain. *Journal of Bacteriology* 187, 2426–2438.

Gillet, Y., Issartel, B., Vanhems, P., Fournet, J.C., Lina, G., Bes, M., Vandenesch, F., Piemont, Y., Brousse, N., Floret, D. and Etienne, J. (2002) Association between *Staphylococcus aureus* strains carrying gene for Panton–Valentine leukocidin and highly lethal necrotising pneumonia in young immunocompetent patients. *Lancet* 359, 753–759.

Goerke, C. and Wolz, C. (2010) Adaptation of *Staphylococcus aureus* to the cystic fibrosis lung. *International Journal of Medical Microbiology* 300, 520–525.

Goerke, C., Wirtz, C., Fluckiger, U. and Wolz, C. (2006) Extensive phage dynamics in *Staphylococcus aureus* contributes to adaptation to the human host during infection. *Molecular Microbiology* 61, 1673–1685.

Goerke, C., Pantucek, R., Holtfreter, S., Schulte, B., Zink, M., Grumann, D., Broker, B.M., Doskar, J. and Wolz, C. (2009) Diversity of prophages in dominant *Staphylococcus aureus* clonal lineages. *Journal of Bacteriology* 191, 3462–3468.

Gorbunova, S.A., Akhverdian, V.S., Cheremukhina, L.V. and Krylov, V.N. (1985) Effective method of transduction with virulent phage pf16 using specific mutants of *Pseudomonas putida* PpG1. *Genetika* 21, 872–874.

Grimont, P.A.D., Weill, F.-X. and WHO Collaborating Center for Reference and Research on Salmonella (2007) *Antigenic formulae of the Salmonella serovars*, 9th edn. Institut Pasteur, Paris, France.

Groman, N.B. (1955) Evidence for the active role of bacteriophage in the conversion of nontoxigenic *Corynebacterium diphtheriae* to toxin production. *Journal of Bacteriology* 69, 9–15.

Groman, N.B. (1984) Conversion by corynephages and its role in the natural history of diphtheria. *Journal of Hygiene* 93, 405–417.

Guinane, C.M., Ben Zakour, N.L., Tormo-Mas, M.A., Weinert, L.A., Lowder, B.V., Cartwright, R.A., Smyth, D.S., Smyth, C.J., Lindsay, J.A., Gould, K.A., Witney, A., Hinds, J., Bollback, J.P., Rambaut, A., Penades, J.R. and Fitzgerald, J.R. (2010) Evolutionary genomics of *Staphylococcus aureus* reveals insights into the origin and molecular basis of ruminant host adaptation. *Genome Biology and Evolution* 2, 454–466.

Hauser, D., Gibert, M., Eklund, M.W., Boquet, P. and Popoff, M.R. (1993) Comparative analysis of C3 and botulinal neurotoxin genes and their environment in *Clostridium botulinum* types C and D. *Journal of Bacteriology* 175, 7260–7268.

Highlander, S.K., Hulten, K.G., Qin, X., Jiang, H., Yerrapragada, S., Mason, E.O. Jr, Shang, Y., Williams, T.M., Fortunov, R.M., Liu, Y., Igboeli, O., Petrosino, J., Tirumalai, M., Uzman, A., Fox, G.E., Cardenas, A.M., Muzny, D.M., Hemphill, L., Ding, Y., Dugan, S., Blyth, P.R., Buhay, C.J., Dinh, H.H., Hawes, A.C., Holder, M., Kovar, C.L., Lee, S.L., Liu, W., Nazareth, L.V., Wang, Q., Zhou, J., Kaplan, S.L. and Weinstock, G.M. (2007) Subtle genetic changes enhance virulence of methicillin resistant and sensitive *Staphylococcus aureus. BMC Microbiology* 7, 99.

Holden, M.T.G., Feil, E.J., Lindsay, J.A., Peacock, S.J., Day, N.P.J., Enright, M.C., Foster, T.J., Moore, C.E., Hurst, L., Atkin, R., Barron, A., Bason, N., Bentley, S.D., Chillingworth, C.,

Chillingworth, T., Churcher, C., Clark, L., Corton, C., Cronin, A., Doggett, J., Dowd, L., Feltwell, T., Hance, Z., Harris, B., Hauser, H., Holroyd, S., Jagels, K., James, K.D., Lennard, N., Line, A., Mayes, R., Moule, S., Mungall, K., Ormond, D., Quail, M.A., Rabbinowitsch, E., Rutherford, K., Sanders, M., Sharp, S., Simmonds, M., Stevens, K., Whitehead, S., Barrell, B.G., Spratt, B.G. and Parkhill, J. (2004) Complete genomes of two clinical *Staphylococcus aureus* strains: evidence for the rapid evolution of virulence and drug resistance. *Proceedings of the National Academy of Sciences USA* 101, 9786–9791.

Holden, M.T., Lindsay, J.A., Corton, C., Quail, M.A., Cockfield, J.D., Pathak, S., Batra, R., Parkhill, J., Bentley, S.D. and Edgeworth, J.D. (2010) Genome sequence of a recently emerged, highly transmissible, multi-antibiotic- and antiseptic-resistant variant of methicillin-resistant *Staphylococcus aureus*, sequence type 239 (TW). *Journal of Bacteriology* 192, 888–892.

Holloway, B.W. and Cooper, G.N. (1962) Lysogenic conversion in *Pseudomonas aeruginosa*. *Journal of Bacteriology* 84, 1324.

Holloway, B.W., Krishnapillai, V. and Stanisich, V. (1971) Pseudomonas genetics. *Annual Review of Genetics* 425–466.

Holmes, A., Ganner, M., McGuane, S., Pitt, T.L., Cookson, B.D. and Kearns, A.M. (2005) *Staphylococcus aureus* isolates carrying Panton-Valentine leucocidin genes in England and Wales: frequency, characterization, and association with clinical disease. *Journal of Clinical Microbiology* 43, 2384–2390.

Holmes, R.K. and Barksdale, L. (1970) Comparative studies with *tox*$^+$ and *tox*$^-$ corynebacteriophages. *Journal of Virology* 5, 783–784.

Holochova, P., Ruzickova, V., Dostalova, L., Pantucek, R., Petras, P. and Doskar, J. (2010) Rapid detection and differentiation of the exfoliative toxin A-producing *Staphylococcus aureus* strains based on φETA prophage polymorphisms. *Diagnostic Microbiology and Infectious Disease* 66, 248–252.

Hyder, S.L. and Streitfeld, M.M. (1978) Transfer of erythromycin resistance from clinically isolated lysogenic strains of *Streptococcus pyogenes* via their endogenous phage. *Journal of Infectious Diseases* 138, 281–286.

Hyman, P. and Abedon, S.T. (2010) Bacteriophage host range and bacterial resistance. *Advances in Applied Microbiology*, 70, 217–248.

Iandolo, J.J., Worrell, V., Groicher, K.H., Qian, Y., Tian, R., Kenton, S., Dorman, A., Ji, H., Lin, S., Loh, P., Qi, S., Zhu, H. and Roe, B.A. (2002) Comparative analysis of the genomes of the temperate bacteriophages ϕ11, ϕ12 and ϕ13 of *Staphylococcus aureus* 8325. *Gene* 289, 109–118.

Inoue, K. and Iida, H. (1968) Bacteriophages of *Clostridium botulinum*. *Journal of Virology* 2, 537–540.

Inoue, K. and Iida, H. (1970) Conversion of toxigenicity in *Clostridium botulinum* type C. *Japanese Journal of Microbiology* 14, 87–89.

Inoue, K. and Iida, H. (1971) Phage conversion of toxigenicity in *Clostridium botulinum* types C and D. *Japanese Journal of Microbiology* 24, 53–56.

Izdebska-Szymona, K., Tyski, S., Drela, N., Sobiczewska, E., Laskowska, M., Kalkhof, D. and Hryniewicz, W. (1992) Staphylococcal lipase affects phagocytosis of *Staphylococcus aureus* by human granulocytes and monocytes. *Acta Microbiologica Polonica* 41, 49–56.

Jin, T., Bokarewa, M., Foster, T., Mitchell, J., Higgins, J. and Tarkowski, A. (2004) *Staphylococcus aureus* resists human defensins by production of staphylokinase, a novel bacterial evasion mechanism. *Journal of Immunology* 172, 1169–1176.

Johnson, W.M. and Lior, H. (1987) Response of Chinese hamster ovary cells to a cytolethal distending toxin (CDT) of *Escherichia coli* and possible misinterpretation as heat-labile (LT) enterotoxin. *FEMS Microbiology Letters* 43, 19–23.

Juhala, R.J., Ford, M.E., Duda, R.L., Youlton, A., Hatfull, G.F. and Hendrix, R.W. (2000) Genetic sequences of bacteriophages HK97 and HK022: pervasive genetic mosaicism in the lambdoid bacteriophages. *Journal of Molecular Biology* 299, 27–51.

Just, I., Mohr, C., Schallehn, G., Menard, L., Didsbury, J.R., Vandekerckhove, J., van Damme, J. and Aktories, K. (1992) Purification and characterization of an ADP-ribosyltransferase produced by *Clostridium limosum*. *Journal of Biological Chemistry* 267, 10274–10280.

Kahankova, J., Pantucek, R., Goerke, C., Ruzickova, V., Holochova, P. and Doskar, J. (2010) Multilocus PCR typing strategy for differentiation of *Staphylococcus aureus* siphoviruses reflecting their modular genome structure. *Environmental Microbiology* 12, 2527–2538.

Kaneko, J., Kimura, T., Narita, S., Tomita, T. and Kamio, Y. (1998) Complete nucleotide sequence and molecular characterization of the temperate staphylococcal bacteriophage ϕPVL carrying Panton–Valentine leukocidin genes. *Gene* 215, 57–67.

Kaper, J.B., Nataro, J.P. and Mobley, H.L. (2004) Pathogenic *Escherichia coli*. *Nature Reviews Microbiology* 2, 123–140.

Karmali, M.A., Petric, M., Lim, C., Fleming, P.C. and Steele, B.T. (1983) *Escherichia coli* cytotoxin, haemolytic–uraemic syndrome, and haemorrhagic colitis. *Lancet* 2, 1299–1300.

Kobayashi, S.D. and DeLeo, F.R. (2009) An update on community-associated MRSA virulence. *Current Opinion in Pharmacology* 9, 545–551.

Kondo, I. and Fujise, K. (1977) Serotype B staphylococcal bacteriophage singly converting staphylokinase. *Infection and Immunity* 18, 266–272.

König, B., Prévost, G., Piémont, Y. and König, W. (1995) Effects of *Staphylococcus aureus* leukocidins on inflammatory mediator release from human granulocytes. *Journal of Infectious Diseases* 171, 607–613.

Kropinski, A.M. (2000) Sequence of the genome of the temperate, serotype-converting, *Pseudomonas aeruginosa* bacteriophage D3. *Journal of Bacteriology* 182, 6066–6074.

Kumagai, R., Nakatani, K., Ikeya, N., Kito, Y., Kaidoh, T. and Takeuchi, S. (2007) Quadruple or quintuple conversion of *hlb*, *sak*, *sea* (or *sep*), *scn*, and *chp* genes by bacteriophages in non-β-hemolysin-producing bovine isolates of *Staphylococcus aureus*. *Veterinary Microbiology* 122, 190–195.

Kuroda, M., Ohta, T., Uchiyama, I., Baba, T., Yuzawa, H., Kobayashi, I., Cui, L., Oguchi, A., Aoki, K., Nagai, Y., Lian, J., Ito, T., Kanamori, M., Matsumaru, H., Maruyama, A., Murakami, H., Hosoyama, A., Mizutani-Ui, Y., Takahashi, N.K., Sawano, T., Inoue, R., Kaito, C., Sekimizu, K., Hirakawa, H., Kuhara, S., Goto, S., Yabuzaki, J., Kanehisa, M., Yamashita, A., Oshima, K., Furuya, K., Yoshino, C., Shiba, T., Hattori, M., Ogasawara, N., Hayashi, H. and Hiramatsu, K. (2001) Whole genome sequencing of meticillin-resistant *Staphylococcus aureus*. *Lancet* 357, 1225–1240.

Kuzio, J. and Kropinski, A.M. (1983) O-antigen conversion in *Pseudomonas aeruginosa* PAO1 by bacteriophage D3. *Journal of Bacteriology* 155, 203–212.

Kwan, T., Liu, J., DuBow, M., Gros, P. and Pelletier, J. (2005) The complete genomes and proteomes of 27 *Staphylococcus aureus* bacteriophages. *Proceedings of the National Academy of Sciences USA* 102, 5174–5179.

Labrie, S.J., Samson, J.E. and Moineau, S. (2010) Bacteriophage resistance mechanisms. *Nature Reviews Microbiology* 8, 317–327.

Ladhani, S., Joannou, C.L., Lochrie, D.P., Evans, R.W. and Poston, S.M. (1999) Clinical, microbial, and biochemical aspects of the exfoliative toxins causing staphylococcal scalded-skin syndrome. *Clinical Microbiology Reviews* 12, 224–242.

Lahteenmaki, K., Edelman, S. and Korhonen, T.K. (2005) Bacterial metastasis: the host plasminogen system in bacterial invasion. *Trends in Microbiology* 13, 79–85.

Lainhart, W., Stolfa, G. and Koudelka, G.B. (2009) Shiga toxin as a bacterial defense against a eukaryotic predator, *Tetrahymena thermophila*. *Journal of Bacteriology* 191, 5116–5122.

Lanyi, B. and Lantos, J. (1976) Antigenic changes in *Pseudomonas aeruginosa in vivo* and after lysogenization *in vitro*. *Acta Microbiology Academy of Science Hungary* 23, 337–351.

Lee, C.Y. and Iandolo, J.J. (1986) Lysogenic conversion of staphylococcal lipase is caused by insertion of the bacteriophage L54a genome into the lipase structural gene. *Journal of Bacteriology* 166, 385–391.

Lee, J.H., Choi, S.Y., Jeon, Y.S., Lee, H.R., Kim, E.J., Nguyen, B.M., Hien, N.T., Ansaruzzaman, M., Islam, M.S., Bhuiyan, N.A., Niyogi, S.K., Sarkar, B.L., Nair, G.B., Kim, D.S., Lopez, A.L., Czerkinsky, C., Clemens, J.D., Chun, J. and Kim, D.W. (2009) Classification of hybrid and altered *Vibrio cholerae* strains by CTX prophage and RS1 element structure. *Journal of Microbiology* 47, 783–788.

Lemire, S., Figueroa-Bossi, N. and Bossi, L. (2008) Prophage contribution to *Salmonella* virulence and diversity. In: Hensel, M. (ed.) *Horizontal Gene Transfer in the Evolution of Pathogenesis*. Cambridge University Press, Cambridge, UK, pp. 159–192.

Leonard, C.G., Colon, A.E. and Cole, R.M. (1968) Transduction in group A *Streptococcus*. *Biochemical and Biophysical Research Communications* 30, 130–135.

Lina, G., Piemont, Y., Godail-Gamot, F., Bes, M., Peter, M.O., Gauduchon, V., Vandenesch, F. and Etienne, J. (1999) Involvement of Panton–Valentine leukocidin-producing *Staphylococcus aureus* in primary skin infections and pneumonia. *Clinical Infectious Diseases* 29, 1128–1132.

Lindsay, J.A. (2010) Genomic variation and evolution of *Staphylococcus aureus*. *International Journal of Medical Microbiology* 300, 98–103.

Lindsay, J.A. and Holden, M.T. (2004) *Staphylococcus aureus*: superbug, super genome? *Trends in Microbiology* 12, 378–385.

Liu, J., Dehbi, M., Moeck, G., Arhin, F., Bauda, P., Bergeron, D., Callejo, M., Ferretti, V., Ha, N., Kwan, T., McCarty, J., Srikumar, R., Williams, D., Wu, J.J., Gros, P., Pelletier, J. and DuBow, M. (2004) Antimicrobial drug discovery through bacteriophage genomics. *Nature Biotechnology* 22, 185–191.

Liu, P.V. (1969) Changes in somatic antigens of *Pseudomonas aeruginosa* induced by bacteriophages. *Journal of Infectious Diseases* 119, 237–246.

Los, M., Kuzio, J., McConnell, M., Kropinski, A.M., Wegrzyn, G. and Christie, G.E. (2010) Lysogenic conversion in bacteria of importance to the food industry. In: Sabour, P. and Griffiths, M. (eds) *Bacteriophages in the Detection and Control of Foodborne Pathogens.* ASM Press, Washington, DC, pp. 157–198.

Lowder, B.V., Guinane, C.M., Ben Zakour, N.L., Weinert, L.A., Conway-Morris, A., Cartwright, R.A., Simpson, A.J., Rambaut, A., Nubel, U. and Fitzgerald, J.R. (2009) Recent human-to-poultry host jump, adaptation, and pandemic spread of *Staphylococcus aureus. Proceedings of the National Academy of Sciences USA* 106, 19545–19550.

Ma, X.X., Ito, T., Chongtrakool, P. and Hiramatsu, K. (2006) Predominance of clones carrying Panton–Valentine leukocidin genes among methicillin-resistant *Staphylococcus aureus* strains isolated in Japanese hospitals from 1979 to 1985. *Journal of Clinical Microbiology* 44, 4515–4527.

Ma, X.X., Ito, T., Kondo, Y., Cho, M., Yoshizawa, Y., Kaneko, J., Katai, A., Higashiide, M., Li, S. and Hiramatsu, K. (2008) Two different Panton–Valentine leukocidin phage lineages predominate in Japan. *Journal of Clinical Microbiology* 46, 3246–3258.

Mathews, A.M. and Novick, R.P. (2005) Staphylococcal phages. In: Waldor, M.K., Friedman, D.I. and Adhya, S.L. (eds) *Phages: Their Role in Bacterial Pathogenesis and Biotechnology.* ASM Press, Washington, DC, pp. 297–318.

McCormick, J.K., Yarwood, J.M. and Schlievert, P.M. (2001) Toxic shock syndrome and bacterial superantigens: an update. *Annual Review of Microbiology* 55, 77–104.

McLeod, S.M., Kimsey, H.H., Davis, B.M. and Waldor, M.K. (2005) CTXϕ and *Vibrio cholerae*: exploring a newly recognized type of phage–host cell relationship. *Molecular Microbiology* 57, 347–356.

Mmolawa, P.T., Willmore, R., Thomas, C.J. and Heuzenroeder, M.W. (2002) Temperate phages in *Salmonella enterica* serovar Typhimurium: implications for epidemiology. *International Journal of Medical Microbiology* 291, 633–644.

Monson, R., Foulds, I., Foweraker, J., Welch, M. and Salmond, G.P. (2011) The *Pseudomonas aeruginosa* generalized transducing phage ϕPA3 is a new member of the ϕKZ-like group of 'jumbo' phages, and infects model laboratory strains and clinical isolates from cystic fibrosis patients. *Microbiology* 157, 859–867.

Morgan, A.F. (1979) Transduction of *Pseudomonas aeruginosa* with a mutant of bacteriophage E79. *Journal of Bacteriology* 139, 137–140.

Moyer, K.E., Kimsey, H.H. and Waldor, M.K. (2001) Evidence for a rolling-circle mechanism of phage DNA synthesis from both replicative and integrated forms of CTXφ. *Molecular Microbiology* 41, 311–323.

Muniesa, M., de Simon, M., Prats, G., Ferrer, D., Panella, H. and Jofre, J. (2003) Shiga toxin 2-converting bacteriophages associated with clonal variability in *Escherichia coli* O157:H7 strains of human origin isolated from a single outbreak. *Infection and Immunity* 71, 4554–4562.

Muniesa, M., Blanco, J.E., de Simon, M., Serra-Moreno, R., Blanch, A.R. and Jofre, J. (2004a) Diversity of stx2 converting bacteriophages induced from Shiga-toxin-producing *Escherichia coli* strains isolated from cattle. *Microbiology* 150, 2959–2971.

Muniesa, M., Serra-Moreno, R. and Jofre, J. (2004b) Free Shiga toxin bacteriophages isolated from sewage showed diversity although the *stx* genes appeared conserved. *Environmental Microbiology* 6, 716–725.

Narita, S., Kaneko, J., Chiba, J., Piemont, Y., Jarraud, S., Etienne, J. and Kamio, Y. (2001) Phage conversion of Panton–Valentine leukocidin in *Staphylococcus aureus*: molecular analysis of a PVL-converting phage, ϕSLT. *Gene* 268, 195–206.

Neoh, H.M., Cui, L., Yuzawa, H., Takeuchi, F., Matsuo, M. and Hiramatsu, K. (2008) Mutated response regulator *graR* is responsible for phenotypic conversion of *Staphylococcus aureus* from heterogeneous vancomycin-intermediate resistance to vancomycin-intermediate resistance. *Antimicrobial Agents and Chemotherapy* 52:45–53.

Newton, G.J., Daniels, C., Burrows, L.L., Kropinski, A.M., Clarke, A.J. and Lam, J.S. (2001) Three-component-mediated serotype conversion in *Pseudomonas aeruginosa* by bacteriophage D3. *Molecular Microbiology* 39, 1237–1247.

Nishifuji, K., Sugai, M. and Amagai, M. (2008) Staphylococcal exfoliative toxins: "molecular scissors" of bacteria that attack the cutaneous defense barrier in mammals. *Journal of Dermatological Science* 49, 21–31.

Novick, R.P. (2006) Staphylococcal pathogenesis and pathogenicity factors: genetics and regulation. In: Fischetti, V.A. (ed.) *Gram-positive Pathogens.* ASM Press, Washington, DC, pp. 496–516.

Novick, R.P. and Subedi, A. (2007) The SaPIs: mobile pathogenicity islands of *Staphylococcus. Chemical Immunology and Allergy* 93, 42–57.

Novick, R.P., Christie, G.E. and Penades, J.R. (2010a) The phage-related chromosomal islands of Gram-positive bacteria. *Nature Reviews Microbiology* 8, 541–551.

Nübel, U., Dordel, J., Kurt, K., Strommenger, B., Westh, H., Shukla, S.K., Zemlickova, H., Leblois, R., Wirth, T., Jombart, T., Balloux, F. and Witte, W. (2010) A timescale for evolution, population expansion, and spatial spread of an emerging clone of methicillin-resistant *Staphylococcus aureus. PLoS Pathogens* 6, e1000855.

O'Brien, A.D., Newland, J.W., Miller, S.F., Holmes, R.K., Smith, H.W. and Formal, S.B. (1984) Shiga-like toxin-converting phages from *Escherichia coli* strains that cause hemorrhagic colitis or infantile diarrhea. *Science* 226, 694–696.

O'Loughlin, E.V. and Robins-Browne, R.M. (2001) Effect of Shiga toxin and Shiga-like toxins on eukaryotic cells. *Microbes and Infection* 3, 493–507.

Oguma, K. (1976) The stability of toxigenicity in *Clostridium botulinum* types C and D. *Journal of General Microbiology* 92, 67–75.

Oguma, K., Iida, H. and Shiozaki, M. (1976) Phage conversion to hemagglutinin production in *Clostridium botulinum* types C and D. *Infection and Immunity* 14, 597–602.

Panton, P.N., Came, M.B. and Valentine, F.C.O. (1932) Staphylococcal toxin. *Lancet* 1, 506–508.

Pantucek, R., Doskar, J., Ruzickova, V., Kaspárek, P., Orácová, E., Kvardova, V. and Rosypal, S. (2004) Identification of bacteriophage types and their carriage in *Staphylococcus aureus. Archives of Virology* 149, 1689–1703.

Pedulla, M.L., Ford, M.E., Karthikeyan, T., Houtz, J.M., Hendrix, R.W., Hatfull, G.F., Poteete, A.R., Gilcrease, E.B., Winn-Stapley, D.A. and Casjens, S.R. (2003) Corrected sequence of the bacteriophage P22 genome. *Journal of Bacteriology* 185, 1475–1477.

Perry, L.L., SanMiguel, P., Minocha, U., Terekhov, A.I., Shroyer, M.L., Farris, L.A., Bright, N., Reuhs, B.L. and Applegate, B.M. (2009) Sequence analysis of *Escherichia coli* O157:H7 bacteriophage ϕV10 and identification of a phage-encoded immunity protein that modifies the O157 antigen. *FEMS Microbiology Letters* 292, 182–186.

Petty, N.K., Foulds, I.J., Pradel, E., Ewbank, J.J. and Salmond, G.P. (2006) A generalized transducing phage (ϕIF3) for the genomically sequenced *Serratia marcescens* strain Db11: a tool for functional genomics of an opportunistic human pathogen. *Microbiology* 152, (Pt 6) 1701–1708.

Petty, N.K., Toribio, A.L., Goulding, D., Foulds, I., Thomson, N., Dougan, G. and Salmond, G.P. (2007) A generalized transducing phage for the murine pathogen *Citrobacter rodentium. Microbiology* 153, 2984–2988.

Proft, T. and Fraser, J.D. (2003) Bacterial superantigens. *Clinical and Experimental Immunology* 133, 299–306.

Proft, T., Moffatt, S.L., Berkahn, C.J. and Fraser, J.D. (1999) Identification and characterization of novel superantigens from *Streptococcus pyogenes. Journal of Experimental Medicine* 189, 89–102.

Proft, T., Sriskandan, S., Yang, L. and Fraser, J.D. (2003a) Superantigens and streptococcal toxic shock syndrome. *Emerging Infectious Diseases* 9, 1211–1218.

Proft, T., Webb, P.D., Handley, V. and Fraser, J.D. (2003b) Two novel superantigens found in both group A and group C *Streptococcus. Infection and Immunity* 71, 1361–1369.

Quinones, M., Kimsey, H.H. and Waldor, M.K. (2005) LexA cleavage is required for CTX prophage induction. *Molecules and Cells* 17, 291–300.

Quinones, M., Davis, B.M. and Waldor, M.K. (2006) Activation of the *Vibrio cholerae* SOS response is not required for intestinal cholera toxin production or colonization. *Infection and Immunity* 74, 927–930.

Rakonjac, J., Bennett, N.J., Spagnuolo, J., Gagic, D. and Russel, M. (2011) Filamentous bacteriophage: biology, phage display and nanotechnology applications. *Current Issues in Molecular Biology* 13, 51–76.

Rappuoli, R., Michel, J.L. and Murphy, J.R. (1983) Restriction endonuclease map of corynebacteriophage ω_c^{tox+} isolated from the Park-Williams no. 8 strain of *Corynebacterium diphtheriae. Journal of Virology* 45, 524–530.

Recktenwald, J. and Schmidt, H. (2002) The nucleotide sequence of Shiga toxin (Stx) 2e-encoding phage ϕP27 is not related to other Stx phage genomes, but the modular genetic structure is conserved. *Infection and Immunity* 70, 1896–1908. [Erratum: 70, 4755.]

Rheinwald, J.G., Chakrabarty, A.M. and Gunsalus, I.C. (1973) A transmissible plasmid controlling camphor oxidation in *Pseudomonas putida. Proceedings of the National Academy of Sciences USA* 70, 885–889.

Riley, L.W., Remis, R.S., Helgerson, S.D., McGee, H.B., Wells, J.G., Davis, B.R., Hebert, R.J., Olcott, E.S., Johnson, L.M., Hargrett, N.T.,

Blake, P.A. and Cohen, M.L. (1983) Hemorrhagic colitis associated with a rare *Escherichia coli* serotype. *New England Journal of Medicine* 308, 681–685.

Roberts, M.D. and Drexler, H. (1981) Isolation and genetic characterization of T1-transducing mutants with increased transduction frequency. *Virology* 112, 662–669.

Rooijakkers, S.H., Ruyken, M., Roos, A., Daha, M.R., Presanis, J.S., Sim, R.B., van Wamel, W.J., van Kessel, K.P. and van Strijp, J.A. (2005) Immune evasion by a staphylococcal complement inhibitor that acts on C3 convertases. *Nature Immunology* 6, 920–927.

Rooijakkers, S.H., Ruyken, M., van Roon, J., van Kessel, K.P., van Strijp, J.A. and van Wamel, W.J. (2006) Early expression of SCIN and CHIPS drives instant immune evasion by *Staphylococcus aureus*. *Cellular Microbiology* 8, 1282–1293.

Rosendal, K., Buelow, P. and Jessen, O. (1964) Lysogen conversion in *Staphylococcus aureus*, to a change in the production of extracellular 'Tween'-splitting enzyme. *Nature*, 204, 1222–1223.

Sachse, S., Seidel, P., Gerlach, D., Gunther, E., Rodel, J., Straube, E. and Schmidt, K.H. (2002) Superantigen-like gene(s) in human pathogenic *Streptococcus dysgalactiae*, subsp *equisimilis*: genomic localisation of the gene encoding streptococcal pyrogenic exotoxin G (*speG*^{dys}). *FEMS Immunology and Medical Microbiology* 34, 159–167.

Sakaguchi, Y., Hayashi, T., Kurokawa, K., Nakayama, K., Oshima, K., Fujinaga, Y., Ohnishi, M., Ohtsubo, E., Hattori, M. and Oguma, K. (2005) The genome sequence of *Clostridium botulinum* type C neurotoxin-converting phage and the molecular mechanisms of unstable lysogeny. *Proceedings of the National Academy of Sciences USA* 102, 17472–17477.

Sako, T. and Tsuchida, N. (1983) Nucleotide sequence of the staphylokinase gene from *Staphylococcus aureus*. *Nucleic Acids Research* 11, 7679–7693.

Sako, T., Sawaki, S., Sakurai, T., Ito, S., Yoshizawa, Y. and Kondo, I. (1983) Cloning and expression of the staphylokinase gene of *Staphylococcus aureus* in *Escherichia coli*. *Molecular and General Genetics* 190, 271–277.

Schallehn, G., Eklund, M.W. and Brandis, H. (1980) [Phage conversion of *Clostridium novyi* type A.] *Zentralblatt für Bakteriologie A* 247, 95–100 (in German).

Schijffelen, M.J., Boel, C.H., van Strijp, J.A. and Fluit, A.C. (2010) Whole genome analysis of a livestock-associated methicillin-resistant *Staphylococcus aureus* ST398 isolate from a case of human endocarditis. *BMC Genomics* 11, 376.

Scott, J., Thompson-Mayberry, P., Lahmamsi, S., King, C.J. and McShan, W.M. (2008) Phage-associated mutator phenotype in group A *Streptococcus*. *Journal of Bacteriology* 190, 6290–6301.

Sebaihia, M., Peck, M.W., Minton, N.P., Thomson, N.R., Holden, M.T., Mitchell, W.J., Carter, A.T., Bentley, S.D., Mason, D.R., Crossman, L., Paul, C.J., Ivens, A., Wells-Bennik, M.H., Davis, I.J., Cerdeno-Tarraga, A.M., Churcher, C., Quail, M.A., Chillingworth, T., Feltwell, T., Fraser, A., Goodhead, I., Hance, Z., Jagels, K., Larke, N., Maddison, M., Moule, S., Mungall, K., Norbertczak, H., Rabbinowitsch, E., Sanders, M., Simmonds, M., White, B., Whithead, S. and Parkhill, J. (2007) Genome sequence of a proteolytic (Group I) *Clostridium botulinum* strain Hall A and comparative analysis of the clostridial genomes. *Genome Research* 17, 1082–1092.

Shimizu, T., Ohta, Y. and Noda, M. (2009) Shiga toxin 2 is specifically released from bacterial cells by two different mechanisms. *Infection and Immunity* 77, 2813–2823.

Shimomura, Y., Okumura, K., Murayama, S.Y., Yagi, J., Ubukata, K., Kirikae, T. and Miyoshi-Akiyama, T. (2011) Complete genome sequencing and analysis of a Lancefield group G *Streptococcus dysgalactiae* subsp. *equisimilis* strain causing streptococcal toxic shock syndrome (STSS). *BMC Genomics* 12, 17.

Simpson, L.L. (1981) The origin, structure, and pharmacological activity of botulinum toxin. *Pharmacological Reviews* 33, 155–188.

Skarin, H., Hafstrom, T., Westerberg, J. and Segerman, B. (2011) *Clostridium botulinum* group III: a group with dual identity shaped by plasmids, phages and mobile elements. *BMC Genomics* 12, 185.

Smith, D.L., James, C.E., Sergeant, M.J., Yaxian, Y., Saunders, J.R., McCarthy, A.J. and Allison, H.E. (2007a) Short-tailed stx phages exploit the conserved YaeT protein to disseminate Shiga toxin genes among enterobacteria. *Journal of Bacteriology* 189, 7223–7233.

Smith, D.L., Wareing, B.M., Fogg, P.C., Riley, L.M., Spencer, M., Cox, M.J., Saunders, J.R., McCarthy, A.J. and Allison, H.E. (2007b) Multilocus characterization scheme for shiga toxin-encoding bacteriophages. *Applied and Environmental Microbiology* 73, 8032–8040.

Smith, T.J., Hill, K.K., Foley, B.T., Detter, J.C., Munk, A.C., Bruce, D.C., Doggett, N.A., Smith, L.A., Marks, J.D., Xie, G. and Brettin, T.S. (2007c)

Analysis of the neurotoxin complex genes in *Clostridium botulinum* A1–A4 and B1 strains: BoNT/A3, /Ba4 and /B1 clusters are located within plasmids. *PLoS ONE* 2, e1271.
Staali, L., Monteil, H. and Colin, D.A. (1998) The staphylococcal pore-forming leukotoxins open Ca^{2+} channels in the membrane of human polymorphonuclear neutrophils. *Journal of Membrane Biology* 162, 209–216.
Stevens, D.L., Tanner, M.H., Winship, J., Swarts, R., Ries, K.M., Schlievert, P.M. and Kaplan, E. (1989) Severe group A streptococcal infections associated with a toxic shock-like syndrome and scarlet fever toxin A. *New England Journal of Medcine.* 321, 1–7.
Strockbine, N.A., Marques, L.R., Newland, J.W., Smith, H.W., Holmes, R.K. and O'Brien, A.D. (1986) Two toxin-converting phages from *Escherichia coli* O157:H7 strain 933 encode antigenically distinct toxins with similar biologic activities. *Infection and Immunity* 53, 135–140.
Su, L.K., Lu, C.P., Wang, Y., Cao, D.M., Sun, J.H. and Yan, Y.X. (2010) Lysogenic infection of a Shiga toxin 2-converting bacteriophage changes host gene expression, enhances host acid resistance and motility. *Molecular Biology (Moscow)* 44, 60–73.
Subedi, A., Ubeda, C., Adhikari, R.P., Penades, J.R. and Novick, R.P. (2007) Sequence analysis reveals genetic exchanges and intraspecific spread of SaPI2, a pathogenicity island involved in menstrual toxic shock. *Microbiology* 153, 3235–3245.
Sugiyama, H. (1980) *Clostridium botulinum* neurotoxin. *Microbiological Reviews* 44, 419–448.
Sumby, P. and Waldor, M.K. (2003) Transcription of the toxin genes present within the staphylococcal phage φSa3ms is intimately linked with the phage's life cycle. *Journal of Bacteriology* 185, 6841–6851.
Szmigielski, S., Sobiczewska, E., Prevost, G., Monteil, H., Colin, D.A. and Jeljaszewicz, J. (1998) Effect of purified staphylococcal leukocidal toxins on isolated blood polymorphonuclear leukocytes and peritoneal macrophages in vitro. *Zentralblatt für Bakteriologie* 288, 383–394.
Tarkowski, T.A., Mooney, D., Thomason, L.C. and Stahl, F.W. (2002) Gene products encoded in the *ninR* region of phage λ participate in Red-mediated recombination. *Genes to Cells* 7, 351–363.
Teel, L.D., Melton-Celsa, A.R., Schmitt, C.K. and O'Brien, A.D. (2002) One of two copies of the gene for the activatable shiga toxin type 2d in *Escherichia coli* O91:H21 strain B2F1 is associated with an inducible bacteriophage. *Infection and Immunity* 70, 4282–4291.
Thierauf, A., Perez, G. and Maloy, A.S. (2009) Generalized transduction. *Methods in Molecular Biology* 501, 267–286.
Thomson, N., Baker, S., Pickard, D., Fookes, M., Anjum, M., Hamlin, N., Wain, J., House, D., Bhutta, Z., Chan, K., Falkow, S., Parkhill, J., Woodward, M., Ivens, A. and Dougan, G. (2004) The role of prophage-like elements in the diversity of *Salmonella enterica* serovars. *Journal of Molecular Biology* 339, 279–300.
Tianiashin, V.I., Zimin, V.I., Boronin, A.M., Tianiashin, V.I., Zimin, V.I. and Boronin, A.M. (2003) [The cotransduction of pET system plasmids by mutants of T4 and RB43 bacteriophages]. *Mikrobiologiia* 72, 785–791 (in Russian).
Todd, J., Fishaut, M., Kapral, F. and Welch, T. (1978) Toxic-shock syndrome associated with phage-group-I staphylococci. *Lancet* 2, 1116–1118.
Toth, I.K., Perombelon, M.C.M. and Salmond, G.P.C. (1993) Bacteriophage ϕKP mediated generalized transduction in *Erwinia carotovora* subsp. *carotovora*. *Journal of General Microbiology* 139, 2705–2709.
Toth, I.K., Mulholland, F., Cooper, V., Bentley, S., Shih, Y.L., Perombelon, M.C.M. and Salmond, G.P.C. (1997) Generalized transduction in the potato blackleg pathogen *Erwinia carotovora* subsp. *atroseptica* by bacteriophage ϕM1. *Microbiology* 143, 2433–2438.
Tranter, H.S. (1990) Foodborne staphylococcal illness. *Lancet* 336, 1044–1046.
Tucker, C.P. and Heuzenroeder, M.W. (2004) ST64B is a defective bacteriophage in *Salmonella enterica* serovar Typhimurium DT64 that encodes a functional immunity region capable of mediating phage-type conversion. *International Journal of Medical Microbiology* 294, 59–63.
Ubeda, C., Maiques, E., Knecht, E., Lasa, I., Novick, R.P. and Penades, J.R. (2005) Antibiotic-induced SOS response promotes horizontal dissemination of pathogenicity island-encoded virulence factors in staphylococci. *Molecular Microbiology* 56, 836–844.
Ubukata, K., Konno, M. and Fujii, R. (1975) Transduction of drug resistance to tetracycline, chloramphenicol, macrolides, lincomycin and clindamycin with phages induced from *Streptococcus pyogenes*. *Journal of Antibiotics (Tokyo)* 28, 681–688.
Vaca, S., Perez, S., Martinez, G. and Enriquez, F. (1993) Partial genetic characterization of FIZ15 bacteriophage of *Pseudomonas aeruginosa*. *Revista Latinoamericana de Microbiologia* 35, 251–257.

Vaca-Pacheco, S., Paniagua-Contreras, G.L., Garcia-Gonzalez, O. and de la Garza, M. (1999) The clinically isolated FIZ15 bacteriophage causes lysogenic conversion in *Pseudomonas aeruginosa* PAO1. *Current Microbiology* 38, 239–243.

van der Vijver, J.C., van Es-Boon, M. and Michel, M.F. (1972) Lysogenic conversion in *Staphylococcus aureus* to leucocidin production. *Journal of Virology* 10, 318–319.

van Wamel, W.J., Rooijakkers, S.H., Ruyken, M., van Kessel, K.P. and van Strijp, J.A. (2006) The innate immune modulators staphylococcal complement inhibitor and chemotaxis inhibitory protein of *Staphylococcus aureus* are located on β-hemolysin-converting bacteriophages. *Journal of Bacteriology* 188, 1310–1315.

Vander Byl, C. and Kropinski, A.M. (2000) Sequence of the genome of *Salmonella* bacteriophage P22. *Journal of Bacteriology* 182, 6472–6481.

Viana, D., Blanco, J., Tormo-Mas, M.A., Selva, L., Guinane, C.M., Baselga, R., Corpa, J.M., Lasa, I., Novick, R.P., Fitzgerald, J.R. and Penades, J.R. (2010) Adaptation of *Staphylococcus aureus* to ruminant and equine hosts involves SaPI-carried variants of von Willebrand factor-binding protein. *Molecular Microbiology* 77, 1583–1594.

Wagner, P.L. and Waldor, M.K. (2002) Bacteriophage control of bacterial virulence. *Infection and Immunity* 70, 3985–3993.

Wagner, P.L., Neely, M.N., Zhang, X., Acheson, D.W., Waldor, M.K. and Friedman, D.I. (2001) Role for a phage promoter in Shiga toxin 2 expression from a pathogenic *Escherichia coli* strain. *Journal of Bacteriology* 183, 2081–2085.

Waldor, M.K. and Mekalanos, J.J. (1996) Lysogenic conversion by a filamentous phage encoding cholera toxin. *Science* 272, 1910–1914.

Wannamaker, L.W., Almquist, S. and Skjold, S. (1973) Intergroup phage reactions and transduction between group C and group A streptococci. *Journal of Experimental Medicine* 137, 1338–1353.

Weiss, B.D., Capage, M.A., Kessel, M. and Benson, S.A. (1994) Isolation and characterization of a generalized transducing phage for *Xanthomonas campestris* pv. *campestris*. *Journal of Bacteriology* 176, 3354–3359.

Wilson, G.G., Young, K.Y., Edlin, G.J. and Konigsberg, W. (1979) High-frequency generalised transduction by bacteriophage T4. *Nature* 280, 80–82.

Winkler, K.C., de Waart, J. and Grootsen, C. (1965) Lysogenic conversion of staphylococci to loss of β-toxin. *Journal of General Microbiology* 39, 321–333.

Yamaguchi, T., Hayashi, T., Takami, H., Nakasone, K., Ohnishi, M., Nakayama, K., Yamada, S., Komatsuzawa, H. and Sugai, M. (2000) Phage conversion of exfoliative toxin A production in *Staphylococcus aureus*. *Molecular Microbiology* 38, 694–705.

Yarwood, J.M., McCormick, J.K., Paustian, M.L., Orwin, P.M., Kapur, V. and Schlievert, P.M. (2002) Characterization and expression analysis of *Staphylococcus aureus* pathogenicity island 3. Implications for the evolution of staphylococcal pathogenicity islands. *Journal of Biological Chemistry* 277, 13138–13147.

Young, K.K., Edlin, G.J. and Wilson, G.G. (1982) Genetic analysis of bacteriophage T4 transducing bacteriophages. *Journal of Virology* 41, 345–347.

Zabicka, D., Mlynarczyk, A., Windyga, B. and Mlynarczyk, G. (1993) Phage-related conversion of enterotoxin A, staphylokinase and β-toxin in *Staphylococcus aureus*. *Acta Microbiologica Polonica* 42, 235–241.

Zou, D., Kaneko, J., Narita, S. and Kamio, Y. (2000) Prophage, ϕPV83-pro, carrying Panton–Valentine leukocidin genes, on the *Staphylococcus aureus* P83 chromosome: comparative analysis of the genome structures of ϕPV83-pro, ϕPVL, ϕ11, and other phages. *Bioscience, Biotechnology and Biochemistry* 64, 2631–2643.

5 The Lion and the Mouse: How Bacteriophages Create, Liberate and Decimate Bacterial Pathogens

Heather Hendrickson[1]
[1]*New Zealand Institute for Advanced Study, Massey University*

'...I had my eyes opened by that microscope, to the fact that there is a world of tiny creatures that you can't see with the naked eye. It's like looking at an African safari park but in miniature...'

Richard Dawkins (Powell, 2011)

The microscopic world is comprised of free-roaming creatures far more complex and impressive in their everyday interactions than what you will observe in a wildlife park. So interdependent are these organisms that defining which is the proverbial lion of the jungle and which the unassuming mouse can be a challenge. Bacterial viruses – bacteriophages or phages – prey on bacteria. On evolutionary timescales, however, phages also can convert bacteria from pussy cats into the human predators we call pathogens. In a surprising twist, it may be the inadvertent action of viruses and the recombination they effect, through horizontal gene transfer (HGT), that enables the continuing existence of bacteria. Phages, as I will discuss below, are the accidental architects of the genetic combinations that we observe in the bacterial world.

Bacteria generally utilize a simple mitotic asexual reproductive cycle. This has profound consequences for these organisms and their evolution. One consequence, of interest here, can be described as Muller's ratchet, the tendency for an asexual organism's DNA to irreversibly acquire increasing loads of deleterious mutations over time (Gordo and Campos, 2008; Allen *et al.*, 2009). Like the ratchet in the handyman's tool box, mutation can be thought of as only permitting change in a single direction, that is, away from the wild type. The absence of sexual reproduction in bacteria means there is no opportunity to reshuffle combinations of genes that sexually reproducing organisms have, which produce both more or less fit combinations. With asexual organisms, by contrast, once a mutation is present, then all future progeny of that individual will probably inherit it, as reversion to wild type will be much rarer than the occurrence of the mutation itself. A continual accumulation of deleterious mutations over time is predicted to eventually destroy asexual lineages and to do so at measured rates that can be as high as 1% of lineages after a mere 1700 generations (Andersson and Hughes, 1996). That bacteria continue to exist suggests that these organisms must have access to recombination of a sort. This takes place through HGT, a phenomenon that adds to genomes by shuffling in new genetic material from other bacterial lineages, both closely and distantly related to the recipient.

In this chapter, I will illustrate the influence of phages in the process of HGT and the impact that HGT has on bacterial evolution with a special interest in the evolution of pathogens. I will begin with a brief word about the day-to-day struggle between phages and bacteria before proceeding to a description of the mechanisms, quantities and impact of HGT in the microbial jungle.

Phages and Bacteria: the Short Game

Over evolutionary timescales, phage-mediated events promote the evolution of bacterial species through the addition of new DNA. The essential relationship between these organisms is however, still properly viewed as antagonistic. Thus phages and bacteria are engaged day to day in a microscopic mêlée, and the mechanisms by which bacteria avoid phage infection can, at times, be indistinguishable from mechanisms by which pathogens avoid the immune responses of potential hosts (Abedon, 2012).

Phages, as a consequence of these antagonistic interactions, can have a significant effect on the short-term evolution of bacteria by modifying their population dynamics. Experiments in controlled competitive environments in which available nutrients are limited have been used to demonstrate the arrival of successive adaptations on both the phage and the bacterial sides in an evolutionary arms race (Bohannan and Lenski, 2000a,b). In population growth experiments where bacterial genetic diversity would normally emerge as a result of niche specialization and trade-offs, the effect of adding predatory phages is to decrease within-population competition, which results in reduced diversity (Buckling and Rainey, 2002a,b). Further studies using soil habitats, rather than liquid cultures, have demonstrated rapid co-evolution between bacteria and their phages, and evidence of fluctuating selection as well as growth costs associated with evolving phage resistance (Gomez and Buckling 2011). Understanding the population dynamics of co-evolution in simple cases such as controlled soil environments will help us to understand these processes better in nature.

Bacteria are engaged in a constant struggle to avoid exploitation by phages. The continued existence of both implies that this effort is matched step for step by continual counter-adaptation on both sides, and antagonistic co-evolution. It is perhaps ironic then that reciprocal evolution in bacteria is mediated, in part, by phage involvement in HGT.

The Long Game: Horizontal Gene Transfer

The bacterial genomic era, starting in 1995 with the complete genome sequence of *Haemophilus influenzae* RD1, was the beginning of fresh insight into the evolutionary processes undergone by bacteria (Fleischmann *et al.*, 1995). This and subsequent sequencing studies revealed that separate genes in single, continuous DNA molecules can have different phylogenetic histories. HGT takes place readily in prokaryotes and leads to organisms with DNA molecules that are chimaeric, with large swathes inherited across species boundaries (Lawrence and Hendrickson, 2003).

A successful HGT event will lead to a recipient bacterial cell with DNA that can be anything from a variant of the genetic material that it already had to a novel section of DNA with genes that are completely new to the bacterium. While occasional transfers of small amounts of genomic content in some fraction of extant lineages would be novel and notable, there are recent suggestions that an average of 81% of bacterial genes might be acquired horizontally (Zhaxybayeva *et al.*, 2006; Dagan *et al.*, 2008). HGT both has and is having a profound impact on the nature of the bacteria that we see around us. The redistribution of genetic material that is implied by such a ubiquitous process gives us cause to reconsider notions that we otherwise rely on in considering bacteria, including key concepts such as 'species', 'pathogen' and 'population'.

Detection of HGT

There are three broad classes of HGT detection methods: parametric mismatch, syntenic disruption and phylogenetic incongruity. In this section, I provide an outline of what these different approaches entail, as well as some of their shortcomings.

Parametric methods posit that bacterial genomes acquire characteristics over evolutionary timescales that are distinctive, much like tool marks acquired in a master sculptor's shop. Various metrics have been proposed for distinguishing between old and new DNA including G+C content, dinucleotide content, higher-order sequences (6mer and 8mer repeats) and the codon adaptation index (CAI), as well as combinations of all of these (Karlin, 1998; Tsirigos and Rigoutsos, 2005). Two major advantages in using parametric methods are that they are computationally simple and that multiple genomes for comparison are not required for transferred genes to be identified.

Many parametric methods have been developed, and these generally identify overlapping but different sets of genes for the same genome. This is due, in part, to the confounding similarity in mutational tendencies among near relatives. In short, atypical regions are identified with the most confidence when donor and recipient genomes are not closely related (Azad and Lawrence, 2007; Arvey *et al.*, 2009; Boc and Makarenkov, 2011). Complicating these determinations, as time passes in a new genomic context the mutational tendencies of the cognate replication machinery in combination with natural selection (acting on spontaneous mutations) will tend to eliminate the distinctiveness of the acquired DNA, a process termed amelioration (Lawrence and Ochman, 1997). It should also be noted that parametric measures, such as CAI, can be misread. Developmental stages in bacterial growth, such as sporulation in *Bacillus subtilis*, for example, can lead to genes with an atypical CAI and mistaken identification, as recently introduced (Moszer *et al.*, 1999).

The second class of HGT detection is syntenic disruption. If closely related organisms are available for genomic comparison, then breaks in synteny, or gene order, can be observed when genomes are aligned. These breaks indicate that a region of DNA has been introduced into or deleted from one of the bacterial genomes (Delcher *et al.*, 2003; Darling *et al.*, 2010). Breaks in synteny can reveal that HGT as well as genomic rearrangements have taken place since the divergence of two organisms.

Phylogenetic incongruity is the third class of HGT detection and involves comparing phylogenetic trees. Homologues in closely related organisms (orthologues) accumulate mutational differences, which can be used as informative character traits. These are in turn used to build phylogenetic trees that represent hypotheses about the relatedness of these genes and the organisms that possess them. The 16S and 18S rRNAs have been proposed as the standard for phylogenetic comparison across the three domains of life because these molecules are present in all branches of life and are molecularly constrained enough to be informative relative to other potential sites (Fox *et al.*, 1977; Olsen *et al.*, 1986). Even this paragon of molecular metrics, however, has been horizontally transferred in some cases (Gogarten *et al.*, 2002). Approaches to discovering the extent to which bacteria are chimaeric using phylogenetic trees have evolved quickly and often lead to differing estimates of the degree to which HGT is influencing the evolution of bacterial lineages (Ochman, 2005).

HGT complicates determinations of a true tree of life. Complete genome sequences are a record of all of the genes that an organism has, as well as a robust source of phylogenetic information regarding each of these genes. Homologous genes from different organisms can therefore be used to construct a tree that represents the history of that gene, but where such a gene tree fails to align with the inferred organismal tree then a transfer event can be inferred (Snel *et al.*, 2005). How then can one find a true organismal tree to use as the benchmark for such comparisons? Multiple methods are being developed to determine the phylogenetic tree of most of genes, using divergence from this norm as the metric by which transfer is judged (Gogarten *et al.*, 2002; Poptsova and Gogarten, 2007; Andam and Gogarten, 2011).

Another route to determining a tree of life is to try to identify a set of genes that are present in as many organisms as possible while avoiding including any transferred genes. Snel and colleagues have made an impressive effort in this regard and generated a tree that includes the three domains of life by using 31 genes from 191 completely sequenced genomes (Ciccarelli *et al.*, 2006). By ignoring the contribution of gene transfer, however, what does this tree represent? By one account, 31 genes represent just 1% of the genes that the average microbial genome contains (Dagan and Martin, 2006). Another method, utilized by Rivera and Lake (2004), allowed transfer and genome fusion to enter into the model of relatedness. This analysis revealed a circular pattern or relatedness at deep branches, evidence of ancient and abundant HGT (Rivera and Lake, 2004). For these reasons, biologists are being encouraged to stop seeking a bifurcating 'tree of life' to represent the nature of relatedness when considering the prokaryotes (Dagan and Martin, 2006; Dagan *et al.*, 2008). Despite the controversy generated by evidence of widespread HGT, there is still strong support for the assertion that all cellular life has arisen from a single ancestor. This hypothesis received support recently through direct testing using a set of ubiquitously conserved gene sequences, which suggested a high-level tree of universal relatedness (Theobald, 2010).

Mechanisms of gene exchange

When HGT is identified in a recipient genome, the donor genome is generally not known, although there are new methods being proposed to overcome this limitation (Azad and Lawrence, 2011; Boc and Makarenkov, 2011). HGT therefore is generally detected solely as novel DNA in a recipient genome, and this acquisition takes place through three processes: transformation, conjugation and transduction (Fig. 5.1).

Transformation is the process by which bacteria take up DNA from the extracellular environment and enact repair-like events that inadvertently bring this linear DNA into the circular bacterial genome. Many naturally competent bacteria have mechanisms in place to ensure that they import DNA from bacteria they are closely related to (Krüger and Stingl, 2011). Gene transfer occurs through transformation as the occasional by-product of bacteria feeding on DNA (Redfield, 2001). Conjugation is the process by which plasmids, small accessory chromosomes with their own genes, control the transfer of DNA, usually the plasmid, between bacteria. Conjugation can be considered an HGT event whether or not the plasmid DNA is subsequently recombined into the bulk of the recipient genome.

In transduction, a virus infects a bacterium and subsequently produces non-infectious viral particles that have accidentally packaged the DNA of the host bacterium rather than the virus's own DNA. These defective phage particles then can inject this bacterial DNA into a recipient bacterium, and

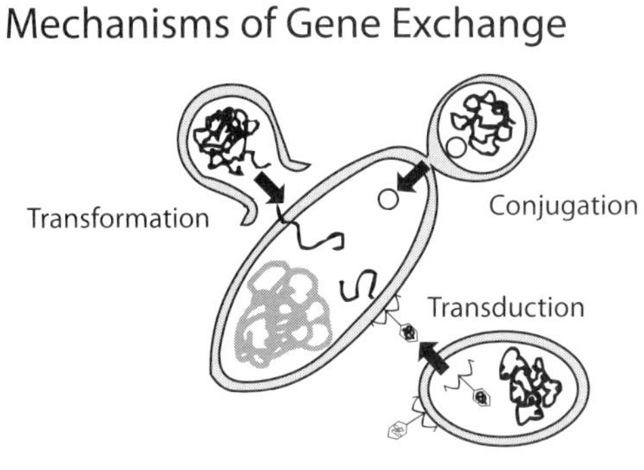

Fig. 5.1. The three major categories of events that can lead to gene transfer: (i) transformation, whereby naked DNA from the extracellular environment can be taken up by bacterial cells; (ii) conjugation, the transmittance from one cell to another of a plasmid; and (iii) transduction, the passage of DNA from one cell to another mediated through the accidental incorporation of bacterial DNA into a phage particle in a prior infection.

it can then be recombined into the recipient genome by repair-like recombination processes. It should be noted that the process of lysogeny or integration of a phage genome into a bacterial chromosome can also be considered a special instance of HGT (see Abedon and Christie *et al.*, Chapters 1 and 4, this volume). Inspection of closely related bacterial genome sequences reveals much alignment in gene order, followed by gaps that can be as short as 1 kb and can range up to many hundreds of kb. Upon more detailed analysis, many of these gaps can be attributed to phages that have integrated in one bacterial genome and not the other. Integrated phage genomes can carry genetic elements that contribute both to the pathogenicity of the strain and to their own retention (Brüssow *et al.*, 2004; see also Christie *et al.*, Chapter 4, this volume).

Constraints on HGT

If all HGT was equally likely, then we might expect the tree of prokaryotic life to be a web of inter-related threads; a panmictic mess. Our notions of bacterial relatedness, however, has structure that is present despite HGT. Organisms that are closely related according to 16S rRNA gene estimates also share unique properties and lifestyles. This family-level cohesion, despite seemingly massive opportunities for HGT, suggests that there are natural constraints that define or structure this process (Gogarten *et al.*, 2002; Andam and Gogarten, 2011). If organisms that are most closely related are those that are exchanging DNA most frequently, then vertical inheritance and HGT may both contribute to the observed relatedness in bacterial families, just as sexual reproduction in higher organisms contributes directly to the species cohesiveness (and, indeed, defines the biological species concept).

Barriers to transfer

Consistent with these ideas of limitations on likelihood of HGT, newly integrated DNA must pass through a series of chance and selective constraints in order to be maintained for long periods of time. These limits to transfer are probably concurrent and, although all are real, many are difficult to measure and none should be considered as a brick wall to transfer. Many natural barriers to transfer in particular have components that make them likely to contribute to enforcing tendencies whereby close relatives exchange DNA most frequently. As such, the following is far from an exhaustive discussion but gives a breadth of mechanisms that constrain gene transfer.

The first real barrier to HGT is the transmittance of genetic material between donor and recipient cells. This is mediated by multiple processes, all of which will have natural host-range limitations to one degree or another (Fig. 5.1). Phages often have limited abilities to infect (Hyman and Abedon, 2010), plasmids are at the mercy of the success of conjugation mechanisms between individual cells, and transformation mechanisms can differentiate between closely and distantly related DNA upon uptake. Next, incoming genetic material will often be encountered in the cytoplasm by recipient enzymes and degraded by nucleases including restriction endonucleases. Recipient-cell DNA has short, specific sequences throughout the genome that are continuously marked by methylation. If a span of DNA enters the cytoplasm and lacks this mark then that DNA is taken to be foreign and is cut. Only if DNA overcomes these barriers can a recombination repair pathway instead mistake the DNA for native damaged DNA and thereby integrate it into the recipient bacterial chromosome or other cytoplasmic replicon such as a native plasmid.

Subsequent to incorporation, new DNA faces a number of immediate challenges that serve to protect the integrity of the recipient bacterium's genome. The mismatch repair system, for example, acts to correct mutations, including small insertions and deletions. It does so by determining, immediately after replication, which is the newly synthesized and which is the older strand of DNA (Harfe and Jinks-Robertson, 2000). If there are differences between the two, then this system directs new strand repair in order to match the older, and therefore trusted, template.

Methylation of a common 4 bp sequence is responsible for this distinction, with methylation lagging behind DNA replication such that new strands are unmethylated for a time. Incoming DNA that does not harbour the native methylation signals is similarly subject to mismatch repair anti-recombination processing, which can contribute to genetic isolation of even closely related bacteria (Berndt *et al.*, 2003).

An additional, sequence-based discrimination mechanism has been recognized in the clustered, regularly interspaced, short palindromic repeats (CRISPR) module. Estimated to be present in up to 70% of eubacterial genomes, CRISPR is essentially an adaptive immune system for bacterial cells wherein the memory of previously encountered foreign DNA is incorporated as small DNA spacers in the CRISPR region of the genome (Deveau *et al.*, 2010). These are used to recognize and target incoming DNA for destruction and have been shown to interfere with lateral gene transfer (Marraffini and Sontheimer, 2008).

Barriers to retention

In the event that a newly inserted region of DNA has both integrated into a replicon and passed the inspection of these various defence mechanisms, existence in the population depends on the fitness effect of new DNA for the host bacterium. Fitness involves many factors. Interruptions of important genes or regulatory regions, for example, tend to be detrimental to fitness. Likewise, newly inserted DNA can interfere with existing chromosomal architecture, which can impact the ability of replicated sister chromosomes to segregate correctly into daughter cells (Bigot *et al.*, 2005; Hendrickson and Lawrence, 2006; Sivanathan *et al.*, 2009). Early evidence of the importance of bacterial chromosome sequence architecture was the observation that integration of phage λ into the *Escherichia coli* chromosome was orientation dependent. As a long-term resident of *E. coli* genomes, phage λ has the same replication-mediated repeat structure as its host genome and integration events are sensitive to matching the host (Corre *et al.*, 2000; Pease *et al.*, 2005).

Constraints on HGT that are mediated by chromosome sequence architecture may be avoided in two primary ways. First, HGT between closely related organisms will tend to bring in novel DNA with compatible sequence structure. This will be less likely to interfere with segregation after integration. Secondly, transfer of entire replicating units such as plasmids will entirely avoid this limitation. Transfer that involves recombination into plasmids will also avoid the constraints of integrating into a structured genome. In this way, conjugation events can create a safe haven for HGT (Hendrickson and Lawrence 2006).

Acquisition of new functions

Newly acquired DNA in an architecturally structured genome is more likely to decrease rather than increase the fitness of a new recombinant. By contrast, the acquired *functions* of the inherited genes have the potential to increase the fitness of the recombinant bacterium. In order for any fitness change to be realized, the genetic material must either include gene operator regions that are correctly read by the transcription factors resident in the host cytoplasm or must have recombined downstream of a native operator region, ideally without disrupting existing functions. Thus, if the genes cannot be transcribed and translated, then the additional DNA is unlikely to be anything more than neutral or may be deleterious to the fitness of the recipient organism. Weakly expressed or misexpressed genes can adapt over time, driving accelerated evolution of regulatory regions in recently acquired DNA (Lercher and Pál, 2008).

If DNA can be transcribed and translated, then new encoded functions will determine whether this HGT event is advantageous in the environments in which this bacterium and its descendants find themselves. If newly acquired DNA is not sufficiently advantageous, then the carrying individual will probably be lost from the population. Alternatively, the new DNA that survives in the population despite its poor contribution to the organism's fitness can become mutated to the point where there is no longer any possibility of function, forming pseudogenes that will eventually be lost. Observations of recently acquired genes may reflect neutral or even disadvantageous acquisitions in the process of being purged, perhaps having been maintained up to that point by genetic drift. Thus, a genome sequence should be considered to be a snapshot of a single bacterium at some point on its evolutionary trajectory, rather than as an organism whose genetic repertoire necessarily represents perfect adaptation to its environment. The combination of gene conservation by vertical descent, frequent HGT and subsequent taxonomically sensitive constraints leads to the observed cohesion between closely related bacteria.

Levels of Horizontal Gene Transfer Observed in Bacterial Genomes

The myriad methods available for detection of HGT and the speed at which the field is advancing means that any author's descriptions of levels of gene transfer for extant organisms are subject to both speculation and fierce debate. Early estimates of the fixation of novel DNA into *E. coli* put the rate at 16 kb per 1 million years (Lawrence and Ochman, 1998). Today, it is generally accepted that most bacterial genomes have experienced recent and frequent gene transfer and that nearly all genes can be transferred (Ochman *et al.*, 2000; Koonin *et al.*, 2001; Gogarten and Townsend, 2005; Sorek *et al.*, 2007).

Bacterial obligate intracellular parasites of eukaryotic cells appear to be a telling exception to this rule. The conditions in which these organisms live, inside their eukaryotic hosts, limits their exposure to the agents of gene transfer: phages, plasmids and extracellular DNA (Moran, 2003; Blanc *et al.*, 2007; Nikoh and Nakabachi, 2009). The absence of recombination through phages or other HGT delivery mechanisms renders these organisms victims of Muller's ratchet and the consequences of continual mutation (Moran, 1996; Andersson and Kurland, 1998). Accumulation of mutations over evolutionary timescales eventually limits their potential to live outside the host cells resulting in their becoming organelles such as mitochondria and chloroplasts (Gray and Doolittle, 1982; Moran, 1996; Lang *et al.*, 1997; Allen *et al.*, 2009). It should be noted, however, that a recent study suggests that rare co-infection events can support genetic exchange in some cases between such obligately intracellular bacteria (Kent *et al.*, 2011).

Genes found on plasmids, in phage genomes or that are associated with other 'mobility elements' are more likely to be transferred than genes that are either intimately involved with large numbers of other genes or genes that are otherwise very specific in their interactions. In addition, a recent study suggested that the number of functional domains a protein has may also positively affect the likelihood of its successful transfer (Gophna and Ofran, 2010). Initially, the notion that some genes are more reticent to transfer led to the classification of some genes as 'mobile', while other genes are considered to be 'core' and therefore good candidates for establishing phylogenetic relationships (Ge *et al.*, 2005). These 'core' genes, however, have been observed to be distributed in non-tree-like manners, suggesting that, while the idea of a set of genes that are resistant to the action of HGT might be appealing, this idea does not appear to be readily demonstrated (Beauregard-Racine *et al.*, 2011). Lastly, it would seem that the more we examine the scope of genes that are present in some but not all members of a species, the higher our estimates of the amounts of transfer become (Dagan *et al.*, 2008). *E. coli* provides an insight into this phenomenon, as it is an organism whose genomic depth we have sampled to a remarkable degree.

E. coli as an Example of Degrees of Horizontal Gene Transfer in Bacteria

Recent genome sequencing projects have expanded our understanding of the speed at which novel DNA is entering and exiting the bacteria that we observe today. *E. coli* is an important bacterial species, strains of which have been differentiating for 25–40 million years (Ochman and Jones, 2000). These organisms are intimately associated with humans and include both commensals and deadly pathogens such as the O157:H7 strain (see Kuhl *et al.*, Christie *et al.*, Williams and LeJeune, Goodridge and Steiner, and Niu *et al.*, Chapters 3, 4, 6, 11 and 16, this volume). A recent 20-genome sequencing project examining *E. coli* diversity revealed that the 'core genome', the genes that appear in nearly all *E. coli*, are approximately 98.3% identical at the sequence level between strains and are a set of approximately 2000 genes. Each genome on average has ~4700 genes and the total pool of genes shared is 10,000+ (Touchon *et al.*, 2009). Thus, each genome on average is ~43% core *E. coli* genes and ~57% other, more mobile genes that have either recently arrived or can easily be lost. This other 57% or 10,000+ genes collectively can be referred to as the *E. coli* mobile-ome (Siefert, 2009). The obvious prophage-associated genes in this study were often found in only a single genome, which suggests that, while flux of phage-borne genes into genomes is high, the retention of these genes is low, perhaps due to deleterious effects (Touchon *et al.*, 2009). Such high levels of fluctuating genomic content imply that the utility of examining single strains as representative of species of bacteria is probably short-sighted.

Another noteworthy finding involves the locations that new DNA has been introduced into the genomes. Touchon *et al.* (2009) found evidence of hot spots for newly acquired DNA, with 71% of the non-core transferred genes located in 133 locations across the 21 genomes. This suggests a natural constraint on HGT that is overcome by inserting into genome regions that themselves have been acquired recently. Logically, these would be locations that are not essential to bacterial fitness and therefore that are less likely to be disrupted upon insertion of new DNA. A subsequent study using 61 completely sequenced *E. coli* and the closely related *Shigella* genomes came to the conclusion that over 80% of the genes in a typical genome could be considered 'accessory' rather than core and that 90% of all of the genes found in these organisms (their pan-genome) were 'accessory' genes. The message from these large-scale genomic studies appears to be that bacterial species are fluid due to fluxes in gene content (Lukjancenko *et al.*, 2010).

Horizontal Gene Transfer: New Capabilities, Pathogenicity, and Beyond

The ultimate effect of the flow of genes into bacterial genomes is that these genes bring about new abilities and potentials for these organisms. It is the sudden acquisition of segments of genetic material that gives bacteria access to new niches resulting, for example, in the spread of antibiotic resistance and the transformation of commensal bacteria into pathogenic lifestyles. These processes have caught our attention in recent years as dangerous new species have arisen, with phages often acting as direct mediators of change.

Antibiotic resistance

Many microorganisms express antibiotics that serve to either kill or slow the growth of local bacteria. Bacteria in turn develop capacities to resist antibiotics. Phages mediate these processes on both sides by mobilizing genetic material in nature that allows both resistance to and production of antibiotics (Fig. 5.2; Davison, 1999). Williams and LeJeune (Chapter 6, this volume) also briefly consider the role that phages might play especially in the transduction of antibiotic resistance genes.

Bacterial evolution and pathogenicity as a process

Within a single lineage of bacteria, evolving over millions of years, we can find the entire spectrum of host interactions: pathogens, commensals and mutualists. This observation alone suggests that these identities are fluid over time, as the last common ancestor of all of these must have fallen into only one of these categories – or did it? The designation of pathogen, commensal or mutualist often

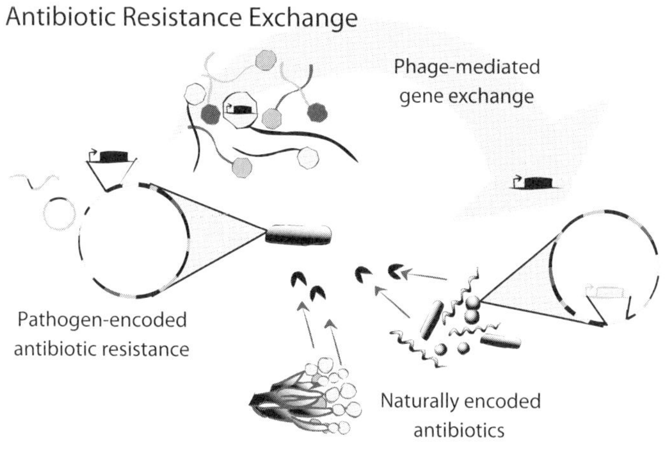

Fig. 5.2. Phage involvement in cycling antibiotic production and resistance genes in the environment. Antibiotics are part of the normal competitive regime of microorganisms in nature and are produced by a wide range of organisms (grey arrows). Bacteria are therefore under pressure to evolve resistance to these (black notched circles). The genes that encode resistance may also be found in organisms that produce the antibiotics (black gene in genome) and are generally beneficial when transferred through phage or other processes. Phages are therefore taking part in the transmission of genetic material that encodes for both antibiotics and antibiotic resistance in nature.

appears to be dependent on the specific host that the bacterium is being observed with (reviewed by Lawrence and Hendrickson, 2008). This is most likely a result of the specific nature of the chemical signals that are transmitted between a host organism and a bacterium.

Bacteriophages are constrained somewhat in the genes that they can carry. On average, these genomes are 60 kb, with a range from a lower limit of 3.5 kb to an upper described limit of ~500 kb (see www.ncbi.nlm.nih.gov/genomes/GenomesGroup.cgi?opt=virus&taxid=10239&host=bacteria; Hendrix, 2009). Classically, bacteriophages must carry a genomic content that encodes structural proteins as well as genome replication and cell-escape functions. In addition, most phage genomes contain non-essential or accessory genes, some of whose functions are only needed in particular host strains (Pope *et al.*, 2011). Bacteriophages none the less contribute in major ways to the genomic fluidity displayed by pathogens. There are numerous cases where the causative agent that makes a bacterium pathogenic is a toxin that is encoded by a prophage (see Christie *et al.*, Chapter 4, this volume). These toxins can then be secreted by the pathogen into the extracellular environment to wreck havoc on the host organism (reviewed by Casas and Maloy, 2011). In addition to mediating pathogenicity in this fashion, bacteriophages at times carry alternative O-antigens that aid bacteria in evading host immune systems or other predation, promote adhesion of cells to tissues and contribute to generalized transduction of antibiotic resistance genes in populations (Firth and Skurray, 2006; Wagner and Waldor, 2006; Wildschutte and Lawrence, 2007).

If bacteria can inherit a suite of deadly pathogenicity genes in a single phage integration event but still look otherwise harmless, then the medically important difference between an inoffensive enteric bacterium and a pathogen is, from a genetic perspective, a small one indeed. For historical reasons, much nomenclature in bacteria is inherently uninformative in this respect. *E. coli* K12 strain MG1655 is a harmless laboratory strain that was obtained from a patient in Palo Alto in 1922 (Blattner *et al.*, 1997; Anon, 2011). On the other hand, beginning 60 years later, *E. coli* O157:H7 strain EDL933 was a deadly pathogen, the causative agent of haemolytic–uraemic syndrome and responsible for fatalities after consumption of contaminated meat in the USA (for more on *E. coli* O157:H7 see Kuhl *et al.*, Christie *et al.*, Williams and LeJeune, Goodridge and Steiner, and Niu *et al.*, Chapters 3, 4, 6, 11 and 16, this volume). These versions of the same species had diverged over an estimated period of 4.5 million years and each had acquired large segments of DNA through HGT in the intervening time (Perna *et al.*, 2001; Welch *et al.*, 2002). These early comparisons of strains of the same species comfortingly led to the conclusion that these additions had been made on a 4.1 Mb backbone of core genes, or roughly 89% of the MG1655 and 72% of the EDL933 total genome size. As stated earlier, however, more recent sequencing efforts put genes shared by all *E. coli* at closer to 43% of total genes in an average representative (Touchon *et al.*, 2009). These efforts miss much of the nearly homologous recombination taking place between more closely related strains, and inclusion of these recombination events as well could send estimates of the percentage of genes in the average genome that have been recently transferred to 81% (±15% ; Dagan *et al.*, 2008). It seems that the more we learn about the amount of transfer taking place, the less certain we can be about species delineations and their relevance to human disease.

Phages can also increase the pathogenicity of bacterial communities without changing the genetic make-up of these bacteria. Bacterial biofilms have become the subject of much speculation recently, and the nature and significance of these microbial conglomerations are being investigated. Microorganisms living in these mature pellicle structures of excreted polysaccharides appear to be less susceptible to antibiotics and phages. In fact, the biofilm community of a cystic fibrosis lung was recently found to be the first phage-free environment recorded (Høiby *et al.*, 2010, 2011; Willner *et al.*, 2011). At the same time, however, it is also becoming clear that phages are a natural part of many of these communities (Abedon, 2011). Biofilm formation has been demonstrated in multiple pathogenic lineages to be bolstered by the presence of extracellular DNA. As phages lyse bacteria, they release this DNA into the environment and this additional extracellular DNA enhances biofilm structural stability, and in some cases the degree of phage-mediated lysis appears to be regulated by the bacteria (Carrolo *et al.*, 2010; Kay *et al.*, 2011; Petrova *et al.*, 2011).

The 'Why' of Bacteriophages and Pathogenicity

The consensus regarding the impact of HGT on the evolution of bacteria has been recognized only in phases. Immediately after completely sequenced bacterial genomes started to be published (1996), HGT was considered to be a novel or rare event. Within a few years, gene transfer began to be appreciated as an ongoing but nevertheless restricted process, where some genes were thought to be mobile but most were not transferred. More recently, by contrast, there has been increasing suggestion that the majority of genes in many genomes have been transferred and that cohesion within families might be driven by a combination of vertical descent and constraints on these transfer frequencies. Continued sequencing efforts and improved methodologies for detection have carried us through these shifts of paradigm. Today, we are still at an impasse as to what observed transfer means for meaningful taxonomy. This becomes a crucial point when nomenclature meets medical practice (Brüssow *et al.*, 2004; Gevers *et al.*, 2005; see also Kuhl *et al.* and Williams and LeJeune, Chapters 3 and 6, this volume).

Evolutionary biologists are often interested in the 'why' of phenomena more than the 'how'. The short answer to why bacteriophages carry genes that enhance pathogenicity is that it helps them to make more bacteriophages. In 1976, Richard Dawkins wrote, in *The Selfish Gene*, that one could consider genes as 'selfish replicators' coding for phenotypes whose sole goal it is to increase their own numbers (Dawkins, 1976). From this perspective, any gene can be seen as contributing to an organism only in order to ensure that more copies of the gene are made. As this concept has influenced modern biological thought, multiple lines of evidence have contributed to the notion that many genetic elements can be thought of in much the same way. Selfish genetic elements have been defined as 'those having characteristics that enhance their own transmission relative to the rest of an individual's genome, and that are either neutral or detrimental to the organism as a whole' (Werren *et al.*, 1988). By this definition, bacteriophages can be thought of as a quintessentially selfish genetic element. These bundles of DNA wrapped in a protein coat travel through the natural world, inert until they happen upon a replicator to prey on. From this view, the success of a temperate bacteriophage is dependent on the genes that it can imbue to a host that it integrates into (Abedon and LeJeune, 2005). If the bacteriophage encodes for sufficiently niche-expanding or fortifying genes, then it can increase its copy number by accompanying a successful host organism in its clonal expansion during an infectious cycle. In some cases, this will propel otherwise harmless commensals into more pathogenic lifestyles but the measure of success for the phage during this outgrowth is simply the production of more phages.

Summary

The relationship between pathogens and bacteriophages is a complex one and the two entities can be thought of as interdependent in many ways. On the one hand, phages can serve as 'catalysts' that impel bacteria, particularly via HGT, into a pathogenic lifestyle. On the other hand, pathogens as well as bacteria in general appear to be doomed, ultimately, to mutational collapse under the pressure of Muller's ratchet when phages and other mediators of HGT are not present to mediate intra-specific re-combination. Nevertheless, although phages apparently are beneficial when viewed over these evolutionary timescales, over short timescales they fundamentally serve as predators that search out and 'devour' bacteria, including the same pathogens that they can play such important roles in creating. In the tale of the lion and the mouse, the lion threatens the mouse one day and is freed by the mouse from a hunter's trap the next. Given the evidence of the eminently complex co-dependencies between pathogens and phages in the microbial world, who is the true king of the jungle? Bacteria perhaps owe deference to their phage predators, which serve simultaneously as creators, sustainers and destroyers of bacterial pathogens?

Acknowledgements

The author would like to acknowledge the generous support of the Human Frontier of Science Program Organization. She is also grateful for helpful discussion with D.I. Andersson and J.G. Lawrence, as well as helpful access to a manuscript in press by S. Maloy.

References

Abedon, S.T. (2011) *Bacteriophages and Biofilms: Ecology, Phage Therapy, Plaques.* Nova Science Publishers, Hauppauge, New York.

Abedon, S.T. (2012) Bacterial 'immunity' against bacteriophages. *Bacteriophage* 2, 50–54.

Abedon, S.T. and LeJeune, J.T. (2005) Why bacteriophage encode exotoxins and other virulence factors. *Evolutionary Bioinformatics Online* 1, 97.

Allen, J.M., Light, J.E., Perotti, M.A., Braig, H.R. and Reed, D.L. (2009) Mutational meltdown in primary endosymbionts: Selection limits Muller's ratchet. *PLoS ONE*, 4, p.e4969.

Andam, C.P. and Gogarten, J.P. (2011) Biased gene transfer in microbial evolution. *Nature Reviews Microbiology* 9, 543–555.

Andersson, D.I. and Hughes, D. (1996) Muller's ratchet decreases fitness of a DNA-based microbe. *Proceedings of the National Academy of Sciences USA* 93, 906–907.

Andersson, S.G. and Kurland, C.G. (1998) Reductive evolution of resident genomes. *Trends in Microbiology* 6, 263–268.

Anon. (2011) Strain Information for *E. coli* K12 MG1655. <http://www.genome.wisc.edu/resources/strains.htm>.

Arvey, A.J., Azad, R.K., Raval, A. and Lawrence, J.G. (2009) Detection of genomic islands via

segmental genome heterogeneity. *Nucleic Acids Research* 37, 5255–5266.
Azad, R.K. and Lawrence, J.G. (2007) Detecting laterally transferred genes: use of entropic clustering methods and genome position. *Nucleic Acids Research* 35, 4629–4639.
Azad, R.K. and Lawrence, J.G. (2011) Towards more robust methods of alien gene detection. *Nucleic Acids Research* 39, e56.
Beauregard-Racine, J., Bicep, C., Schliep, K., Lopez, P., Lapointe, F-J. and Bapteste, E. (2011) Of woods and webs: possible alternatives to the tree of life for studying genomic fluidity in *E. coli*. *Biology Direct* 6, 39.
Berndt, C., Meier, P. and Wackernagel, W. (2003) DNA restriction in a barrier to natural transformation in *Pseudomonas stutzeri* JM300. *Microbiology* 149, 895–901.
Bigot, S., Saleh, O.A., Lesterlin, C., Pages, C., El Karoui, M., Dennis, C., Grigoriev, M., Allemand, J-F., Barre, F-X. and Cornet, F. (2005) KOPS: DNA motifs that control *E. coli* chromosome segregation by orienting the FtsK translocase. *EMBO Journal* 24, 3770–3780.
Blanc, G., Ogata, H., Robert, C., Audic, S., Claverie, J-M. and Raoult, D. (2007) Lateral gene transfer between obligate intracellular bacteria: Evidence from the *Rickettsia massiliae* genome. *Genome Research* 17, 1657–1664.
Blattner, F.R., Plunkett, G. III, Bloch, C.A., Perna, N.T., Burland, V., Riley, M., Collado-Vides, J., Glasner, J.D., Rode, C.K., Mayhew, G.F., Gregor, J., Davis, N.W., Kirkpatrick, H.A., Goeden, M.A., Rose, D.J., Mau, B. and Shao, Y. (1997) The Complete Genome Sequence of *Escherichia coli* K-12. *Science* 277, 1453–1462.
Boc, A. and Makarenkov, V. (2011) Towards an accurate identification of mosaic genes and partial horizontal gene transfers. *Nucleic Acids Research* 39, e144.
Bohannan, B.J.M. and Lenski, R.E. (2000a) Linking genetic change to community evolution: insights from studies of bacteria and bacteriophage. *Ecology Letters* 3, 362–377.
Bohannan, B.J.M. and Lenski, R.E. (2000b) The relative importance of competition and predation varies with productivity in a model community. *American Naturalist* 156, 329–340.
Brüssow, H., Canchaya, C. and Hardt, W-D. (2004) Phages and the evolution of bacterial pathogens: from genomic rearrangements to lysogenic conversion. *Microbiology and Molecular Biology Reviews* 68, 560–602.
Buckling, A. and Rainey, P.B. (2002a) Antagonistic coevolution between a bacterium and a bacteriophage. *Proceedings of the Royal Society of London B: Biological Sciences* 269, 931–936.
Buckling, A. and Rainey, P.B. (2002b) The role of parasites in sympatric and allopatric host diversification. *Nature* 420, 496–499.
Carrolo, M., Frias, M.J., Pinto, F.R., Melo-Cristino, J. and Ramirez, M. (2010) Prophage spontaneous activation promotes DNA release enhancing biofilm formation in *Streptococcus pneumoniae*. *PLoS One* 5, e15678.
Casas, V. and Maloy, S. (2011) Role of bacteriophage encoded exotoxins in the evolution of bacterial pathogens. *Future Microbiology* 6, 1461–1473.
Ciccarelli, F.D., Doerks, T., von Mering, C., Creevey, C.J., Snel, B. and Bork, P. (2006) Toward automatic reconstruction of a highly resolved tree of life. *Science* 311, 1283–1287.
Corre, J., Patte, J. and Louarn, J.M. (2000) Prophage λ induces terminal recombination in *Escherichia coli* by inhibiting chromosome dimer resolution: an orientation-dependent *cis*-effect lending support to bipolarization of the terminus. *Genetics* 154, 39–48.
Dagan, T. and Martin, W. (2006) The tree of one percent. *Genome Biology* 7.
Dagan, T., Artzy-Randrup, Y. and Martin, W. (2008) Modular networks and cumulative impact of lateral transfer in prokaryote genome evolution. *Proceedings of the National Academy of Sciences USA* 105, 10039–10044.
Darling, A.E., Mau, B. and Perna, N.T. (2010) progressiveMauve: multiple genome alignment with gene gain, loss and rearrangement. *PLoS One* 5, e11147.
Davison, J. (1999) Genetic exchange between bacteria in the environment. *Plasmid* 42, 73–91.
Dawkins, R. (1976) *The Selfish Gene*. Oxford University Press, Oxford, UK.
Delcher, A.L., Salzberg, S.L. and Phillippy, A.M. (2003) Using MUMmer to identify similar regions in large sequence sets. *Current Protocols in Bioinformatics* 10.3.1–10.3.18.
Deveau, H., Garneau, J.E. and Moineau, S. (2010) CRISPR/Cas system and its role in phage–bacteria interactions. *Annual Review of Microbiology* 64, 475–493.
Firth, N. and Skurray, R.A. (2006) Genetics: accessory elements and genetic exchange. In: Fischetti, V.A. (ed.) *Gram-positive Pathogens*. ASM Press, Washington, DC, pp. 326–338.
Fleischmann, R.D., Adams, M.D., White, O., Clayton, R.A., Kirkness, E.F., Kerlavage, A.R., Bult, C.J., Tomb, J.-F., Dougherty, B.A., Merrick, J.M., McKenney, K., Suton, G., FitzHugh, W., Fields, C., Gocayne, J.D., Scot, J., Shirley, R., Liu, L., Glodek, A., Keley, J.M., Weidman, J.F., Philips, C.A., Sprigs, T., Hedblom, E., Cotton, M.D., Utterback, T.R., Hanna, M.C., Nguyen, D.T., Saudek, D.M., Brandon, R.C., Fine, L.D.,

Fritchman, J.L., Fuhrmann, J.L., Geoghagen, N.S.M., Gnehm, C.L., McDonald, L.A., Smal, K.V., Fraser, C.M., Smith, H.O. and Venter, J.C. (1995) Whole-genome random sequencing and assembly of *Haemophilus influenzae* Rd. *Science* 269, 496–512.

Fox, G.E., Pechman, K.R. and Woese, C.R., (1977) Comparative cataloging of 16S ribosomal ribonucleic acid: molecular approach to procaryotic systematics. *International Journal of Systematic Bacteriology* 27, 44–57.

Ge, F., Wang, L.-S. and Kim, J. (2005) The cobweb of life revealed by genome-scale estimates of horizontal gene transfer. *PLoS Biol* 3, e316.

Gevers, D., Cohan, F.M., Lawrence, J.G., Spratt, B.G., Coenye, T., Feil, E.J., Stackebrandt, E., van de Peer, Y., Vandamme, P., Thompson, F.L. and Swings, J. (2005) Re-evaluating prokaryotic species. *Nature Reviews Microbiology* 3, 733–739.

Gogarten, J.P. and Townsend, J.P. (2005) Horizontal gene transfer, genome innovation and evolution. *Nature Reviews Microbiology* 3, 679–687.

Gogarten, J.P., Doolittle, W.F. and Lawrence, J.G. (2002) Prokaryotic evolution in light of gene transfer. *Molecular Biology and Evolution* 19, 2226–2238.

Gomez, P. and Buckling, A. (2011) Bacteria–phage antagonistic coevolution in soil. *Science* 332, 106–109.

Gophna, U. and Ofran, Y. (2010) Lateral acquisition of genes is affected by the friendliness of their products. *Proceedings of the National Academy of Sciences USA* 108, 343–348.

Gordo, I. and Campos, P.R.A. (2008) Sex and deleterious mutations. *Genetics* 179, 621–626.

Gray, M.W. and Doolittle, W.F. (1982) Has the endosymbiont hypothesis been proven? *Microbiological Reviews* 46, 1–42.

Harfe, B.D. and Jinks-Robertson, S. (2000) DNA mistmatch repair and genetic instability. *Annual Review of Genetics* 34, 359–399.

Hendrickson, H. and Lawrence, J.G. (2006) Selection for chromosome architecture in bacteria. *Journal of Molecular Evolution* 62, 615–629.

Hendrix, R. (2009) Lesser known large dsDNA viruses, *Current Topics in Microbiology and Immunology* 328, 229–240.

Høiby, N., Ciofu, O. and Bjarnsholt, T. (2010) *Pseudomonas aeruginosa* biofilms in cystic fibrosis. *Future Microbiology* 5, 1663–1674.

Høiby, N., Ciofu, O., Johansen, H.K., Song, Z.-J., Moser, C., Jensen, P.O., Molin, S., Givskov, M., Tolker-Nielsen, T. and Bjarnsholt, T. (2011) The clinical impact of bacterial biofilms. *International Journal of Oral Science* 3, 55–65.

Hyman, P. and Abedon, S.T. (2010) Bacteriophage host range and bacterial resistance. *Advances in Applied Microbiology* 70, 217–248.

Karlin, S. (1998) Global dinucleotide signatures and analysis of genomic heterogeneity. *Current Opinion in Microbiology* 1, 598–610.

Kay, M.K., Erwin, T.C., McLean, R.J.C. and Aron, G.M. (2011) Bacteriophage ecology in *Escherichia coli* and *Pseudomonas aeruginosa* mixed-biofilm communities. *Applied and Environmental Microbiology* 77, 821–829.

Kent, B.N., Salichos, L., Gibbons, J.G., Rokas, A., Newton, I.L.G., Clark, M.E. and Bordenstein, S.R. (2011) Complete bacteriophage transfer in a bacterial endosymbiont (*Wolbachia*) determined by targeted genome capture. *Genome Biology and Evolution* 3, 209–218.

Koonin, E.V., Makarova, K.S. and Aravind, L. (2001) Horizontal gene transfer in prokaryotes: quantification and classification. *Annual Review of Microbiology* 55, 709–742.

Krüger, N. and Stingl, K., (2011) Two steps away from novelty – principles of bacterial DNA uptake. *Molecular Microbiology* 80, 860–867.

Lang, B.F., Burger, B., O'Kelly, C.J., Cedergren, R., Golding, G.B., Lemieux, C., Sankoff, D., Turmel, M. and Gray, M.W. (1997) An ancestral mitochondrial DNA resembling a eubacterial genome in miniature. *Nature* 387, 493–497.

Lawrence, J.G. and Hendrickson, H. (2003) Lateral gene transfer: when will adolescence end? *Molecular Microbiology* 50, 739–749.

Lawrence, J.G. and Hendrickson, H. (2008) Genomes in motion: gene transfer as a catalyst for genome change. In: Hensel, M. and Schmidt, H. (eds) *Horizontal Gene Transfer in the Evolution of Pathogenesis*. Cambridge University Press, Cambridge, UK, pp. 3–22.

Lawrence, J.G. and Ochman, H. (1997) Amelioration of bacterial genomes: rates of change and exchange. *Journal of Molecular Evolution* 44, 383–397.

Lawrence, J.G. and Ochman, H. (1998) Molecular archaeology of the *Escherichia coli* genome. Proceedings *of the National Academy of Sciences USA* 95, 9413–9417.

Lercher, M.J. and Pál, C. (2008) Integration of horizontally transferred genes into regulatory interaction networks takes many million years. *Molecular Biology and Evolution* 25, 559–567.

Lukjancenko, O., Wassenaar, T.M. and Ussery, D.W. (2010) Comparison of 61 sequenced *Escherichia coli* genomes. *Microbial Ecology* 60, 708–720.

Marraffini, L.A. and Sontheimer, E.J. (2008) CRISPR interference limits horizontal gene transfer in staphylococci by targeting DNA. *Science* 322, 1843–1845.

Moran, N.A. (1996) Accelerated evolution and Muller's rachet in endosymbiotic bacteria. *Proceedings of the National Academy of Sciences USA* 93, 2873–2878.
Moran, N.A. (2003) Tracing the evolution of gene loss in obligate bacterial symbionts. *Current Opinion in Microbiology* 6, 512–518.
Moszer, I., Rocha, E.P. and Danchin, A. (1999) Codon usage and lateral gene transfer in *Bacillus subtilis*. *Current Opinion in Microbiology* 2, 524–528.
Nikoh, N. and Nakabachi, A. (2009) Aphids acquired symbiotic genes via lateral gene transfer. *BMC Biology* 7, 12.
Ochman, H. (2005) Examining bacterial species under the specter of gene transfer and exchange. *Proceedings of the National Academy of Sciences USA* 102, 6595–6599.
Ochman, H. and Jones, I.B. (2000) Evolutionary dynamics of full genome content in *Escherichia coli*. *EMBO Journal* 19, 6637–6643.
Ochman, H., Lawrence, J.G. and Groisman, E.A. (2000) Lateral gene transfer and the nature of bacterial innovation. *Nature* 405, 299–304.
Olsen, G.J., Lane, D.J., Giovannoni, S.J., Pace, N.R. and Stahl, D.A. (1986) Microbial ecology and evolution: a ribosomal RNA approach. *Annual Review of Microbiology* 40, 337–365.
Pease, P.J., Levy, O., Cost, G.J., Gore, J., Ptacin, J.L., Sherratt, D., Bustamante, C. and Cozzarelli, N.R. (2005) Sequence-directed DNA translocation by purified FtsK. *Science* 307, 586–590.
Perna, N.T., Plunkett, G. III., Burland, V., Mau, B., Glasner, J.D., Rose, D.J., Mayhew, G.F., Evans, P.S., Gregor, J., Kirkpatrick, H.A., Pósfai, G., Hackett, J., Klink, S., Boutin, A., Shao, Y., Miller, L., Grotbeck, E.J., Davis, N.W., Lim, A., Dimalanta, E.T., Potamousis, K.D., Apodaca, J., Anantharaman, T.S., Lin, J., Yen, G., Schwartz, D.C., Welch, R.A. and Blattner, F.R. (2001) Genome sequence of enterohaemorrhagic *Escherichia coli* O157:H7. *Nature* 409, 529–533.
Petrova, O.E., Schurr, J.R., Schurr, M.J. and Sauer, K. (2011) The novel *Pseudomonas aeruginosa* two-component regulator BfmR controls bacteriophage-mediated lysis and DNA release during biofilm development through PhdA. *Molecular Microbiology* 81, 767–783.
Pope, W.H., Jacobs-Sera, D., Russell, D.A., Peebles, C.L., Al-Atrache, Z., and others. (2011) Expanding the diversity of mycobacteriophages: insights into genome architecture and evolution. *PLoS ONE* 6, e16329.
Poptsova, M.S. and Gogarten, J.P. (2007) The power of phylogenetic approaches to detect horizontally transferred genes. *BMC Evolutionary Biology* 7, 45.
Powell, M. (2011) Richard Dawkins: a knack for bashing orthodoxy. The New York Times <http:/ www.nytimes.com/2011/09/20/ science/20dawkins.html>.
Redfield, R.J. (2001) Do bacteria have sex? *Nature Reviews Genetics* 2, 634–639.
Rivera, M.C. and Lake, J.A. (2004) The ring of life provides evidence for a genome fusion origin of eukaryotes. *Nature* 431, 152–155.
Siefert, J.L. (2009) Defining the mobilome. *Methods in Molecular Biology* 532, 13–27.
Sivanathan, V., Emerson, J.E., Pages, C., Cornet, F., Sherratt, D.J. and Arciszewska, L.K. (2009) KOPS-guided DNA translocation by FtsK safeguards *Escherichia coli* chromosome segregation. *Molecular Microbiology* 71, 1031–1042.
Snel, B., Huynen, M.A. and Dutilh, B.E. (2005) Genome trees and the nature of genome evolution. *Annual Review of Microbiology* 59, 191–209.
Sorek, R., Zhu, Y., Creevey, C.J., Francino, M.P., Bork, P. and Rubin, E.M. (2007) Genome-wide experimental determination of barriers to horizontal gene transfer. *Science* 318, 1449–1452.
Theobald, D.L. (2010) A formal test of the theory of universal common ancestry. *Nature* 465, 219–222.
Touchon, M., Hoede, C., Tenaillon, O., Barbe, V., Baeriswyl, S., Bidet, P., Bingen, E., Bonacorsi, S., Bouchier, C., Bouvet, O., Calteau, A., Chiapello, H., Clermont, O., Cruveiller, S., Danchin, A., Diard, M., Dossat, C., Karoui, M.E., Frapy, E., Garry, L., Ghigo, J.M., Gilles, A.M., Johnson, J., Le Bouguénec, C., Lescat, M., Mangenot, S., Martinez-Jéhanne, V., Matic, I., Nassif, X., Oztas, S., Petit, M.A., Pichon, C., Rouy, Z., Ruf, C.S., Schneider, D., Tourret, J., Vacherie, B., Vallenet, D., Médigue, C., Rocha, E.P. and Denamur, E. (2009) Organised genome dynamics in the *Escherichia coli* species results in highly diverse adaptive paths *PLoS Genetics* 5, e1000344.
Tsirigos, A. and Rigoutsos, I. (2005) A new computational method for the detection of horizontal gene transfer events. *Nucleic Acids Research* 33, 922–933.
Wagner, P.L. and Waldor, M.K. (2006) Bacteriophages and bacterial pathogenesis In: Calendar, R. and Abedon, S.T. (eds) *The Bacteriophages*, 2nd edn. Oxford University Press, Oxford, UK, pp. 710–724.
Welch, R.A., Burland, V., Plunkett, G. III, Redford, P., Roesch, P., Rasko, D., Buckles, E.L., Liou, S.R., Boutin, A., Hackett, J., Stroud, D., Mayhew, G.F., Rose, D.J., Zhou, S., Schwartz, D.C.,

Perna, N.T., Mobley, H.L., Donnenberg, M.S. and Blattner, F.R. (2002) Extensive mosaic structure revealed by the complete genome sequence of uropathogenic *Escherichia coli*. *Proceedings of the National Academy of Sciences USA* 99, 17020–17024.

Werren, J.H., Nur, U. and Wu, C.I. (1988) Selfish genetic elements. *Trends in Ecology and Evolution* 3, 297–302.

Wildschutte, H. and Lawrence, J.G. (2007) Differential *Salmonella* survival against communities of intestinal amoebae. *Microbiology* 153, 1781–1789.

Willner, D., Haynes, M.R., Furlan, M., Hanson, N., Kirby, B., Lim, Y.W., Rainey, P.B., Schmieder, R., Youle, M., Conrad, D. and Rohwer, F. (2011) Case studies of the spatial heterogeneity of DNA viruses in the cystic fibrosis lung. *American Journal of Respiratory Cell and Molecular Biology* 46, 127–131.

Zhaxybayeva, O., Gogarten, J.P., Charlebois, R.L., Doolittle, W.F. and Papke, R.T. (2006) Phylogenetic analyses of cyanobacterial genomes: quantification of horizontal gene transfer events. *Genome Research* 16, 1099–1108.

6 Phages and Bacterial Epidemiology

Michele L. Williams[1] and Jeffrey T. LeJeune[1]
[1]*Food Animal Health Research Program, The Ohio Agricultural Research and Development Center.*

The frequent, complex and intimate interactions between phages and their bacterial hosts have influenced the evolution of both groups of organisms. The consequences of their relationship have significantly impacted human, animal and plant health. As discussed in Kuhl *et al.*, Christie *et al.* and Hendrickson (Chapters 3–5, this volume), some organisms have acquired genes to become more virulent. Notwithstanding this, ever since the discovery of phages, researchers have been attempting to harness and exploit the biological and genetic properties of phages to advance science and to enhance health. The use of phages in molecular biology studies (see Abedon, Chapter 1, this volume) and as agents for therapeutic (see Olszowska-Zaremba *et al.*, Loc-Carrillo *et al.*, Burrowes and Harper, and Abedon, Chapters 12, 13, 14 and 17, this volume) or diagnostic (see Cox, Chapter 10, this volume) purposes, and for food processing (see Niu *et al.*, Chapter 16, this volume) has clearly provided benefits to society. Another important way that the association between phages and bacteria has been exploited is in the tracking of outbreaks and epidemiological research related to disease.

The differentiation of bacterial isolates at a greater resolution than the species level, in order to identify strains or subspecies, provides valuable information in the investigation of disease outbreaks and research on the epidemiology and ecology of pathogens (Moorman *et al.*, 2010). The essential and preferred criteria for such bacterial typing methods have evolved little since their advent (Anderson and Williams, 1956; van Belkum *et al.*, 2007). These include inherent characteristics of the test: (i) stability (consistency over time); (ii) typeability (the ability of the test to describe a unique phage type); (iii) discriminatory power (the ability to differentiate strains, (iv) epidemiological concordance (grouping of related isolates similarly); and (v) reproducibility (the ability to get the same result each time the test is run). They also include practical aspects of use such as: (i) ease of use; (ii) applicability across a broad range of bacterial species (flexibility); (iii) speed or turnaround time; (iv) accessibility; and (v) cost (van Belkum *et al.*, 2007). In this chapter, the methodologies and application of phages and prophages in the tracking and subtyping of bacterial strains is reviewed.

Phage Typing: a Foundation for Epidemiological Investigations

In 1922, Callow reported host-range differences among a collection of staphylococcal phages (Callow, 1922). A few years later, Craigie and Yen (1938) were able to develop the first phage-based typing method by exploiting the receptor specificity of phages for the Vi antigen present on the surface of some *Salmonella typhi*. Briefly, phage typing requires a collection of phages lytic for at least some of the members of the bacterial group being typed. The principles of phage typing reflect the fundamental concepts governing the relationship between phages and their bacterial hosts, notably receptor binding, restriction enzyme systems and immunity as mediated by prophages (Hyman and Abedon, 2010; Labrie *et al.*, 2010). Both obligately lytic and temperate phages may be employed. Using the highest possible dilution known to produce areas of lysis on lawns of its strain type (the bacterial host used to propagate the phage), the bacterial strain to be typed is challenged with a panel of phages. The type of lysis that results is recorded (confluent, opaque, individual clear plaques, individual opaque plaques or no lysis). It is not expected that every phage will result in lysis of every test strain. Instead, it is the pattern of lysis caused by the panel of phages that defines the particular phage type (PT) of the test strain. In effect, phage typing can be viewed as a phenotypic assessment tool (spot formation) that represents or is governed by genetic elements present in the host.

The initial experiments, reported in 1938 by Craigie and Yen, became the foundation for all subsequent phage typing schemes including those still in use today. Many bacteria are still commonly typed using phages (Table 6.1). The methods and specific phages employed are standardized for each typing scheme. The International Federation of Enteric Phage Typing oversees and coordinates phage typing in 55 countries around the globe.

Issues relevant to phage typing utility: inherent characteristics

Despite the longstanding history and value of phage typing in epidemiological investigations, this methodology performs poorly when evaluated for several of the aforementioned desirable criteria for typing methods.

Stability

Host susceptibility to specific phages is subject to change. Prophages, stably integrated into the host genome, can confer immunity to

Table 6.1. Bacteria commonly typed using bacteriophages.

Bacteria	Reference
Acinetobacter spp.	Santos *et al.* (1984)
Bacillus spp.	Ahmed *et al.* (1995)
Campylobacter spp.	Frost *et al.* (1999)
Clostridium difficile	Sell *et al.* (1983)
Clostridium perfringens	Yan (1989)
Corynebacterium diphtheriae	Andronescu *et al.* (1997)
Erwinia spp.	Toth *et al.* (1999)
Escherichia coli	Pearce *et al.* (2009)
Klebsiella spp.	Sechter *et al.* (2000)
Listeria monocytogenes	McLauchlin *et al.* (1996)
Mycobacterium spp.	Schurch and van Soolingen (2011)
Pasteurella multocida	Nielsen and Rosdahl (1990)
Proteus spp.	Sekaninova *et al.* (1999)
Pseudomonas spp.	Bergmans *et al.* (1997)
Salmonella spp.	Kafatos *et al.* (2009)
Serratia spp.	Hamilton and Brown (1972)
Shigella spp.	Bentley *et al.* (1996)
Streptococcus spp.	Domelier *et al.* (2009)
Vibrio cholera	Sarkar *et al.* (2011)
Yersinia spp.	Kawaoka et al. (1987)

superinfection. Under certain circumstances, however, these prophages may be excised and lost. This can result in the potential for the phage type of a given bacterial strain to change. Prophage excision, also known as curing, may also occur either during the course of an outbreak (diphtheria) or during laboratory propagation, such as following the initial isolation of Shiga toxin-producing *Escherichia coli* (Bielaszewska *et al.*, 2007). In addition to the loss of prophages, bacteria may become lysogenized by temperate phages, again potentially changing their phage type (Rankin and Platt, 1995). Although it would be unusual for such an event to happen in the laboratory after the initial isolation of the bacterial strain, lysogenization occurs naturally in the environment (Revathi *et al.*, 2011).

In addition to phage-mediated transfer of DNA and the loss of prophages, the gain or loss of other mobile genetic elements, notably plasmids, can also result in phage type conversion (Frost *et al.*, 1989). Loss of phage receptors can also result in changes in phage types. For example, the sporadic transition from smooth to rough strains of bacteria associated with the loss of lipopolysaccharide can result a change in the susceptibility to phages used in typing, resulting, for example, in the conversion of *Salmonella enterica* serovar Enteritidis PT 4 to PT 7 (Chart *et al.*, 1989).

Typeability

Typeability refers to the ability of the testing methodology to assign a subtype to the organism being evaluated. Because the typing system relies on the pattern of lysis for a collection of individual phages, it is imperative that the reference phage typing set contains one or more phages capable of causing lysis in the tested organism, otherwise the organism is considered 'untypeable'. Depending on the bacterial genus and species being tested, the origin of the samples and the diversity as well as size of the typing set of phages, phage typeability may be as low as 0% and as high as 100% (Kuramasu *et al.*, 1967; Valdezate *et al.*, 2000).

Discriminatory power

Discriminatory power refers to the ability of the typing system to distinguish between unrelated isolates, that is, the probability that two organisms randomly selected from a population of genetically unrelated organisms will produce different subtypes (Hunter and Gaston, 1988). As with other typing systems, the discriminatory power of phage typing is dependent on the population (genus and species) of the bacterial strain being tested and is highly variable, but indices between 80 and 90% are typical and can be useful for epidemiological investigations. For example, the discriminatory power of phage typing for *S. enterica* serovar Enteritidis is 85% (Boxrud *et al.*, 2007) and for *Listeria monocytogenes* it is 87% (Boerlin *et al.*, 1995). The problem associated with phage typing tests of low discriminatory power is that isolates with the same phage type could be genetically unrelated, resulting in a misclassification of unrelated cases into an outbreak definition.

Epidemiological concordance

It is desirable that the typing method groups all of the outbreak strains from the same source together. Epidemiological concordance (E) is a measure of how well the typing method agrees with the epidemiological data obtained from other sources. E can be calculated by determining the fraction of epidemiologically linked sets (five to ten isolates per set) of strains recovered from outbreaks that are indistinguishable (van Belkum *et al.*, 2007). This attribute can be considered the counterbalancing factor to discriminatory power. High discriminatory power with low epidemiological concordance would result in a failure to cluster epidemiologically related isolates from the same outbreak together. The problem associated with low epidemiological concordance is that two or more isolates of similar clonal origin that have been modified, for example by a recent prophage incorporation or phage mobilization event, will not be considered to be part of the same outbreak when in fact they are.

Reproducibility

Reproducibility is the ability of the typing method to generate the same result when repeated on the same isolate on different occasions in different laboratories. It differs from stability (the first criterion, above) in that it is influenced by factors extrinsic to the bacteria. Specifically, intra- and inter-laboratory inconsistencies, including operator error, contribute to variations in analysis and interpretation, thereby resulting in poor reproducibility. Phage typing can be highly reproducible (0.985; Marquet-Van der Mee and Audurier, 1995), but it requires standardized panels of phages and should be conducted only in a limited number of reference laboratories by experienced personnel. Without tight controls on laboratory methods and reagents, reproducibility is poor.

Issues relevant to phage typing utility: practicality

In addition to the abovementioned performance criteria, there are several practical aspects of phage typing that are also considered when evaluating the usefulness of bacterial typing schemes. These include: (i) ease of use; (ii) applicability; (iii) cost; (iv) accessibility; and (v) speed. Phage typing falls short in several of these areas when compared with other typing methods. For example, each bacterial species requires a different set of test phages for typing, thereby limiting its applicability or flexibility in terms of usefulness for multiple bacterial species. Although the laboratory equipment required for test performance is not highly specialized, it does require a fair amount of technical expertise to perform the analysis, ranking phage typing lower on the 'ease-of-use' scale. The technical expertise required and the maintenance of a wide range of phage libraries add to the cost and restrict the availability of phage typing to a limited number of highly specialized reference laboratories. The speed at which the assay can be completed is dependent on the time required to detect spots on bacterial lawns. This may be quicker than some other typing methods (e.g. pulsed-field gel electrophoresis), similar to methods of phenotypic characterization (e.g. biochemical tests) or slower than some more modern molecular-based methods such as PCR.

Choosing the most appropriate typing method is dependent on the question at hand and the resources available. Despite the abovementioned limitations, the use of phage typing is still considered useful in epidemiological surveillance and investigations (Kafatos *et al.*, 2009; Baggesen *et al.*, 2010). In many instances, using a combination of typing methods including traditional phage typing can add valuable molecular insight into the epidemiological investigation of outbreaks (Bhowmick *et al.*, 2009; Cho *et al.*, 2010).

The Next Frontier in Phage Typing

Fundamentally, phage susceptibility and ultimately phage typing are phenotypic representations of the genetic composition of the bacterial host, although as yet unidentified environmental factors governing phage susceptibility may also exist. Host-cell susceptibility to specific phages is viewed primarily as a reflection of: (i) the encoding and expression of specific phage receptors; and (ii) pre-existing prophage lysogenization potentially resulting in immunity (Stephan *et al.*, 2008). There additionally exist a substantial number of anti-phage resistance mechanisms that are encoded by bacteria and which, in principle, could be correlated with the failure of specific typing phages to recognize the appropriate phage. These mechanisms include the familiar restriction-modification mechanisms, as well as abortive infection systems. Certainly, of greatest current interest are clustered, regularly interspaced, short palindromic repeat (CRISPR)/Cas systems, which not only can bestow anti-phage resistance but do so in such a manner that resistant bacteria and resisted phages display small regions of high genetic identity (Deveau *et al.*, 2010; Hyman and Abedon, 2010; Labrie *et al.*, 2010). Thus, although phage typing is characterized as a phenotypic subtyping method, like other phenotypic methods it too could mature methodologically – as more information becomes available – into genotypic methods, that is, if the genetic elements underlying phage-resistance or phage-sensitivity phenotypes are resolved. A number of investigators are exploring this line of research.

Phages often use outer-membrane proteins as receptors. Thus, characterization of the diversity of receptors present on bacterial hosts capable of binding a variety of phages, including those used in phage typing, could provide a novel phage-centric tool to discriminate and classify bacteria. Unfortunately, not all phage receptors are known, and it is expected that new receptors will be discovered. Nevertheless, this approach has been piloted for the discrimination of *S. enterica* serovar Enteritidis phage types (Preisner *et al.*, 2010). In these studies, Fourier transform infrared spectroscopy was used to analyse the outer-membrane proteins present in extracts of cell membranes of various phage types of *Salmonella* Enteritidis. Although the exact components in the extract that were correlated with the phage typing classification were not identified, their analysis demonstrated that this rapid method of differentiation of phage types was highly discriminatory. Presumably, the factors that contribute to the strength of the association were either the receptors for phages themselves or other membrane components that were genetically linked with susceptibility of host lysis by specific groups of phages.

Another approach that has been used to move away from phenotypic methods is the identification of correlates between phage typing systems and genetics. In *Salmonella* Typhimurium, Lan *et al.* (2007) identified ten sequences using amplified fragment length polymorphism, which could be used for discriminating eight different phage types of this organism. A similar approach, PCR of phage-derived open reading frames, was proposed by O'Sullivan *et al.* (2010) to differentiate methicillin-resistant *Staphylococcus aureus* (MRSA) subtypes. More recently, in experiments conducted using microarray-based technology, specifically diversity arrays technology, Hackl *et al.* (2010) described a set of DNA probes that could be used to differentiate between *Salmonella enterica* serovar Enteritidis and Typhimurium strains based on phage types. Many of the genetic markers were subsequently sequenced and in a number of cases demonstrated high identity to hypothetical or putative proteins from phages common to *Salmonella* (Fels-2, Gifsy-1 and ST64B).

Studies such as the ones described here demonstrate the possibility of linking rapid molecular analysis of bacterial proteins or genome sequences with established bacterial groups defined by phage typing. This would allow the application of rapid analysis methods for new strain identification with strong epidemiological concordance to existing groups and bypassing the limitations of traditional phage typing described above. See Cox (Chapter 10, this volume) for additional consideration of phage-based methods that may be employed to distinguish bacterial species or strains.

Beyond Typing: Additional Epidemiological Insights

Many bacterial genomes are pervaded by cryptic phages and prophages. As described above, the public health arena has exploited this phenomenon to develop and refine tools to differentiate strains of bacteria to determine which cases of illness may be attributed to a single source during outbreak investigations. Prophages and their respective chromosomal insertion sites, however, may also provide valuable insights into bacterial evolution, ecology and the pathogenesis of disease. Although not fully understood, lysogeny by temperate phages may impact on the ecology and physiology of the bacterial host (Chen *et al.*, 2005). Moreover, several prophages encode cardinal virulence factors necessary for disease (e.g. botulism toxin, diphtheria toxin, cholera toxin and Shiga toxin) or elements for the regulation of virulence among pathogenic bacteria (Abedon and Lejeune, 2005; see Christie *et al.*, Chapter 4, this volume). Further elucidation of the role of phage-encoded genes, or at least identifying associations between their presence/absence and epidemiological niches, may help to explain the distribution of pathogens in the environment and their relevance to human disease – information that is critically needed to develop enhanced prevention and control strategies to improve human health.

One example of a phage-encoded gene used for epidemiological investigation is the phage λ anti-terminator *Q* gene present in Shiga toxin-producing *E. coli* O157. A *Q* allele identical to that present in bacteriophage 933W is associated with high Shiga toxin production and with most cases of clinical disease in humans (LeJeune *et al.*, 2004). Additional typing systems have been described that differentiate between human clinical isolates of *E. coli* O157 and the larger, more diverse pool of genotypes present in the cattle population, a predominant reservoir host for *E. coli* O157. Notably, the lineage-specific polymorphism assay (LSPA) targeted a number of phage-encoded genes (Zhang *et al.*, 2007). In contrast to LSPA, the Shiga toxin-encoding bacteriophage insertion (SBI) sites method screens putative phage insertion sites to determine whether the sites are accessible for lysogeny or whether they are occupied, and, if occupied, with what size of genetic insertion (Shaikh and Tarr, 2003). Collectively, all of these phage-based methods concordantly identify distribution differences in *E. coli* O157 genotypes recovered from human clinical cases and the bovine reservoir (Whitworth *et al.*, 2010). These genetic tools can be used to track subtypes of bacteria that are over- or under-represented in different environments (Besser *et al.*, 2007). Our understanding of factors governing lysogeny in *E. coli* O157, however, is incomplete. Stephan *et al.* (2009) were unable to transduce two Shiga toxin-negative *E. coli* O157 strains despite the presence of intact SBI sites. Hence, the presence of intact SBI sites in bacterial genomes is not a clear indication of potential for lysogenic conversion and the emergence of novel pathogenic strains (for more on *E. coli* O157:H7, see Kuhl *et al.*, Goodridge and Steiner, and Niu *et al.*, Chapters 3, 11 and 16, this volume).

Antibiotic resistance

Another important role bacteriophages may play in the epidemiology of human, animal and plant health is in the emergence of antibiotic resistance. The emergence of and increase in resistance to antibiotics among bacterial pathogens and commensal organisms are important and growing public health concerns (Bush *et al.*, 2011), particularly because antimicrobial-resistant infections are often difficult to treat and result in increased healthcare costs (McGowan, 2001; Merz *et al.*, 2010). Although much attention has been placed on bacterial conjugation as a mechanism for the emergence and dissemination of antimicrobial resistance (AMR) genes, bacteriophages encoding AMR genes are widespread in the environment (Colomer-Lluch *et al.*, 2011a,b), and the potential for transduction of AMR genes has been often overlooked.

Examples of how phages may impact on the epidemiology of human and animal diseases include the emergence of the penta-resistant *S. enterica* serovar Typhimurium, definite type (DT) 104 in the 1990s. The genes that confer resistance to ampicillin, chloramphenicol, streptomycin/pectinomycin, sulfonamides and tetracycline in this organism are clustered chromosomally (Schmieger and Schicklmaier, 1999) and are not located on a plasmid. These genes, however, are transducible by several phages endemic to DT104 isolates (Schmieger and Schicklmaier, 1999). Similarly, our laboratory has demonstrated the transduction of genes encoding extended spectrum β-lactamase (*cmy-2*) and tetracycline (*tetA* and *tetB*) resistance from *S. enterica* subsp. *enterica* serovar Heidelberg to *S. enterica* serovar Typhimurium (Zhang and LeJeune, 2008). There is also growing evidence that MRSA may have also evolved by the transduction of the *mecA* gene to *S. aureus* from other staphylococci. For additional discussion of transduction and the evolution of bacterial genomes, including the genomes of bacterial pathogens, see Christie *et al.* and Hendrickson (Chapters 4 and 5, this volume).

Summary and Outlook

Clearly bacteriophages have shaped the landscape of infectious agents with significant impacts on the epidemiology and ecology of bacterial pathogens (see Hendrickson, Chapter 5, this volume). Prophages along with various cryptic elements present in microorganisms provide a trail of genetic 'footprints' that can be used to track the evolution of pathogens and the short-term epidemiology of disease. At present, despite its limitations, classical phage typing – which operates in part by phenotypically detecting these elements – provides a valuable tool in outbreak identification and investigation. The discovery of the specific genotypic elements underpinning the traditional phenotypic methods of phage typing will provide more robust and reliable methods to track outbreaks. As more information is acquired and the associations among phage types, genotypes and diseases are better understood, it is expected that the use of phage-based molecular methods will supplement, if not replace, current phage typing in epidemiological investigations and research.

References

Abedon, S.T. and Lejeune, J.T. (2005) Why bacteriophage encode exotoxins and other virulence factors. *Evolutionary Bioinformatics Online* 1, 97–110.

Ahmed, R., Sankar-Mistry, P., Jackson, S., Ackermann, H.W. and Kasatiya, S.S. (1995) *Bacillus cereus* phage typing as an epidemiological tool in outbreaks of food poisoning. *Journal of Clinical Microbiology* 33, 636–640.

Anderson, E.S. and Williams, R.E.O. (1956) Bacteriophage typing of enteric pathogens and staphylococci, and its use in epidemiology. A review. *Journal of Clinical Pathology* 9, 94–127.

Andronescu, C., Diaconescu, A., Marin, B. and Petric, A. (1997) Phage typing, an accessible and useful method in the epidemiological surveillance of diphtheria. *Romanian Archives of Microbiology and Immunology* 56, 139–146.

Baggesen, D.L., Sorensen, G., Nielsesn, E. and Wegener, H.C. (2010) Phage typing of *Salmonella* Typhimurium – is it still a useful tool for surveillance and outbreak investigation? *Eurosurveillance* 15, Article 4.

Bentley, C.A., Frost, J.A. and Rowe, B. (1996) Phage typing and drug resistance of *Shigella sonnei* isolated in England and Wales. *Epidemiology and Infection* 116, 295–302.

Bergmans, D., Bonten, M., van Tiel, F., Gaillard, C., London, N., van der Geest, S., de Leeuw, P. and Stobberingh, E. (1997) Value of phenotyping methods as an initial screening of *Pseudomonas aeruginosa* in epidemiologic studies. *Infection* 25, 350–354.

Besser, T.E., Shaikh, N., Holt, N.J., Tarr, P.I., Konkel, M.E., Malik-Kale, P., Walsh, C.W., Whittam, T.S., *et al.* (2007) Greater diversity of Shiga toxin-encoding bacteriophage insertion sites among *Escherichia coli* O157:H7 isolates from cattle than in those from humans. *Applied and Environmental Microbiology* 73, 671–679.

Bhowmick, T.S., Das, M., Ruppitsch, W., Stoeger, A., Pietzka, A.T., Allerberger, F., Rodrigues, D.P. and Sarkar, B.L. (2009) Detection of virulence-associated and regulatory protein genes in association with phage typing of human *Vibrio cholerae* from several geographical regions of the world. *Journal of Medical Microbiology* 58, 1160–1167.

Bielaszewska, M., Prager, R., Köck, R., Mellmann, A., Zhang W., Tschäpe, H., Tarr, P., Karch, H. (2007) Shiga toxin gene loss and transfer in vitro and in vivo during enterohemorrhagic *Escherichia coli* O26 infection in humans. *Applied and Environmental Microbiology* 10, 3144–3150.

Boerlin, P., Bannerman, E., Ischer, F., Rocourt, J. and Bille, J. (1995) Typing *Listeria monocytogenes*: a comparison of random amplification of polymorphic DNA with 5 other methods. *Research in Microbiology* 146, 35–49.

Boxrud, D., Pederson-Gulrud, K., Wotton, J., Medus, C., Lyszkowicz, E., Besser, J. and Bartkus, J.M. (2007) Comparison of multiple-locus variable-number tandem repeat analysis, pulsed-field gel electrophoresis, and phage typing for subtype analysis of *Salmonella enterica* serotype Enteritidis. *Journal of Clinical Microbiology* 45, 536–543.

Bush, K., Courvalin, P., Dantas, G., Davies, J., Eisenstein, B., Huovinen, P., Jacoby, G.A., Kishony, R., Kreiswirth, B.N., Kutter, E., Lerner, S.A., Levy, S., Lewis, K., Lomovskaya, O., Miller, J.H., Mobashery, S., Piddock, L.J., Projan, S., Thomas, C.M., Tomasz, A., Tulkens, P.M., Walsh, T.R., Watson, J.D., Witkowski, J., Witte, W., Wright, G., Yeh, P. and Zgurskaya, H.I. (2011) Tackling antibiotic resistance. *Nature Reviews Microbiology* 9, 894–896.

Callow, B. (1922) Bacteriophage phenomena with *Staphylococcus aureus*. *Journal of Infectious Diseases* 30, 643–650.

Chart, H., Row, B., Threlfall, E.J. and Ward, L.R. (1989) Conversion of *Salmonella enteritidis* phage type 4 to phage type 7 involves loss of lipopolysaccharide with concomitant loss of virulence. *FEMS Microbiology Letters* 51, 37–40.

Chen, Y., Golding, I., Sawai, S., Guo, L. and Cox, E.C. (2005) Population fitness and the regulation of *Escherichia coli* genes by bacterial viruses. *PLoS Biology* 3, e229.

Cho, S., Whittam, T.S., Boxrud, D.J., Bartkus, J.M., Rankin, S.C., Wilkins, M.J., Somsel, P., Downes, F.P., Musser K.A., Root, T.P., Warnick, L.D., Wiedemann, M. and Saeed, A.M. (2010) Use of multiple-locus variable number tandem repeat analysis and phage typing for subtyping of *Salmonella* Enteritidis from sporadic human cases in the United States. *Journal of Applied Microbiology* 108, 859–867.

Colomer-Lluch, M., Imamovic, L., Joffe, D. and Muniesa, M. (2011a) Bacteriophages carrying antibiotic resistance genes in fecal waste from cattle, pigs, and poultry. *Antimicrobial Agents and Chemotherapy* 55, 4908–4911.

Colomer-Lluch, M., Jofre, J. and Muniesa, M. (2011b) Antibiotic resistance genes in the bacteriophage DNA fraction of environmental samples. *PLoS One* 6, e17549.

Craigie, J. and Yen, C. (1938) The demonstration of types of *B. typhosa* by means of preparations of type II Vi phage. I. Principles and technique. *Canadian Journal of Public Health* 29, 448–484.

Deveau H., Garneau, J., and Moineau S. (2010) CRISPR/Cas system and its role in phage–bacteria interactions. *Annual Review of Microbiology* 64, 475–493.

Domelier, A.S., van der Mee-Marquet, N., Sizaret, P.Y., Hery-Arnaud, G., Lartigue, M.F., Mereghetti, L. and Quentin, R. (2009) Molecular characterization and lytic activities of *Streptococcus agalactiae* bacteriophages and determination of lysogenic-strain features. *Journal of Bacteriology* 191, 4776–4785.

Frost, J.A., Kramer, J.M. and Gillanders, S.A. (1999) Phage typing of *Campylobacter jejuni* and *Campylobacter coli* and its use as an adjunct to serotyping. *Epidemiology and Infection* 123, 47–55.

Frost, J.A., Ward, L.R. and Rowe, B. (1989) Acquisition of a drug resistance plasmid converts *Salmonella enteritidis* phage type 4 to phage type 24. *Epidemiology and Infection* 103, 243–248.

Hackl, E., Konrad-Köszler, M., Kilian, A., Wenzl, P., Kornschober, C. and Sessitsch, A. (2010) Phage-type specific markers identified by Diversity Arrays Technology (DArT) analysis of *Salmonella enterica* spp. *enterica* serovars Enteritidis and Typhimurium. *Journal of Microbiological Methods* 80, 100–105.

Hamilton, R.L. and Brown, W.J. (1972) Bacteriophage typing of clinically isolated *Serratia marcescens*. *Applied Microbiology* 24, 899–906.

Hunter, P.R. and Gaston, M.A. (1988) Numerical index of the discriminatory ability of typing systems: an application of Simpson's index of diversity. *Journal of Clinical Microbiology* 26, 2465–2466.

Hyman, P. and Abedon, S.T. (2010) Bacteriophage host range and bacterial resistance. *Advances in Applied Microbiology* 70, 217–248.

Kafatos, G., Andrews, N., Gillespie, I.A., Charlett, A., Adak, G.K., de Pinna, E. and Threlfall, E.J. (2009) Impact of reduced numbers of isolates phage-typed on the detection of *Salmonella* outbreaks. *Epidemiology and Infection* 137, 821–827.

Kawaoka, Y., Mitani, T., Otsuki, K. and Tsubokura, M. (1987) Isolation and use of eight phages for typing *Yersinia enterocolitica* O3. *Journal of Medical Microbiology* 23, 349–352.

Kuramasu, S., Imamura, Y., Takizawa, T., Oguchi, F. and Tajima, Y. (1967) Studies on staphylococcosis in chickens. I. Outbreaks of staphylococcal infection on poultry farms and characteristics of *Staphylococcus aureus* isolated from chickens. *Journal of Veterinary Medicine Series B* 14, 646–656.

Labrie, S.J., Samson, J.E. and Moineau, S. (2010) Bacteriophage resistance mechanisms. *Nature Reviews Microbiology* 8, 317–327.

Lan, R., Stevenson, G., Donohoe, K., Ward, L. and Reeves, P.R. (2007) Molecular markers with potential to replace phage typing for *Salmonella enterica* serovar typhimurium. *Journal of Microbiological Methods* 68, 145–156.

LeJeune, J.T., Abedon, S.T., Takemura, K., Christie, N.P. and Sreevatsan, S. (2004) Human *Escherichia coli* O157:H7 genetic marker in isolates of bovine origin. *Emerging Infectious Diseases* 10, 1482–1485.

Marquet-Van der Mee, N. and Audurier , A. (1995) Proposals for optimization of the international phage typing system for *Listeria monocytogenes*: combined analysis of phage lytic spectrum and variability of typing results. *Applied and Environmental Microbiology* 61, 303–306.

McGowan, J.E. Jr (2001) Economic impact of antimicrobial resistance. *Emerging Infectious Diseases* 7, 286–292.

McLauchlin, J., Audurier, A., Frommelt, A., Gerner-Smidt, P., Jacquet, C., Loessner, M.J., van der Mee-Marquet, N., Rocourt, J., Shah, S. and Wilhelms, D. (1996) WHO study on subtyping *Listeria monocytogenes*: results of phage-typing. *International Journal of Food Microbiology* 32, 289–299.

Merz, L.R., Guth, R.M. and Fraser, V.J. (2010). Cost of antimicrobial resistance in healthcare settings: a critical review. *Issues in Infectious Diseases* 6, 102–119.

Moorman, M., Pruette, P. and Weidman, M. (2010) Value and methods for molecular subtyping of bacteria. In: Kornacki, J. (ed.) *Principles of Microbiological Troubleshooting in the Industrial Food Processing Environment Food Microbiology and Food Safety.* Springer, New York, NY, pp. 157–174.

Nielsen, J.P. and Rosdahl, V.T. (1990) Development and epidemiological applications of a bacteriophage typing system for typing *Pasteurella multocida*. *Journal of Clinical Microbiology* 28, 103–107.

O'Sullivan, M.V., Kong, F., Sintchenko, V. and Gilbert, G.L. (2010) Rapid identification of methicillin-resistant *Staphylococcus aureus* transmission in hospitals by use of phage-derived open reading frame typing enhanced by multiplex PCR and reverse line blot assay. *Journal of Clinical Microbiology* 48, 2741–2748.

Pearce, M.C., Chase-Topping, M.E., McKendrick, I.J., Mellor, D.J., Locking, M.E., Allison, L., Ternent, H.E., Matthews, L., Knight, H.I., Smith, A.W., Synge, B.A., Reilly, W., Low, J.C., Reid, S.W., Gunn, G.J. and Woolhouse, M.E. (2009) Temporal and spatial patterns of bovine *Escherichia coli* O157 prevalence and comparison of temporal changes in the patterns of phage types associated with bovine shedding and human *E. coli* O157 cases in Scotland between 1998–2000 and 2002–2004. *BMC Microbiology* 9, 276.

Preisner, O., Guiomar, R., Machado, J., Menezes, J.C. and Lopes, J.A. (2010) Application of Fourier transform infrared spectroscopy and chemometrics for differentiation of *Salmonella enterica* serovar Enteritidis phage types. *Applied and Environmental Microbiology* 76, 3538–3544.

Rankin, S. and Platt, D.J. (1995) Phage conversion in *Salmonella enterica* serotype Enteritidis: implications for epidemiology. *Epidemiology and Infection* 114, 227–236.

Revathi, G., Fralick, J.A. and Rolfe, R.D. (2011) *In vivo* lysogenization of a *Clostridium difficile* bacteriophage ϕCD 119. *Anaerobe* 17, 125–129.

Santos Ferreira, M.O., Vieu, J.F. and Klein, B. (1984) Phage-types and susceptibility to 26 antibiotics of nosocomial strains of *Acinetobacter* isolated in Portugal. *Journal of International Medical Research* 12, 364–368.

Sarkar, B.L., Bhowmick, T.S., Das, M., Rajendran, K. and Nair, G.B. (2011) Phage types of *Vibrio cholerae* O1 and O139 in the past decade in India. *Japanese Journal of Infectious Diseases* 64, 312–315.

Schmieger, H. and Schicklmaier, P. (1999) Transduction of multiple drug resistance of *Salmonella enterica* serovar Typhimurium DT104. *FEMS Microbiology Letters* 170, 251–256.

Schurch, A.C. and van Soolingen, D. (2011) DNA fingerprinting of *Mycobacterium tuberculosis*: from phage typing to whole-genome sequencing. *Infection, Genetics and Evolution* doi: 10.1016/j.meegid.2011.08.032 (Epub ahead of print).

Sechter, I., Mestre, F. and Hansen, D.S. (2000) Twenty-three years of *Klebsiella* phage typing: a review of phage typing of 12 clusters of nosocomial infections, and a comparison of phage typing with K serotyping. *Clinical Microbiology and Infection* 6, 233–238.

Sekaninova, G., Kolarova, M., Pillich, J., Semenka, J., Slavikova, H., Kubickova, D. and Zajicova, V. (1999) *Pseudomonas aeruginosa* phage lysate as an immunobiological agent. 1. Selection of *Pseudomonas aeruginosa* clinical strains for phage lysate preparation. *Folia Microbiologica* 44, 93–97.

Sell, T.L., Schaberg, D.R. and Fekety, F.R. (1983) Bacteriophage and bacteriocin typing scheme for *Clostridium difficile*. *Journal of Clinical Microbiology* 17, 1148–1152.

Shaikh, N. and Tarr, P.I. (2003) *Escherichia coli* O157:H7 Shiga toxin-encoding bacteriophages: integrations, excisions, truncations, and evolutionary implications. *Journal of Bacteriology* 185, 3596–3605.

Stephan, R., Schumacher, S., Corti, S., Krause, G., Danuser, J. and Beutin, L. (2008) Prevalence and characteristics of Shiga toxin-producing *Escherichia coli* in Swiss raw milk cheeses collected at producer level. *Journal of Dairy Science* 91, 2561–2565.

Stephan, R., Zhang, W., Bielaszewska, M., Mellmann, A. and Karch, H. (2009) Phenotypic and genotypic traits of Shiga toxin-negative *E. coli* O157:H7/H$^-$ bovine and porcine strains. *Foodborne Pathogens and Disease* 6, 235–243.

Toth, I.K., Bertheau, Y., Hyman, L.J., Laplaze, L., Lopez, M.M., McNicol, J., Niepold, F., Persson, P., Salmond, G.P., Sletten, A., van der Wolf, J.M. and Pérombelon, M.C. (1999) Evaluation of phenotypic and molecular typing techniques for determining diversity in *Erwinia carotovora* subspp. *atroseptica*. *Journal of Applied Microbiology* 87, 770–781.

Valdezate, S., Echeita, A., Diez, R. and Usera, M.A. (2000) Evaluation of phenotypic and genotypic markers for characterisation of the emerging gastroenteritis pathogen *Salmonella* Hadar. *European Journal of Clinical Microbiology and Infectious Diseases* 19, 275–281.

van Belkum, A., Tassios, P.T., Dijkshoorn, L., Haeggman, S., Cookson, B., Fry, N.K., Fussing, V., Green, J., Feil, E., Gerner-Smidt, P., Brisse, S., Struelens, M. and the European Society of Clinical Microbiology and Infectious Diseases (ESCMID) Study Group on Epidemiological Markers (ESGEM) (2007) Guidelines for the validation and application of typing methods for use in bacterial epidemiology. *Clinical Microbiology and Infection* 13 (Suppl. 3), 1–46.

Whitworth, J., Zhang, Y., Bono, J., Pleydell, E., French, N. and Besser, T. (2010) Diverse genetic markers concordantly identify bovine origin *Escherichia coli* O157 genotypes under-represented in human disease. *Applied and Environmental Microbiology* 76, 361–365.

Yan, W.K. (1989) Use of host modified bacteriophages in development of a phage typing scheme for *Clostridium perfringens*. *Medical Laboratory Sciences* 46, 186–193.

Zhang, Y. and LeJeune, J.T. (2008) Transduction of $bla_{CMY\text{-}2}$, tet(A), and tet(B) from *Salmonella enterica* subspecies *enterica* serovar Heidelberg to *S.* Typhimurium. *Veterinary Microbiology* 129, 418–425.

Zhang, Y., Laing, C., Steele, M., Ziebell, K., Johnson, R., Benson, A., Taboada, E. and Gannon, V. (2007) Genome evolution in major *Escherichia coli* O157:H7 lineages. *BMC Genomics* 8, 121.

7 Phages as Therapeutic Delivery Vehicles

Jason Clark[1], Stephen T. Abedon[2] and Paul Hyman[3]
[1]*BigDNA Ltd, Roslin BioCentre, Roslin;* [2]*Department of Microbiology, The Ohio State University;* [3]*Department of Biology/Toxicology, Ashlands University*

Because of their small genomes, relative ease of genome isolation and the more direct link between genotype and phenotype, bacteriophages were some of the first organisms upon which modern genetic engineering techniques were performed. This relative ease of manipulation means that phages also have a versatility in biotechnology applications that is distinct from other platforms. One aspect of that versatility can be summarized in terms of phages serving as delivery vehicles. When used for this purpose, bacteriophage virions can be considered as inert, nanoscale particles that have specific activities when they interact with their targets.

Examples of such applications include: the direct use of phages as killing agents (phage therapy; see Olszowska-Zaremba *et al.*, Loc-Carrillo *et al.*, Burrowes and Harper, and Abedon, Chapters 12, 13, 14 and 17, this volume); delivery of nucleic acid to mammalian cells (gene therapy and vaccines, this chapter); immunomodulation via phage proteins (vaccines/chemotherapy; see Olszowska-Zaremba *et al.*, Chapter 12, this volume); cell binding of phages linked to toxins (chemotherapy, this chapter); and linking bacteria to detectors (biosensors; see Cox, Chapter 10, this volume). These various processes can be categorized in two ways: (i) by considering which physical/chemical part of the phage is considered the 'cargo' being delivered; and (ii) by considering effects on the recipient cells. Table 7.1 summarizes the classification of some applications according to these categories. For reason of space limitations, we have avoided providing a comprehensive review of the literature and will instead concentrate on an illustrative subset of those studies we have identified.

Variations on the Theme

The phage genome can be modified to include heterologous sequences designed to be expressed in or to otherwise modify a target cell. Gene expression can be under the control of either prokaryotic or eukaryotic sequences. Phages can be modified to deliver proteins directly, by displaying the proteins as coat protein fusions (phage display; see Siegel, Chapter 8, this volume) or by artificial conjugation of proteins to the intact phage capsid (see Cox, Chapter 10, this volume). Other cargos may include non-protein molecules that are attached to phages for delivery to targets using the phage proteins to provide targeting via specific binding sequences. Non-DNA cargo may be subject to control by a human operator or modified in response to the target entity, such as being released (i.e. solubilized) and thereby activated.

Table 7.1. Characteristics of material being delivered by phage in various technologies.

Application	Type of cargo delivered[a]	Type of target	Effect on host cell[b]	Result of modification[c]	Duration of effect[d]	Whole phage uptake needed[e]
Normal phage infection	Nucleic acid	Bacteria	Toxic	Not applicable	Permanent	No
Phage therapy	Nucleic acid	Bacteria	Toxic	Manipulation	Permanent	No
Chemotherapy	Exogenous toxin	Cancer cell	Toxic	Manipulation	Permanent	Yes/no (depends on toxin type)
Vaccine	Nucleic acid or protein	Immune or other cells	Non-toxic	Manipulation	Temporary (on cells taking up phage)	Yes
Gene therapy	Nucleic acid	Multiple cell types	Non-toxic	Manipulation	Permanent	Yes
Imaging agent	Exogenous	Multiple cell types	Non-toxic	Signalling	Temporary	No
Bacterial biosensor	Protein or nucleic acid	Bacteria	Non-toxic	Signalling	Temporary	No

[a] Is the phage protein, nucleic acid sequence or some molecule bound to the phage virion the active material being transported by the phage?
[b] Is the material being delivered toxic to the target cell or not?
[c] Target cells may be altered in some way (manipulation) or detected (signalling).
[d] Target cells may be permanently changed or some property may be altered temporarily.
[e] Does the whole phage need to be taken into the target cell for the desired effect to take place?

The target for phage-delivered material can be either prokaryotic or eukaryotic. The obvious targets for phages are bacteria and the natural specificity of phages has been harnessed for a variety of biotechnological applications. Alternatively, phage display or the conjugation of non-phage molecules can be used to target phages to cells (see Siegel, Chapter 8, this volume), either prokaryotic or eukaryotic, for which there otherwise is no natural tropism. In these instances, the phages act as agents that attach to the cell surface, with this attachment sometimes followed by endocytosis if the target cell is eukaryotic. For some applications, the interaction can be limited to simply binding to the target, co-locating the phage's cargo with its target. While antibodies can serve similar purposes, and indeed may be responsible for homing in on non-host targets, phages have the advantage of providing a larger, more flexible platform with which cargo may be delivered.

Phage-delivered cargo can be benign or toxic to target cells. Benign treatments can include gene therapy as well as the use of phages as DNA vaccines, which are intended to leave the cell metabolically active, and producing phage-DNA encoded protein. Alternatively, phages may be modified genetically, phenotypically or both so that they deliver toxins or genes expressing damaging products. The latter can include a form of phage therapy where prokaryotic cells are the target. Phages can be designed to damage or kill non-bacterial targets such as tumour cells. Phage modification of targets can also result in the generation of a signal for detection (Galikowska *et al.*, 2011; see Cox, Chapter 10, this volume).

Phage-delivered cargo can result in only temporary expression or impact such as for the sake of bacterial identification (see Cox, Chapter 10, this volume). The use of phages as carriers of DNA vaccines is also a temporary phenomenon that induces a long-term effect on the host. Phage therapy or phage carriage of toxic genes or materials to any target also, ideally, is a short-term effect. Employment of phages as vectors for either gene cloning or gene therapy will be relatively long lasting, not just in its effect but also in terms of retention and expression of constructs. Some therapies may also rely on long-term expression in a relatively small number of cells (e.g. gene therapy), whereas others may rely on high-level expression over short timescales (e.g. treatment of cancers).

There are a number of technologies involving phage applications intended to modify mammalian physiology either by generating immune responses (vaccines; see Olszowska-Zaremba *et al.*, Chapter 12, this volume), modifying the permanent or semi-permanent genetic complement of human cells (gene therapy) or killing unwanted tissues such as tumour or white fat cells (cytotoxic agents). Phages that can bind to mammalian tissues can become endocytosed. Phage clearance from circulation to sites of antigen presentation can result in increased expression from eukaryotic expression cassettes (Lankes *et al.*, 2007) in a manner that has been shown to be antibody dependent and Fc receptor mediated, and to enhance transfer of genes in mammalian cells both *in vivo* and *in vitro* (Sapinoro *et al.*, 2008). Phages also can serve as scaffolds for altering the pharmacokinetic properties of other molecules.

Phages as Vaccine Delivery Vehicles

The ease of production of large numbers of phages on relatively inexpensive media, combined with the proposed ease of purification (Gill and Hyman, 2010), suggests that bacteriophages could serve as ideal vectors for vaccine delivery. Phages also may have natural adjuvant properties with, for example, unmethylated CpG motifs in phages with double-stranded DNA genomes being potent stimulators of Toll-like receptor 9 (Senti *et al.*, 2009). Similarly, the protein component of phages can provide T-cell epitopes that enhance antibody response and improve immune memory. It has also been reported that T4 phages are potent stimulators of interferon (IFN) production (Kleinschmidt *et al.*, 1970), as supported by the observation that high levels of IFN-γ are produced in mouse cells following immunization with phage λ (J. Clark and J.B. March, unpublished data). Similarly, it has been shown that treatment of blood cultures with phages results in increased levels of tumour necrosis factor-α and interleukin-6 (Weber-Dabrowska *et al.*, 2001; see also Olszowska-Zaremba *et al.*, Chapter 12, this volume). The natural adjuvant effect of phages is also implied by results suggesting that phage vaccines give rise to similar immune responses in the presence or absence of adjuvant, whether they are protein (Greenwood *et al.*, 1991; Minenkova *et al.*, 1993; Willis *et al.*, 1993) or DNA (i.e. genetic) vaccines (Clark and March, 2004b). We differentiate phage-based delivery of vaccines into two basic types: those where DNA is the cargo and those where protein instead serves as the cargo.

Phages as DNA vaccines

In phage-mediated DNA vaccination, a eukaryotic expression cassette is cloned into the genome of a phage vector and whole phage particles are used to immunize the host. This has several potential advantages over standard vaccination including the generation of strong type 1 immune responses along with the production of vaccine antigen within host cells so that the resultant proteins are correctly post-translationally modified. Additionally, the phage coat protects the DNA from degradation and the phage particles themselves can be stable over a wide range of temperatures and pH values (Jepson and March, 2004). As particulate antigens, phages should be targeted to sites of antigen presentation where the vaccine component can efficiently raise an immune response.

The majority of research into phage-delivered DNA vaccination has been performed using phage λ, with the first published report of this method using a λ gt-11 vector to deliver the hepatitis B small surface antigen (HBsAg) gene, under the control of the cytomegalovirus promoter, to mice (Clark and March, 2004b). Antibody responses against the HBsAg protein were detected after two immunizations by a number of different methodologies and in some instances anti-HBsAg responses were in excess of 10 mIU ml^{-1}, the internationally recognized level required for protection, with the highest level being approximately 150 mIU ml^{-1}. A λ phage containing an expression cassette expressing enhanced green fluorescent protein (EGFP) also generated antibody responses in mice after two immunizations, with sera recognizing the EGFP protein in Western blots (March, 2002; March *et al.*, 2004). Subsequent research in rabbits confirmed that the same phage λ-delivered expression cassette was able to raise significant immune responses in rabbits after intramuscular immunization (March *et al.*, 2004). More recently, the same phage λ-delivered HBsAg expression cassette was tested in rabbits and the responses compared with Engerix B, a commercially available recombinant protein vaccine (Clark *et al.*, 2011). In this study, three out of five phage-vaccinated rabbits responded after one immunization, with the remaining two responding after a booster immunization. By contrast, after three vaccinations, one of the Engerix B-vaccinated rabbits had not responded. By 2 weeks after the second immunization, responses in the phage-vaccinated group were higher by all tests compared with the Engerix B-vaccinated group.

In an interesting modification to the standard phage-mediated DNA vaccine technique, March *et al.* (2006) screened a whole-genome library of the bovine pathogen *Mycoplasma mycoides* subsp. *mycoides* (small colony biotype) for immunodominant clones using a ZAP Express phage λ containing both prokaryotic and eukaryotic promoters. Immune sera from both rabbits and cows were used to probe plaques with the immunodominant clones selected. After sequencing and analysis, two of these clones were tested in a mouse challenge model. Immunization with both clones resulted in significant antibody responses, with one clone also giving significant lymphocyte proliferation after stimulation with whole bacteria. Vaccination with this same clone gave protection, as measured by reduced levels of mycoplasmaemia after challenge. This approach to vaccination is particularly interesting as protection was obtained after screening with no initial knowledge of the protective antigens.

In other studies using phage λ as a vaccine delivery vehicle, Ghaemi and co-workers demonstrated the ability of phage-delivered human papillomavirus genes to express both in bacteria (Ghaemi *et al.*, 2010) and in cell lines (Ghaemi *et al.*, 2011). In the latter study, mice were shown to be protected against challenge with a mouse tumour cell line expressing the same protein. The release of IFN-γ and granzyme B, along with proliferation of spleen cells, suggested that a cell-mediated response was at least partly responsible for protection.

Recently, protection against challenge in a mouse model has been demonstrated following phage-mediated DNA vaccination (Ling *et al.*, 2011). Phage λ containing a eukaryotic cassette expressing the major outer-membrane protein of *Chlamydophila abortus* was used to immunize mice. Three immunizations of the phage vaccine were compared with three immunizations of a homologous DNA vaccine and a single immunization of a live attenuated vaccine strain. Although antibody responses were higher in the group immunized with the live attenuated strain, the phage DNA vaccine group showed antibody responses significantly higher than the DNA vaccine group, and levels of IFN-γ and interleukin-2 were higher in the phage-vaccinated group than in the other two groups. Protection was measured as clearance of chlamydiae from spleens after challenge. The phage-vaccinated group gave significantly better protection than the DNA-vaccinated group or the control group, although mice immunized with the live attenuated strain showed the highest levels of protection.

Other phages have been tested. Hashemi *et al.* (2010) used filamentous phages to deliver a genetic vaccine against sherpes simplex virus type 1 (HSV-1), based on HSV-1 glycoprotein D. In this case, both humoral and cellular responses were found to be similar to those from a plasmid DNA vaccine control group. Contrastingly, in the published literature, phage λ-delivered DNA vaccines are reported to be more effective than plasmid DNA vaccines. The reduced response observed with filamentous phage may be due to the single-stranded nature of the genome. Additionally, there is a report of a phage-delivered mRNA vaccine (Sun *et al.*, 2011). Virus-like particles (VLPs) derived from the bacteriophage MS2, which has a positive-sense, single-stranded RNA genome were used. These particles self-assemble and can package foreign nucleic acid. VLPs packaging human immunodeficiency virus type 1 (HIV-1) *gag* mRNAs were produced in yeast cells and used to immunize mice. Following three immunizations with Freund's adjuvant Gag-specific antibody responses over 2 logs higher than in the control groups were detected.

Phages as protein vaccines

Bacteriophages have also been used for the delivery of vaccines in the form of proteins or peptides displayed as capsid protein fusions or chemically conjugated to the surface of the phage. Phages displaying protective antigens have been tested in animals (Irving *et al.*, 2001; Wang and Yu, 2004). Specific antibodies (e.g. di Marzo *et al.*, 1994) and protection against challenge (e.g. Bastien *et al.*, 1997) have been demonstrated. Peptides/proteins that are displayed on the surface of phage particles can be chosen based on previous knowledge of the specific disease, particularly of protective epitopes. An alternative approach is to screen a phage display library and select from random peptides for those that bind to serum raised against a given immunogen or pathogenic organism (Folgori *et al.*, 1994). The utility of this latter approach is that, rather than guess at which linear epitopes are most protective, one instead can begin with an immune response that is known to be protective. This latter approach, however, is limited by an incomplete knowledge of which aspects of the humoral immune response provide protection and by the fact that, in some cases, peptides can be antigenic in that they react with the products of an immune response but may not be immunogenic, that is, they are unable to generate an immune response themselves. Similarly, Zhong *et al.* (1994) were able to identify randomly mutagenized peptides that bound to antibodies known to recognize an epitope found in its native conformation.

Meola *et al.* (1995) compared display of peptides with delivery of the same peptides as multiple antigenic peptides or fused to hepatitis B virus core protein and human H ferritin. They found the phage-delivered peptides to be more immunogenic. It has also been demonstrated that peptides displayed on phages can reach both MHC class I and class II compartments, suggesting that peptides presented in this way can stimulate both humoral and cellular responses (Wan *et al.*, 2001; de Berardinis *et al.*, 2003; Gaubin *et al.*, 2003). Filamentous phages have also been shown to stimulate both $CD8^+$ and $CD4^+$ T cells inducing both short-term and long-term cytotoxic T-lymphocyte (CTL) responses (Mascolo *et al.*, 2007; del Pozzo *et al.*, 2010).

A variety of different phages have been used to deliver protein vaccines to animals, including the filamentous phages M13 and fd, T4, λ, T7 and the RNA phages Qβ, MS2 and PP7. Each of these phage vectors has advantages and disadvantages in terms of display capacity and ease of production. Phage display by filamentous phages is covered in substantial detail by Siegel (Chapter 8, this volume). While the high density of display possible with filamentous phages can enhance immune responses (Minenkova *et al.*, 1993), the limit on the size of peptides that can be displayed can limit immunogenicity. Such constructs may be more suited to generating protective CTLs, rather than antibody responses. It has been noted that CTL responses could be improved 1000- to 10,000-fold if peptides are displayed on large carriers (Raychaudhuri and Rock, 1998). Some researchers have been able to show improved immune responses to vaccine phages by co-administration of T-cell epitopes displayed on a second filamentous phage (di Marzo *et al.*, 1994; Perham *et al.*, 1995; Guardiola *et al.*, 2001). Although filamentous phage-display vaccines have usually been delivered via the intraperitoneal or sub-cutaneous routes, some researchers have also shown immune responses after delivery via oral (Delmastro *et al.*, 1997; Zuercher *et al.*, 2000) or nasal (Delmastro *et al.*, 1997) routes. Interestingly, Zuercher *et al.* (2000) observed that only clones displaying a high-copy-number peptide remained active after oral delivery, with low-copy-number display phages being rapidly inactivated. Table 7.2 gives a summary of antigens used with filamentous phage display as vaccines.

Bacteriophage T4 is probably the second

Table 7.2. Filamentous phage display vaccines.

Antigen	Type of display	Reference
Circumsporozoite protein of *Plasmodium falciparum*	pVIII by complementation on plasmid with inducible promoter	Greenwood *et al.* (1991); Willis *et al.* (1993)
Human immunodeficiency virus (HIV-1) Gag p17	pVIII	Minenkova *et al.* (1993); Loktev *et al.* (1996)
HIV-1 gp120 V3 loop	pVIII (complemented)	di Marzo *et al.* (1994)
HIV-1 proteins	pVIII (complemented)	Lundin *et al.* (1996); Scala *et al.* (1999)
HIV-1	pVIII (complemented)	de Berardinis *et al.* (1999, 2000, 2003); Guardiola *et al.* (2001)
HIV-1	pVIII (complemented)	Chen *et al.* (2001)
Chlamydia trachomatis	pVIII (complemented)	Zhong *et al.* (1994)
Hepatitis B	pVIII and pIII	Meola *et al.* (1995); Delmastro *et al.* (1997)
	pVIII (complemented)	Wan *et al.* (2001)
Plasmodium vivax	pVIII	Demangel *et al.* (1996)
Human respiratory	pIII	Bastien *et al.* (1997)
syncytial virus	pVIII (complemented)	Grabowska *et al.* (2000)
Pig cysticercosis	pIII and pVIII (complemented)	Manoutcharian *et al.* (2004); Morales *et al.* (2008)
Enterotoxigenic *Escherichia coli*	pIII	van Gerven *et al.* (2008)
Trichinella spiralis	pIII	Gu *et al.* (2008)
Alzheimer's disease	pIII and pVIII (complemented)	Frenkel *et al.* (2000, 2001, 2004)
Melanoma	pVIII (complemented)	Fang *et al.* (2005); Sartorius *et al.* (2008)

most well-characterized phage in terms of phage display for vaccine purposes. The capsid of T4 contains two non-essential proteins, both of which have been used for display purposes; highly antigenic outer capsid protein (HOC) 39 kDa with 155 copies per phage and small outer capsid protein (SOC) at 10 kDa with 810 copies per phage (for further discussion of phage T4 and its immunogenicity, see Olszowska-Zaremba *et al.*, Chapter 12, this volume). The main advantage of T4 over filamentous phages for display purposes is that large peptides/proteins can be displayed at high copy numbers, which in many cases will lead to improved immune responses. A number of studies have employed phage T4 as a carrier of protein antigens for vaccination, most recently for example by an intramuscularly delivered T4 vaccine that conferred complete protection to Dutch-belted rabbits against anthrax spores (Peachman *et al.*, 2012) and in non-human primates, delivered with or without a monophosphoryl lipid A adjuvant (Rao *et al.*, 2011). Overall, although filamentous phages are the most well-characterized vaccine delivery vehicles, the ability to display large proteins and peptides with significant copy numbers continues to make T4 an attractive option for vaccine development.

Several groups have used RNA phages (e.g. MS2, Qβ, AP205 and PP7) to deliver peptide or protein vaccines. These particles spontaneously assemble if expressed in *Escherichia coli*, are highly amenable to genetic engineering and can display peptides at a relatively high density. They often do not contain genetic material and thus are often used as VLPs rather than as phages. As a consequence, they may be safer for use in humans, although removing the ability to replicate may make production of vaccine particles more difficult. Vaccine delivery vehicles based on both display of antigens as coat protein fusions and chemical conjugates of antigens to coat proteins have been described; for more recent efforts, see Zou *et al.* (2010) and Tissot *et al.* (2010). An additional advantage of RNA phage VLPs for the delivery of vaccine products is that, in some cases, the particles can be loaded with CpG oligodeoxynucleotides, which act as potent immunostimulators in mammalian hosts. For example, phage Qβ VLPs were loaded with CpG oligodeoxynucleotides and used as an adjuvant in a co-immunization with house dust mite allergen in trials that demonstrated a significant reduction in the symptoms of rhinitis and allergic asthma (Senti *et al.*, 2009).

Use of RNA phage VLPs as vaccine delivery vehicles is perhaps the most developed example of a phage-derived product for use in humans. Cytos Biotechnology Ltd (Schlieren, Switzerland; www.cytos.com) is currently conducting phase II clinical trials with a number of products. These trials include nicotine addiction (Maurer *et al.*, 2005; Cornuz *et al.*, 2008), hypertension (Tissot *et al.*, 2008; Maurer and Bachmann, 2010), allergic rhinoconjunctivitis (Klimek *et al.*, 2011), allergic asthma (Senti *et al.*, 2009) and Alzheimer's disease (Wiessner *et al.*, 2011). A full discussion of this delivery technology is beyond the scope of this chapter, but the Cytos website has a full list of relevant references (www.cytos.com/?id=158#candidates).

Although Cytos Biotechnology has targeted relatively high-value indications for vaccine development, it has also been suggested that the potential ease of production of some phages could make them particularly attractive for the development of vaccines against orphan diseases (Rao *et al.*, 2011). It is also important to note that only a very few phages have been tested as vaccine delivery vehicles. There is a growing understanding of the role that shape, size and charge play in the interaction of foreign particles with the immune system (Wang *et al.*, 2011), and it is possible that other phage vectors could have advantages in terms of tissue tropism or how they interact with the host to generate an immune response. Further information on phages as vaccines is given by Benhar (2001); Irving *et al.* (2001); Wang and Yu (2004); Clark and March (2004a) and Gao *et al.* (2010).

Phages as Gene-Therapy Vectors

The most commonly used vectors for gene therapy – the addition of DNA to eukaryotic cells to correct a genetic defect or augment the cell's phenotype – are viruses of eukaryotes. Although not considered here, a description of gene therapy of prokaryotic organisms has also been proposed to combat antibiotic-resistant infections (e.g. Norris *et al.*, 2000). The problems with employing animal viruses include difficulties controlling their inherent virulence, at worst resulting in subject death, as well as difficulties associated with modifying tissue tropism so that the desired cells are preferentially targeted. Therefore, there is a strong need for gene-delivery technologies that are both more benign and more effective. An important consideration is that not all gene therapy is delivered *in vivo*, as it is possible in certain instances to remove specific cell types temporarily from the body where they may be treated *ex vivo*. The now genetically modified cells are then returned to the body where, ideally, relatively long-term expression of the introduced gene may occur. Thus, not all gene therapy requires exposure of systemic body tissues to DNA-delivering vectors.

The advantages of using phages as gene-therapy vectors are similar to the advantages of using phages as DNA vaccines: phages are safe, stable, inexpensive and easy to manipulate genotypically as well as phenotypically. Indeed, modification of the tissue tropism of phages is an almost trivial exercise involving phage display technologies (see Siegel, Chapter 8, this volume) and cloning of the desired genes into phages. In short, phages are perhaps ideal cloning vectors combined with facile trophic targeting, with shortcomings found only in terms of the size of DNA that may be cloned, a lack of inherent nucleus-homing mechanisms and the lack of natural sequences in vectors able to function in eukaryotes, although some of these shortcomings can be addressed by modification of the phage particle. In filamentous phages, there is the additional problem that the genome is single stranded and therefore presumably requires conversion to double-stranded DNA before transcription can occur. As Hajitou *et al.* (2007, p. 523) noted, '...phage-based vectors have inherently been considered poor gene delivery vehicles. As a working hypothesis, we proposed that the rate-limiting step might be mechanistically related to the post-targeting fate of the single-stranded DNA of the phage genome.' This hypothesis is further supported by research indicating that genotoxic treatment increases the rate of transduction of phagemid particles containing single-stranded DNA (Burg *et al.*, 2002; Liang *et al.*, 2006).

Initial studies in transfection with phage particles used chemical agents to facilitate *in vitro* uptake. Such agents included DEAE-dextran (Yokoyama-Kobayashi and Kato, 1993) and cationic lipids (Yokoyama-Kobayashi and Kato, 1994; Aujame *et al.*, 2000) with filamentous phages and calcium phosphate with phage λ (Ishiura *et al.*, 1982; Okayama and Berg, 1985). These initial studies proved that phage particles can express eukaryotic genes contained in the phage genome if uptake is sufficiently efficient.

Non-specific uptake of phage particles has relatively limited biotechnological applications, and several groups have attempted to modify phages to provide tissue tropism. This specific targeting of phage particles has been shown to increase transfection efficiencies *in vitro* from 1–2% to as much as 45% (Larocca *et al.*, 2002). In many cases, relatively small peptides can give tissue tropism to phage particles, or in some instances antibodies displayed on phages can be used to increase cellular uptake of phage particles (Becerril *et al.*, 1999; Chung *et al.*, 2008), and antibody libraries can also be screened to select phage clones with tropism for specific cell types, such as tumour cells (Poul *et al.*, 2000). Another advantage of filamentous phages in the context of gene therapy is the ability to pan phage libraries to select for phages displaying peptides that are selectively taken up by specific cell types either *in vitro* (Ivanenkov *et al.*, 1999) or *in vivo* (Molenaar *et al.*, 2002). See Siegel (Chapter 8, this volume) for additional discussion of phage display and biopanning. A number of reviews are available that further discuss the use of modified phages as gene-therapy vectors (Uppala and Koivunen, 2000; Monaci *et al.*, 2001; Sergeeva *et al.*, 2006; Hajitou, 2010).

Phages as cytotoxic agents

Phages may be employed not just to 'do good' in a gene-therapy or gene-vaccination sense but also to 'do good' in a cytotoxic sense. Thus, not all cell types found within bodies are desirable or even healthy to possess. Undesirable cells include tumour cells but also can include white fat cells, against which a large number of strategies (often ineffective) exist aimed at reducing their size or number. Phages as cytotoxic agents against such cells may be differentiated in a number of different ways. Phages can deliver toxins to a cell, after targeting and either co-localization or endocytosis. Toxins can be physically delivered by the phage or encoded by appropriate expression cassettes cloned into the phage genome. Alternatively, an immune response against the phage itself (facilitated by the adjuvant effect of the phage particle) can result in killing of target cells by the host immune system. In all cases, the key to proper functioning is phage targeting to specific cell types, which typically is effected by capsid modification based on phage display technologies.

The most common research using phages as cytotoxic agents is in the targeting and destruction of cancer cells. Generally, the phage particles are targeted to cancerous cells and either a toxin is released or the immune response against the phage itself promotes a cell-killing effect. The most concerted programme of work in using phages as cytotoxic agents is that of Yacoby, Benhar and others, who initially targeted bacterial pathogens. These authors demonstrated a technology where a filamentous phage displays a cell-binding ligand along with a cytotoxin that is subject to controlled release (Yacoby *et al.*, 2006, 2007; Yacoby and Benhar, 2008). These phages display peptides that specifically target the bacterial pathogen of interest, even if the bacterium is not a natural target for the phage. After targeting, an antibacterial compound is released at a locally very high concentration, which results in more efficient bacterial killing. The same group has developed a related technology for application against eukaryotic cells that is dependent on endocytosis. A filamentous phage displaying a eukaryotic cell-binding ligand along with a cytotoxin, which is subject to controlled release, is used to target a specific cell type, such as a tumour cell (Bar *et al.*, 2008; Yacoby and Benhar, 2008). The phages are endocytosed, resulting in both concentration and delivery of the cytotoxin only within the targeted cells.

Chung *et al.* (2008) proposed an antibody-targeted phage construct to target tumour cells, specifically for use in inducing apoptosis in Hodgkin's-derived cell lines. In this initial proof-of-principle report, they used an *in vitro* system based on GFP expression as a measure of phage uptake, rather than demonstrating apoptosis *per se*. In an alternative strategy, Eriksson *et al.* (2007; 2009) targeted filamentous phages to tumour cells but these phages did not specifically express or carry a cytotoxic molecule. In this case, tumour-cell killing was promoted by an immune response against the phage particle, specifically the activation of tumour-associated macrophages (Eriksson *et al.*, 2009).

Other Therapies Associated with Phage-mediated Delivery

Phages that display specific molecules can also be used to effect protein distribution to locations where the protein alone would have difficulty penetrating. This approach is exemplified by the work of Solomon and colleagues (e.g. Frenkel and Solomon, 2002) who used phage display to deliver monoclonal antibodies to the brains of mice. The antibodies bound to an epitope found in the amyloid plaques associated with Alzheimer's disease, and antibody binding gave rise to plaque clearing. While antibody alone is difficult to deliver to the central nervous system, this barrier apparently may readily be overcome by phage particles. As a consequence, the phage essentially alters the pharmacokinetic properties of the displayed antibody, allowing penetration to amyloid plaques, where the therapeutic effect is required. In addition, the same group found that the phage construct could penetrate the central nervous system following intranasal delivery. Brain localization by phages has been found by others (Pasqualini and Ruoslahti, 1996), in this case to brain-specific blood vessels. See Olszowska-Zaremba *et al.* (Chapter 12, this volume) for additional perspectives on this idea of phage penetration through anatomical barriers.

Additional work by the Solomon group targeting Alzheimer's disease via the immune system (Frenkel *et al.*, 2000, 2003) found that filamentous phages could serve as an adjuvant for a vaccine consisting of an epitope found on amyloid plaques. Here, phages were repeatedly injected intraperitoneally into mice at a dosage of 10^{11} phages ml^{-1} every 2 weeks for a total of six injections. The resulting humoral immune response produced antibody, which had the effect of reducing amyloid plaques in the transgenic mouse Alzheimer's model.

Carrera *et al.* (2000, 2004) provided a similar example of intranasal phage delivery to the central nervous system. They produced a phage construct, based on the filamentous phage fd, which displayed cocaine-binding antibodies, which can sequester cocaine molecules penetrating the brain. Using a rat model, the phage dosage was 5×10^{12} per naris (50 μl of 10^{14} phages ml^{-1} suspended in PBS) delivered twice daily for 3 days. A serum immune response (antibody titre) was not observed against the administered phage, while substantial phage titres were localized to the brain, as many as 2.5×10^{13} (units indicated solely as 'phage titre' and a total of 10^{15} phages were reportedly administered to each rat prior to this determination). In all, administration of phages displaying anti-cocaine antibodies reduced cocaine-associated behavioural symptoms to a statistically significant extent. For a more detailed discussion, see also Dickerson *et al.* (2005).

As discussed previously, RNA phage VLPs have been used to deliver CpG-containing nucleic acids as vaccine adjuvants. A modification of this technique is the use of bacteriophage MS2 VLPs displaying transferrin on their surface to encapsulate antisense oligonucleotides and deliver them to leukaemia cells (Wu *et al.*, 2005). These targeted antisense oligonucleotide delivery vehicles were shown to increase killing of leukaemia cells. In related research, transferrin was also chemically conjugated to Qβ particles (Banerjee *et al.*, 2010). It was demonstrated that the efficiency of uptake was dependent on ligand density, implying that chemical conjugation (as opposed to coat protein fusion) of some targeting molecules to phage VLPs could increase the efficiency of gene transfer.

Conclusions

Bacteriophages have evolved capsids that can arguably be described as the most efficient known delivery vehicle of a genome payload to their natural target cells. The development of genetic engineering techniques allows the facile manipulation of these highly efficient delivery systems to provide a tropism towards non-natural targets to specifically deliver therapeutic or toxic molecules. Phages can also be modified physically or genetically to deliver immunogenic molecules for the purpose of vaccination. Techniques to modify the capsid of phages have existed for over 25 years, but their use in the various applications described here remains largely at the proof-of-principle and early developmental stages, although a few systems are already in clinical trials and others are advancing towards this stage. Well-defined and consistent programmes of work geared towards the end point of human clinical trials are required to transform the technologies described here from research into viable products.

References

Aujame, L., Seguin, D., Droy, C. and Hessler, C. (2000) Experimental design optimization of filamentous phage transfection into mammalian cells by cationic lipids. *Biotechniques* 28, 1202–6, 1208, 1210.

Banerjee, D., Liu, A.P., Voss, N.R., Schmid, S.L. and Finn, M.G. (2010) Multivalent display and receptor-mediated endocytosis of transferrin on virus-like particles. *ChemBioChem* 11, 1273–1279.

Bar, H., Yacoby, I. and Benhar, I. (2008) Killing cancer cells by targeted drug-carrying phage nanomedicines. *BMC Biotechology* 8, 37.

Bastien, N., Trudel, M. and Simard, C. (1997) Protective immune responses induced by the immunization of mice with a recombinant bacteriophage displaying an epitope of the human respiratory syncytial virus. *Virology* 234, 118–122.

Becerril, B., Poul, M.A. and Marks, J.D. (1999) Toward selection of internalizing antibodies from phage libraries. *Biochemistry and Biophysics Research Communications* 255, 386–393.

Benhar, I. (2001) Biotechnological applications of phage and cell display. *Biotechnology Advances* 19, 1–33.

Burg, M.A., Jensen-Pergakes, K., Gonzalez, A.M., Ravey, P., Baird, A. and Larocca, D. (2002) Enhanced phagemid particle gene transfer in camptothecin-treated carcinoma cells. *Cancer Research* 62, 977–981.

Carrera, M.R., Ashley, J.A., Zhou, B., Wirsching, P., Koob, G.F. and Janda, K.D. (2000) Cocaine vaccines: antibody protection against relapse in a rat model. *Proceedings of the National Academy of Sciences USA* 97, 6202–6206.

Carrera, M.R., Meijler, M.M. and Janda, K.D. (2004) Cocaine pharmacology and current pharmacotherapies for its abuse. *Bioorganic and Medicinal Chemistry* 12, 5019–5030.

Chen, X., Scala, G., Quinto, I., Liu, W., Chun, T.W., Justement, J.S., Cohen, O.J., vanCott, T.C., Iwanicki, M., Lewis, M.G., Greenhouse, J., Barry, T., Venzon, D. and Fauci, A.S. (2001) Protection of rhesus macaques against disease progression from pathogenic SHIV-89.6PD by vaccination with phage-displayed HIV-1 epitopes. *Nature Medicine* 7, 1225–1231.

Chung, Y.S., Sabel, K., Kronke, M. and Klimka, A. (2008) Gene transfer of Hodgkin cell lines via multivalent anti-CD30 scFv displaying bacteriophage. *BMC Molecular Biology* 9, 37.

Clark, J.R. and March, J.B. (2004a) Bacterial viruses as human vaccines? *Expert Reviews in Vaccines* 3, 463–476.

Clark, J.R. and March, J.B. (2004b) Bacteriophage-mediated nucleic acid immunisation. *FEMS Immunology and Medical Microbiology* 40, 21–26.

Clark, J.R., Bartley, K., Jepson, C.D., Craik, V. and March, J.B. (2011) Comparison of a bacteriophage-delivered DNA vaccine and a commercially available recombinant protein vaccine against hepatitis B. *FEMS Immunology and Medical Microbiology* 61, 197–204.

Cornuz, J., Zwahlen, S., Jungi, W.F., Osterwalder, J., Klingler, K., van Melle, G., Bangala, Y., Guessous, I., Muller, P., Willers, J., Maurer, P., Bachmann, M.F. and Cerny, T. (2008) A vaccine against nicotine for smoking cessation: a randomized controlled trial. *PLoS One* 3, e2547.

de Berardinis, P., d'Apice, L., Prisco, A., Ombra, M.N., Barba, P., del Pozzo, G., Petukhov, S., Malik, P., Perham, R.N. and Guardiola, J. (1999) Recognition of HIV-derived B and T cell epitopes displayed on filamentous phages. *Vaccine* 17, 1434–1441.

de Berardinis, P., Sartorius, R., Fanutti, C., Perham, R.N., del Pozzo, G. and Guardiola, J. (2000) Phage display of peptide epitopes from HIV-1 elicits strong cytolytic responses. *Nature Biotechnology* 18, 873–876.

de Berardinis, P., Sartorius, R., Caivano, A., Mascolo, D., Domingo, G.J., del Pozzo, G., Gaubin, M., Perham, R.N., Piatier-Tonneau, D. and Guardiola, J. (2003) Use of fusion proteins and procaryotic display systems for delivery of HIV-1 antigens: development of novel vaccines for HIV-1 infection. *Current HIV Research* 1, 441–446.

del Pozzo, G., Mascolo, D., Sartorius, R., Citro, A., Barba, P., d'Apice, L. and de Berardinis, P. (2010) Triggering DTH and CTL activity by fd filamentous bacteriophages: role of $CD4^+$ T cells in memory responses. *Journal of Biomedicine and Biotechnology* 2010, 894971.

Delmastro, P., Meola, A., Monaci, P., Cortese, R. and Galfre, G. (1997) Immunogenicity of filamentous phage displaying peptide mimotopes after oral administration. *Vaccine* 15, 1276–1285.

Demangel, C., Lafaye, P. and Mazie, J.C. (1996) Reproducing the immune response against the *Plasmodium vivax* merozoite surface protein 1 with mimotopes selected from a phage-displayed peptide library. *Molecular Immunology* 33, 909–916.

di Marzo, V., Willis, A.E., Boyer-Thompson, C., Appella, E. and Perham, R.N. (1994) Structural mimicry and enhanced immunogenicity of peptide epitopes displayed on filamentous bacteriophage. The V3 loop of HIV-1 gp120. *Journal of Molecular Biology* 243, 167–172.

Dickerson, T.J., Kaufmann, G.F. and Janda, K.D. (2005) Bacteriophage-mediated protein delivery into the central nervous system and its application in immunopharmacotherapy. *Expert Opinion on Biological Therapy* 5, 773–781.

Eriksson, F., Culp, W.D., Massey, R., Egevad, L., Garland, D., Persson, M.A. and Pisa, P. (2007) Tumor specific phage particles promote tumor regression in a mouse melanoma model. *Cancer Immunology and Immunotherapy* 56, 677–687.

Eriksson, F., Tsagozis, P., Lundberg, K., Parsa, R., Mangsbo, S.M., Persson, M.A., Harris, R.A. and Pisa, P. (2009) Tumor-specific bacteriophages induce tumor destruction through activation of tumor-associated macrophages. *Journal of Immunology* 182, 3105–3111.

Fang, J., Wang, G., Yang, Q., Song, J., Wang, Y. and Wang, L. (2005) The potential of phage display virions expressing malignant tumor specific antigen MAGE-A1 epitope in murine model. *Vaccine* 23, 4860–4866.

Folgori, A., Tafi, R., Meola, A., Felici, F., Galfre, G., Cortese, R., Monaci, P. and Nicosia, A. (1994) A general strategy to identify mimotopes of pathological antigens using only random peptide libraries and human sera. *EMBO Journal* 13, 2236–2243.

Frenkel, D. and Solomon, B. (2002) Filamentous phage as vector-mediated antibody delivery to the brain. *Proceedings of the National Academy of Sciences USA* 99, 5675–5679.

Frenkel, D., Katz, O. and Solomon, B. (2000) Immunization against Alzheimer's β-amyloid plaques via EFRH phage administration. *Proceedings of the National Academy of Sciences USA* 97, 11455–11459.

Frenkel, D., Kariv, N. and Solomon, B. (2001) Generation of auto-antibodies towards Alzheimer's disease vaccination. *Vaccine* 19, 2615–2619.

Frenkel, D., Dewachter, I., van Leuven, F. and Solomon, B. (2003) Reduction of β-amyloid plaques in brain of transgenic mouse model of Alzheimer's disease by EFRH-phage immunization. *Vaccine* 21, 1060–1065.

Frenkel, D., Dori, M. and Solomon, B. (2004) Generation of anti-β-amyloid antibodies via phage display technology. *Vaccine* 22, 2505–2508.

Galikowska, E., Kunikowska, D., Tokarska-Pietrzak, E., Dziadziuszko, H., Los, J.M., Golec, P., Wegrzyn, G. and Los, M. (2011) Specific detection of *Salmonella enterica* and *Escherichia coli* strains by using ELISA with bacteriophages as recognition agents. *European Journal of Clinical Microbiology and Infectious Diseases* 30, 1067–1073.

Gao, J., Wang, Y., Liu, Z. and Wang, Z. (2010) Phage display and its application in vaccine design. *Annals of Microbiology* 60, 13–19.

Gaubin, M., Fanutti, C., Mishal, Z., Durrbach, A., de Berardinis, P., Sartorius, R., del Pozzo, G., Guardiola, J., Perham, R.N. and Piatier-Tonneau, D. (2003) Processing of filamentous bacteriophage virions in antigen-presenting cells targets both HLA class I and class II peptide loading compartments. *DNA and Cell Biology* 22, 11–18.

Ghaemi, A., Soleimanjahi, H., Gill, P., Hassan, Z., Jahromi, S.R. and Roohvand, F. (2010) Recombinant λ-phage nanobioparticles for tumor therapy in mice models. *Genetic Vaccines and Therapy* 8, 3.

Ghaemi, A., Soleimanjahi, H., Gill, P., Hassan, Z.M., Razeghi, S., Fazeli, M. and Razavinikoo, S.M. (2011) Protection of mice by a λ-based therapeutic vaccine against cancer associated with human papillomavirus type 16. *Intervirology* 54, 105–112.

Gill, J.J. and Hyman, P. (2010) Phage choice, isolation and preparation for phage therapy. *Current Pharmaceutical Biotechnology* 11, 2–14.

Grabowska, A.M., Jennings, R., Laing, P., Darsley, M., Jameson, C.L., Swift, L. and Irving, W.L. (2000) Immunisation with phage displaying peptides representing single epitopes of the glycoprotein G can give rise to partial protective immunity to HSV-2. *Virology* 269, 47–53.

Greenwood, J., Willis, A.E. and Perham, R.N. (1991) Multiple display of foreign peptides on a filamentous bacteriophage. Peptides from *Plasmodium falciparum* circumsporozoite protein as antigens. *Journal of Molecular Biology* 220, 821–827.

Gu, Y., Li, J., Zhu, X., Yang, J., Li, Q., Liu, Z., Yu, S. and Li, Y. (2008) *Trichinella spiralis*: characterization of phage-displayed specific epitopes and their protective immunity in BALB/c mice. *Experimental Parasitology* 118, 66–74.

Guardiola, J., de Berardinis, P., Sartorius, R., Fanutti, C., Perham, R.N. and del Pozzo, G. (2001) Phage display of epitopes from HIV-1 elicits strong cytolytic responses *in vitro* and *in vivo*. *Advances in Experimental Medicine and Biology* 495, 291–298.

Hajitou, A. (2010) Targeted systemic gene therapy and molecular imaging of cancer contribution of the vascular-targeted AAVP vector. *Advances in Genetics* 69, 65–82.

Hajitou, A., Rangel, R., Trepel, M., Soghomonyan, S., Gelovani, J.G., Alauddin, M.M., Pasqualini, R. and Arap, W. (2007) Design and construction of targeted AAVP vectors for mammalian cell transduction. *Nature Protocols* 2, 523–531.

Hashemi, H., Bamdad, T., Jamali, A., Pouyanfard, S. and Mohammadi, M.G. (2010) Evaluation of humoral and cellular immune responses against HSV-1 using genetic immunization by filamentous phage particles: a comparative approach to conventional DNA vaccine. *Journal of Viological Methods* 163, 440–444.

Irving, M.B., Pan, O. and Scott, J.K. (2001) Random-peptide libraries and antigen-fragment libraries for epitope mapping and the development of vaccines and diagnostics. *Current Opinion in Chemical Biology* 5, 314–324.

Ishiura, M., Hirose, S., Uchida, T., Hamada, Y., Suzuki, Y. and Okada, Y. (1982) Phage particle-mediated gene transfer to cultured mammalian cells. *Molecular and Cellular Biology* 2, 607–616.

Ivanenkov, V.V., Felici, F. and Menon, A.G. (1999) Targeted delivery of multivalent phage display vectors into mammalian cells. *Biochimica et Biophysica Acta* 1448, 463–472.

Jepson, C.D. and March, J.B. (2004) Bacteriophage lambda is a highly stable DNA vaccine delivery vehicle. *Vaccine* 22, 2413–2419.

Kleinschmidt, W.J., Douthart, R.J. and Murphy, E.B. (1970) Interferon production by T4 coliphage. *Nature* 228, 27–30.

Klimek, L., Willers, J., Hammann-Haenni, A., Pfaar, O., Stocker, H., Mueller, P., Renner, W.A. and Bachmann, M.F. (2011) Assessment of clinical efficacy of CYT003-QbG10 in patients with allergic rhinoconjunctivitis: a phase IIb study. *Clinical and Experimental Allergy* 41, 1305–1312.

Lankes, H.A., Zanghi, C.N., Santos, K., Capella, C., Duke, C.M. and Dewhurst, S. (2007) *In vivo* gene delivery and expression by bacteriophage lambda vectors. *Journal of Applied Microbiology* 102, 1337–1349.

Larocca, D., Burg, M.A., Jensen-Pergakes, K., Ravey, E.P., Gonzalez, A.M. and Baird, A. (2002) Evolving phage vectors for cell targeted gene delivery. *Current Pharmaceutical Biotechnology* 3, 45–57.

Liang, Y., Shi, B., Zhang, J., Jiang, H., Xu, Y., Li, Z. and Gu, J. (2006) Better gene expression by (–) gene than by (+)gene in phage gene delivery systems. *Biotechnology Progress*, 22, 626–630.

Ling, Y., Liu, W., Clark, J.R., March, J.B., Yang, J. and He, C. (2011) Protection of mice against *Chlamydophila abortus* infection with a bacteriophage-mediated DNA vaccine expressing the major outer membrane protein. *Veterinary Immunology and Immunopathology* 144, 389–395.

Loktev, V.B., Ilyichev, A.A., Eroshkin, A.M., Karpenko, L.I., Pokrovsky, A.G., Pereboev, A.V., Svyatchenko, V.A., Ignat'ev, G.M., Smolina, M.I., Melamed, N.V., Lebedeva, C.D. and Sandakhchiev, L.S. (1996) Design of immunogens as components of a new generation of molecular vaccines. *Journal of Biotechnology* 44, 129–137.

Lundin, K., Samuelsson, A., Jansson, M., Hinkula, J., Wahren, B., Wigzell, H. and Persson, M.A. (1996) Peptides isolated from random peptide libraries on phage elicit a neutralizing anti-HIV-1 response: analysis of immunological mimicry. *Immunology* 89, 579–586.

Manoutcharian, K., Díaz-Orea, A., Gevorkian, G., Fragoso, G., Acero, G., Gonzalez, E., De, Aluja, A., Villalobos, N., Gomez-Conde, E. and Sciutto, E. (2004) Recombinant bacteriophage-based multiepitope vaccine against *Taenia solium* pig cysticercosis. *Veterinary Immunology and Immunopathology* 99, 11–24.

March, J.B. (2002) Bacteriophage-mediated immunisation. World Patent WO 02/076498.

March, J.B., Clark, J.R. and Jepson, C.D. (2004) Genetic immunisation against hepatitis B using whole bacteriophage λ particles. *Vaccine* 22, 1666–1671.

March, J.B., Jepson, C.D., Clark, J.R., Totsika, M. and Calcutt, M.J. (2006) Phage library screening for the rapid identification and *in vivo* testing of candidate genes for a DNA vaccine against *Mycoplasma mycoides* subsp. *mycoides* small colony biotype. *Infection and Immunity* 74, 167–174.

Mascolo, D., Barba, P., de Berardinis, P., di Rosa, F. and del Pozzo, G. (2007) Phage display of a CTL epitope elicits a long-term *in vivo* cytotoxic response. *FEMS Immunology and Medical Microbiology* 50, 59–66.

Maurer, P. and Bachmann, M.F. (2010) Immunization against angiotensins for the treatment of hypertension. *Clinical Immunology* 134, 89–95.

Maurer, P., Jennings, G.T., Willers, J., Rohner, F., Lindman, Y., Roubicek, K., Renner, W.A., Muller, P. and Bachmann, M.F. (2005) A therapeutic vaccine for nicotine dependence: preclinical efficacy, and Phase I safety and immunogenicity. *European Journal of Immunology* 35, 2031–2040.

Meola, A., Delmastro, P., Monaci, P., Luzzago, A., Nicosia, A., Felici, F., Cortese, R. and Galfre, G. (1995) Derivation of vaccines from mimotopes. Immunologic properties of human hepatitis B virus surface antigen mimotopes displayed on filamentous phage. *Journal of Immunology* 154, 3162–3172.

Minenkova, O.O., Ilyichev, A.A., Kishchenko, G.P. and Petrenko, V.A. (1993) Design of specific immunogens using filamentous phage as the carrier. *Gene* 128, 85–88.

Molenaar, T.J., Michon, I., de Haas, S.A., van Berkel, T.J., Kuiper, J. and Biessen, E.A. (2002) Uptake and processing of modified bacteriophage M13 in mice: implications for phage display. *Virology* 293, 182–191.

Monaci, P., Urbanelli, L. and Fontana, L. (2001) Phage as gene delivery vectors. *Current Opinion in Molecular Therapeutics* 3, 159–169.

Morales, J., Martinez, J.J., Manoutcharian, K., Hernandez, M., Fleury, A., Gevorkian, G., Acero, G., Blancas, A., Toledo, A., Cervantes, J., Maza, V., Quet, F., Bonnabau, H., de Aluja, A.S., Fragoso, G., Larralde, C. and Sciutto, E. (2008) Inexpensive anti-cysticercosis vaccine: S3Pvac expressed in heat inactivated M13 filamentous phage proves effective against naturally acquired *Taenia solium* porcine cysticercosis. *Vaccine* 26, 2899–2905.

Norris, J.S., Westwater, C. and Schofield, D. (2000) Prokaryotic gene therapy to combat multidrug resistant bacterial infection. *Gene Therapy* 7, 723–725.

Okayama, H. and Berg, P. (1985) Bacteriophage lambda vector for transducing a cDNA clone library into mammalian cells. *Molecular and Cellular Biology* 5, 1136–1142.

Pasqualini, R. and Ruoslahti, E. (1996) Tissue targeting with phage peptide libraries. *Molecular Psychiatry* 1, 423.

Peachman, K.K., Li, Q., Matyas, G.R., Shivachandra, S.B., Lovchik, J., Lyons, R.C., Alving, C.R., Rao, V.B. and Rao, M. (2012) Anthrax vaccine antigen-adjuvant formulations completely protect New Zealand white rabbits against challenge with *Bacillus anthracis* Ames strain spores. *Clinical and Vaccine Immunology* 19, 11–16.

Perham, R.N., Terry, T.D., Willis, A.E., Greenwood, J., di Marzo Veronese, F. and Appella, E. (1995) Engineering a peptide epitope display system on filamentous bacteriophage. *FEMS Microbiology Reviews* 17, 25–31.

Poul, M.A., Becerril, B., Nielsen, U.B., Morisson, P. and Marks, J.D. (2000) Selection of tumor-specific internalizing human antibodies from phage libraries. *Journal of Molecular Biology* 301, 1149–1161.

Rao, M., Peachman, K.K., Li, Q., Matyas, G.R., Shivachandra, S.B., Borschel, R., Morthole, V.I., Fernandez-Prada, C., Alving, C.R. and Rao, V.B. (2011) Highly effective generic adjuvant systems for orphan or poverty-related vaccines. *Vaccine*, 29, 873–877.

Raychaudhuri, S. and Rock, K.L. (1998) Fully mobilizing host defense: building better vaccines. *Nature Biotechnology* 16, 1025–1031.

Sapinoro, R., Volcy, K., Rodrigo, W.W., Schlesinger, J.J. and Dewhurst, S. (2008) Fc receptor-mediated, antibody-dependent enhancement of bacteriophage lambda-mediated gene transfer in mammalian cells. *Virology* 373, 274–286.

Sartorius, R., Pisu, P., d'Apice, L., Pizzella, L., Romano, C., Cortese, G., Giorgini, A., Santoni, A., Velotti, F. and de Berardinis, P. (2008) The use of filamentous bacteriophage *fd* to deliver MAGE-A10 or MAGE-A3 HLA-A2-restricted peptides and to induce strong antitumor CTL responses. *Journal of Immunology* 180, 3719–3728.

Scala, G., Chen, X., Liu, W., Telles, J.N., Cohen, O.J., Vaccarezza, M., Igarashi, T. and Fauci, A.S. (1999) Selection of HIV-specific immunogenic epitopes by screening random peptide libraries with HIV-1-positive sera. *Journal of Immunology* 162, 6155–6161.

Senti, G., Johansen, P., Haug, S., Bull, C., Gottschaller, C., Muller, P., Pfister, T., Maurer, P., Bachmann, M.F., Graf, N. and Kundig, T.M. (2009) Use of A-type CpG oligodeoxynucleotides as an adjuvant in allergen-specific immunotherapy in humans: a phase I/IIa clinical trial. *Clinical and Experimental Allergy* 39, 562–570.

Sergeeva, A., Kolonin, M.G., Molldrem, J.J., Pasqualini, R. and Arap, W. (2006) Display technologies: application for the discovery of drug and gene delivery agents. *Advances in Drug Delivery Reviews* 58, 1622–1654.

Sun, S., Li, W., Sun, Y., Pan, Y. and Li, J. (2011) A new RNA vaccine platform based on MS2 virus-like particles produced in *Saccharomyces cerevisiae*. *Biochemistry and Biophysics Research Communications* 407, 124–128.

Tissot, A.C., Maurer, P., Nussberger, J., Sabat, R., Pfister, T., Ignatenko, S., Volk, H.D., Stocker, H., Muller, P., Jennings, G.T., Wagner, F. and Bachmann, M.F. (2008) Effect of immunisation against angiotensin II with CYT006-AngQb on ambulatory blood pressure: a double-blind, randomised, placebo-controlled phase IIa study. *Lancet* 371, 821–827.

Tissot, A.C., Renhofa, R., Schmitz, N., Cielens, I., Meijerink, E., Ose, V., Jennings, G.T., Saudan, P., Pumpens, P. and Bachmann, M.F. (2010) Versatile virus-like particle carrier for epitope based vaccines. *PLoS One* 5, e9809.

Uppala, A. and Koivunen, E. (2000) Targeting of phage display vectors to mammalian cells. *Combinatorial Chemistry and High Throughput Screening* 3, 373–392.

van Gerven, N., de Greve, H. and Hernalsteens, J. (2008) Presentation of the functional receptor-binding domain of the bacterial F17a-G adhesin on bacteriophage M13 induces an immune response. *Antonie van Leeuwenhoek* 93, 219–226.

Wan, Y., Wu, Y., Bian, J., Wang, X.Z., Zhou, W., Jia, Z.C., Tan, Y. and Zhou, L. (2001) Induction of hepatitis B virus-specific cytotoxic T lymphocytes response *in vivo* by filamentous phage display vaccine. *Vaccine* 19, 2918–2923.

Wang, L.F. and Yu, M. (2004) Epitope identification and discovery using phage display libraries: applications in vaccine development and diagnostics. *Current Drug Targets* 5, 1–15.

Wang, J., Byrne, J.D., Napier, M.E. and DeSimone, J.M. (2011) More effective nanomedicines through particle design. *Small* 7, 1919–1931.

Weber-Dabrowska, B., Mulczyk, M. and Górski, A. (2001) Bacteriophage therapy for infections in cancer patients. *Clinical and Applied Immunology Reviews* 1, 131–134.

Wiessner, C., Wiederhold, K.H., Tissot, A.C., Frey, P., Danner, S., Jacobson, L.H., Jennings, G.T., Luond, R., Ortmann, R., Reichwald, J., Zurini, M., Mir, A., Bachmann, M.F. and Staufenbiel, M. (2011) The second-generation active Aβ immunotherapy CAD106 reduces amyloid accumulation in APP transgenic mice while minimizing potential side effects. *Journal of Neuroscience* 31, 9323–9331.

Willis, A.E., Perham, R.N. and Wraith, D. (1993) Immunological properties of foreign peptides in multiple display on a filamentous bacteriophage. *Gene* 128, 79–83.

Wu, M., Sherwin, T., Brown, W.L. and Stockley, P.G. (2005) Delivery of antisense oligonucleotides to leukemia cells by RNA bacteriophage capsids. *Nanomedicine* 1, 67–76.

Yacoby, I. and Benhar, I. (2008) Targeted filamentous bacteriophages as therapeutic agents. *Expert Opinion on Drug Delivery* 5, 321–329.

Yacoby, I., Shamis, M., Bar, H., Shabat, D. and Benhar, I. (2006) Targeting antibacterial agents by using drug-carrying filamentous bacteriophages. *Antimicrobial Agents and Chemotherapy* 50, 2087–2097.

Yacoby, I., Bar, H. and Benhar, I. (2007) Targeted drug-carrying bacteriophages as antibacterial nanomedicines. *Antimicrobial Agents and Chemotherapy* 51, 2156–2163.

Yokoyama-Kobayashi, M. and Kato, S. (1993) Recombinant f1 phage particles can transfect monkey COS-7 cells by DEAE dextran method. *Biochemistry and Biophysics Research Communications* 192, 935–939.

Yokoyama-Kobayashi, M. and Kato, S. (1994) Recombinant f1 phage-mediated transfection of mammalian cells using lipopolyamine technique. *Annals of Biochemistry* 223, 130–134.

Zhong, G., Smith, G.P., Berry, J. and Brunham, R.C.

(1994) Conformational mimicry of a chlamydial neutralization epitope on filamentous phage. *Journal of Biological Chemistry* 269, 24183–24188.

Zou, Y., Sonderegger, I., Lipowsky, G., Jennings, G.T., Schmitz, N., Landi, M., Kopf, M. and Bachmann, M.F. (2010) Combined vaccination against IL-5 and eotaxin blocks eosinophilia in mice. *Vaccine* 28, 3192–3200.

Zuercher, A.W., Miescher, S.M., Vogel, M., Rudolf, M.P., Stadler, M.B. and Stadler, B.M. (2000) Oral anti-IgE immunization with epitope-displaying phage. *European Journal of Immunology* 30, 128–135.

8 Clinical Applications of Phage Display

Don L. Siegel[1]

[1]Department of Pathology and Laboratory Medicine, University of Pennsylvania.

As with many of the other uses for bacteriophages described in the accompanying chapters of this volume, phage display represents an ingenious application of bacteriophages that is completely unrelated to their natural role as a bacterial pathogen – their use as a tool for the discovery of novel peptide and protein binders to molecular targets of basic and clinical interest. Since its initial description over 20 years ago, phage-display technology has been used to understand complex biological signalling pathways, to study protein–protein interactions and to develop diagnostic and therapeutic agents important to human health. These methods have been the subject of a number of recent comprehensive reviews (Hoogenboom, 2005; Lonberg, 2008; Bratkovic, 2010; Pande *et al.*, 2010; Bradbury *et al.*, 2011), and several excellent laboratory manuals are available that provide step-by-step protocols for their use (Barbas *et al.*, 2000; Clackson and Lowman, 2004; Sidhu, 2005; Aitken, 2009). This chapter will focus on a description of the technology and then provide an overview of key clinical areas in which phage display has had its largest impact.

What is Phage Display?

In theory, if one were interested in finding a ligand for a particular receptor of interest, e.g. a 7 amino acid peptide (7mer), one could synthesize every one of the more than 1 billion possible 7mers (~20^7) and perform a seemingly infinite series of biological assays to determine which peptide(s) bound to the target. Such a task would be impractical not only because of the expense and time required, but because of the technical difficulty in identifying the amino acid sequence(s) of positive peptide binders, especially because the binders would be all mixed together. If, however, each peptide were in some way physically connected to the nucleic acid that encodes that particular peptide, then a single receptor-binding assay could be performed with a mixture of all 1 billion peptide–DNA complexes. Unbound complexes could be washed away, and the nucleic acid sequences of the binders could be cloned in bacteria. Sequencing the peptide-encoding DNA in individual bacterial clones would then provide the amino acid sequence(s) of the receptor-binding ligands one was originally interested in determining.

Conceptually, the reason why this hypothetical approach seems feasible is because there is a physical connection between the genotype and phenotype of the ligand – the binding of the peptide carries along with it the nucleic acid needed for its unique identification (its sequence) and its ability to replicate. This 'linkage' of genotype and phenotype of peptides and proteins is the hallmark of phage display.

Although there have been a number of different types of bacteriophage used for phage display, typically the filamentous bacteriophage M13 is employed. M13 produces virions approximately 7 nm in diameter and up to 1 μm in length (Webster, 2001). As shown in Fig. 8.1(a), the genome of phage M13 consists of 11 genes including those that encode the major coat protein, pVIII, for which there are thousands of copies per virion, and the minor coat protein, pIII, which is required for bacterial infectivity (it binds to the F′ pilus of *Escherichia coli*) and is present in three to five copies per phage particle.

In a seminal observation made over 25 years ago, George Smith (1985) likened phage virions (M13 or others) as biological particles that physically link the phenotype of a protein (i.e. the viral coat protein) with the protein's genotype (i.e. the DNA encoding the coat protein contained within the phage genome). He demonstrated that foreign DNA inserted into the M13 phage genome, just upstream and in frame with the minor coat protein gene, *gIII*, would result in the production of phage in which each pIII copy found on the phage virion had the expressed peptide fused to its N terminus (Fig. 8.1b). In his original experiment, Smith used DNA encoding a portion of a bacterial restriction enzyme. Remarkably, the presence of this exogenous stretch of enzyme amino acids, at the tip of each pIII, did not

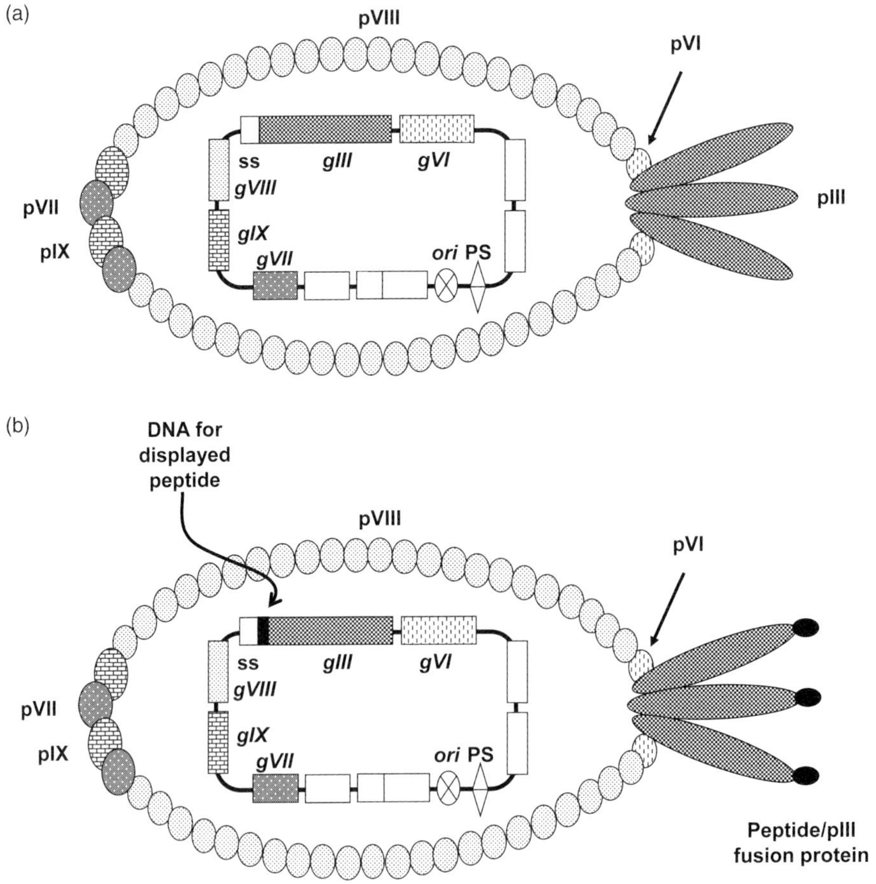

Fig. 8.1. Diagram of simplified M13 filamentous phage genomic DNA and assembled virions in (a) a wild-type and (b) a peptide-displaying phage. Note that, in (b), exogenous peptide DNA is inserted upstream of the minor coat protein gene, *gIII*, resulting in expression of exogenous peptide as fusion protein with all copies of coat protein pIII. SS, signal sequence; PS, phage packaging signal sequence; *ori*, phage DNA origin of replication.

adversely affect phage infectivity or other biological properties of the phage in any significant way. Furthermore, antibodies that he had in hand that recognized the native restriction enzyme were found to bind to the phage particles via the pIII fusion protein.

Although expressing a piece of a bacterial enzyme on the surface of a bacteriophage in and of itself was not necessarily of any practical importance, what Smith did was to demonstrate how the phenotype and genotype of a protein could be physically connected in a self-replicating biological particle. What this led to was the ability to construct massive *libraries* of M13 phage particles generated by the cloning of billions of different random peptide sequences of the desired length (typically seven to 20 codons) upstream of *gIII*. Through a process called 'panning' (discussed in detail later), such libraries could be incubated with a target, and phage-displaying peptides of sufficient binding affinity could be eluted from the target and amplified in bacteria. Reselection of the propagated libraries against the target for several additional times would eventually yield phages, all with specificity for the target. Phage DNA could be isolated from individual bacterial clones, sequenced and used to determine the amino acid sequence of the peptide(s) that bound the target. This was the beginning of what became known as phage display.

Technical Aspects of Peptide and Protein Phage-display Libraries

The peptide discovery approach described above has become a fairly routine laboratory procedure, and pre-made M13 peptide phage-display libraries are commercially available. For example, New England Biolabs (Ipswich, MA) provides linear 7mer and 12mer or cyclic 7mer libraries (displayed 7mers are flanked by cysteine residues to provide secondary structure). Phage vectors for designing one's own peptide library are also commercially available and require only basic molecular biology skills to construct.

A typical phage-display experiment with such a peptide library comprises three or four rounds of selection against a target, which can be accomplished in about a week and are technically relatively simple to perform – a round of selection with a target immobilized to a plastic well of a microtitre plate is essentially like performing an ELISA except that phages are added to wells for incubation with immobilized target, not analytes or antibodies, and instead of enzyme-conjugated antibodies added to the wells for ELISA development, bound phages are eluted from the wells with a dissociation agent (typically low pH buffer) and used to infect bacteria. Subsequent analyses of derived peptides through nucleic acid sequencing of individual phage clones and then production of positive clones as soluble peptides (i.e. unlinked to phage) to verify binding can take an additional few weeks depending on the nature of the project.

Such commercially available or research laboratory-constructed peptide libraries have been used for epitope mapping of monoclonal antibodies (mAbs), developing enzyme inhibitors, mapping protein–protein contacts, designing vaccines, discovering peptides that mimic non-peptide carbohydrate ligands and many other applications (reviewed by Brissette and Goldstein, 2007). A set of particularly clinically relevant uses of peptide phage display in the areas of molecular imaging, tumour targeting and diagnosis of cancer has been reviewed recently (Deutscher, 2010).

The ability to create M13 phages displaying repertoires of molecules larger than a peptide, such as immunoglobulin fragments with molecular weights of 30–50 kDa (~275–450 amino acids) derived from a human's or animal's immune system, has provided leads for the development of therapeutic antibodies for treating a number of health conditions including cancer, autoimmune disorders and others. After selection of phages from such libraries, the DNA encoding positive-binding immunoglobulin fragments can be subcloned into other vectors for the production of full-length IgG suitable for diagnostic or therapeutic use. One of the major driving forces for the development of phage-display methods for antibody discovery was a need for a source of *human* mAbs that would not induce anti-drug immune reactions, as can be seen with murine or even human/murine chimaeric antibodies (Weiner, 2006). Up until the development of phage display, the cloning of mAbs was accomplished by hybridoma technology, which is a viable method for the production of murine antibodies (Kohler and Milstein, 1975) but problematic for the production of human antibodies because immortalizing human B cells is technically challenging (Winter and Milstein, 1991). As shown below, phage display offered a way of cloning human antibody genes in bacteria rather than requiring the ability to immortalize the B cells from which the antibody clones originated.

Cloning the larger-than-peptide-sized stretches of exogenous DNA into the M13 phage genome, such as those required to encode antibody fragments, became technically challenging, as phages containing longer stretches of cloned DNA have a growth disadvantage and their expressed proteins may be toxic to their bacterial hosts. Furthermore, as illustrated in Fig. 8.1(b), when cloning exogenous DNA into the M13 genome, all copies of pIII per virion display the desired ligand. Although such 'multivalent display' can be advantageous in certain applications, for the selection of phages displaying antibody fragments that bind to their target through affinity alone, not influenced by avidity effects, 'monovalent display' of single-chain Fv (scFv) or Fab fragments is most often preferred (Fig. 8.2). This requires the co-expression of wild-type pIII coat protein along with pIII coat protein fused with the displayed antibody fragment. Cloning libraries of antibody fragment DNA into phagemid vectors, rather than into the actual phage genome, can be used to achieve monovalent display, as shown stepwise in Fig. 8.3. Phagemids engineered for this

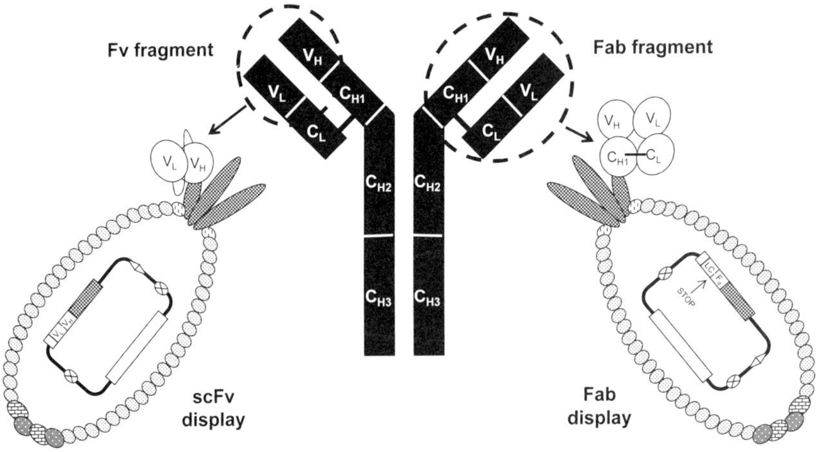

Fig. 8.2. Expression of antibody fragments on a filamentous phage. Monovalent display of an scFv fragment (left) or Fab fragment (right) is illustrated using a conventional diagram of IgG (middle) to indicate parts of IgG expressed as a fusion with the phage pIII coat protein. An Fv fragment of IgG is defined as an antigen-binding molecule comprising just the variable regions of the immunoglobulin heavy (V_H) and light (V_L) chains. In order for V_H and V_L to assemble into one protein, overlap PCR is used to assemble individual V_H and V_L amplification products into a 'single-chain Fv' in which a flexible glycine/serine linker of approximately 18 amino acids permits alignment of the N termini of heavy and light variable regions. A Fab fragment is defined as an antigen-binding molecule comprising the Fd fragment of the heavy chain (Fd = V_H and C_{H1} domains of the heavy chain constant region) paired with the entire light chain (V_L plus the constant region of the light chain, C_L). For the construction of Fab phage-display libraries, the Fd and light-chain polypeptides are produced as separate molecules off a dicistronic message (the stop codon in the right-hand illustration); the Fd fragment is produced as a fusion with pIII and the light chain associates with the Fd fragment (as it would do naturally in a B lymphocyte), and a covalent disulfide bond forms between C_{H1} and C_L. Note that each phage particle comprises both wild-type pIII and pIII antibody fusion protein, and DNAs for displayed proteins are cloned into phagemid vectors rather than the phage genome, as described in the text.

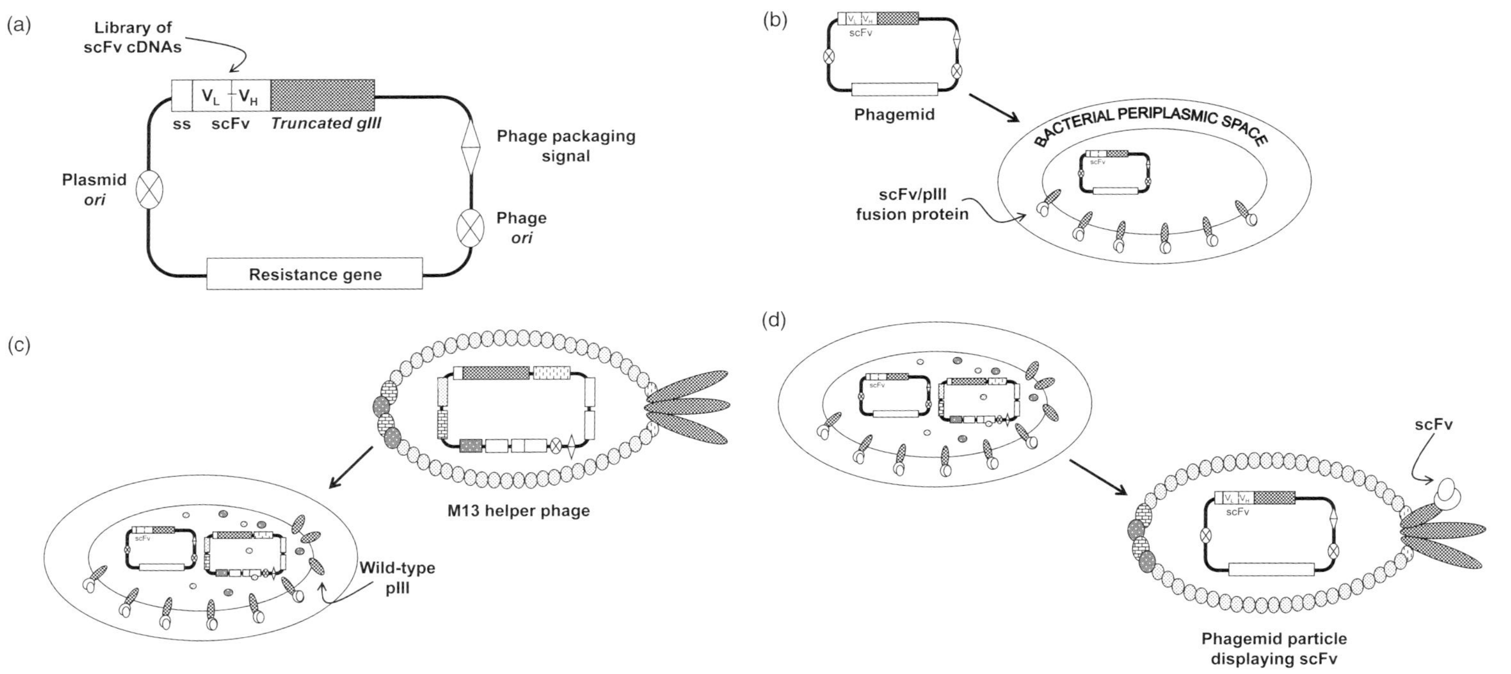

Fig. 8.3. Overview of the construction of phagemid-based antibody fragment (scFv) display libraries. (a) A large array of scFv constructs (e.g. those derived from RT-PCR of immune cells such as peripheral blood lymphocytes, splenocytes or bone marrow mononuclear cells) is introduced upstream of a truncated *gIII* (see text) in a phagemid vector comprising an M13 phage origin of replication (phage *ori*), a phage packaging signal, a resistance gene for selection and a plasmid origin of replication (plasmid *ori*). (b) The library of scFv constructs is electroporated into a culture of *Escherichia coli* and scFv/*gIII* fusion proteins are expressed and sent to the bacterial periplasmic space by the signal sequence (ss). (c) The bacterial culture is superinfected with M13 helper phage (essentially wild-type phage with an antibiotic resistance gene and a defective packaging signal sequence) and M13 phage proteins, including wild-type pIII and proteins required for phage assembly, are expressed. (d) Through dual antibiotic selection for both phagemid- and helper phage-containing bacteria, phagemid particles are released into culture that contain scFv plasmid DNA within and a mixture of wild-type and scFv-fused pIII coat proteins on the surface. Because the phage packaging signal sequence is defective in the helper phage, the plasmid DNA is preferentially packaged by the helper phage DNA-encoded proteins. Through the optimization of various parameters (e.g. level of expression of scFv/pIII, timing of various steps including addition of helper phage), each phage particle, on average, will have one scFv/pIII molecule per virion to approximate monovalent display.

purpose have cloning sites for antibody fragment DNAs just upstream of a form of *gIII* that lacks an N-terminal domain (Scott and Barbas, 2001). This so-called 'truncated' *gIII* lacks a string of nucleotides that encodes a region of pIII that signals to the bacterium to cease production of pili. This natural control mechanism carried out by the N-terminal portion of pIII serves to prevent an individual bacterium from being infected by more than one phage particle (super-infection immunity). When working with phage as vectors, there is no need for pilus formation after infection, so it does not matter that pilus formation is inhibited. In fact, inhibition of pilus formation is absolutely essential when working with phage vectors, otherwise reinfection would lead to a 'scrambling' of the correct linkages between displayed peptides and their encoding DNAs. For antibody libraries, however, as phagemid vectors require subsequent co-infection with helper phages to provide genes for wild-type pIII as well as for all other phage genes required for phage particle assembly and secretion, continued pilus expression is required after infection of bacterial cultures with phagemids. The use of the truncated *gIII* allows this necessary ongoing pilus expression (Fig. 8.3c).

Selection Strategies for Phage-display Libraries

Whether a peptide or antibody phage-display library, selection of phages displaying a target-binding moiety is accomplished through a series of enrichment steps referred to as 'panning'. Typically, three to four rounds of panning are required for most, if not all, of the resulting phages to display peptides or proteins with affinity for the target. After panning rounds are completed, the resulting preparation of 'polyclonal' phages are titrated on *E. coli* to produce either individual plaques or bacterial colonies, each having been infected with a single bacteriophage. Typically, several dozen clones are randomly selected from agar plates and grown for the purpose of preparing phage (or phagemid) DNA from which the peptide (or antibody) sequences can be determined. This analysis will provide an estimate of the diversity in binders obtained following the given number of rounds of selection.

Depending on the nature of the project, there may be only one or perhaps a few different peptide or antibody sequences, or the collection may be very diverse. Typically in a peptide phage-display library experiment, a series of different peptide sequences will be obtained within which one can identify a common shorter continuous or discontinuous amino acid motif. For an antibody phage-display experiment (see below), the diversity in sequence of the resulting binders will depend on a number of factors including whether the antibody library was constructed from an immunized animal, in which there may be a relatively small number of antibodies that bind with particularly high affinity to overshadow (outcompete for target binding) other binders. Conversely, if the antibody library was constructed from a non-immunized or synthetic source of V_H and V_L segments, there may be a relatively larger number of equal but weaker binders that may require '*in vitro* affinity maturation', as discussed later. Quite often, one may find a series of clonally related antibodies comprising heavy and/or light chains that can be shown to have been derived *in vivo* from a single-original B cell, yet may differ in their exact nucleotide sequences due to somatic mutation that had occurred in the human or animal from which the library was constructed (Chang and Siegel, 1998; Roark *et al.*, 2002).

Panning against purified, immobilized antigen is the most common and straight-forward type of phage library selection and is illustrated in Fig. 8.4 for an antibody phagemid library. As briefly described above, this form of panning is very much like carrying out an ELISA except that binders are eluted from microplate wells rather than detected with a secondary antibody. Elution is most often carried out with low pH (~3.0) buffer for about 10 min, which is generally sufficient to denature the target and/or displayed antibody, resulting in release of the phage from the well, but remarkably does not harm the integrity of the phage so that reinfection and amplification of eluted phage can take place. Elution can also be accomplished by alkaline pH or by other conditions that are known specifically to affect the particular antibody–target interaction with which one is working, such as reduction with dithiothreitol, exposure to certain detergents or proteases, or an elevation in temperature, or by competition by the addition of excess soluble target or other antibodies to the target. Because the eluted phages contain phagemid DNA capable of producing scFv/pIII fusion protein but no other phage components, helper phages are required during the panning process to rescue the phagemids, as in the process of library construction outlined in Fig. 8.3. After overnight incubation, the phage particles are harvested from the supernatant of the bacterial culture and applied to a fresh well of target for the subsequent round of panning.

Although 'solid-phase' panning shown in Fig. 8.4 is the most straightforward method, a number of other approaches have been described that may eliminate potential problems, for example if adsorption of the particular target to plastic distorts its native conformation or results in blocking the binding surfaces (epitopes) of interest, rendering them inaccessible to the displayed ligand on the phage particles. To get around these potential pitfalls, the use of biotinylated targets permits incubation in solution with the phage library and the use of streptavidin-coated magnetic beads for separation of the target for the washing and elution steps (Hawkins *et al.*, 1992). Other advantages of the biotinylated antigen approach are that it allows the investigator to control precisely the effective in-solution concentration of the target during selection and thus influence the expected affinity of the captured phage-displayed ligand. In fact, a strategy that is often used is to sequentially decrease the concentration of biotinylated target from round to round of panning so as to create competition between binders and eventually

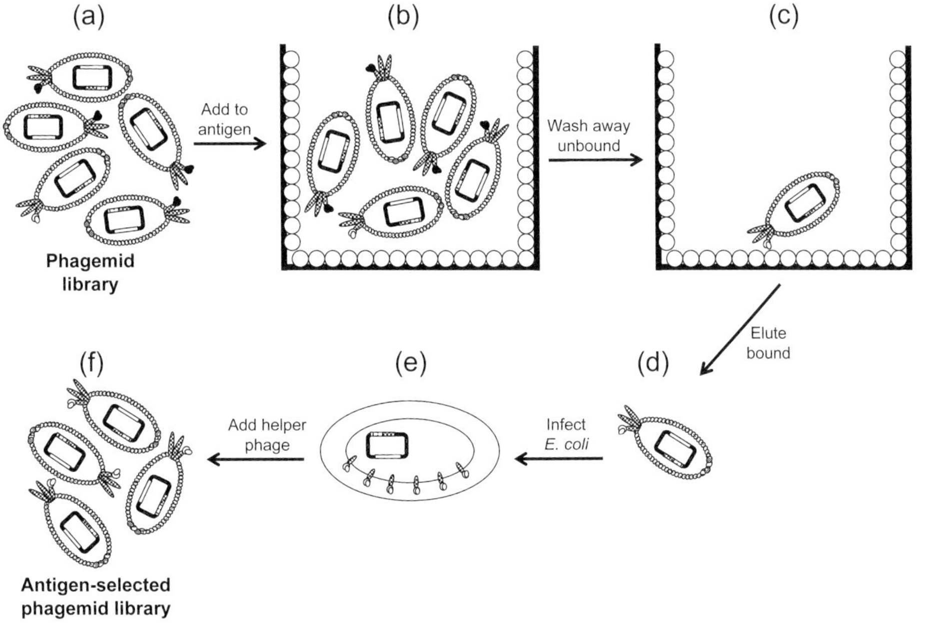

Fig. 8.4. Antibody phage-display library selection by the solid-phage panning approach. A phage antibody library (a) is applied to the well of a microplate coated with the target antigen of interest (b) and unbound phage are washed away (c). Bound phage are eluted (see text) (d) and used to infect *E. coli* (e). Phagemid DNA is rescued by subsequent superinfection with M13 helper phage to produce a second-generation antibody phage-display library selected on the antigen.

get those with the highest affinity to 'win out' during progressive rounds of selection as the amount of target becomes limiting.

There is a wide range of other selection approaches, illustrating the enormous versatility of phage-display approaches for discovering binding partners for targets of interest (reviewed by Griffiths and Duncan, 1998; Hoogenboom, 2005). For example, phage libraries may be panned on intact live cell surfaces to isolate binders to an antigen or receptor that is either unknown (e.g. in the case of a putative tumour-specific antigen) or unpurifiable in its native state (e.g. in the case of many transmembrane proteins that would denature if solubilized in detergent). Most often, such cell-surface panning approaches include initial 'subtractive selections' of phage libraries with suspensions of cells otherwise identical to the target cells but that specifically lack the antigen of interest (e.g. the normal cellular counterpart of the tumour cells). Such 'negative selections' will adsorb pan-reactive 'generic' phages, leaving behind target-specific phages that can be enriched through positive selection on antigen-positive cells using magnetically activated cell sorting, as illustrated in Fig. 8.5 (Siegel *et al.*, 1997; Siegel, 2001; McWhirter *et al.*, 2006). Other methods for differential cell-surface panning of antibody phage-display libraries have utilized fluorescently activated cell sorting (de Kruif *et al.*, 1995a,b; van Ewijk *et al.*, 1997; Huls *et al.*, 1999; Lekkerkerker and Logtenberg, 1999; van der Vuurst de Vries and Logtenberg, 1999).

Panning of phage libraries can also take place on immunoblots of SDS-PAGE-separated polypeptides (Liu and Marks, 2000) or on tissue sections (van Ewijk *et al.*, 1997) to isolate phage binders to particular components within complex mixtures of proteins or cells. For example, if all that is known about a target of interest is its molecular weight and thus its position as a protein band on an SDS-PAGE gel, then a suitable phage-display library can be incubated with an immunoblot of the gel as though one were performing a Western blot. The band of interest can then be cut out of the sheet of nitrocellulose with a razor blade, the phages eluted with acid and the phage eluate used to reinfect bacteria for amplification. By analogy, the panning of phage-display libraries on tissue sections combined with laser microdissection techniques has permitted the cloning of antibodies directed against nanogram amounts of proteins contained within intracellular structures or against particular cells (e.g. tumour cells) within a heterogeneous slice of tissue (Tanaka *et al.*, 2002). These types of selection procedures illustrate the power of antibody phage display when compared with traditional tissue culture hybridoma methods in which an impractical number (thousands) of assays (by Western blot or immunohistochemistry) would need to be performed to screen each hybridoma clone one by one.

'*In vivo* panning' is perhaps one of the most interesting forms of panning in which phage-display libraries have been injected into living animals including humans (reviewed by Sergeeva *et al.*, 2006). Following a brief time circulating in the bloodstream, tissues are removed and bound phages are eluted and then amplified in bacteria. Such approaches have attempted to identify peptides or other ligands that home in on vascular structures specific to particular organs or tumours.

Use of Phage Display for Antibody Discovery

Overview

Figure 8.6 presents an overview of the ways in which phage display has been used to discover new diagnostic reagents and assays for their use, as well as reagents that have served as leads for the development of novel pharmaceuticals. The process begins with a source of immunoglobulin DNA, the construction of a phage-display library, the selection of binders through panning and the isolation of target-specific phage particles (Fig. 8.6a). Immunoglobulin DNA can then be extracted from individual phages and subcloned into other systems to produce full-length IgG in which antibody Fc constant region domains are appended (Fig. 8.6b), or to produce antibody fragments as soluble

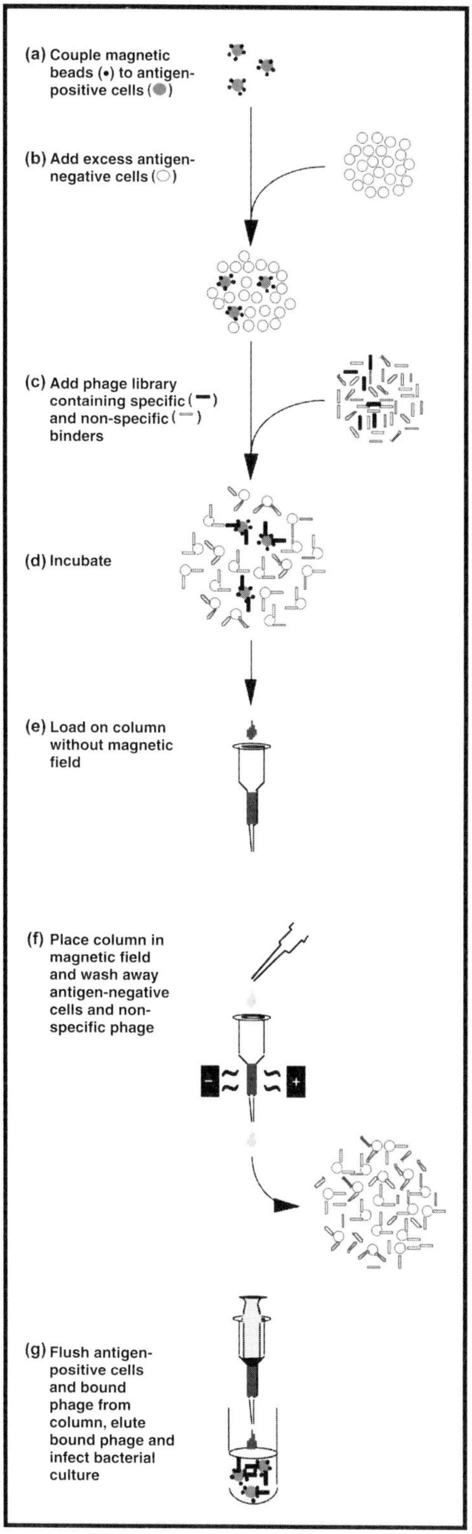

Fig. 8.5. Cell-surface panning of phage-display libraries using magnetically activated cell sorting. (a) Target cells (in this case, red blood cells bearing the human Rh(D) blood group antigen) are cell-surface biotinylated and coupled to streptavidin-conjugated magnetic beads. (b) A large excess of antigen-negative distractor cells (in this case, Rh(D)-negative red cells) are added and (c, d) the cell admixture is incubated with a phage library constructed from the peripheral blood lymphocytes of an Rh(D)-negative patient previously sensitized to the Rh(D) antigen. (e) The incubation mix is added to the top of a column used for magnetically activated cell sorting. The volume of the incubation mix is smaller than the included volume of the column and the column is initially kept out of a magnetic field so that the cells spread out evenly within the column. (f) The column is then placed within a magnetic field and antigen-negative cells along with non-specific phage-displayed antibodies are washed away. (g) Antigen-positive cells are then flushed out of the column, bound phage antibodies are eluted and clones are amplified in bacterial culture. Amplified phage are then subjected to several additional identical rounds of competitive cell-surface selection, after which essentially all eluted phage are specific for the Rh(D) antigen. (Adapted from Siegel *et al.*, 1997.)

molecules for applications in which natural antibody effector function or long circulation half-life is not necessary (Fig. 8.6c). Alternatively, the phage particles can be used directly as highly sensitive and specific immunological detection reagents (Fig. 8.6d).

Types of antibody phage-display libraries

Phage-display libraries can be classified as 'immune', 'naïve' or 'synthetic' depending on the source of immunoglobulin DNA. 'Immune libraries' are derived from the RT-PCR of B-cell-containing material of an animal (e.g. peripheral blood lymphocytes, spleen, bone marrow) that has been immunized with a target of interest. In theory, a phage-display library can be made from the immune system of any animal for which PCR primers can be designed that will amplify the repertoire of heavy chain and light chain variable region sequences for that species. Though the ability to do this at first may seem counterintuitive, given that variable regions of antibodies are 'variable' in sequence and unknown *a priori*. It should be appreciated that the very beginning 5′ end of heavy and light chain genes comprises sequences referred to as 'framework regions' for which there are a relatively limited number of possible sequences. For example, in the human, any one of the billions of heavy-chain antibody sequences in a given individual will fall into one of only seven families of V_H germline genes (de Bono *et al.*, 2004). To PCR amplify that repertoire of heavy chains, only half a dozen or so forward PCR primers are needed, each paired with the same 3′ primer that anneals to the beginning of the antibody constant region. Human κ and λ light-chain variable-region genes also fall into families with limited diversity in the 5′-most framework region, thus permitting their PCR

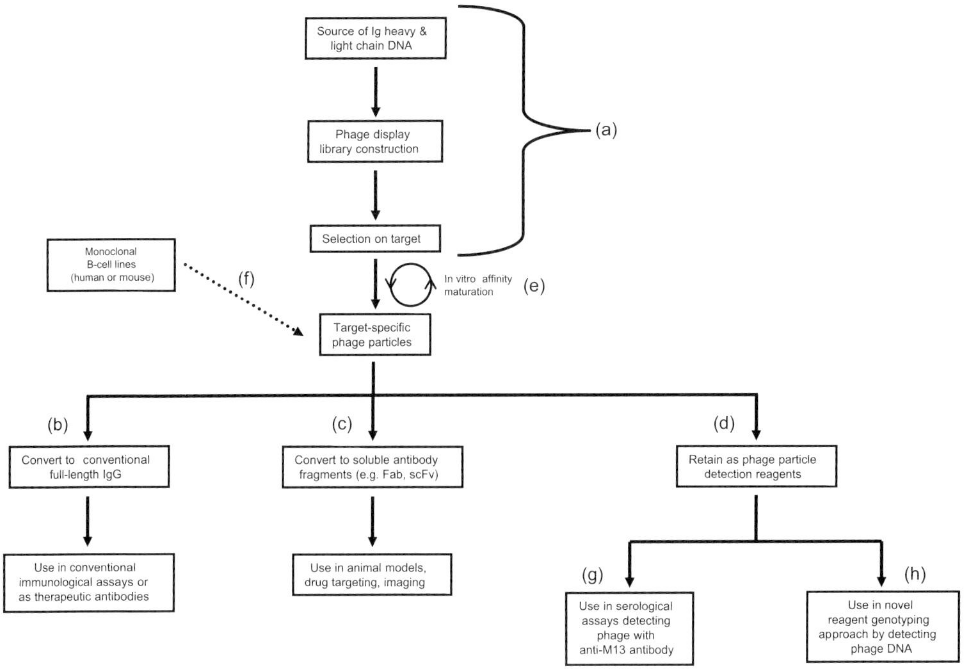

Fig. 8.6. Overview of phage-display applications for antibody discovery, production and use (see text for details).

amplification with a manageable number of amplification reactions (~20–40 depending on the phage-display system used).

Species for which phage-display libraries have been constructed have included human, non-human primate, mouse, rabbit, sheep, chicken, camel, shark, dog and others (Marks *et al.*, 1991; Siegel *et al.*, 1999; Andris-Widhopf *et al.*, 2000; Andris-Widhopf *et al.*, 2001; Popkov *et al.*, 2003; Harmsen and de Haard, 2007; Braganza *et al.*, 2011; Finlay *et al.*, 2011). Using human material as a source of immunoglobulin cDNA is clearly advantageous if the ultimate goal is to produce a therapeutic antibody that will be most compatible with the immune systems of the patients who will be treated with that drug. However, for obvious ethical reasons, humans cannot be immunized at will with any substance to which one might want to generate an antibody. Consequently, human immune phage-display libraries are generally limited to those constructed from individuals who are found to already make potentially clinically useful antibodies. Such scenarios include post-vaccination (e.g. with tetanus toxoid or anthrax; Barbas *et al.*, 1991; Wild *et al.*, 2003), post-infection (e.g. with human immunodeficiency virus; Barbas *et al.*, 1993), post-treatment for underlying disease (e.g. from sensitization to blood group antigens following transfusion; Chang and Siegel, 1998) or in certain autoimmune diseases in which subjects make antibodies to human self-antigens as part of their disorder (McIntosh *et al.*, 1996; Graus *et al.*, 1997; Roark *et al.*, 2002; Payne *et al.*, 2005).

Although there are phage-display methods for 'humanizing' mouse antibodies (Osbourn *et al.*, 2005), animals are most often used for developing antibodies for diagnostic or research purposes where the species of origin is not relevant. Although a comprehensive discussion of the relative merits of using one animal over another for a given phage-display project is beyond the scope of this chapter, chickens and camels as sources for the generation of phage-display-derived antibodies are worth noting. Chickens offer two advantages. First, from a biological perspective, as birds and being evolutionarily distinct from mammals, their immune systems may mount vigorous immune reactions to highly important and thus highly conserved mammalian proteins to which the immune system of a mammal (e.g. human, rabbit or mouse) will not respond due to tolerance mechanisms (Andris-Widhopf *et al.*, 2001). Secondly, from a technical perspective, chicken immune systems create their antibody diversity through gene conversion events rather than through the inheritance of multiple heavy- and light-chain immunoglobulin genes, as described for humans and other animals (McCormack *et al.*, 1993). In fact, chicken immune systems comprise only one heavy-chain germline gene and one light-chain germline gene, which limits the number of PCR amplification steps to a total of two, thus simplifying phage-display library construction. Antibodies derived by phage display from camel immune systems (Arbabi Ghahroudi *et al.*, 1997) have been shown to offer certain interesting advantages over those of other species due to their unusual stability in the presence of detergent or extremes in temperature. For example, phage-display libraries made from immunized camelids (camels and llamas) have produced 'anti-dandruff' antibodies that are stable in shampoo (Dolk *et al.*, 2005), as well as dip stick devices to detect the presence of caffeine to differentiate between caffeinated versus decaffeinated hot coffee (Ladenson *et al.*, 2006).

In contrast to the 'immune' phage-display libraries discussed above, 'naïve' and 'synthetic' phage-display libraries are designed to get around the ethical issues precluding the at-will immunization of human subjects with clinically relevant target antigens. Naïve libraries are generally constructed from large pools of B cells collected from normal healthy individuals (Vaughan *et al.*, 1996; Lloyd *et al.*, 2009). Their construction was originally designed to exploit the observation, originally made with murine hybridomas, that naïve IgM antibodies (i.e. those that generally have not undergone somatic mutation) can recognize a variety of antigens. The theoretical advantage of these libraries is that a single 'universal' library could be constructed and used to provide mAbs with any desired specificity. Although this has been the prevailing view, the affinity of the binders may, in some cases, be suboptimal. Nevertheless, once target-specific phage particles have been isolated, there are a number of methods – including chemical mutagenesis, the use of mutagenic strains of bacteria, incorporation of degenerate oligonucleotides and error-prone PCR – that can be used to generate random or targeted nucleotide mutations in antibody gene segments (Martineau, 2002; Thie *et al.*, 2009). Reselection of those daughter libraries on target antigens can be used to select for those mutations that have improved the affinity of the original parent antibody (Fig. 8.6e). Along these lines, phage display can be used to improve the affinity or other properties of pre-existing mAbs whether derived from mouse hybridomas, human–mouse heterohybridomas, or human Epstein–Barr virus-transformed B cells. By cloning the heavy and light gene segments into a phage vector, sequences with improved binding properties can be selected (Fig. 8.6f).

'Synthetic' phage-display libraries differ from the immune and naïve types in that no B cells are involved in the generation of library diversity. Rather, the library is composed of 'human' antibody genes that have been constructed *in vitro* using PCR or other strategies to randomize the hypervariable regions of one or more human germline-encoded variable-region genes (Hoogenboom and Winter, 1992; de Kruif *et al.*, 1995a; Knappik *et al.*, 2000; Krebs *et al.*, 2001). Although antibodies isolated from such synthetic repertoires may also suffer from suboptimal affinity and require *in vitro* affinity maturation, as described above, a significant advantage over immune and naïve libraries is that the antibody composition of synthetic libraries is not constrained by the mechanisms at play *in vivo* that normally prevent the generation of antibodies to human self-antigens. Therefore, the synthetic phage library approach inherently possesses the ability to provide 'human antibodies' to clinically important human proteins, excess production of which or expression on cells may be responsible for human disease.

Use of phage particles as immunological reagents

The vast majority of studies have used phage display as a tool to isolate novel antibody DNA that would then be subcloned into another recombinant antibody expression system to produce soluble immunoglobulin. A number of studies have shown, however, that antibody-displaying phage particles themselves can also be used as immunological reagents, ones that are capable of self-replication (Fig. 8.6d). For example, phage particles expressing antibody fragments for red blood cell antigens were able to phenotype cells using an indirect agglutination assay analogous to the traditional Coombs reaction (which utilizes rabbit or mouse anti-human IgG) but instead using antibodies to the phage particles as a secondary antibody (Fig. 8.6g) (Siegel *et al.*, 1997). The advantages of using phage particles for these types of assays instead of conventional IgM- or IgG-containing antisera is sensitivity – as few as 10–25 anti-red cell phage per red cell will produce a positive reaction – and cost. Indeed, an inexpensive, 1 litre overnight bacterial culture of phage can provide enough reagent for over 100,000 diagnostic assays. The enormous sensitivity of these assays is due in large part to the enormous size of a phage particle when compared with a conventional antibody; thus, virions provide a large surface area to which secondary antibodies can bind (Fig. 8.7).

Similarly, antibody-displaying phage particles can be used in other immunological assays such as ELISAs (Steinberger *et al.*, 2001) and flow cytometry (Siegel and Silberstein, 1994) by covalently modifying the phage particles with colour-developing enzymes, fluorescent molecules or biotin, or by using an appropriately conjugated anti-M13 secondary antibody. However, more recently, the potential use of intact phage particles in immunological assays has been extended further by exploiting the fact that unique sequences of DNA are present within the particles (Fig. 8.6h; Jaye *et al.*, 2003; Siegel, 2005). The concept here is that the presence of a target antigen can be determined by binding target-specific phages and then genotyping the phage particle. The advantages of this

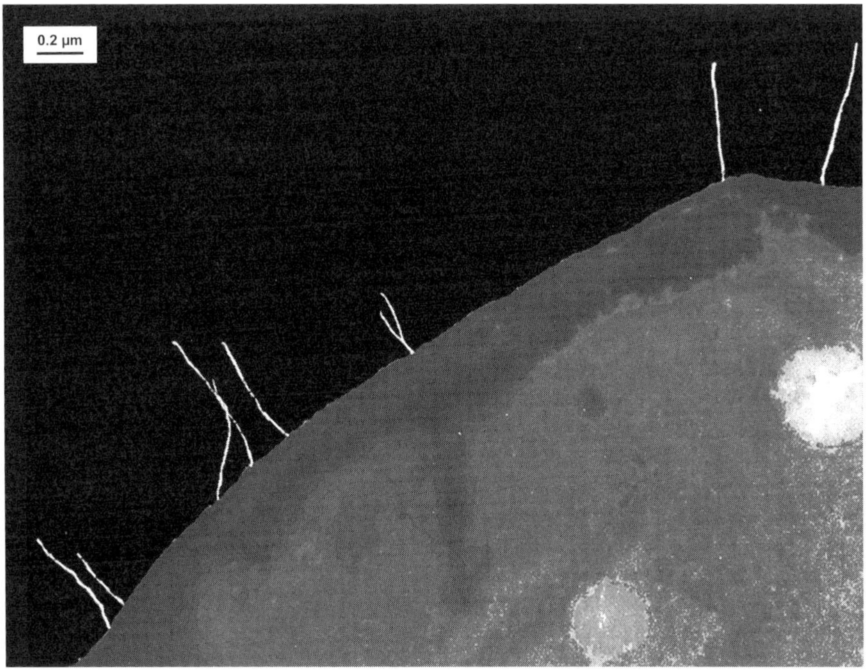

Fig. 8.7. Transmission electron micrographic images of anti-Rh(D)-expressing M13 bacteriophage (white filaments) bound to the surface of a human Rh(D)-positive red blood cell. Binding occurs at the tips of the phage where the recombinant antibody is fused to the pIII phage coat protein. Original magnification ×44,000. Bar, 0.2 μm. (Adapted from Siegel and Silberstein, 1994.)

approach are that methods that use nucleic acid detection schemes offer extraordinarily high sensitivity and specificity, require minute amounts of reagents and are adaptable to automation. In addition, nucleic acid-based assays lend themselves to multiplexing strategies, which, in the case of phenotyping cells, would offer the possibility of simultaneously determining the presence or absence of an entire set of cell-surface antigens by adding mixtures of phage, each expressing a different antibody specificity on the outside of the phage particle and bearing a unique DNA 'sequence tag' that had been engineered into the phagemid DNA (Siegel, 2005).

Current Status of Phage-display-derived Therapeutic Antibodies

The market for therapeutic mAbs has increased to more than US$30 billion within the last few years with approximately two dozen mAbs currently approved for human use (Bratkovic, 2010: Nelson *et al.*, 2010). The majority (approximately two-thirds) of these antibodies comprise molecules that were derived using conventional mouse hybridoma methods developed in the 1970s (Kohler and Milstein, 1975), which then underwent some form of 'humanization' by which either the murine antibody variable regions were spliced on to a human constant region ('chimaeric' mAbs) or the complementarity-determining regions, which are the most variable domains of the variable regions, were grafted on to human framework regions equipped with human constant regions ('humanized' mAbs). However, as noted earlier, such antibodies are not entirely derived from human genes and may still elicit anti-drug immune responses that can affect their efficacy as well as cause potentially serious allergic or anaphylactic reactions (Weiner, 2006). With the development during the 1990s of phage display along with one other technology that immunizes transgenic mice engineered to express human immunoglobulin germline genes (Green *et al.*, 1994; Lonberg *et al.*, 1994; Lonberg, 2005), it became possible to produce so-called 'fully human' mAbs that in theory would be better tolerated when used therapeutically. Of the approximately 90 mAbs currently being evaluated for clinical use that were derived using one of these two 'human' methods, to date phage-display has contributed approximately 35 candidates targeting cancer, inflammatory/autoimmune disorders, neurodegenerative diseases, infectious diseases and others (Lonberg, 2008; Thie *et al.*, 2008; Nelson *et al.*, 2010).

The anti-tumour necrosis factor-α (TNF-α) antibody adalimumab (Humira®) was the first of approximately 35 phage-display-developed mAbs to be approved by the US Food and Drug Administration (FDA) in 2002, and is indicated for a number of inflammatory conditions including rheumatoid arthritis, psoriatic arthritis, ankylosing spondylitis and Crohn's disease (van de Putte *et al.*, 2004). Interestingly, this antibody was derived by employing both conventional murine hybridoma methods and phage-display approaches. Initially, the heavy- and light-chain gene segments of a hybridoma-derived murine anti-human TNF-α neutralizing antibody were cloned into the M13 phage-display system (Fig. 8.6f). This was followed by the use of 'guided selection' (Jespers *et al.*, 1994; Osbourn *et al.*, 2005) in which the heavy chain of the mouse antibody was paired with a repertoire of random human light chains to display human–mouse antibody fragments on phage. This phage-display library was then panned on TNF-α. In parallel, the light chain from the original murine antibody was paired with a repertoire of random human heavy chains, displayed on phage and then panned against TNF-α. Finally, a set of 'guided' human heavy and light chains 'selected' from each chimaeric library were combined together to form yet another phage library (now completely 'human'), which was panned on TNF-α. *In vitro* affinity maturation (Fig. 8.6e) of the best binder was then used to generate a higher affinity, higher potency, anti-TNF-α mAb (D2E7) that became the therapeutic drug adalimumab.

In contrast to adalimumab, belimumab (Benlysta®), approved by the FDA in 2011 for the treatment of systemic lupus erythematosus, was discovered *de novo* from a human naïve phage-display library selected on B-Lys (a B-lymphocyte stimulator) and then affinity matured to improve binding (Baker *et al.*, 2003; Halpern *et al.*, 2006). B-Lys, a naturally occurring protein important in B-cell maturation, has been implicated in the development of autoantibodies leading to certain forms of autoimmune disease, and belimumab is thought to work, in part, by antagonizing the action of B-Lys on B-cell stimulation.

Future Perspectives

Currently celebrating its 25th anniversary, phage-display technology has proved to be an enormously powerful approach for the discovery of new diagnostic reagents and assays for their use, as well as leads for the development of novel molecules for clinical use. One of its most significant advantages in the design of therapeutic targeting agents is its ability to create antibodies that appear to be less immunogenic than those produced using conventional methods. Although the great majority of therapeutic antibodies currently approved for human use have not been derived using phage-display approaches, it is believed that the difficult-to-navigate intellectual property climate surrounding phage-display technology since its inception has been in large part responsible for restricting its proliferation in the therapeutic market (Bradbury *et al.*, 2011). With a number of core patents covering key aspects of phage display now expiring, using these methods for commercial use may not be a limiting factor in exploiting the enormous capabilities of phage display (Storz, 2010). In addition to these applications, the visibility of phage-display technologies in the fields of nanotechnology (Sarikaya *et al.*, 2003; Hemminga *et al.*, 2010) and nanomedicine (Souza *et al.*, 2006, 2008, 2010) will increase greatly, as will their contribution to the interface between clinical genomics and proteomics.

References

Aitken, R. (2009) *Antibody Phage Display: Methods and Protocols.* Springer-Verlag, New York, NY.

Andris-Widhopf, J., Rader, C., Steinberger, P., Fuller, R. and Barbas, C.F. III (2000) Methods for the generation of chicken monoclonal antibody fragments by phage display. *Journal of Immunological Methods* 242, 159–181.

Andris-Widhopf, J., Steinberger, P., Fuller, R., Rader, C. and Barbas, C.F. (2001) Generation of antibody libraries: PCR amplification and assembly of light- and heavy-chain coding sequences. In: Barbas, C.F., Burton, D.R., Scott, J.K. and Silverman, G.J. (eds) *Phage Display: a Laboratory Manual.* Cold Spring Harbor Press, Cold Spring Harbor, NY, pp. 9.1–9.113.

Arbabi Ghahroudi, M., Desmyter, A., Wyns, L., Hamers, R. and Muyldermans, S. (1997) Selection and identification of single domain antibody fragments from camel heavy-chain antibodies. *FEBS Letters* 414, 521–526.

Baker, K.P., Edwards, B.M., Main, S.H., Choi, G.H., Wager, R.E., Halpern, W.G., Lappin, P.B., Riccobene, T., Abramian, D., Sekut, L., Sturm, B., Poortman, C., Minter, R.R., Dobson, C.L., Williams, E., Carmen, S., Smith, R., Roschke, V., Hilbert, D.M., Vaughan, T.J. and Albert, V.R. (2003) Generation and characterization of LymphoStat-B, a human monoclonal antibody that antagonizes the bioactivities of B lymphocyte stimulator. *Arthritis and Rheumatism* 48, 3253–3265.

Barbas, C., Burton, D., Silverman, G. and Scott, J. (eds) (2000) *Phage Display of Proteins and Peptides: a Laboratory Manual.* Cold Spring Harbor Press, Cold Spring Harbor, NY.

Barbas, C.F. III, Kang, A.S., Lerner, R.A. and Benkovic, S.J. (1991) Assembly of combinatorial antibody libraries on phage surfaces: the gene III site. *Proceedings of the National Academy of Sciences USA* 88, 7978–7982.

Barbas, C.F., Collet, T.A., Amberg, W., Roben, P., Binley, J.M., Hoekstra, D., Cababa, D., Jones, T.M., Williamson, R.A., Pilkington, G.R., Haigwood, N.L., Cabezas, E., Satterthwait, A.C., Sanz, I. and Burton, D.R. (1993) Molecular profile of an antibody response to HIV-1 as probed by combinatorial libraries. *Journal of Molecular Biology* 230, 812–823.

Bradbury, A.R.M., Sidhu, S., Dubel, S. and Mccafferty, J. (2011) Beyond natural antibodies: the power of *in vitro* display technologies. *Nature Biotechnology* 29, 245–254.

Braganza, A., Wallace, K., Pell, L., Parrish, C.R., Siegel, D.L. and Mason, N.J. (2011) Generation and validation of canine single chain variable fragment phage display libraries. *Veterinary Immunology and Immunopathology* 139, 27–40.

Bratkovic, T. (2010) Progress in phage display: evolution of the technique and its application. *Cellular and Molecular Life Sciences* 67, 749–767.

Brissette, R. and Goldstein, N.I. (2007) The use of phage display peptide libraries for basic and translational research. *Methods in Molecular Biology* 383, 203–213.

Chang, T.Y. and Siegel, D.L. (1998) Genetic and immunological properties of phage-displayed human anti-Rh(D) antibodies: implications for Rh(D) epitope topology. *Blood* 91, 3066–3078.

Clackson, T. and Lowman, H. (eds.) (2004) *Phage Display: a Practical Approach.* Oxford University Press, New York, NY.

de Bono, B., Madera, M. and Chothia, C. (2004) V_H gene segments in the mouse and human genomes. *Journal of Molecular Biology* 342, 131–143.

de Kruif, J., Boel, E. and Logtenberg, T. (1995a) Selection and application of human single chain Fv antibody fragments from a semi-synthetic phage antibody display library with designed CDR3 regions. *Journal of Molecular Biology* 248, 97–105.

de Kruif, J., Terstappen, L., Boel, E. and Logtenberg, T. (1995b) Rapid selection of cell subpopulation-specific human monoclonal antibodies from a synthetic phage antibody library. *Proceedings of the National Academy of Sciences USA* 92, 3938–3942.

Deutscher, S.L. (2010) Phage display in molecular imaging and diagnosis of cancer. *Chemical Reviews* 110, 3196–3211.

Dolk, E., van der Vaart, M., Lutje Hulsik, D., Vriend, G., de Haard, H., Spinelli, S., Cambillau, C., Frenken, L. and Verrips, T. (2005) Isolation of llama antibody fragments for prevention of dandruff by phage display in shampoo. *Applied and Environmental Microbiology* 71, 442–450.

Finlay, W.J.J., Bloom, L. and Cunningham, O. (2011) Phage display: a powerful technology for the generation of high specificity affinity reagents from alternative immune sources. *Methods in Molecular Biology* 681, 87–101.

Graus, Y.F., de Baets, M.H., Parren, P.W., Berrih-Aknin, S., Wokke, J., van Breda Vriesman, P.J. and Burton, D.R. (1997) Human anti-nicotinic acetylcholine receptor recombinant Fab fragments isolated from thymus-derived phage display libraries from myasthenia gravis patients reflect predominant specificities in serum and block the action of pathogenic serum antibodies. *Journal of Immunology* 158, 1919–1929.

Green, L.L., Hardy, M.C., Maynard-Currie, C.E.,

Tsuda, H., Louie, D.M., Mendez, M.J., Abderrahim, H., Noguchi, M., Smith, D.H. and Zeng, Y. (1994) Antigen-specific human monoclonal antibodies from mice engineered with human Ig heavy and light chain YACs. *Nature Genetics* 7, 13–21.

Griffiths, A.D. and Duncan, A.R. (1998) Strategies for selection of antibodies by phage display. *Current Opinion in Biotechnology* 9, 102–108.

Halpern, W.G., Lappin, P., Zanardi, T., Cai, W., Corcoran, M., Zhong, J. and Baker, K.P. (2006) Chronic administration of belimumab, a BLyS antagonist, decreases tissue and peripheral blood B-lymphocyte populations in cynomolgus monkeys: pharmacokinetic, pharmacodynamic, and toxicologic effects. *Toxicological Sciences* 91, 586–599.

Harmsen, M.M. and de Haard, H.J. (2007) Properties, production, and applications of camelid single-domain antibody fragments. *Applied Microbiology and Biotechnology* 77, 13–22.

Hawkins, R.E., Russell, S.J. and Winter, G. (1992) Selection of phage antibodies by binding affinity mimicking affinity maturation. *Journal of Molecular Biology* 226, 889–896.

Hemminga, M.A., Vos, W.L., Nazarov, P.V., Koehorst, R.B.M., Wolfs, C.J.A.M., Spruijt, R.B. and Stopar, D. (2010) Viruses: incredible nanomachines. New advances with filamentous phages. *European Biophysics Journal* 39, 541–550.

Hoogenboom, H.R. (2005) Selecting and screening recombinant antibody libraries. *Nature Biotechnology* 23, 1105–1116.

Hoogenboom, H.R. and Winter, G. (1992) By-passing immunisation: human antibodies from synthetic repertoires of germline V_H gene segments rearranged *in vitro*. *Journal of Molecular Biology* 227, 381–388.

Huls, G.A., Heijnen, I.A., Cuomo, M.E., Koningsberger, J.C., Wiegman, L., Boel, E., van der Vuurst de Vries, A.R., Loyson, S. A., Helfrich, W., van Berge Henegouwen, G.P., van Meijer, M., de Kruif, J. and Logtenberg, T. (1999) A recombinant, fully human monoclonal antibody with antitumor activity constructed from phage-displayed antibody fragments. *Nature Biotechnology* 17, 276–281.

Jaye, D.L., Nolte, F.S., Mazzucchelli, L., Geigerman, C., Akyildiz, A. and Parkos, C.A. (2003) Use of real-time polymerase chain reaction to identify cell- and tissue-type-selective peptides by phage display. *American Journal of Pathology* 162, 1419–1429.

Jespers, L.S., Roberts, A., Mahler, S.M., Winter, G. and Hoogenboom, H.R. (1994) Guiding the selection of human antibodies from phage display repertoires to a single epitope of an antigen. *Bio/Technology* 12, 899–903.

Knappik, A., Ge, L., Honegger, A., Pack, P., Fischer, M., Wellnhofer, G., Hoess, A., Wolle, J., Pluckthun, A. and Virnekas, B. (2000) Fully synthetic human combinatorial antibody libraries (HuCAL) based on modular consensus frameworks and CDRs randomized with trinucleotides. *Journal of Molecular Biology* 296, 57–86.

Kohler, G. and Milstein, C. (1975) Continuous cultures of fused cells secreting antibody of predefined specificity. *Nature* 256, 495–497.

Krebs, B., Rauchenberger, R., Reiffert, S., Rothe, C., Tesar, M., Thomassen, E., Cao, M., Dreier, T., Fischer, D., Hoss, A., Inge, L., Knappik, A., Marget, M., Pack, P., Meng, X.Q., Schier, R., Sohlemann, P., Winter, J., Wolle, J. and Kretzschmar, T. (2001) High-throughput generation and engineering of recombinant human antibodies. *Journal of Immunological Methods* 254, 67–84.

Ladenson, R.C., Crimmins, D.L., Landt, Y. and Ladenson, J.H. (2006) Isolation and characterization of a thermally stable recombinant anti-caffeine heavy-chain antibody fragment. *Analytical Chemistry* 78, 4501–4508.

Lekkerkerker, A. and Logtenberg, T. (1999) Phage antibodies against human dendritic cell subpopulations obtained by flow cytometry-based selection on freshly isolated cells. *Journal of Immunological Methods* 231, 53–63.

Liu, B. and Marks, J.D. (2000) Applying phage antibodies to proteomics: selecting single chain Fv antibodies to antigens blotted on nitrocellulose. *Analytical Biochemistry* 286, 119–128.

Lloyd, C., Lowe, D., Edwards, B., Welsh, F., Dilks, T., Hardman, C. and Vaughan, T. (2009) Modelling the human immune response: performance of a 10^{11} human antibody repertoire against a broad panel of therapeutically relevant antigens. *Protein Engineering, Design and Selection* 22, 159–168.

Lonberg, N. (2005) Human antibodies from transgenic animals. *Nature Biotechnology* 23, 1117–1125.

Lonberg, N. (2008) Fully human antibodies from transgenic mouse and phage display platforms. *Current Opinion in Immunology* 20, 450–459.

Lonberg, N., Taylor, L.D., Harding, F.A., Trounstine, M., Higgins, K.M., Schramm, S.R., Kuo, C.C., Mashayekh, R., Wymore, K. and McCabe, J.G. (1994) Antigen-specific human antibodies from mice comprising four distinct genetic modifications. *Nature* 368, 856–859.

Marks, J.D., Hoogenboom, H.R., Bonnert, T.P.,

McCafferty, J., Griffiths, A.D. and Winter, G. (1991) By-passing immunization. Human antibodies from V-gene libraries displayed on phage. *Journal of Molecular Biology* 222, 581–597.

Martineau, P. (2002) Error-prone polymerase chain reaction for modification of scFvs. *Methods in Molecular Biology* 178, 287–294.

McCormack, W.T., Tjoelker, L.W. and Thompson, C.B. (1993) Immunoglobulin gene diversification by gene conversion. *Progress in Nucleic Acid Research and Molecular Biology* 45, 27–45.

McIntosh, R.S., Asghar, M.S., Watson, P.F., Kemp, E.H. and Weetman, A.P. (1996) Cloning and analysis of IgGκ and IgGλ anti-thyroglobulin autoantibodies from a patient with Hashimoto's thyroiditis: evidence for *in vivo* antigen-driven repertoire selection. *Journal of Immunology* 157, 927–935.

McWhirter, J., Kretz-Rommel, A., Saven, A., Maruyama, T., Potter, K., Mockridge, C., Ravey, E., Qin, F. and Bowdish, K. (2006) Antibodies selected from combinatorial libraries block a tumor antigen that plays a key role in immunomodulation. *Proceedings of the National Academy of Sciences USA* 103, 1041–1046.

Nelson, A.L., Dhimolea, E. and Reichert, J.M. (2010) Development trends for human monoclonal antibody therapeutics. *Nature Reviews Drug Discovery* 9, 767–774.

Osbourn, J., Groves, M. and Vaughan, T. (2005) From rodent reagents to human therapeutics using antibody guided selection. *Methods* 36, 61–68.

Pande, J., Szewczyk, M.M. and Grover, A.K. (2010) Phage display: concept, innovations, applications and future. *Biotechnology Advances* 28, 849–858.

Payne, A.S., Ishii, K., Kacir, S., Lin, C., Li, H., Hanakawa, Y., Tsunoda, K., Amagai, M., Stanley, J.R. and Siegel, D.L. (2005) Genetic and functional characterization of human pemphigus vulgaris monoclonal autoantibodies isolated by phage display. *Journal of Clinical Investigation* 115, 888–899.

Popkov, M., Mage, R.G., Alexander, C.B., Thundivalappil, S., Barbas, C.F. III and Rader, C. (2003) Rabbit immune repertoires as sources for therapeutic monoclonal antibodies: the impact of kappa allotype-correlated variation in cysteine content on antibody libraries selected by phage display. *Journal of Molecular Biology* 325, 325–335.

Roark, J.H., Bussel, J.B., Cines, D.B. and Siegel, D.L. (2002) Genetic analysis of autoantibodies in ITP reveals evidence of clonal expansion and somatic mutation. *Blood* 100, 1388–1398.

Sarikaya, M., Tamerler, C., Jen, A.K.Y., Schulten, K. and Baneyx, F. (2003) Molecular biomimetics: nanotechnology through biology. *Nature Materials* 2, 577–585.

Scott, J.K. and Barbas, C.F. (2001) Phage-display vectors. In: Barbas, C.F., Burton, D.R., Silverman, G.J. and Scott, J.K. (eds) *Phage Display of Proteins and Peptides: a Laboratory Manual.* Cold Spring Harbor Press, Cold Spring Harbor, NY, pp. 2.1–2.19.

Sergeeva, A., Kolonin, M.G., Molldrem, J.J., Pasqualini, R. and Arap, W. (2006) Display technologies: application for the discovery of drug and gene delivery agents. *Advanced Drug Delivery Reviews* 58, 1622–1654.

Sidhu, S.S. (2005) *Phage Display in Biotechnology and Drug Discovery.* CRC Press, Boca Raton, FL.

Siegel, D.L. (2001) Cell-surface panning of phage-display libraries. In: Barbas, C.F., Burton, D.R., Silverman, G.J. and Scott, J.K. (eds) *Phage Display of Proteins and Peptides: a Laboratory Manual.* Cold Spring Harbor Press, Cold Spring Harbor, NY, pp 23.1–23.32.

Siegel, D.L. (2005) Phage-display tools for blood typing. *Current Hematology Reports* 4, 459–464.

Siegel, D.L. and Silberstein, L.E. (1994) Expression and characterization of recombinant anti-Rh(D) antibodies on filamentous phage: a model system for isolating human red blood cell antibodies by repertoire cloning. *Blood* 83, 2334–2344.

Siegel, D.L., Chang, T.Y., Russell, S.L. and Bunya, V.Y. (1997) Isolation of cell surface-specific human monoclonal antibodies using phage display and magnetically-activated cell sorting: applications in immunohematology. *Journal of Immunological Methods* 206, 73–85.

Siegel, D.L., Reid, M.E., Lee, H. and Blancher, A. (1999) Production of large repertories of macaque mAbs to human RBCs using phage display. *Transfusion* 39, 92S.

Smith, G.P. (1985) Filamentous fusion phage: novel expression vectors that display cloned antigens on the virion surface. *Science* 228, 1315–1317.

Souza, G.R., Christianson, D.R., Staquicini, F.I., Ozawa, M.G., Snyder, E.Y., Sidman, R.L., Miller, J.H., Arap, W. and Pasqualini, R. (2006) Networks of gold nanoparticles and bacteriophage as biological sensors and cell-targeting agents. *Proceedings of the National Academy of Sciences USA* 103, 1215–1220.

Souza, G.R., Yonel-Gumruk, E., Fan, D., Easley, J., Rangel, R., Guzman-Rojas, L., Miller, J. H., Arap, W. and Pasqualini, R. (2008) Bottom-up assembly of hydrogels from bacteriophage and

Au nanoparticles: the effect of cis- and trans-acting factors. *PloS One* 3, e2242.

Souza, G.R., Staquicini, F.I., Christianson, D.R., Ozawa, M.G., Miller, J.H., Pasqualini, R. and Arap, W. (2010) Combinatorial targeting and nanotechnology applications. *Biomedical Microdevices* 12, 597–606.

Steinberger, P., Rader, C. and Barbas, C.F. (2001) Analysis of selected antibodies. In: Barbas, C.F., Burton, D.R., Scott, J.K. and Silverman, G.J. (eds) *Phage Display: a Laboratory Manual.* Cold Spring Harbor Press, Cold Spring Harbor, NY, pp. 11.1–11.24.

Storz, U. (2010) IP issues in the therapeutic antibody industry. In: Kontermann, R. and Dubel, S. (eds) *Antibody Enginneering.* Springer-Verlag, New York, NY, pp. 517–581.

Tanaka, T., Ito, T., Furuta, M., Eguchi, C., Toda, H., Wakabayashi-Takai, E. and Kaneko, K. (2002) *In situ* phage screening. A method for identification of subnanogram tissue components *in situ*. *Journal of Biological Chemistry* 277, 30382–30387.

Thie, H., Meyer, T., Schirrmann, T., Hust, M. and Dubel, S. (2008) Phage display derived therapeutic antibodies. *Current Pharmaceutical Biotechnology* 9, 439–446.

Thie, H., Voedisch, B., Dubel, S., Hust, M. and Schirrmann, T. (2009) Affinity maturation by phage display. *Methods in Molecular Biology* 525, 309–322.

van de Putte, L.B.A., Atkins, C., Malaise, M., Sany, J., Russell, A.S., van Riel, P.L.C.M., Settas, L., Bijlsma, J.W., Todesco, S., Dougados, M., Nash, P., Emery, P., Walter, N., Kaul, M., Fischkoff, S. and Kupper, H. (2004) Efficacy and safety of adalimumab as monotherapy in patients with rheumatoid arthritis for whom previous disease modifying antirheumatic drug treatment has failed. *Annals of the Rheumatic Diseases* 63, 508–516.

van der Vuurst de Vries, A. and Logtenberg, T. (1999) Dissecting the human peripheral B-cell compartment with phage display-derived antibodies. *Immunology* 98, 55–62.

van Ewijk, W., de Kruif, J., Germeraad, W.T., Berendes, P., Ropke, C., Platenburg, P.P. and Logtenberg, T. (1997) Subtractive isolation of phage-displayed single-chain antibodies to thymic stromal cells by using intact thymic fragments. *Proceedings of the National Academy of Sciences USA* 94, 3903–3908.

Vaughan, T.J., Williams, A.J., Pritchard, K., Osbourn, J.K., Pope, A.R., Earnshaw, J.C., Mccafferty, J., Hodits, R.A., Wilton, J. and Johnson, K.S. (1996) Human antibodies with sub-nanomolar affinities isolated from a large non-immunized phage display library. *Nature Biotechnology* 14, 309–314.

Webster, R. (2001) Filamentous phage biology. In: Barbas, C.F., Burton, D.R., Silverman, G.J. and Scott, J.K. (ed.) *Phage Display of Proteins and Peptides: a Laboratory Manual.* Cold Spring Harbor Press, Cold Spring Harbor, NY, pp. 1.1–1.37.

Weiner, L.M. (2006) Fully human therapeutic monoclonal antibodies. *Journal of Immunotherapy* 29, 1–9.

Wild, M.A., Xin, H., Maruyama, T., Nolan, M.J., Calveley, P.M., Malone, J.D., Wallace, M.R. and Bowdish, K.S. (2003) Human antibodies from immunized donors are protective against anthrax toxin *in vivo*. *Nature Biotechnology* 21, 1305–1306.

Winter, G. and Milstein, C. (1991) Man-made antibodies. *Nature* 349, 293–299.

9 Phages and Their Hosts: a Web of Interactions – Applications to Drug Design

Jeroen Wagemans[1] and Rob Lavigne[1]
[1]Department of Biosystems, Katholieke Universiteit Leuven

The development of antibiotic resistance in virtually all clinically important pathogens is driving the search for alternatives to conventional antibiotics. Currently, drug-resistant bacteria for almost every available antibiotic class are seen in hospital and community settings (Clatworthy *et al.*, 2007; Falconer and Brown, 2009; Fernebro, 2011; see also Kuhl *et al.*, Chapter 3, this volume). The growing problem of antibiotic resistance has been exacerbated by the use of new drugs that are merely variants of older, overused broad-spectrum antibiotics. While it is naïve to expect to restrain the spread of resistance without controlling antibacterial usage, the desperate need for drugs with novel modes of action has been recognized by health organizations, industry and academia (Lock and Harry, 2008). In general, the molecular target specificity of currently used antimicrobials is very limited, as most interfere with the biosynthesis of DNA, proteins or cell walls (Brown, 2006; Zlitni and Brown, 2009). In fact, commercially available antimicrobial agents target fewer than 20 families of proteins (Lange *et al.*, 2007; Silver, 2011), whereas genome-scale mutagenesis assays suggest that most bacteria encode at least ten times more essential proteins, indispensable for *in vivo* growth and crucial to pathogenesis (Akerley *et al.*, 2002; Forsyth *et al.*, 2002; Mecsas, 2002; Thanassi *et al.*, 2002; Kobayashi *et al.*, 2003; Knuth *et al.*, 2004; Baba *et al.*, 2008). Hence, it would appear that there is a plethora of unexploited targets for antibacterial drug discovery. Success in exploiting these targets, however, relies on selecting those possessing the highest potential for commercial success (Brown, 2004).

Large pharmaceutical companies have put great efforts into screening small-molecule chemical libraries against novel essential protein targets that are widely conserved among clinically relevant bacterial species. There is increasing evidence, however, that this high-throughput screening has failed to produce the influx into the antibacterial discovery pipeline that was hoped for, in contrast to screening campaigns in other therapeutic areas, which ordinarily yield multiple leads for each screen. Out of 70 screens conducted by GlaxoSmithKline from 1995 to 2001, for example, only five compounds were identified that were able to inhibit target activity at a low micromolar concentration and that had also antibacterial activity (So *et al.*, 2011). Two of the main problems with antibacterial screens are the need for a proper target selection, targets that are not prone to rapid resistance development and the limitation of chemical diversity in currently used libraries. Marketed antibacterials apparently do not generally follow Lipinski's rule of five, which defines drug-likeness of chemical compounds by certain solubility and permeability characteristics (Lipinski *et al.*, 2001). Nevertheless, most modern libraries are based on this rule of thumb. These findings and the limited profit and extensive timelines associated with drug development make antibacterial high-throughput screenings less attractive for many large pharmaceutical companies, although advances in small-molecule libraries may make these screenings more cost-effective and fruitful (Payne *et al.*, 2007; Gwynn *et al.*, 2010; Silver, 2011; So *et al.*, 2011).

Proper target-molecule selection in high-throughput screening can be enhanced using the knowledge we have gained from the study of bacterial natural enemies, the bacteriophages. Drug resistance has been an identified issue for over 50 years (Finland *et al.*, 1959), and ever since its emergence, there has been particular interest in re-examining bacteriophages, which have shown great promise in combating pathogenic bacteria. Besides using phages themselves (see Olszowska-Zaremba *et al.*, Loc-Carrillo *et al.*, Burrowes and Harper, and Abedon, Chapters 12, 13, 14 and 17, this volume) or phage antibacterial enzymes (see Shen *et al.*, Chapter 15, this volume) as antibacterial treatments, their non-structural metabolic proteins could also be used to validate new bacterial targets and even to develop novel chemical antibacterials (Liu *et al.*, 2004), as discussed further in this chapter.

After injection of phage DNA into the host bacterium, intracellular phage development entails a number of tightly programmed steps (see Cox, Chapter 10, this volume, for an overview). The transition to a phage-oriented metabolism early in infection is crucial (Miller *et al.*, 2003; Guttman *et al.*, 2005; Roucourt and Lavigne, 2009). Bacterial viruses consequently have evolved novel proteins that either inhibit or co-opt critical bacterial metabolic processes to ensure proper viral development. Intriguingly, many phage genes can be deleted under standard laboratory conditions without seriously affecting phage production, suggesting that the corresponding phage proteins are only necessary under specific environmental conditions, for infecting specific hosts or have functions that are redundant with those of other proteins expressed by the same phage (Miller *et al.*, 2003).

This chapter specifically addresses interactions between phage-encoded and bacterial proteins within the phage-infected cell, hereafter referred to as (bacterio)phage–host interactions. Although important to different stages of infection, other phage–host interactions, such as phage–receptor interactions or protein–DNA interactions, are not discussed. From known bacteriophage–host interactions, crucial examples are described, with emphasis placed on those that could readily be exploited in drug design. A general phage genomics-based strategy for drug discovery is also discussed.

Phages and Their Host: an Intriguing Web of Interactions

Known bacteriophage–host interactions demonstrate that phages influence their hosts with regard to different metabolic aspects (Sau *et al.*, 2008; Roucourt and Lavigne, 2009). Besides protecting the bacteriophage from bacterial defence mechanisms, such as restriction enzymes and proteases, several crucial interactions could be utilized in drug discovery. For example, the replication and transcription apparatus is a common target among phages. Furthermore, there are also indications that phages directly affect the translation machinery and maybe even the global host metabolism. The following sections describe in more detail these different phage strategies and their molecular background to 'hijack' their host-cell macromolecular machinery and how they can be applied in the drug-design field.

Phages taking over the host transcription apparatus

All bacterial RNA polymerases are closely related and consist of five subunits, designated α, β, β' and σ, with α present in two identical copies. These subunits interact to form the active enzyme, called the RNA polymerase holoenzyme, but the σ specificity factor is not as tightly bound as the others, thus easily dissociating. RNA polymerase minus this σ factor is described as the core enzyme, consisting of $\alpha_2\beta\beta'$. This core enzyme alone can catalyse the formation of RNA, but a σ subunit is required for recognition of the appropriate promoter sites on the DNA and the initiation of RNA synthesis.

The RNA polymerase holoenzyme distinguishes between different types of promoters, which consist of various consensus sequences, on the basis of which type of σ factor is attached. The most common promoters are those recognized by the RNA polymerase-σ^{70} holoenzyme. These promoter sequences have two important regions: a short AT-rich region about 10 bp upstream of the transcription start site, known as the –10 sequence, and a region about 35 bp upstream of the start site, called the –35 sequence. The σ^{70} factor must bind to both sequences to initiate transcription. Apart from σ^{70}, *Escherichia coli* uses several other specificity factors for special circumstances: for example, σ^{54} exerts its function in nitrogen metabolism-related gene transcription, while σ^{32} is required for expression of heat-shock genes (Snyder and Champness, 2003). Bacteriophages can also encode their own σ factors to co-opt the host RNA polymerase to express phage genes. This is most common for genes expressed later in the infection cycle – the 'middle' and 'late' phage genes. Conversely, 'early' phage genes, transcribed immediately after phage genome injection, have promoters that are more like bacterial promoters (Kutter *et al.*, 2005).

To begin transcription, the RNA polymerase binds to the promoter (forming the 'closed complex') and separates the DNA strands, exposing the bases ('open complex'). Once the DNA strands have been pulled apart, RNA synthesis begins, complementary to the transcribed strand. As the RNA chain begins to grow, the holoenzyme releases its σ factor, and the core enzyme continues RNA polymerization in the 5′-to-3′ direction, until it encounters a transcription termination site in the DNA (Snyder and Champness, 2003; Madigan *et al.*, 2006).

As they tend to favour their own transcription and reduce that of their host, most phages exploit the bacterial transcription apparatus at some stage during the infection cycle. Consequently, the RNA polymerase complex is a common target for early phage proteins. Phages especially can shut down host transcription by: (i) modifying the bacterial RNA polymerase, thus recruiting it for their own transcription, as in the case of T4; or (ii) inhibiting the host RNA polymerase and taking over transcription with a phage-encoded RNA polymerase, as in the case of T7 (Nechaev and Severinov, 2003; Sau *et al.*, 2008). Table 9.1 lists known phage–host interactions affecting the transcription apparatus. For each described bacteriophage–host interaction, the bacteriophage protein and the phage it originates from as well as the host protein are shown. If known, the effect of the interaction is summarized. Interactions between coliphages T4 and T7 and the *E. coli* RNA polymerase have been reviewed widely elsewhere (Roucourt and Lavigne, 2009), and a few additional interesting examples that could be useful for drug discovery are discussed below.

As a first example, *Thermus thermophilus* phage P23-45 fully depends on the host transcription machinery for transcription of its middle and late genes. Two P23-45 proteins, gp76 and gp39, both bind the host RNA polymerase core enzyme as well as the RNA polymerase-σ^{A} holoenzyme independently and at separate sites. As a result, both gp76 and gp39 inhibit *in vitro* transcription of the bacterial –35/–10 promoters. Moreover, the inhibitory effect is even greater when both phage proteins are present, reducing host transcription to a minimum. In contrast, –10 promoters, predominantly located in front of the middle and late phage P23-45 genes, are not inhibited (Berdygulova *et al.*, 2011). Although this bacterium does not cause any infections, this inhibition mechanism by phage P23-45 might also be used in other phages infecting human pathogens.

In contrast to *T. thermophilus*, the rice pathogen *Xanthomonas oryzae* siphovirus Xp10 was shown to encode its own RNA polymerase (Yuzenkova *et al.*, 2003). After transcription of the early genes by the host RNA polymerase, Xp10 transcription regulator p7 stimulates late viral gene

Table 9.1. Phage–host interactions affecting the transcription machinery.

Phage	Phage protein	Host protein	Effect	Reference
G1	gp67	σ^{SA} factor[a]	Anti-σ factor	Dehbi *et al.* (2009)
λ	N	RNA polymerase core, NusA, NusG	Anti-termination, delayed early transcription	Mason and Greenblatt (1991)
	Q	RNA polymerase holoenzyme (core + σ^{70})	Anti-termination, late transcription	Marr *et al.* (2001)
N4	SSB	RNA polymerase β' subunit	Activation of late transcription	Miller *et al.* (1997)
T7	gp0.7 C-term	RNA polymerase β' subunit	Host transcription shutoff	Michalewicz and Nicholson (1992)
	gp0.7 N-term	RNA polymerase β and β' subunit	Efficiency of termination	Severinova and Severinov (2006)
		RNase III	Processing of T7 mRNA	Mayer and Schweiger (1983)
		RnaseE and RhlB	Protection of T7 mRNA from degradation	Marchand *et al.* (2001)
	gp2	RNA polymerase β' subunit	Inhibition of transcription initiation	Nechaev and Severinov (1999)
T4	Alt	One RNA polymerase α subunit	Preferential expression of T4 early genes	Sommer *et al.* (2000)
	Alc	RNA polymerase β subunit (postulated)	Host transcription shutoff	Kashlev *et al.* (1993)
	ModA	Both RNA polymerase α subunits	Lower expression of T4 early and host genes	Skorko *et al.* (1977)
	AsiA	σ^{70} factor	Inhibits recognition of σ^{70} promoters	Ouhammouch *et al.* (1995)
	MotA	σ^{70} factor	Recognition of T4 middle promoters	Ouhammouch *et al.* (1995)
	gp55	RNA polymerase core enzyme	σ factor for expression from late promoters	Kassavetis *et al.* (1983)
	gp33	RNA polymerase core enzyme	Helper of gp55	Kassavetis *et al.* (1983)
	RpbA	RNA polymerase core enzyme	Favours expression of T4 late genes (postulated)	Kolesky *et al.* (1999)
	Mrh	σ^{32} factor	Decoy of σ^{32} from RNA polymerase core (postulated)	Mosig *et al.* (1998)
	Srh	RNA polymerase core enzyme (postulated)	Decoy of σ^{32} from RNA polymerase core (postulated)	Mosig *et al.* (1998)
	Srd	RNA polymerase core enzyme (postulated)	Decoy of σ^{70}/σ^{38} from RNA polymerase core (postulated)	Mosig *et al.* (1998)
P2	Org	RNA polymerase α subunit	Initiation of late transcription	Wood *et al.* (1997)
P4	Psu	Rho factor	Inhibition of Rho-dependent termination	Pani *et al.* (2009)
P23-45	gp39	RNA polymerase holoenzyme (core + σ^{A})	Shutoff of host transcription	Berdygulova *et al.* (2011)
	gp76	RNA polymerase holoenzyme (core + σ^{A})	Shutoff of host transcription	Berdygulova *et al.* (2011)
phiEco32	gp36	RNA polymerase core enzyme	σ factor (recognition of phage promoters)	Savalia *et al.* (2008)
	gp79	RNA polymerase core enzyme	Inhibition of σ^{70}-dependent transcription	Savalia *et al.* (2008)
SPO1	gp44 (E3)	RNA polymerase β subunit	Inhibition of the host RNA polymerase	Wei and Stewart (1995)
	gp28	RNA polymerase core enzyme	σ factor for middle transcription	Losick and Pero (1981)
	gp33	RNA polymerase core enzyme	σ factor for late transcription	Losick and Pero (1981)
	gp34	RNA polymerase core enzyme	Helper protein for late transcription	Losick and Pero (1981)
Unknown	gp67	Component of RNA polymerase	Shutoff of host transcription	Liu *et al.* (2004)
Xp10	p7	RNA polymerase β' subunit	Inhibition and anti-termination of transcription	Nechaev *et al.* (2002)
YuA	Unknown	σ^{54} factor	Expression of late genes	Nechaev and Severinov (2008)

[a]SA = *Staphylococcus aureus*.

transcription in two ways: p7 not only blocks transcription from most bacterial and early viral promotors by binding the β′ subunit of the RNA polymerase, it also stimulates late gene transcription by inhibiting termination of host RNA polymerase transcription from a p7-resistant promoter (Nechaev *et al.*, 2002; Djordjevic *et al.*, 2006; Yuzenkova *et al.*, 2008).

One last example of inhibition of host transcription through direct interaction between a phage-encoded protein and the RNA polymerase complex is the anti-staphylococcal polypeptide G1ORF67 (Dehbi *et al.*, 2009). This phage G1 protein binds to the primary σ factor of *Staphylococcus aureus*, σ^{SA}. As a result, transcription of host house-keeping genes is shut down during the phage's exponential growth phase as transcription of these genes is σ^{SA} dependent. Dehbi *et al.* (2009) concluded that G1ORF67 is a phage-encoded anti-σ factor, given both its interaction with σ^{SA} and its ability to inhibit σ^{SA} function. As inhibiting transcription unavoidably leads to cell death, non-proteinaceous chemical analogues that mimic these phage proteins and therefore their inhibitory effect could serve as new antibacterial compounds.

Phages hijacking host genome replication

Although less extensively documented, bacteriophages also directly target their host's replication system. Table 9.2 summarizes phage–host interactions influencing host replication, of which some examples will be discussed below.

Whereas most obligately lytic phages, such as T4, encode most proteins for their own genome replication (Mueser *et al.*, 2010), temperate phages and small lytic phages often depend heavily upon the host replisome (Friedman *et al.*, 1984). To prioritize their own DNA biosynthesis while abolishing host genome replication, lytic phages usually block several host DNA metabolism proteins through direct bacteriophage–host interactions. In addition, the host nucleoid is often degraded to provide precursors for phage genome replication. Temperate phages, on the other hand, redirect the host replisome through phage–host interactions, thereby recruiting it for the replication of the phage genome (Guttman *et al.*, 2005). These differences between lytic and temperate phages in completely shutting down or instead redirecting the host replication machinery, respectively, demonstrate the possibility of identifying a wide diversity of susceptible drug targets within their host and different inhibition mechanisms.

Liu *et al.* (2004) showed that the DNA replication pathway of bacteria is indeed very vulnerable to inhibition. They demonstrated that among 26 diverse *Staphylococcus aureus* phages, a total of seven unrelated phage polypeptide families target and inhibit four proteins of the replication machinery of the host cell. The DNA primase DnaG, which is necessary for replication initiation, and the uncharacterized replication protein PT-R14 are targeted by only one phage protein each, while DnaI, which is the helicase loader, is directed by two phage proteins from two different phages. Moreover, three distinct phage polypeptides from three different phages target DnaN, the sliding clamp, which tethers the DNA polymerase to its template after the clamp loading process and therefore confers processivity on that enzyme. Except for phage Twort gp168 and G1 gp240, the molecular mechanisms of these interactions have not been studied. These Twort and G1 proteins have a bactericidal effect by directly blocking the DNA sliding clamp and therefore may be ideal candidates for the development of mimicking drugs (Belley *et al.*, 2006).

Both coliphages λ and P2 encode polypeptides, proteins P and B, respectively, which interact with the DNA helicase DnaB, which unwinds the DNA prior to replication, of their host (Klein *et al.*, 1980; Odegrip *et al.*, 2000). As a consequence, phage λ protein P strongly inhibits the ATPase activity of the helicase as well as its ability to assist the *E. coli* primase, thereby blocking host DNA replication. Moreover, both protein P and phage P2 protein B serve as DNA helicase loaders, directing the helicase to the phage origin of replication for commencing replication (Mallory *et al.*, 1990). The specific actions of both proteins could be mimicked by chemical analogues.

Table 9.2. Phage–host interactions involved in host replication shutoff.

Phage	Phage protein	Host protein	Effect	Reference
77	gp104	DnaI (helicase loader)	Shutoff of host replication	Liu *et al.* (2004)
G1	gp240	DnaN (DNA polymerase III β subunit)	Shutoff of host replication	Liu *et al.* (2004)
λ	P	DnaB (DNA helicase)	Initiation of λ genome replication	Klein *et al.* (1980)
	O	RNA polymerase β subunit	Initiation of λ DNA replication	Szambowska *et al.* (2011)
	Gam	RecBCD	Inhibits nuclease and helicase activity of RecBCD	Court *et al.* (2007)
	CII	DnaB (DNA helicase) and DnaC (helicase loader)	Inhibition host replication	Sau *et al.* (2008)
N4	gp8	HolA (DNA polymerase III δ subunit)	Shutoff host replication	Yano and Rothman-Denes (2011)
T7	gp5	Thioredoxine	Processivity factor of the DNA polymerase	Tabor *et al.* (1987)
	gp5.9	RecBCD nuclease	Inhibits nuclease and ATPase activity	Molineux (2005)
P2	B	DnaB (DNA helicase)	Initiation of P2 genome replication	Odegrip *et al.* (2000)
PA16	gp106	DnaN (DNA polymerase III β subunit)	Shutoff of host replication (postulated)	Belley *et al.* (2006)
Twort	gp168	DnaN (DNA polymerase III β subunit)	Shutoff of host replication	Liu *et al.* (2004)
Unknown	gp016	DnaI (helicase loader)	Shutoff of host replication	Liu *et al.* (2004)
Unknown	gp025	DnaN (DNA polymerase III β subunit)	Shutoff of host replication	Liu *et al.* (2004)
Unknown	gp078	DnaG (DNA primase)	Shutoff of host replication	Liu *et al.* (2004)
Unknown	gp140	PT-R14 (involved in replication)	Shutoff of host replication	Liu *et al.* (2004)

Recently, Yano and Rothman-Denes (2011) reported a different mechanism of arrest of host DNA replication, which favours phage genome replication. *E. coli* phage N4 specifically and effectively inhibits host replication within 5 min of infection (Schito, 1973). Yano and Rothman-Denes (2011) revealed the interaction between N4 gp8 and the δ-subunit of the DNA polymerase III, which is the clamp loader of this ten-subunit complex. Gene product 8 was demonstrated to be the specific inhibitor of *E. coli* DNA replication elongation, responsible for host DNA synthesis shutoff. N4 gp8 does not eliminate host genome replication, but without loading the DNA clamp, the processivity of the DNA polymerase III holoenzyme is extremely low. The authors also illustrated the bacteriostatic effect of N4 gp8, which again could be exploited in antibacterial research. Interestingly, gp8 is not conserved between N4 and any of the currently known N4-like phages, such as *Yersinia ruckeri* phage NC10 or *Pseudomonas aeruginosa* phages LIT1 and LUZ7 (Yano and Rothman-Denes, 2011). Consequently, application of gp8 derivatives may lead to a narrow-host-range antibiotic.

Phages manipulating the translation machinery

The ubiquitous availability of cellular ribosomes allows some phages to utilize the unmodified host translation machinery to translate their mRNA (Nechaev and Severinov, 2008). Several findings, however, suggest a direct influence of infection on translation by some phages. These are summarized in this section.

Both T4 and T7 modify proteins directly involved in translation (Robertson *et al.*, 1994; Depping *et al.*, 2005). For example, T4 ADP-ribosyltransferases Alt and ModB interact with 27 and eight different host proteins, respectively, some of which are involved in translation. Nevertheless, the individual effects of these ribosylations remain to be fully elucidated (Depping *et al.*, 2005). Although it might be difficult or even impossible to imitate ribosylation or other modifications with small chemical compounds, these findings indicate that there may well be other phage mechanisms to modify or inhibit the host translation system.

T. thermophilus phage ϕYS40 also seems to influence the host translation machinery. This phage might control the expression of its genes through a novel mechanism that directs the cellular translation apparatus towards mRNAs lacking a leader sequence. This leader sequence contains the ribosome-binding site and is required for correct translation initiation. As most transcripts of ϕYS40 middle and late genes are leaderless, this strategy would favour translation of bacteriophage mRNA over host mRNA. Sevostyanova *et al.* (2007) suggested this translation manipulation because phage ϕYS40 appears to specifically upregulate the levels of translation initiation factor IF2, which is implicated in translation of leaderless RNA (Grill *et al.*, 2001). The exact molecular basis of this influence and its possible application in drug development is still unknown.

Do phages impact on global host metabolism?

Inhibiting host replication, transcription and translation after infection are clearly favourable for a phage in that it allows a shift to phage processes from host ones. Similarly, it is not unimaginable that phages also might be able to influence the bacterial energy metabolism or any other essential pathway, as superfluous and energy-wasting steps could be a burden to efficient intracellular phage development. On the other hand, inhibiting an important energy source might help a phage in blocking the host response to infection.

Examples are the interactions between phage T4 ADP-ribosyltransferase Alt and host pyruvate kinase I and ribose 5-phosphate isomerase A (Depping *et al.*, 2005). These are important enzymes in glycolysis and the non-oxidative branch of the pentose phosphate pathway, respectively. Although the effect caused by these modifications remains unclear, this provides a first indication for the influence of phages on host energy metabolism. Similarly, a systematic yeast two-hybrid screening for bacteriophage–host interactions between *P. aeruginosa* and podovirus ϕKMV revealed three interactions: again an enzyme of the central energy metabolism (malate synthase G), a regulator of a secretion system and a regulator of nitrogen metabolism (Roucourt *et al.*, 2009a,b). These first examples suggest that there might also be some deleterious influences on global metabolic pathways due to direct phage–host interactions, potentially useful in drug discovery.

Indeed, recently, Rybniker *et al.* (2011) reported the interaction between the cytotoxic early protein 77 of mycobacteriophage L5 and MSMEG_3532, an L-serine dehydratase of *Mycobacterium smegmatis*. Minute amounts of intracellular protein 77 act bacteriostatically on the host of phage L5 (Rybniker *et al.*, 2008). In *E. coli*, deficiency of L -serine dehydratase has been shown to impair growth through the inhibition of cell division (Zhang and Newman, 2008). Rybniker *et al.* (2008) could not confirm an *in vitro* effect of gp77 on dehydratase activity, however. The specific interaction of phage protein and dehydratase may also play a role in providing amino acids or products of the amino acid metabolism required for phage maturation (Rybniker *et al.*, 2011). In terms of drug development, the hope would be that the host target (or targets) whose presumed inhibition resulted in the bacteriostatic effect would be present as well in *M. smegmatis*-related pathogens such as *Mycobacterium tuberculosis*.

Phage-directed inhibition of cell-wall biosynthesis?

Cell-wall biosynthesis is a crucial step associated with bacterial growth. Not surprisingly, many antibiotics in the past have been shown to inhibit this process directly. Some findings suggest a direct inhibition of cell-wall biosynthesis by several phage-encoded proteins, which could lead to new inhibition mechanisms of this crucial step in cell division (Sau *et al.*, 2008). For example, the Kil protein of defective prophage Rac was shown to arrest *E. coli* cell division severely. This detrimental effect was reversed when both Kil and cell division protein FtsZ were overexpressed, suggesting a direct or indirect interaction between the two proteins (Conter *et al.*, 1996). Like the Kil gene product of the Rac prophage, the Kil proteins of Mu (Waggoner *et al.*, 1989), P22 (Semerjian *et al.*, 1989), λ (Sergueev *et al.*, 2001) and Sf6 (Casjens *et al.*, 2004) also have been reported to affect cell division, illustrating this common Kil mechanism among phages. So far, however, no exclusive proof for a direct effect on cell-wall biosynthesis or cell division has been provided.

From interactions towards new antibacterials

Beyond identifying new antibacterial targets, non-lytic phage molecules could also serve as antibacterial compounds themselves or they can form the basis for a phage protein-derived small-molecule antibiotic, as demonstrated by Liu and colleagues (2004) and further discussed in the next paragraph. Although not all phage–host interactions are detrimental to the host cell, some inhibit the host macromolecular machinery or other critical cellular processes, leading to cell-cycle arrest or even host lethality. These particular bacteriophage–host interactions thus might be used towards development of new selectively bacteriostatic or bactericidal agents (Liu *et al.*, 2004; Projan, 2004).

Bacteriophage–host interaction-based development of new antibacterial compounds is a multistage process, as depicted in Fig. 9.1. First, bacteriophages are isolated and their genomes sequenced and annotated. Subsequently, predicted or selected genes are cloned into an inducible expression vector and transferred to the host cell. By inducing expression, toxic or inhibitory phage products are identified by inhibition of host-cell growth. In the next phase, the bacterial interaction partner of the phage-encoded inhibitor is revealed by protein–protein interaction analysis, such as yeast two-hybrid or affinity purification. As interaction analysis often contains both false-positive and false-negative interactions, it is important to use a combination of different techniques to find and confirm all bacteriophage–host interactions (Braun *et al.*, 2009; Chen *et al.*, 2010). Once the interaction partner is known, it can be validated as a suitable target for drug development. Apart from identifying the new antibacterial target, the interaction itself between the phage protein and the bacterial protein can serve as the basis for a screening assay in the development of a new antibacterial compound. Presumably, small chemical molecules capable of disrupting the phage–host interaction by binding to the host protein would mimic the inhibitory effect of the bacteriophage protein. These assays, using, for example, time-resolved fluorescence resonance energy transfer (TR-FRET) to demonstrate disruption of interaction, facilitate high-throughput screening of large libraries of small synthetic compounds. The chemical equivalent of the phage protein then could provide the basis for a novel antibacterial (Liu *et al.*, 2004).

Liu and colleagues (2004) demonstrated the effectiveness of this bacteriophage-inspired platform to identify new targets and antibacterials in the Gram-positive *S. aureus*. Out of 26 fully sequenced *S. aureus*-specific phages, 31 novel polypeptide families of early proteins were identified as abolishing growth upon expression in the bacterial host. Using affinity chromatography of *S. aureus* lysates on phage protein-coated columns and subsequent identification by mass spectrometry, the cellular targets for some of the inhibitory proteins were identified and several were shown to be essential components of the host DNA replication and

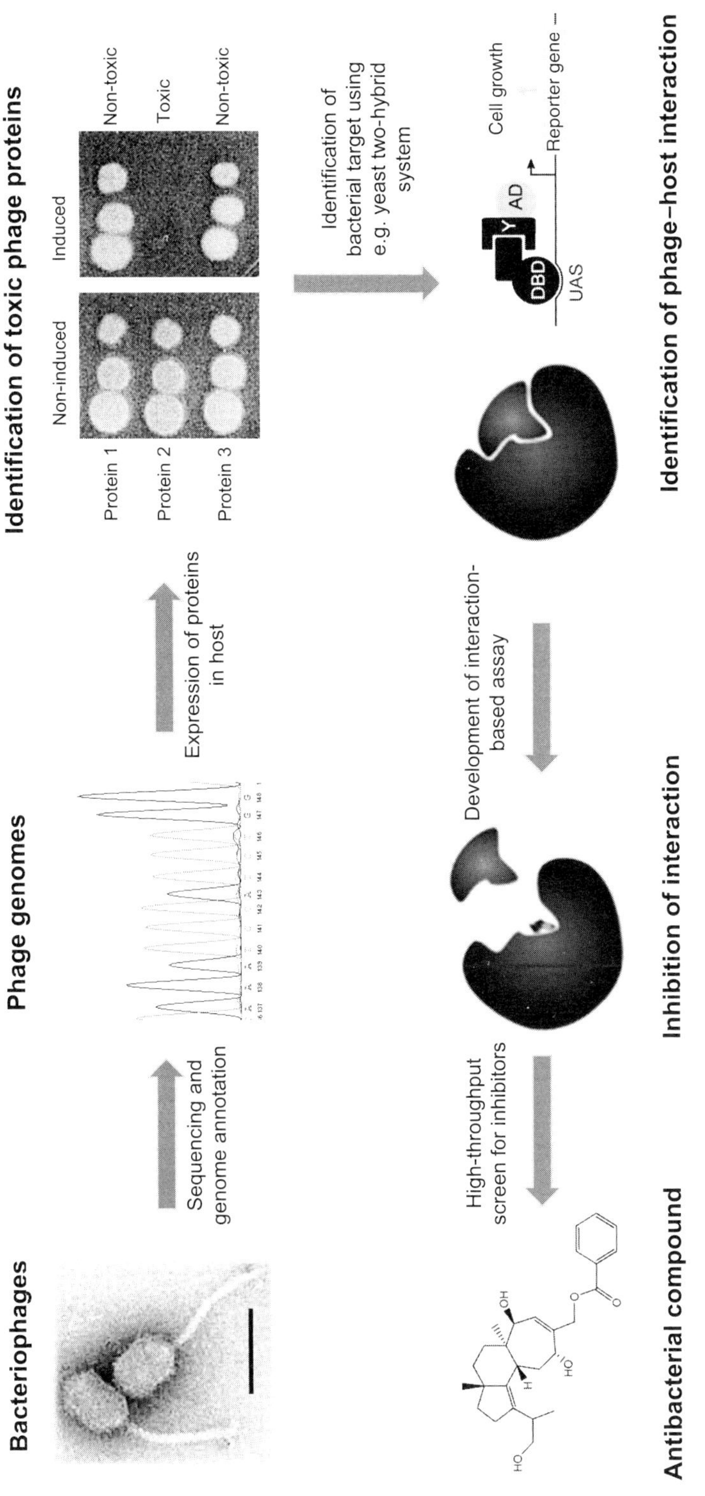

Fig. 9.1. Bacteriophage–host interaction-based development of new antibacterial compounds.

transcription machineries. One of these was the interaction between ORF104 of phage 77 and the helicase loader DnaI, which is essential for DNA replication initiation. This phage–host interaction was confirmed to be specific and direct using four independent assays: yeast two-hybrid analysis, far Western analysis with immobilized DnaI and ^{32}P-labelled gp104, biomolecular interaction analysis (Biacore) and TR-FRET. Moreover, DnaI was shown to be essential for viability of the host cell. These results were then used to screen for small synthetic molecule inhibitors, mimicking the phage protein. Out of 125,000 compounds from commercially available libraries, 36 were able to inhibit the phage–host interaction with a 50% inhibitory concentration (IC_{50}) of less than 10 μM. Among these, 11 compounds were found to inhibit bacterial growth of *S. aureus* on their own with a minimum inhibitory concentration of less than 16 μg ml^{-1}. Furthermore, neither compound was significantly cytotoxic to human primary hepatocytes or to the cell lines Hep G2 and HeLa. These lead compounds were proposed to form a basis for a new phage-based antibiotic (Liu *et al.*, 2004).

Conclusion

Although microbial resistance is probably an unavoidable consequence of antibiotic therapy, a bacteriophage-based platform has great potential with respect to identifying novel mechanisms and targets to treat bacterial infections (Falconer and Brown, 2009). In fact, known phage–host interactions illustrate the potential for phage systems to be used for the identification of points in host metabolism that may be susceptible to small-molecule inhibitors. The most efficient and vulnerable targets are selected and validated through billions of years of co-evolution between phages and their hosts (Brown, 2004). As there is no dearth of bacteriophages in nature, the quest for lethal phage proteins as well as their cognate bacterial targets should be continued in order to expedite the research on antibacterial drug discovery (Sau *et al.*, 2008). Despite the body of available research and the vast untapped potential, implementations by industry have remained limited, however. This can be explained by a number of factors.

First, the bacteriophage research field remains strongly genome oriented and only a limited number of research groups analyse the intracellular phage–host interactions (several of them focusing on model hosts and not pathogens). Secondly, the step from fundamental research to product development is great, as an extensive body of evidence is necessary to obtain a suitable bacterial target candidate and relevant phage-based inhibitors. Efforts to crystallize and solve the structure of phage-encoded proteins involved in influencing the host metabolism should be encouraged to tackle this large and diverse, at least at the primary sequence level, group of proteins and provide a solid basis for rational antibacterial design. The strategy proposed by Liu *et al.* (2004) (Fig. 9.1) appears straightforward but contains genome and expression-based assays, protein interaction assays, crystallography and small-molecule design. Although high-throughput sequencing and protein expression can now be considered commonplace, protein interaction analyses (like yeast two-hybrid analyses, affinity purification and FRET) and (rational) small-molecule development remain technically challenging. Thirdly, from an industry perspective, one could state that major pharmaceutical companies are cautious towards new antibacterial strategies in view of the limited profit margin and highly competitive market for antibiotics. This implies the necessity of small biotech company initiatives (vulnerable to economic circumstances) to generate a substantial (and costly) body of material for product development.

The development of phage-inspired antibiotics nevertheless remains a strong option for the future. One can anticipate a shift in the general research focus from genome-based research to proteome- and even metabolomics-based research. We can anticipate that technological developments of protein interaction and mass spectrometry approaches will be revolutionized further, in much the same way as we have seen at the genomics level, with the advent of next-generation and single-molecule sequencing techniques. Proteome- and metabolomics-based research will rely on these developments. In addition, the bacteriophage domain has an extensive body of front-line researchers with experience in the structural analysis of virion-associated proteins, which provides a large body of experience to tackle the dark matter within phage genomes. From an industry perspective, we can anticipate an increasing need within our society towards novel antibacterials with a related increased profit potential, which will hopefully motivate large pharmaceutical companies and investors in small companies as well. It will be the responsibility of the scientists, however, to generate the fundamental knowledge that will help bridge the gap towards product development in a sustainable manner.

Acknowledgements

J.W. holds a predoctoral fellowship of the *'Agentschap voor Innovatie door Wetenschap en Technologie in Vlaanderen'* (IWT, Belgium).

References

Akerley, B.J., Rubin, E.J., Novick, V.L., Amaya, K., Judson, N. and Mekalanos, J.J. (2002) A genome-scale analysis for identification of genes required for growth or survival of *Haemophilus influenzae*. *Proceedings of the National Academy of Sciences USA* 99, 966–971.

Baba, T., Huan, H.C., Datsenko, K., Wanner, B.L. and Mori, H. (2008) The applications of systematic in-frame, single-gene knockout mutant collection of *Escherichia coli* K-12. *Methods in Molecular Biology* 416, 183–194.

Belley, A., Callejo, M., Arhin, F., Dehbi, M., Fadhil, I., Liu, J., McKay, G., Srikumar, R., Bauda, P., Ha, N., DuBow, M., Gros, P., Pelletier, J. and Moeck, G. (2006) Competition of bacteriophage polypeptides with native replicase proteins for binding to the DNA sliding clamp reveals a novel mechanism for DNA replication arrest in *Staphylococcus aureus*. *Molecular Microbiology* 62, 1132–1143.

Berdygulova, Z., Westblade, L.F., Florens, L., Koonin, E.V., Chait, B.T., Ramanculov, E., Washburn, M.P., Darst, S.A., Severinov, K. and Minakhin, L. (2011) Temporal regulation of gene expression of the *Thermus thermophilus* bacteriophage P23-45. *Journal of Molecular Biology* 405, 125–142.

Braun, P., Tasan, M., Dreze, M., Barrios-Rodiles, M., Lemmens, I., Yu, H., Sahalie, J.M., Murray, R.R., Roncari, L., de Smet, A.S., Venkatesan, K., Rual, J.-F., Vandenhaute, J., Cusick, M.E., Pawson, T., Hill, D.E., Tavernier, J., Wrana, J.L., Roth, F.P. and Vidal, M., (2009) An experimentally derived confidence score for binary protein–protein interactions. *Nature Methods* 6, 91–97.

Brown, E.D. (2004) Drugs against superbugs: private lessons from bacteriophages. *Trends in Biotechnology* 22, 434–436.

Brown, E.D. (2006) Microbiology: antibiotic stops 'ping-pong' match. *Nature* 441, 293–294.

Casjens, S., Winn-Stapley, D.A., Gilcrease, E.B., Morona, R., Kuhlewein, C., Chua, J.E., Manning, P.A., Inwood, W. and Clark, A.J. (2004) The chromosome of *Shigella flexneri* bacteriophage Sf6: complete nucleotide sequence, genetic mosaicism, and DNA packaging. *Journal of Molecular Biology* 339, 379–394.

Chen, Y.C., Rajagopala, S.V., Stellberger, T. and Uetz, P. (2010) Exhaustive benchmarking of the yeast two-hybrid system. *Nature Methods* 7, 667–668; author reply 668.

Clatworthy, A.E., Pierson, E. and Hung, D.T. (2007) Targeting virulence: a new paradigm for antimicrobial therapy. *Nature Chemical Biology* 3, 541–548.

Conter, A., Bouche, J.P. and Dassain, M. (1996) Identification of a new inhibitor of essential division gene *ftsZ* as the *kil* gene of defective prophage Rac. *Journal of Bacteriology* 178, 5100–5104.

Court, R., Cook, N., Saikrishnan, K. and Wigley, D. (2007) The crystal structure of l-Gam protein suggests a model for RecBCD inhibition. *Journal of Molecular Biology* 371, 25–33.

Dehbi, M., Moeck, G., Arhin, F.F., Bauda, P., Bergeron, D., Kwan, T., Liu, J., McCarty, J., Dubow, M. and Pelletier, J. (2009) Inhibition of transcription in *Staphylococcus aureus* by a primary sigma factor-binding polypeptide from phage G1. *Journal of Bacteriology* 191, 3763–3771.

Depping, R., Lohaus, C., Meyer, H.E. and Ruger, W. (2005) The mono-ADP-ribosyltransferases Alt and ModB of bacteriophage T4: target proteins identified. *Biochemical and Biophysical Research Communications* 335, 1217–1223.

Djordjevic, M., Semenova, E., Shraiman, B. and Severinov, K. (2006) Quantitative analysis of a virulent bacteriophage transcription strategy. *Virology* 354, 240–251.

Falconer, S.B. and Brown, E.D. (2009) New screens and targets in antibacterial drug discovery. *Current Opinion in Microbiology* 12, 497–504.

Fernebro, J. (2011) Fighting bacterial infections – future treatment options. *Drug Resistance Updates* 14, 125–139.

Finland, M., Jones, W.F. Jr and Barnes, M.W. (1959) Occurrence of serious bacterial infections since introduction of antibacterial agents. *Journal of the American Medical Association* 170, 2188–2197.

Forsyth, R.A., Haselbeck, R.J., Ohlsen, K.L., Yamamoto, R.T., Xu, H., Trawick, J.D., Wall, D., Wang, L., Brown-Driver, V., Froelich, J.M., Kedar, G.C., King, P., McCarthy, M., Malone, C., Misiner, B., Robbins, D., Tan, Z., Zhu Z.Y., Carr, G., Mosca, D.A., Zamudio, C., Foulkes, J.G. and Zyskind, J.W. (2002) A genome-wide strategy for the identification of essential genes in *Staphylococcus aureus*. *Molecular Microbiology* 43, 1387–1400.

Friedman, D.I., Olson, E.R., Georgopoulos, C., Tilly, K., Herskowitz, I. and Banuett, F. (1984) Interactions of bacteriophage and host macromolecules in the growth of bacteriophage l. *Microbiological Reviews* 48, 299–325.

Grill, S., Moll, I., Hasenohrl, D., Gualerzi, C.O. and Blasi, U. (2001) Modulation of ribosomal recruitment to 5′-terminal start codons by translation initiation factors IF2 and IF3. *FEBS Letters* 495, 167–171.

Guttman, B., Raya, R. and Kutter, E. (2005). Basic phage biology. In: Kutter, E. and Sulakvelidze, A. (eds) *Bacteriophages Biology and Application*. CRC Press, Boca Rota, FL, pp. 29–66.

Gwynn, M.N., Portnoy, A., Rittenhouse, S.F. and Payne, D.J. (2010) Challenges of antibacterial discovery revisited. *Annals of the New York Academy of Sciences* 1213, 5–19.

Kashlev, M., Nudler, E., Goldfarb, A., White, T. and Kutter, E. (1993) Bacteriophage T4 Alc protein: a transcription termination factor sensing local modification of DNA. *Cell* 75, 147–154.

Kassavetis, G.A., Elliott, T., Rabussay, D.P. and Geiduschek, E.P. (1983) Initiation of transcription at phage T4 late promoters with purified RNA polymerase. *Cell* 33, 887–897.

Klein, A., Lanka, E. and Schuster, H. (1980) Isolation of a complex between the P protein of phage l and the *dnaB* protein of *Escherichia coli*. *European Journal of Biochemistry* 105, 1–6.

Knuth, K., Niesalla, H., Hueck, C.J. and Fuchs, T.M. (2004) Large-scale identification of essential *Salmonella* genes by trapping lethal insertions. *Molecular Microbiology* 51, 1729–1744.

Kobayashi, K., Ehrlich, S.D., Albertini, A., Amati, G., Andersen, K.K., Arnaud, M., Asai, K., Ashikaga, S., Aymerich, S., Bessieres, P., Boland, F., Brignell, S.C., Bron, S., Bunai, K., Chapuis, J., Christiansen, L.C., Danchin, A., Débarbouille, M., Dervyn, E., Deuerling, E., Devine, K., Devine, S.K., Dreesen, O., Errington, J., Fillinger, S., Foster, S.J., Fujita, Y., Galizzi, A., Gardan, R., Eschevins, C., Fukushima, T., Haga, K., Harwood, C.R., Hecker, M., Hosoya, D., Hullo, M.F., Kakeshita, H., Karamata, D., Kasahara, Y., Kawamura, F., Koga, K., Koski, P., Kuwana, R., Imamura, D., Ishimaru, M., Ishikawa, S., Ishio, I., Le Coq, D., Masson, A., Mauël, C., Meima, R., Mellado, R.P., Moir, A., Moriya, S., Nagakawa, E., Nanamiya, H., Nakai, S., Nygaard, P., Ogura, M., Ohanan, T., O'Reilly, M., O'Rourke, M., Pragai, Z., Pooley, H.M., Rapoport, G., Rawlins, J.P., Rivas, L.A., Rivolta, C., Sadaie, A., Sadaie, Y., Sarvas, M., Sato, T., Saxild, H.H., Scanlan, E., Schumann, W., Seegers, J.F., Sekiguchi, J., Sekowska, A., Séror, S.J., Simon, M., Stragier, P., Studer, R., Takamatsu, H., Tanaka, T., Takeuchi, M., Thomaides, H.B., Vagner, V., van Dijl, J.M., Watabe, K., Wipat, A., Yamamoto, H., Yamamoto, M., Yamamoto, Y., Yamane, K., Yata, K., Yoshida, K., Yoshikawa, H., Zuber, U. and Ogasawara, N. (2003) Essential *Bacillus subtilis* genes. *Proceedings of the National Academy of Sciences USA* 100, 4678–4683.

Kolesky, S., Ouhammouch, M., Brody, E.N. and Geiduschek, E.P. (1999) Sigma competition: the contest between bacteriophage T4 middle and late transcription. *Journal of Molecular Biology* 291, 267–281.

Kutter, E., Raya, R. and Carlson, K. (2005). Molecular mechanisms of phage infection. In: Kutter, E. and Sulakvelidze, A. (eds) *Bacteriophages: Biology and Applications*. CRC Press, Boca Raton, FL, pp. 165–222.

Lange, R.P., Locher, H.H., Wyss, P.C. and Then, R.L. (2007) The targets of currently used antibacterial agents: lessons for drug discovery. *Current Pharmacological Design* 13, 3140–3154.

Lipinski, C.A., Lombardo, F., Dominy, B.W. and Feeney, P.J. (2001) Experimental and computational approaches to estimate solubility and permeability in drug discovery and development settings. *Advances in Drug Delivery Reviews* 46, 3–26.

Liu, J., Dehbi, M., Moeck, G., Arhin, F., Bauda, P., Bergeron, D., Callejo, M., Ferretti, V., Ha, N., Kwan, T., McCarty, J., Srikumar, R., Williams, D., Wu, J.J., Gros, P., Pelletier, J. and DuBow, M. (2004) Antimicrobial drug discovery through bacteriophage genomics. *Nature Biotechnology* 22, 185–191.

Lock, R.L. and Harry, E.J. (2008) Cell-division inhibitors: new insights for future antibiotics. *Nature Reviews Drug Discovery* 7, 324–338.

Losick, R. and Pero, J. (1981) Cascades of sigma factors. *Cell* 25, 582–584.

Madigan, M.T., Martinko, J.M. and Brock, T.D. (2006). *Brock Biology of Microorganisms*, 11th edn. Pearson Prentice Hall, Upper Saddle River, NJ.

Mallory, J.B., Alfano, C. and McMacken, R. (1990) Host virus interactions in the initiation of bacteriophage l DNA replication. Recruitment of *Escherichia coli* DnaB helicase by λ P replication protein. *Journal of Biological Chemistry* 265, 13297–13307.

Marchand, I., Nicholson, A.W. and Dreyfus, M. (2001) Bacteriophage T7 protein kinase phosphorylates RNase E and stabilizes mRNAs synthesized by T7 RNA polymerase. *Molecular Microbiology* 42, 767–776.

Marr, M.T., Datwyler, S.A., Meares, C.F. and Roberts, J.W. (2001) Restructuring of an RNA polymerase holoenzyme elongation complex by lambdoid phage Q proteins. *Proceedings of the National Academy of Sciences USA* 98, 8972–8978.

Mason, S.W. and Greenblatt, J. (1991) Assembly of transcription elongation complexes containing the N protein of phage λ and the *Escherichia coli* elongation factors NusA, NusB, NusG, and S10. *Genes and Development* 5, 1504–1512.

Mayer, J.E. and Schweiger, M. (1983) RNase III is positively regulated by T7 protein kinase. *Journal of Biological Chemistry* 258, 5340–5343.

Mecsas, J. (2002) Use of signature-tagged mutagenesis in pathogenesis studies. *Current Opinion on Microbiology* 5, 33–37.

Michalewicz, J. and Nicholson, A.W. (1992) Molecular cloning and expression of the bacteriophage T7 0.7 (protein kinase) gene. *Virology* 186, 452–462.

Miller, A., Wood, D., Ebright, R.H. and Rothman-Denes, L.B. (1997) RNA polymerase β′ subunit: a target of DNA binding-independent activation. *Science* 275, 1655–1657.

Miller, E.S., Kutter, E., Mosig, G., Arisaka, F., Kunisawa, T. and Ruger, W. (2003) Bacteriophage T4 genome. *Microbiology and Molecular Biology Reviews* 67, 86–156.

Molineux, I.J. (2005). The T7 group. In: Calendar, R. and Abedon, S.T. (eds) *The Bacteriophages.* Oxford University Press, Oxford, UK, pp. 277–301.

Mosig, G., Colowick, N.E. and Pietz, B.C. (1998) Several new bacteriophage T4 genes, mapped by sequencing deletion endpoints between genes 56 (dCTPase) and dda (a DNA-dependent ATPase-helicase) modulate transcription. *Gene* 223, 143–155.

Mueser, T.C., Hinerman, J.M., Devos, J.M., Boyer, R.A. and Williams, K.J. (2010) Structural analysis of bacteriophage T4 DNA replication: a review in the Virology Journal series on bacteriophage T4 and its relatives. *Virology Journal* 7, 359.

Nechaev, S. and Severinov, K. (1999) Inhibition of *Escherichia coli* RNA polymerase by bacteriophage T7 gene 2 protein. *Journal of Molecular Biology* 289, 815–826.

Nechaev, S. and Severinov, K. (2003) Bacteriophage-induced modifications of host RNA polymerase. *Annual Review of Microbiology* 57, 301–322.

Nechaev, S. and Severinov, K. (2008) The elusive object of desire – interactions of bacteriophages and their hosts. *Current Opinion on Microbiology* 11, 186–193.

Nechaev, S., Yuzenkova, Y., Niedziela-Majka, A., Heyduk, T. and Severinov, K. (2002) A novel bacteriophage-encoded RNA polymerase binding protein inhibits transcription initiation and abolishes transcription termination by host RNA polymerase. *Journal of Molecular Biology* 320, 11–22.

Odegrip, R., Schoen, S., Haggard-Ljungquist, E., Park, K. and Chattoraj, D.K. (2000) The interaction of bacteriophage P2 B protein with *Escherichia coli* DnaB helicase. *Journal of Virology* 74, 4057–4063.

Ouhammouch, M., Adelman, K., Harvey, S.R., Orsini, G. and Brody, E.N. (1995) Bacteriophage T4 MotA and AsiA proteins suffice to direct *Escherichia coli* RNA polymerase to initiate transcription at T4 middle promoters. *Proceedings of the National Academy of Sciences USA* 92, 1451–1455.

Pani, B., Ranjan, A. and Sen, R. (2009) Interaction surface of bacteriophage P4 protein Psu required for complex formation with the transcription terminator Rho. *Journal of Molecular Biology* 389, 647–660.

Payne, D.J., Gwynn, M.N., Holmes, D.J. and Pompliano, D.L. (2007) Drugs for bad bugs: confronting the challenges of antibacterial discovery. *Nature Reviews Drug Discovery* 6, 29–40.

Projan, S. (2004) Phage-inspired antibiotics? *Nature Biotechnology* 22, 167–168.

Robertson, E.S., Aggison, L.A. and Nicholson, A.W. (1994) Phosphorylation of elongation factor G and ribosomal protein S6 in bacteriophage T7-infected *Escherichia coli*. *Molecular Microbiology* 11, 1045–1057.

Roucourt, B. and Lavigne, R. (2009) The role of interactions between phage and bacterial proteins within the infected cell: a diverse and puzzling interactome. *Environmental Microbiology* 11, 2789–2805.

Roucourt, B., Lecoutere, E., Chibeu, A., Hertveldt, K., Volckaert, G. and Lavigne, R. (2009a) A procedure for systematic identification of bacteriophage–host interactions of *P. aeruginosa* phages. *Virology* 387, 50–58.

Roucourt, B., Minnebo, N., Augustijns, P., Hertveldt, K., Volckaert, G. and Lavigne, R. (2009b) Biochemical characterization of malate synthase G of *P. aeruginosa*. *BMC Biochemistry* 10, 20.

Rybniker, J., Plum, G., Robinson, N., Small, P.L. and Hartmann, P. (2008) Identification of three cytotoxic early proteins of mycobacteriophage L5 leading to growth inhibition in *Mycobacterium smegmatis*. *Microbiology* 154, 2304–2314.

Rybniker, J., Krumbach, K., van Gumpel, E., Plum, G., Eggeling, L. and Hartmann, P. (2011) The cytotoxic early protein 77 of mycobacteriophage L5 interacts with MSMEG_3532, an L-serine dehydratase of *Mycobacterium smegmatis*. *Journal of Basic Microbiology* 51, 515–522.

Sau, S., Chattoraj, P., Ganguly, T., Chanda, P.K. and Mandal, N.C. (2008) Inactivation of indispensable bacterial proteins by early proteins of bacteriophages: implication in antibacterial drug discovery. *Current Protein and Peptide Science* 9, 284–290.

Savalia, D., Westblade, L.F., Goel, M., Florens, L., Kemp, P., Akulenko, N., Pavlova, O., Padovan, J.C., Chait, B.T., Washburn, M.P., Ackermann, H.-W., Mushegian, A., Gabisonia, T., Molineux, I. and Severinov, K. (2008) Genomic and proteomic analysis of phiEco32, a novel *Escherichia coli* bacteriophage. *Journal of Molecular Biology* 377, 774–789.

Schito, G.C. (1973) The genetics and physiology of coliphage N4. *Virology* 55, 254–265.

Semerjian, A.V., Malloy, D.C. and Poteete, A.R. (1989) Genetic structure of the bacteriophage P22 P_L operon. *Journal of Molecular Biology* 207, 1–13.

Sergueev, K., Yu, D., Austin, S. and Court, D. (2001) Cell toxicity caused by products of the p_L operon of bacteriophage lambda. *Gene* 272, 227–235.

Severinova, E. and Severinov, K. (2006) Localization of the *Escherichia coli* RNA polymerase β′ subunit residue phosphorylated by bacteriophage T7 kinase Gp0.7. *Journal of Bacteriology* 188, 3470–3476.

Sevostyanova, A., Djordjevic, M., Kuznedelov, K., Naryshkina, T., Gelfand, M.S., Severinov, K. and Minakhin, L. (2007) Temporal regulation of viral transcription during development of *Thermus thermophilus* bacteriophage ϕYS40. *Journal of Molecular Biology* 366, 420–435.

Silver, L.L. (2011) Challenges of antibacterial discovery. *Clin Microbiological Reviews* 24, 71–109.

Skorko, R., Zillig, W., Rohrer, H., Fujiki, H. and Mailhammer, R. (1977) Purification and properties of the NAD^+:protein ADP-ribosyltransferase responsible for the T4-phage-induced modification of the α subunit of DNA-dependent RNA polymerase of *Escherichia coli*. *European Journal of Biochemistry* 79, 55–66.

Snyder, L. and Champness, W. (2003). *Molecular Genetics of Bacteria*, 2nd edn. ASM Press, Washington, DC.

So, A.D., Gupta, N., Brahmachari, S.K., Chopra, I., Munos, B., Nathan, C., Outterson, K., Paccaud, J.P., Payne, D.J., Peeling, R.W., Spigelman, M. and Weigelt, J., (2011) Towards new business models for R&D for novel antibiotics. *Drug Resistance Updates* 14, 88–94.

Sommer, N., Salniene, V., Gineikiene, E., Nivinskas, R. and Ruger, W. (2000) T4 early promoter strength *probed in vivo* with unribosylated and ADP-ribosylated *Escherichia coli* RNA polymerase: a mutation analysis. *Microbiology* 146, 2643–2653.

Szambowska, A., Pierechod, M., Wegrzyn, G. and Glinkowska, M. (2011) Coupling of transcription and replication machineries in λ DNA replication initiation: evidence for direct interaction of *Escherichia coli* RNA polymerase and the λO protein. *Nucleic Acids Research* 39, 168–177.

Tabor, S., Huber, H.E. and Richardson, C.C. (1987) *Escherichia coli* thioredoxin confers processivity on the DNA polymerase activity of the gene 5 protein of bacteriophage T7. *Journal of Biological Chemistry* 262, 16212–16223.

Thanassi, J.A., Hartman-Neumann, S.L., Dougherty, T.J., Dougherty, B.A. and Pucci, M.J. (2002) Identification of 113 conserved essential genes using a high-throughput gene disruption system in *Streptococcus pneumoniae*. *Nucleic Acids Research* 30, 3152–3162.

Waggoner, B.T., Sultana, K., Symonds, N., Karlok, M.A. and Pato, M.L. (1989) Identification of the bacteriophage Mu *kil* gene. *Virology* 173, 378–389.

Wei, P. and Stewart, C.R. (1995) Genes that protect against the host-killing activity of the E3 protein of *Bacillus subtilis* bacteriophage SPO1. *Journal of Bacteriology* 177, 2933–2937.

Wood, L.F., Tszine, N.Y. and Christie, G.E. (1997) Activation of P2 late transcription by P2 Ogr protein requires a discrete contact site on the C terminus of the α subunit of *Escherichia coli* RNA polymerase. *Journal of Molecular Biology* 274, 1–7.

Yano, S.T. and Rothman-Denes, L.B. (2011) A phage-encoded inhibitor of *Escherichia coli* DNA replication targets the DNA polymerase clamp loader. *Molecular Microbiology* 79, 1325–1338.

Yuzenkova, J., Nechaev, S., Berlin, J., Rogulja, D., Kuznedelov, K., Inman, R., Mushegian, A. and Severinov, K. (2003) Genome of *Xanthomonas oryzae* bacteriophage Xp10: an odd T-odd phage. *Journal of Molecular Biology* 330, 735–748.

Yuzenkova, Y., Zenkin, N. and Severinov, K. (2008) Mapping of RNA polymerase residues that interact with bacteriophage Xp10 transcription antitermination factor p7. *Journal of Molecular Biology* 375, 29–35.

Zhang, X. and Newman, E. (2008) Deficiency in L-serine deaminase results in abnormal growth and cell division of *Escherichia coli* K-12. *Molecular Microbiology* 69, 870–881.

Zlitni, S. and Brown, E.D. (2009) Drug discovery: not as fab as we thought. *Nature* 458, 39–40.

10 Bacteriophage-based Methods of Bacterial Detection and Identification

Christopher R. Cox[1]

[1]Department of Chemistry and Geochemistry, Colorado School of Mines.

The need for rapid and inexpensive methods of bacterial detection and identification has become an increasingly common theme among the medical, agricultural, food and water processing, military, homeland security and academic fields. Despite the permeation of humanity by disease-causing and resource-contaminating bacterial agents, conventional methods fall short of widespread practicality for bacterial detection and identification, particularly due to a lack of rapidity or cost-effectiveness. Consequently, an abundance of biochemical and genotypic methods have been developed, although with varied levels of success.

Some newer methodologies include 16S–23S rRNA gene sequencing, which has become the gold standard of genotypic bacterial identification (Bottger, 1989; Fredericks and Relman, 1996; Clarridge, 2004), but also multilocus enzyme electrophoresis (Selander *et al.*, 1986; Duan *et al.*, 1991), multilocus sequence typing (Maiden *et al.*, 1998), multilocus variable tandem repeat analysis (Ross *et al.*, 2011), pulsed-field gel electrophoresis (Durmaz *et al.*, 2009), amplified fragment length polymorphism (AFLP; Jayarao *et al.*, 1992; Koeleman *et al.*, 1998), repetitive sequence-based PCR (Jayarao *et al.*, 1992; Grisold *et al.*, 2010) and Raman spectroscopy-based systems (Maquelin *et al.*, 2006; Patel *et al.*, 2008). All of these conventional methods, however, rely on labour-intensive microbiological culturing practices or costly and time-consuming DNA isolation, amplification and/or sequencing protocols utilizing highly specialized equipment. This makes them impractical for deployment as rapid, cost-effective point-of-care or field detection and identification methods. In addition, most require 24–50 h for positive bacterial identification. While these various methods have been used effectively as tools for laboratory analysis, there is growing interest in the development of much faster, simpler approaches, ideally ones that are much more cost-effective, that do not rely on complex equipment and that allow on-site positive pathogen identification without the need for extensive personnel training.

Exploitation of the bacteriophage (or phage) life cycle and its accompanying species specificity can eliminate much of the time and cost associated with these methods while rivalling the detection capabilities of even the most sensitive of current technologies. The use of phages in combination with existing and emerging technologies consequently has the potential to overcome many of the shortcomings of current detection and identification methodologies and presents a promising opportunity to develop the next generation of truly rapid, tractable bacterial detection techniques. Newly developed applications centring on the use of phage amplification coupled to matrix-assisted laser desorption ionization time-of-flight mass spectrometry (MALDI-TOF MS) or easy-to-use lateral flow immunochromatography now allow rapid bacterial detection with total time for sampling and analysis in 2–12 h timeframes. Advancements in techniques using immobilized phages, labelled phages, individual phage components or phages as vehicles for transferring reporter genes to a bacterial host have also opened the door for a multitude of new phage-based bacterial detection and identification methods.

In this chapter, both traditional and newer approaches to phage-based bacterial detection and identification are reviewed. For additional consideration of traditional as well as modern approaches to phage-based bacterial detection technologies, see Williams and LeJeune (Chapter 6, this volume), who in part provide a discussion of phage typing techniques (bacterial classification based on susceptibility to phage infection), and Goodridge and Steiner (Chapter 11, this volume), who consider phage detection as a means of inferring bacterial presence within environments, with an emphasis on faecal contamination of water.

Phage Amplification

Phages possess two key features that make them ideal 'reagents' for bacterial detection and identification: their specificity and their ability to amplify in number in the presence of target bacteria. These two properties are considered in this section within the context of a traditional approach to bacterial identification – the plaque assay. A third key feature, phage genetic malleability, is considered in subsequent sections.

The plaque assay

The phage plaque serves as a means of visualizing phage amplification during growth in a Petri dish. This technique consists simply of infection of a suspected bacterial host with a well-characterized and species-specific phage, followed by plating (Cherry *et al.*, 1954; Stewart *et al.*, 1998). If the bacterial target is phage susceptible, then infection is initiated, resulting in lysis of the host and release or burst of progeny phage. This is followed by continued infection of remaining uninfected bacteria in the vicinity (Kutter *et al.*, 2005). The result of these actions is observed on the bacterial lawn as areas of lysis or clearing called phage plaques (Fig. 10.1a). Traditional plaque assays have been used routinely to identify numerous human bacterial pathogens including *Bacillus anthracis* (Thal and Nordberg, 1968; Abshire *et al.*, 2005), *Campylobacter jejuni* (Grajewski *et al.*, 1985), *Enterococcus faecalis* (Pleceas and Brandis, 1974), *Escherichia coli* (Nicolle *et al.*, 1952), *Listeria monocytogenes* (Shcheglova and Neidbailik, 1968), *Pseudomonas aeruginosa* (Postic and Finland, 1961), *Salmonella typhimurium* (Felix, 1956) and *Staphylococcus aureus* (Wallmark and Laurell, 1952).

Plaque development possesses a number of limitations that can result in false negatives during bacterial identification. Of particular concern is the potential for bacterial strains to differ in their specific mechanisms of phage resistance, that is, failure of plaque formation on hosts that otherwise are the same species as strains that can support plaque growth. A number of reviews of these mechanisms exist (Abedon and Yin, 2009; Hyman and Abedon, 2010; Labrie *et al.*, 2010; Stern and Sorek, 2011). In addition, plaques typically require at least overnight incubation as well as bacterial cultures that are both pure and themselves amplified overnight, problems also seen with more elaborate phage identification techniques (see Williams and LeJeune, Chapter 6, this volume). These characteristics detract from the utility of the conventional plaque assay in most cases as an approach to rapid bacterial identification.

An important exception is seen when considering *Mycobacteria* spp. for which the rapidly growing *Mycobacterium smegmatis* host may be employed as a surrogate indicator during phage amplification on pathogenic *Mycobacterium* strains (Stanley *et al.*, 2007). Such an approach can be particularly useful for characterizing the antibiotic susceptibility of these pathogens. Notwithstanding the

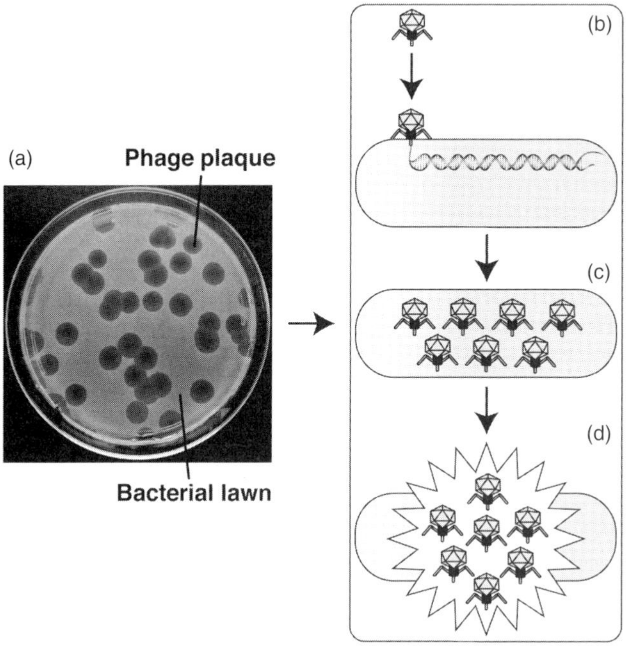

Fig. 10.1. A plaque assay performed by mixing a species-specific phage with an unknown bacteria and growing the bacteria on an agar plate (a). If the target bacteria are susceptible to infection, then phage attachment to the host is followed by insertion of phage genetic material (b), which reprogrammes the bacterial replication machinery to produce numerous progeny phage (c). This is followed by lysis of the host bacterium and release of new phages for subsequent infection (d), resulting in the development of zones of clearance in the lawn, called phage plaques, as shown in (a).

known caveats, the ability of phages to form plaques is illustrative of the utility of phage properties towards bacterial detection and identification properties. These are considered below, along with more modern and sophisticated approaches to phage-based bacterial detection and identification.

Phage properties useful to detection technologies

Plaque formation involves the prototypical phage infection cycle (Fig. 10.1). Familiarity with this infection cycle is useful for understanding other phage-based bacterial detection methods, some of which utilize only certain portions of the cycle.

The phage infection cycle begins with virion adsorption to the bacterial cell surface. This step involves attachment, for example, to Gram-negative outer-membrane porins (Hashemolhosseini *et al.*, 1994) or transporter proteins (Hantke, 1978), or attachment to bacterial cell-wall components such as lipopolysaccharide (Heller, 1992) and Gram-positive peptidoglycan (Valyasevi *et al.*, 1990; Duplessis and Moineau, 2001). These virion–host interactions tend to be highly precise, resulting in adsorption by specific phages only to specific hosts. All phage-based bacterial detection and identification methodologies consequently take advantage of the specificity of adsorption to distinguish among different bacterial types.

The next step of the infection cycle involves transfer of phage genetic material into the host, which occurs by a number of possible mechanisms depending on the phage (Letellier *et al.*, 1999; Perez *et al.*, 2009). This is followed by transcription and translation of phage genes by host cellular machinery (Kutter *et al.*, 2005). Although phage transcription and translation, as well as other molecular processes, can be highly dependent on the biochemistry of specific bacterial hosts, thereby providing additional levels of specificity, the generality of these mechanisms also allows expression of reporter genes from genetically engineered phages. In particular, reporter genes, as considered below, can be more likely to be expressed, as well as more rapid in their expression, than plaque formation, thereby potentially contributing to reductions in both detection times and the occurrence of false negatives.

The final steps in the phage infection cycle are assembly of new virions and virion release, typically via host-cell lysis. Although not all phage-based techniques involve or require phage release, such release is crucial for virion amplification. In addition, lysis can release non-phage bacterial contents, which can also be employed as a means of bacterial detection. Phage amplification in particular can be coupled with modern detection instrumentation to exploit the phage infection process to intensify a detectable signal that is attributable to the presence of the targeted bacteria (Madonna *et al.*, 2007).

Phage Amplification and MALDI-TOF MS

MALDI-TOF MS is an effective tool for analysis and identification of a large range of analyte molecules including peptides, proteins, carbohydrates and polymers. The principal mechanism of MALDI-TOF MS centres on the rapid transfer of target peptides, proteins or other molecular targets into the gas phase by laser ablation of samples embedded in a UV-absorbing matrix, combined with TOF MS analysis. Of particular interest with regard to the application of MALDI-TOF MS to biological detection and identification is its potential for rapid analysis of intact bacteria and viruses (Karas and Hillenkamp, 1988; Tanaka *et al.*, 1988; Holland *et al.*, 1996). Phage amplification-coupled MALDI-TOF MS requires species-specific bacterial infection and allows bacterial detection and identification in as little as 2 h. When combined with disulfide bond reduction methods to enhance MALDI-TOF MS resolution (McAlpin *et al.*, 2010), this method lowers the limit of detection by several orders of magnitude over conventional biochemical detection methods. As such, MALDI-TOF MS protein profiling or mass fingerprinting has emerged as a powerful method of bacterial detection and identification (Holland *et al.*, 1996; Madonna *et al.*, 2003) and is readily combined with phage amplification for rapid bacterial identification (Fig. 10.2).

Mechanisms of MALDI-TOF MS

Typical MALDI-TOF MS applications involve mixing of an analyte with an organic matrix that catalyses sample crystallization and facilitates ionization. Samples are exposed to a laser pulse under vacuum, resulting in the liberation of ions. Surface-ejected ions are accelerated by an electrical field down a TOF tube towards a mass analyser. Smaller analytes are accelerated to higher velocities than larger ones and thus reach the analyser sooner. The TOF for a given target analyte is then proportional to its relative mass to charge ratio (m/z). The charge (z) of an ionized molecule, if given a value of 1, thus makes the m/z value equal to the mass of the molecule. In the case of biological analysis, the output of each detection event can then be compiled in a series of readings to generate a spectral profile for each analyte present in a sample. When plotted graphically, each analyte present within the detection limits of the device is described by its m/z ratio along the horizontal axis and by a peak-intensity value on the vertical axis to arrive at a spectral analyte profile, as exemplified in Fig. 10.3.

MALDI-TOF MS is useful for direct analysis of enriched bacterial cultures. Phage amplification, by virtue of the production of many copies of the phages from each bacterial infection, increases sensitivity by effectively lowering the amount of bacteria necessary to elicit a detectable signal. This concept was used to develop tractable MALDI-TOF MS detection methods for *E. coli* (Madonna *et al.*, 2003, 2007) and *Salmonella enterica* (subsp.

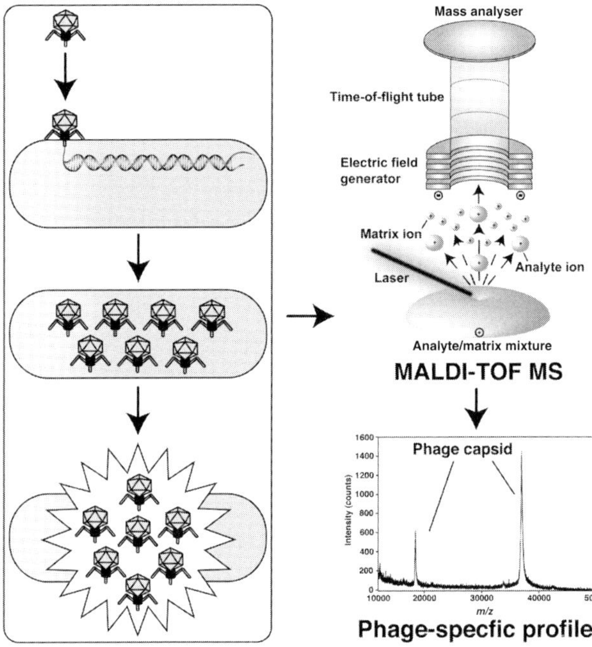

Fig. 10.2. Phage amplification-coupled MALDI-TOF MS for bacterial detection and identification. See text for full description. Shown is a phage-specific protein profile, seen here as a 37.8 kDa major capsid protein peak and its 18.9 kDa doubly charged ion.

enterica serovar Typhimurium) (Rees, 2005). Madonna *et al.* (2003) determined that a typical MALDI-TOF MS limit of detection for *E. coli* of 1.0×10^5 colony-forming units (CFU) ml^{-1} was decreased by two orders of magnitude when coupled with MS2 phage amplification, while Rees and Voorhees (2005) extended this work to demonstrate, with similar sensitivity, the simultaneous detection of *E. coli* and *S. enterica* with MS2 and MPSS1 phage amplification. As an example of MALDI-TOF MS phage amplification analysis, spectra resulting from the infection of a *L. monocytogenes* culture with the *Listeria* typing phage A511 are shown in Fig. 10.3.

A pure *L. monocytogenes* culture yields a mass spectrum showing multiple bacterial proteins (Fig. 10.3a). In contrast, purified A511 phage yields a mass spectrum consisting of one main peak representative of a previously described 48.7 kDa major capsid protein, which is present in many copies (Loessner and Scherer, 1995; Fig. 10.3b). Immediately following infection with phage and bacterial concentrations below the limit of MALDI-TOF MS detection, mass spectra show no discernible bacterial or viral peaks (Fig. 10.3c). After a 12 h incubation at room temperature, however, the same infection reveals the presence of the A511 major capsid protein among the peaks generated by lysed bacterial cells, effectively demonstrating the use of phage amplification and MALDI-TOF MS analysis for bacterial detection without the need for overnight (16–20 h) bacterial culture (Fig. 10.3d).

Phage Amplification and Lateral Flow Immunoassays

Lateral flow immunoassays (LFIs) have been adapted in many formats for detection of a large number of analytes and have proven useful for rapid, phage amplification-based detection and identification using an easy-to-use handheld strip (Fig. 10.4; see also Goodridge and Steiner, Chapter 11, this volume). First described in 1980 and later

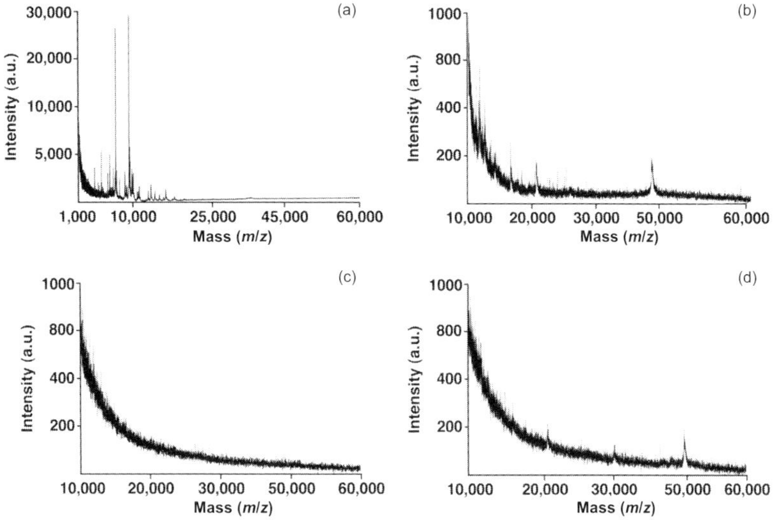

Fig. 10.3. MALDI-TOF MS phage amplification analysis illustrating phage protein profiling for bacterial identification. (a) Representative *Listeria monocytogenes* bacterial protein profile obtained after overnight culture. The intensity scale is expanded to compensate for a very high concentration of bacterial proteins. (b) *L. monocytogenes* phage A511 spectrum showing the abundant 48.7 kDa major capsid protein. (c) Initial infection at phage and bacterial input concentrations below the limit of MALDI-TOF MS detection. (d) Phage amplification at 12 h results in the formation of the distinct capsid protein. This is the earliest time this peak is reliably detected. The slight shift in the mass scale in (b) and (d) are due to inherent drift in MALDI-TOF MS instrument calibration over the course of the experiment.

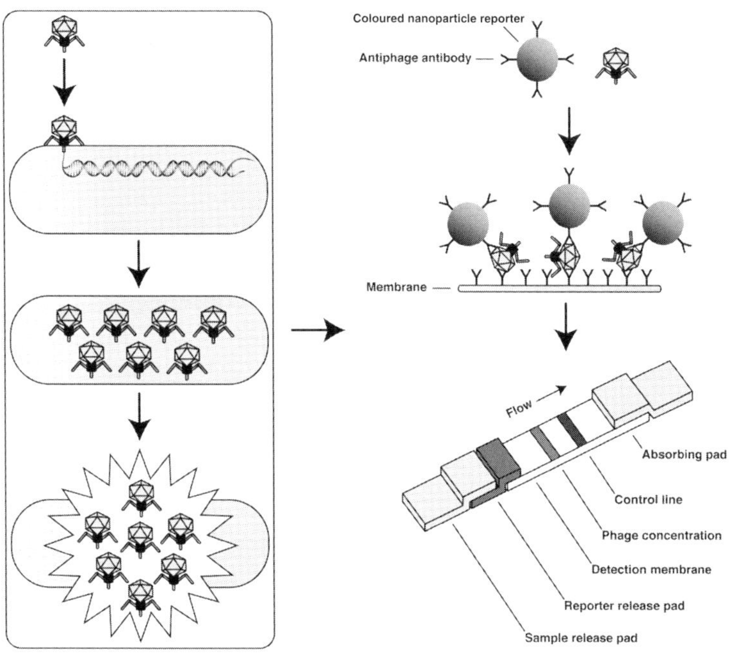

Fig. 10.4. Phage amplification coupled to LFI for bacterial detection and identification. See text for details.

developed as a commercial pregnancy test, the technique was originally termed the 'sol particle immunoassay' (Leuvering *et al.*, 1980). Since then, LFI devices have been utilized with various levels of success for the detection of numerous human bacterial pathogens, including *B. anthracis* (Carter and Cary, 2007), *Brucella abortus* (Clavijo *et al.*, 2003), *E. coli* (Aldus *et al.*, 2003), *Helicobacter pylori* (Kato *et al.*, 2004), *S. aureus* (Fong *et al.*, 2000), *Streptococcus pneumoniae* (Zuiderwijk *et al.*, 2003) and *Treponema pallidum* (Oku *et al.*, 2001). LFIs have also been successfully applied to the detection of a number of human (Al-Yousif *et al.*, 2002; Kikuta *et al.*, 2008; Chaiyaratana *et al.*, 2009; Li *et al.*, 2009), plant (Kusano *et al.*, 2007) and animal (Joon Tam *et al.*, 2004; Lyoo *et al.*, 2005; Sithigorngul *et al.*, 2007) viral pathogens.

Mechanisms of LFIs

While LFIs have been applied to detection of a myriad of analytes, the basic concept remains constant. In the case of phage amplification-coupled LFIs, a patient sample (e.g. urine, serum, skin swab), hospital surface swab or other sample suspected of containing a target bacterial species is mixed with a previously prepared species-specific phage and incubated to allow phage infection and amplification. Following incubation, a primary reporter consisting of coloured nanoparticles conjugated to phage-specific antibodies is added to the infection reaction. An aliquot of this is applied to a test strip (at S in Fig. 10.5) where the liquid sample wicks from the sample pad into a secondary control reporter conjugate release pad (containing a second colour of nanoparticles conjugated to biotin). The secondary reporter and any phage–primary reporter complexes are carried on to a nitrocellulose detection membrane striped with a detection line composed of immobilized, anti-phage antibodies. In the case of a positive test, phage–primary reporter complexes concentrate along this line resulting in a visible, coloured line (T in Fig. 10.5). At the same time, the secondary control reporter is carried across a second detection zone (consisting of a stripe of streptavidin) and concentrates there, to indicate that the sample was completely transported across the detection zone (C on the device in Fig. 10.5). A positive test is thus indicated by the formation of two separate coloured lines – one at the detection line and one at the control line as illustrated in Figs 10.4 and 10.5(a). It follows that a negative test is indicated by the formation of a single control line only.

While effective as a relatively inexpensive, portable means of bacterial detection, conventional LFI applications all depend on lengthy pre-enrichment of target bacteria (24 h minimum) – an undesirable attribute for truly rapid, point-of-care detection and pathogen identification. By coupling species-specific phage amplification with an LFI, rather than bacterial amplification, rapid bacterial detection can be achieved in hours rather than days. As an example, Fig. 10.5 depicts a phage amplification-coupled *B. anthracis* detection device.

Figure 10.5(a) shows a positive result following the addition of *B. anthracis*-specific phage γ to a liquid bacterial sample. A sample

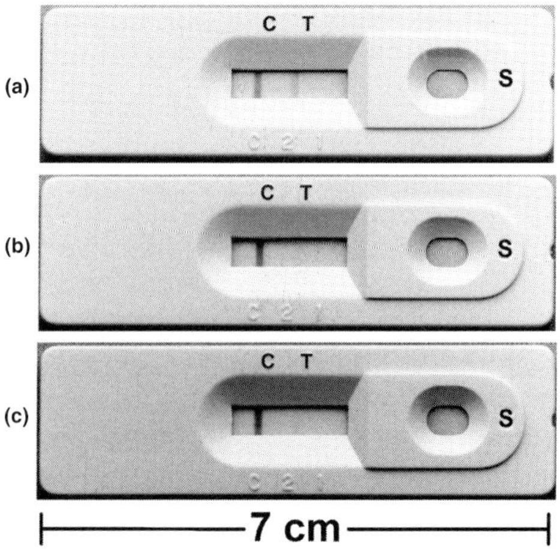

Fig. 10.5. Phage amplification applied to LFI-based detection of *B. anthracis.* (a) A positive test resulting from phage amplification in the presence of the target bacterial species as indicated by the development of two lines. (b) A phage-free, bacteria-free control showing a single control line, which indicates a negative result. (c) A phage-free bacterial culture control showing a single negative control line. C, control line; T, test line; S, sample applicator pad.

(in this case an aliquot of a known bacterial solution at a concentration of 5×10^5 CFU ml^{-1}) was infected with a high-titre phage. After 5 h incubation, an aliquot of the phage infection reaction was applied to the pictured handheld cassette on the sample applicator pad (S) and allowed to wick across the test window. A positive result is indicated by the formation of two lines – one at the test line T and another at the control line C. In order to verify that this positive test was the result of phage amplification, accompanying control tests were run at the same time. Figure 10.5(b) shows the result of tests run without either phage or a bacterial target, verifying the absence of a false positive, as indicated by the formation of only a single control line. The test in Fig. 10.5(c) was conducted with the same bacterial sample used in Fig. 10.5(a) but without the addition of phage. A single test line in the absence of phage was observed only at the control line. Handheld LFI devices such as this allow rapid, user-friendly, species-specific detection of any number of phage–bacterial host pairs. In doing so, this platform bypasses the need for laborious microbial culturing and/or costly DNA amplification methods, or complex detection instrumentation associated with most commonly employed microbial identification methods.

In 2003, Voorhees *et al.* patented the use of phage amplification and LFIs for rapid bacterial identification and antibiotic resistance determination. This led to the founding of Microphage, Inc. (Longmont, CO) for commercialization of this technology. In 2011, Microphage gained approval from the US Food and Drug Administration and a European CE mark for use of phage amplification-based LFI clinical detection and antibiotic resistance determination of methicillin-resistant *S. aureus* (MRSA). The Microphage KeyPath™ MRSA/methicillin-sensitive *S. aureus* (MSSA) blood culture test utilizes phage amplification for handheld, point-of-care *S. aureus* detection and methicillin resistance determination with a sensitivity of 5×10^5 CFU ml^{-1} (after 5.5 h enrichment) and a total analysis time of 5.5 h. By comparison, conventional MRSA identification methods require a minimum of 48 h for analysis.

Phage Immobilization for Bacterial Detection

Phages can be attached to a solid surface such that they can adsorb to their bacterial host while remaining fixed to that surface. In this way, phages have been used as a species-specific biosorbent to capture or separate bacteria from liquid samples for further analysis by secondary methods such as PCR, ELISA and microscopy. The first reported use of immobilized phages for bacterial capture and detection focused on separation and concentration of *Salmonella* from liquid cultures (Bennett *et al.*, 1997). The 'Sapphire' phage was passively attached to microtitre plates or polystyrene dipsticks and used to capture *Salmonella* from pure and mixed bacterial cultures. While requiring only 2 h for concentration of bacteria from an overnight culture, poor capture efficiency (1%) was observed, requiring a minimum of 10^5 CFU ml^{-1} for post-capture PCR detection. Non-specific binding of bacterial cells was observed in phage-free controls, further hindering the usefulness of this approach, but improvements using biotinylated phages have since followed (Fig. 10.6).

In one such approach, *S. enteritidis* phage SJ2 was chemically biotinylated and attached to streptavidin-coated magnetic microparticles (Sun *et al.*, 2001). Phage-coated particles were reacted with overnight cultures of a bioluminescent *lux*-expressing *S. enteritidis* strain, and phage–microparticle complexes were collected with a magnetic particle separator and assayed with a luminometer. This method allowed the capture of 19.3% of cells, with a limit of detection of 2×10^6 CFU ml^{-1}, requiring 20 min for post-adsorption analysis. In a similar approach, coliphage T4 was genetically engineered to express a biotinylated capsid protein to allow attachment to a streptavidin-coated gold electrode (Gervais *et al.*, 2007). *E. coli* growth was monitored following adsorption to immobilized phages by electric cell-substrate impedance (ECIS) at the gold surface. This method demonstrated an approximately 15-fold improvement in biotinylated phage attachment compared with passive phage attachment to untreated gold. A recent

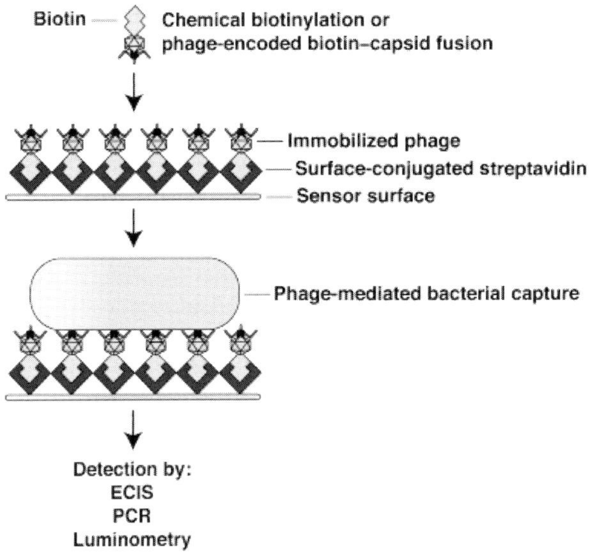

Fig. 10.6. Phage immobilization for bacterial detection. See main text for description. ECIS, electric cell-substrate impedance.

advancement with T4 biotin–capsid fusion and streptavidin-coated magnetic beads was shown to capture 99% of *E. coli* within 10 min, with a detection limit of 8 × 10^2 CFU ml^{-1} measured by real-time PCR (Tolba *et al.*, 2010). Other approaches involve covalent attachment of *Salmonella* phages to glass for bacterial capture (Handa *et al.*, 2008), as well as incorporation of *Salmonella* and *B. anthracis* phages into immobilized magnetoelastic sensors (Guntupalli *et al.*, 2007; Lakshmanan *et al.*, 2007; Huang *et al.*, 2009), although with varied levels of success and detection limits in the 10^3 CFU ml^{-1} range. The use of immobilized phages has also been applied to ELISA-based bacterial detection (Galikowska *et al.*, 2011). By this approach, several *S. enterica* and *E. coli* phages were investigated as primary antibody replacements and non-specifically bound to ELISA plates to facilitate bacterial capture and subsequent secondary detection by conventional alkaline phosphatase-mediated absorbance. Assay sensitivity was reported at 10^6 CFU ml^{-1} and required approximately 5 h. Despite these marked improvements over the last decade, phage immobilization as a component of bacterial detection systems remains at the proof-of-principle stage and has not yet been incorporated into any commercial applications.

Phage-encoded Reporter Genes for Bacterial Detection

Numerous recombinant reporter phages have been created for visual bacterial detection utilizing a range of bioluminescence, chemiluminescence and fluorescence-based approaches. The basic principle centres on incorporation of reporter genes such as bacterial or firefly luciferase, or green fluorescent protein (GFP) into a phage genome. As a result of phage infection, these genes are transferred to a bacterial host and are ultimately expressed to produce a detectible photonic emission that can in turn be used as an indication of the presence of a given bacterial target.

lux-based reporter phages

The well-characterized bacterial luciferase genes, encoded by the *lux* operon from the aquatic Gram-negative *Vibrio fischeri* (Meighen, 1994), are perhaps the most frequently used reporters for incorporation into recombinant phages. The *lux* operon is composed of seven individual open reading frames. The genes *luxA* and *luxB* encode the two-component luciferase holoenzyme responsible for light emission. The *lux* system, in various configurations, has been utilized for construction of recombinant reporter phages. The first description of its use in bacterial detection involved insertion of the complete *lux* operon (Fig. 10.7) into a phage λ-based cloning vector for detection of *E. coli* (Ulitzur and Kuhn, 1987; Ulitzur and Kuhn, 1989) with a reported detection limit of as few as 10 CFU ml^{-1} in contaminated milk within 1 h.

Numerous other iterations using *lux* genes from either *V. fischeri* or *Vibrio harveyi* have since been reported with similar success and include the use of recombinant *luxAB* (in the absence of *luxCDE*) for detection of *L. monocytogenes* (Loessner *et al.*, 1996, 1997), *Salmonella* spp. (Chen and Griffiths, 1996; Kuhn *et al.*, 2002; Thouand *et al.*, 2008) and *E. coli* (Ulitzur and Kuhn, 2000; Waddell and Poppe, 2000). Another recent approach for *E. coli* detection (Ripp *et al.*, 2006; Birmele *et al.*, 2008) involves the use of a phage λ recombinant encoding only *luxI*, which encodes the production of acyl-homoserine lactone (AHL), the freely diffusible component of the quorum-sensing aspect of *lux* activity, and an *E. coli* bioreporter strain carrying *luxCDABE* and *luxR* (also involved in *lux*-associated quorum sensing). The latter strain is added in tandem with the recombinant phage to the test sample. By this method, the phage-infected target bacteria release *luxI*-encoded AHL, which induces bioluminescence in the bacterial reporter strain as an indication of the presence of target *E. coli*. This approach was shown to be particularly sensitive when used with pure cultures, demonstrating detection of 1 CFU ml^{-1} within 10 h, while higher target bacterial concentrations reduced detection times to as little as 1.5 h. Because of its reliance on freely diffusible AHL for activation of reporter bioluminescence, however, this system could be susceptible to non-specific activation by autoinducers produced by other bacterial phylotypes present within mixed bacterial samples.

luc-based reporter phages

The *luc*-encoded luciferase of the firefly, *Photinus pyralis*, is a well-characterized bioluminescent reporter (McElroy and DeLuca, 1983; Viviani, 2002). It has been adapted

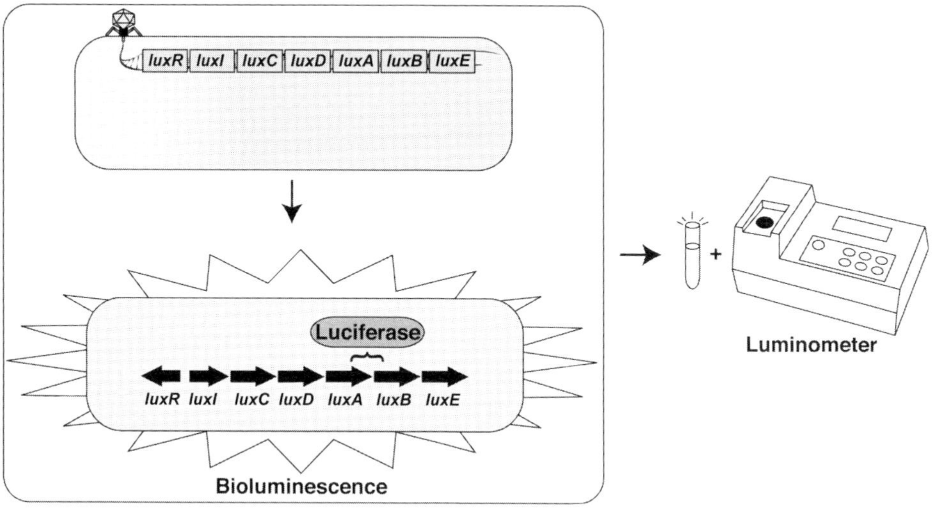

Fig. 10.7. Phage-based bioluminescence for bacterial detection. See text for description.

much in the same way as the *lux* operon for recombinant bioluminescence and is commercially available for use in a wide range of applications (for example, the Luc-Screen® Firefly Luciferase Reporter Gene Assay System; Applied Biosystems). Luc bioluminescence is brought about by ATP-dependent, luciferase-catalysed oxidation of a luciferin substrate (benzothiazole) resulting in light emission (DeLuca and McElroy, 1978). As is the case with many *lux*-based reporter systems, *luc* relies on the exogenous application of the luciferin substrate to a phage-infected culture to bring about light emission.

The first use of firefly luciferase for phage-based bacterial targeting was reported for detection and drug susceptibility determination of *M. smegmatis* and *Mycobacterium tuberculosis* (Jacobs *et al.*, 1993). Recombinant mycobacteriophages carrying the *luc* gene allowed the detection of 5×10^2–5×10 CFU ml^{-1} of *M. smegmatis* within minutes of infection and were capable of distinguishing between antibiotic-resistant and -susceptible *M. smegmatis* and *M. tuberculosis*. Due to the inherently slow growth rate of mycobacteria, however, lengthy pre-enrichment was required to obtain sufficient bacterial concentrations for detection, resulting in total analysis times in excess of 9 days. Since then, several other mycobacteriophage constructs have been engineered to exploit *luc*-based bioluminescence with the aim of increasing assay sensitivity and reducing analysis time (Sarkis *et al.*, 1995; Bardarov *et al.*, 2003; Dusthackeer *et al.*, 2008). While these studies show promise for mycobacterial detection, they are still hampered by extensive pre-enrichment requirements, making it unlikely that these systems will be tractable for rapid bacterial detection and identification in a truly rapid diagnostic capacity.

Fluorescent protein-based reporter phages

GFP and its many derivatives are used in numerous applications and have been expressed in organisms ranging from bacteria and nematodes (Chalfie *et al.*, 1994), insects (Wang and Hazelrigg, 1994), yeast (Kahana *et al.*, 1995) and plants (Haseloff and Amos, 1995) to mammalian cell cultures (Olson *et al.*, 1995). Originally isolated from the bioluminescent hydrozoan jellyfish *Aequorea victoria* (Shimomura *et al.*, 1962), a plethora of commercial GFP-based cloning and expression tools are available from multiple manufacturers and include enhanced green, red, cyan, yellow and blue-shifted variants. While dependent on the presence of oxygen (Heim *et al.*, 1994) and often hampered by low signal-to-noise ratios because of bacterial, tissue or other sample autofluorescence, the advantages offered by GFP-based fluorescence include that it does not require the addition of an enzyme substrate, is relatively resistant to photobleaching in comparison with other fluorophores and produces a signal over a fairly wide emission timeframe.

The first use of GFP-encoding phages for bacterial detection involved cloning GFP into the phage λ genome followed by infection, bacterial expression and epifluorescence microscopy-based detection of *E. coli* (Funatsu *et al.*, 2002). This strategy allowed detection of GFP-expressing bacteria within 4 h of infection but suffered from reduced sensitivity as a result of phage-mediated cell lysis, which resulted in dispersal of the GFP-based signal. Several studies aimed at addressing this and improving the target host range followed. One strategy centred on the use of a GFP-encoding, lysozyme-deficient, broad host range coliphage T4 mutant designed to prevent host-cell destruction (Tanji *et al.*, 2004). Using similar approaches, lytic (Oda *et al.*, 2004) and lysozyme-deficient (Awais *et al.*, 2006) GFP-encoding PP01 mutants were independently engineered. These iterations allowed rapid, highly specific detection and differentiation between O157:H7 and K12 strains within 10 min of infection of 24 h cultures. Interestingly, unlike many methods that require living target cells for phage infection, these approaches, because they relied upon expression of a GFP–minor capsid protein fusion (an externally GFP-tagged phage), allowed the detection of viable but non-culturable, as well as heat-killed (pasteurized), *E. coli* by way of phage adsorption. It should be noted, however, that viable cells gave the most intense fluorescent signal.

Recently, GFP and yellow fluorescence protein (YFP)-encoding mycobacteriophages were reported for fluorescence-based detection and antibiotic resistance determination of *M. tuberculosis* and *M. smegmatis* (Piuri *et al.*, 2009; Rondon *et al.*, 2011). Using epifluorescence microscopy and flow cytometry, these techniques were capable of detecting approximately 100 CFU of phage-infected, GFP-expressing mycobacteria. Total analysis time, including bacterial pre-enrichment was similar to that observed for *luc*-based mycobacterial detection, requiring weeks to obtain sufficient optical densities for experimentation. Fluorescence was observable within 4 h of phage infection, however, which is a significant improvement over conventional, culture-based detection methods commonly employed by clinical laboratories. The use of phage-mediated fluorescence for antibiotic resistance determination of established liquid cultures required 24–48 h, while the same analysis conducted on mycobacteria derived from single colonies required 2–3 weeks.

Phage-conjugated Quantum Dots

Quantum dots (QDs) are colloidal, semiconducting nanocrystals with electronic characteristics, such as fluorescent emission wavelengths, that are closely related to and can be controlled by their crystal diameter (size tunability; Murray *et al.*, 2000). With respect to their applicability to imaging and detection strategies, they exhibit superior photostability in comparison with conventional fluorophores, are more reactive across a broad spectrum of excitation wavelengths and fluoresce at narrow, size-dependent emission spectra. These attributes, in combination with recent advances in surface conjugation chemistry, make QDs ideal candidates as probes for imaging of biological systems and bacterial detection (Dubertret *et al.*, 2002; Hahn *et al.*, 2005; Li *et al.*, 2007). Moreover, the tunability of QDs over a wide emission bandwidth makes multiplexed detection of numerous targets in a single sample a favourable possibility.

The use of QDs in combination with species-specific phages was first explored using *E. coli* and biotinylated coliphage T7 (Edgar *et al.*, 2006). T7 was genetically engineered to express a small peptide fused to its major capsid protein, which, following infection and production of progeny phages, was post-translationally biotinylated by host cell biotin ligases. Once released from host cells, biotinylated progeny phages were then tagged by the addition of streptavidin-conjugated QDs. Phage–QD complexes extracted from mixed bacterial cultures were then visualized by flow cytometry and fluorescence microscopy 1 h after infection with detection limits of around 10 CFU ml^{-1}. A similar approach used biotinylatable phage λ and QDs in combination with conventional and image-based flow cytometry to achieve phage-based signal quantification (Yim *et al.*, 2009).

Fluorescent Phage Endolysin Cell Wall-binding Domains

Phage-encoded endolysins are peptidoglycan hydrolases that serve as one component of the phage lytic system (Fishetti, 2005). They are produced in the host-cell cytoplasm during phage infection but target the host cell-wall peptidoglycan, resulting in weakening and rupture of the bacterial cell along with subsequent release of phage progeny. Structurally, endolysins are composed of an enzymatically active N-terminal domain and a cell wall-binding C-terminal domain (CBD) (Moak and Molineux, 2004). To date, numerous endolysins have been sequenced, purified and applied from without to destroy bacteria and explore their use as potential alternatives to antibiotics (Loeffler *et al.*, 2001; Nelson *et al.*, 2001; Schuch *et al.*, 2002; see also Shen *et al.*, Chapter 15, this volume). Their use is largely limited to detecting Gram-positive bacteria, however, because the Gram-negative outer membrane serves as a protective barrier against exogenously applied endolysins. Nevertheless, endolysins are potentially powerful tools, not only for biocontrol but also for highly specific detection.

In one of the most recent of the new approaches to phage-based bacterial detection, phage endolysin CBDs were paired with green, blue, cyan, yellow and red fluorescent reporter proteins, magnetic separation and fluorescence microscopy to allow the capture and detection of *Listeria* (Schmelcher *et al.*, 2010). Using several different CBD–fluorescent protein fusion-coated magnetic particles for bacterial separation, multiplexed detection of *Listeria* strains in mixed cultures as well as in spiked milk and cheese samples were successfully carried out. Bacterial recovery rates of greater than 90% were observed following 24 h of enrichment and 16 h of selective plating of the captured cells.

Conclusions

Despite the growing need for rapid, inexpensive methods of bacterial detection and identification, most current methods fall short as user-friendly, point-of-care diagnostics due to their complexity, lack of sensitivity and reliance on expensive equipment and highly specialized facilities and expertise, as well as their dependence on time-consuming microbiological pre-enrichment. By exploiting the highly species-specific phage life cycle and accompanying amplification, the adsorption or attachment affinity for their bacterial hosts, their ability to take up and transfer reporter genes and display surface-expressed molecular tags, phage-based detection methods, depending on the application, bypass nearly all of these pitfalls. Phage amplification-coupled MALDI-TOF MS, for example, allows instrumental advances in bacterial detection via rapid, species-specific phage protein profiling. This novel signal amplification method allows bacterial detection and identification in as little as 2 h, while lowering the limits of detection by several orders of magnitude over conventional spectroscopy and phage-typing methods.

With the goal of bypassing the need for complex instrumentation and highly trained personnel, methods for the use of phage amplification-coupled LFI provide a combination of both rapid and sensitive bacterial detection in an inexpensive, portable configuration, which is ideal for point-of-care applications. The first of such technologies is the recently commercialized Microphage KeyPath^{TM} system, which takes advantage of phage amplification for patient screening for MRSA and provides a rapid, sensitive method for point-of-care diagnostic bacterial identification in a fraction of the time required by conventional laboratory-based techniques.

Newly refined phage immobilization techniques continue to advance our ability to quickly and specifically isolate and concentrate target bacteria for analysis with existing detection instrumentation. Fluorescence-based bacterial detection by targeted infection and delivery of fluorescent reporter systems by recombinant phages show great promise as multiplexed bioluminescent and fluorescent detection tools, and the emerging use of highly refined phage endolysins show strong potential in high-efficiency bacterial capture and identification. In all these approaches, phage-based bacterial detection methods have proven to be equal, if not superior, to conventional methods in every category and stand to transform diagnostic microbiology in the very near future.

Acknowledgements

Portions of this chapter describing work conducted by the author were supported by the Defense Threat Reduction Agency of the United States Department of Defense.

References

Abedon, S.T. and Yin, J. (2009) Bacteriophage plaques: theory and analysis. *Methods in Molecular Biology* 501, 161–174.

Abshire, T.G., Brown, J.E. and Ezzell, J.W. (2005) Production and validation of the use of gamma phage for identification of *Bacillus anthracis*. *Journal of Clinical Microbiology* 43, 4780–4788.

Al-Yousif, Y., Anderson, J., Chard-Bergstrom, C. and Kapil, S. (2002) Development, evaluation, and application of lateral-flow immunoassay (immunochromatography) for detection of rotavirus in bovine fecal samples. *Clinical Diagnostic and Laboratory Immunology* 9, 723–725.

Aldus, C.F., van Amerongen, A., Ariens, R.M., Peck, M.W., Wichers, J.H. and Wyatt, G.M. (2003) Principles of some novel rapid dipstick methods for detection and characterization of verotoxigenic *Escherichia coli*. *Journal of Applied Microbiology* 95, 380–389.

Awais, R., Fukudomi, H., Miyanaga, K., Unno, H. and Tanji, Y. (2006) A recombinant bacteriophage-based assay for the discriminative detection of culturable and viable but nonculturable *Escherichia coli* O157:H7. *Biotechnology Progress* 22, 853–859.

Bardarov, S. Jr, Dou, H., Eisenach, K., Banaiee, N., Ya, S., Chan, J., Jacobs, W.R. Jr and Riska, P.F. (2003) Detection and drug-susceptibility testing of *M. tuberculosis* from sputum samples using luciferase reporter phage: comparison with the Mycobacteria Growth Indicator Tube (MGIT) system. *Diagnostic Microbiology and Infectious Disease* 45, 53–61.

Bennett, A.R., Davids, F.G., Vlahodimou, S., Banks, J.G. and Betts, R.P. (1997) The use of bacteriophage-based systems for the separation and concentration of *Salmonella*. *Journal of Applied Microbiology* 83, 259–265.

Birmele, M., Ripp, S., Jegier, P., Roberts, M.S., Sayler, G. and Garland, J. (2008) Characterization and validation of a bioluminescent bioreporter for the direct detection of *Escherichia coli*. *Journal of Microbiological Methods* 75, 354–356.

Bottger, E.C. (1989) Rapid determination of bacterial ribosomal RNA sequences by direct sequencing of enzymatically amplified DNA. *FEMS Microbiology Letters* 53, 171–176.a

Carter, D.J. and Cary, R.B. (2007) Lateral flow microarrays: a novel platform for rapid nucleic acid detection based on miniaturized lateral flow chromatography. *Nucleic Acids Research* 35, e74.

Chaiyaratana, W., Chuansumrit, A., Pongthanapisith, V., Tangnararatchakit, K., Lertwongrath, S. and Yoksan, S. (2009) Evaluation of dengue nonstructural protein 1 antigen strip for the rapid diagnosis of patients with dengue infection. *Diagnostic Microbiology and Infectious Disease* 64, 83–84.

Chalfie, M., Tu, Y., Euskirchen, G., Ward, W.W. and Prasher, D.C. (1994) Green fluorescent protein as a marker for gene expression. *Science* 263, 802–805.

Chen, J. and Griffiths, M.W. (1996) Luminescent *Salmonella* strains as real time reporters of growth and recovery from sublethal injury in food. *International Journal of Food Microbiology* 31, 27–43.

Cherry, W.B., Davis, B.R., Edwards, P.R. and Hogan, R.B. (1954) A simple procedure for the identification of the genus *Salmonella* by means of a specific bacteriophage. *Journal of Laboratory and Clinical Medicine* 44, 51–55.

Clarridge, J.E. III (2004) Impact of 16S rRNA gene sequence analysis for identification of bacteria on clinical microbiology and infectious diseases. *Clinical Microbiological Reviews* 17, 840–862.

Clavijo, E., Diaz, R., Anguita, A., Garcia, A., Pinedo, A. and Smits, H.L. (2003) Comparison of a dipstick assay for detection of *Brucella*-specific immunoglobulin M antibodies with other tests for serodiagnosis of human brucellosis. *Clinical Diagnostic and Laboratory Immunology* 10, 612–615.

DeLuca, M. and McElroy, W.D. (1978) Purification and properties of firefly luciferase. *Methods in Enzymology* 57, 3–15.

Duan, G., Liu, Y. and Qi, G. (1991) Use of multilocus enzyme electrophoresis for bacterial population genetics, classification and molecular epidemiology. *Zhonghua Liu Xing Bing Xue Za Zhi* 12, 177–181.

Dubertret, B., Skourides, P., Norris, D.J., Noireaux, V., Brivanlou, A.H. and Libchaber, A. (2002) *In vivo* imaging of quantum dots encapsulated in phospholipid micelles. *Science* 298, 1759–1762.

Duplessis, M. and Moineau, S. (2001) Identification of a genetic determinant responsible for host specificity in *Streptococcus thermophilus* bacteriophages. *Molecular Microbiology* 41, 325–336.

Durmaz, R., Otlu, B., Koksal, F., Hosoglu, S., Ozturk, R., Ersoy, Y., Aktas, E., Gursoy, N.C. and Caliskan, A. (2009) The optimization of a rapid pulsed-field gel electrophoresis protocol for the typing of *Acinetobacter baumannii*, *Escherichia coli* and *Klebsiella* spp. *Japanese Journal of Infectious Disease* 62, 372–377.

Dusthackeer, A., Kumar, V., Subbian, S., Sivaramakrishnan, G., Zhu, G., Subramanyam, B., Hassan, S., Nagamaiah, S., Chan, J. and Paranji Rama, N. (2008) Construction and evaluation of luciferase reporter phages for the detection of active and non-replicating tubercle bacilli. *Journal of Microbiological Methods* 73, 18–25.

Edgar, R., McKinstry, M., Hwang, J., Oppenheim, A.B., Fekete, R.A., Giulian, G., Merril, C., Nagashima, K. and Adhya, S. (2006) High-sensitivity bacterial detection using biotin-tagged phage and quantum-dot nanocomplexes. *Proceedings of the National Academy of Sciences USA* 103, 4841–4845.

Felix, A. (1956) Phage typing of *Salmonella typhimurium*: its place in epidemiological and epizootiological investigations. *Journal of General Microbiology* 14, 208–222.

Fishetti, V.A. (2005). The use of phage lytic enzymes to control bacterial intections In: Kutter, E. and Sulakvelidze, A. (eds) *Bacteriophages Biology and Applications.* CRC Press, New York, NY, pp. 321–334.

Fong, W.K., Modrusan, Z., McNevin, J.P., Marostenmaki, J., Zin, B. and Bekkaoui, F. (2000) Rapid solid-phase immunoassay for detection of methicillin-resistant *Staphylococcus aureus* using cycling probe technology. *Journal of Clinical Microbiology* 38, 2525–2529.

Fredericks, D.N. and Relman, D.A. (1996) Sequence-based identification of microbial pathogens: a reconsideration of Koch's postulates. *Clinical Microbiological Reviews* 9, 18–33.

Funatsu, T., Taniyama, T., Tajima, T., Tadakuma, H. and Namiki, H. (2002) Rapid and sensitive detection method of a bacterium by using a GFP reporter phage. *Microbiology and Immunology* 46, 365–369.

Galikowska, E., Kunikowska, D., Tokarska-Pietrzak, E., Dziadziuszko, H., Los, J.M., Golec, P., Wegrzyn, G. and Los, M. (2011) Specific detection of *Salmonella enterica* and *Escherichia coli* strains by using ELISA with bacteriophages as recognition agents. *European Journal of Clinical Microbiology and Infectious Disease* 30, 1067–1073.

Gervais, L., Gel, M., Allain, B., Tolba, M., Brovoko, L., Zourob, M., Mandeville, R., Griffiths, M. and Evoy, S. (2007) Immobilization of biotinylated bacteriophages on biosensor surfaces. *Sensors and Actuators* B125, 615–621.

Grajewski, B.A., Kusek, J.W. and Gelfand, H.M. (1985) Development of a bacteriophage typing system for *Campylobacter jejuni* and *Campylobacter coli. Journal of Clinical Microbiology* 22, 13–18.

Grisold, A.J., Zarfel, G., Strenger, V., Feierl, G., Leitner, E., Masoud, L., Hoenigl, M., Raggam, R.B., Dosch, V. and Marth, E. (2010) Use of automated repetitive-sequence-based PCR for rapid laboratory confirmation of nosocomial outbreaks. *Journal of Infection* 60, 44–51.

Guntupalli, R., Lakshmanan, R.S., Hu, J., Huang, T.S., Barbaree, J.M., Vodyanoy, V. and Chin, B.A. (2007) Rapid and sensitive magnetoelastic biosensors for the detection of *Salmonella typhimurium* in a mixed microbial population. *Journal of Microbiological Methods* 70, 112–118.

Hahn, M.A., Tabb, J.S. and Krauss, T.D. (2005) Detection of single bacterial pathogens with semiconductor quantum dots. *Analytical Chemistry* 77, 4861–4869.

Handa, H., Gurczynski, S., Jackson, M.P., Auner, G. and Mao, G. (2008) Recognition of *Salmonella Typhimurium* by immobilized phage P22 monolayers. *Surface Science* 602, 1392–1400.

Hantke, K. (1978) Major outer membrane proteins of *E. coli* K12 serve as receptors for the phages T2 (protein Ia) and 434 (protein Ib). *Molecular and General Genetics* 164, 131–135.

Haseloff, J. and Amos, B. (1995) GFP in plants. *Trends in Genetics* 11, 328–329.

Hashemolhosseini, S., Montag, D., Kramer, L. and Henning, U. (1994) Determinants of receptor specificity of coliphages of the T4 family. A chaperone alters the host range. *Journal of Molecular Biology* 241, 524–533.

Heim, R., Prasher, D.C. and Tsien, R.Y. (1994) Wavelength mutations and posttranslational autoxidation of green fluorescent protein. *Proceedings of the National Academy of Sciences USA* 91, 12501–12504.

Heller, K.J. (1992) Molecular interaction between bacteriophage and the Gram-negative cell envelope. *Archives of Microbiology* 158, 235–248.

Holland, R.D., Wilkes, J.G., Rafii, F., Sutherland, J.B., Persons, C.C., Voorhees, K.J. and Lay, J.O. Jr (1996) Rapid identification of intact whole bacteria based on spectral patterns using matrix-assisted laser desorption/ionization with time-of-flight mass spectrometry. *Rapid Communications in Mass Spectrometry* 10, 1227–1232.

Huang, S., Yang, H., Lakshmanan, R.S., Johnson, M.L., Wan, J., Chen, I.H., Wikle, H.C. III, Petrenko, V.A., Barbaree, J.M. and Chin, B.A. (2009) Sequential detection of *Salmonella typhimurium* and *Bacillus anthracis* spores using magnetoelastic biosensors. *Biosensors and Bioelectronics* 24, 1730–1736.

Hyman, P. and Abedon, S.T. (2010) Bacteriophage host range and bacterial resistance. *Advances in Applied Microbiology* 70, 217–248.

Jacobs, W.R. Jr, Barletta, R.G., Udani, R., Chan, J., Kalkut, G., Sosne, G., Kieser, T., Sarkis, G.J., Hatfull, G.F. and Bloom, B.R. (1993) Rapid assessment of drug susceptibilities of *Mycobacterium tuberculosis* by means of luciferase reporter phages. *Science* 260, 819–822.

Jayarao, B.M., Dore, J.J. Jr and Oliver, S.P. (1992) Restriction fragment length polymorphism analysis of 16S ribosomal DNA of *Streptococcus* and *Enterococcus* species of bovine origin. *Journal of Clinical Microbiology* 30, 2235–2240.

Joon Tam, Y., Mohd Lila, M.A. and Bahaman, A.R. (2004) Development of solid-based paper strips for rapid diagnosis of pseudorabies infection. *Tropical Biomedicine* 21, 121–134.

Kahana, J.A., Schnapp, B.J. and Silver, P.A. (1995) Kinetics of spindle pole body separation in budding yeast. *Proceedings of the National Academy of Sciences USA* 92, 9707–9711.

Karas, M. and Hillenkamp, F. (1988) Laser desorption ionization of proteins with molecular masses exceeding 10,000 daltons. *Analytical Chemistry* 60, 2299–2301.

Kato, S., Ozawa, K., Okuda, M., Nakayama, Y., Yoshimura, N., Konno, M., Minoura, T. and Iinuma, K. (2004) Multicenter comparison of rapid lateral flow stool antigen immunoassay and stool antigen enzyme immunoassay for the diagnosis of *Helicobacter pylori* infection in children. *Helicobacter* 9, 669–673.

Kikuta, H., Sakata, C., Gamo, R., Ishizaka, A., Koga, Y., Konno, M., Ogasawara, Y., Sawada, H., Taguchi, Y., Takahashi, Y., Yasuda, K., Ishiguro, N., Hayashi, A., Ishiko. H. and Kobayashi, K. (2008) Comparison of a lateral-flow immunochromatography assay with real-time reverse transcription-PCR for detection of human metapneumovirus. *Journal of Clinical Microbiology* 46, 928–932.

Koeleman, J.G., Stoof, J., Biesmans, D.J., Savelkoul, P.H. and Vandenbroucke-Grauls, C.M. (1998) Comparison of amplified ribosomal DNA restriction analysis, random amplified polymorphic DNA analysis and amplified fragment length polymorphism fingerprinting for identification of *Acinetobacter* genomic species and typing of *Acinetobacter baumannii*. *Journal of Clinical Microbiology* 36, 2522–2529.

Kuhn, J., Suissa, M., Wyse, J., Cohen, I., Weiser, I., Reznick, S., Lubinsky-Mink, S., Stewart, G. and Ulitzur, S. (2002) Detection of bacteria using foreign DNA: the development of a bacteriophage reagent for *Salmonella*. *International Journal of Food Microbiology* 74, 229–238.

Kusano, N., Hirashima, K., Kuwahara, M., Narahara, K., Imamura, T., Mimori, T., Nakahira, K. and Torii, K. (2007) Immunochromatographic assay for simple and rapid detection of Satsuma dwarf virus and related viruses using monoclonal antibodies. *Journal of General Plant Pathology* 73, 66–71.

Kutter, E., Raya, R. and Carlson, K. (2005). Molecular mechanisms of phage infection. In: Kutter, E. and Sulakvelidze, A. (eds) *Bacteriophages Biology and Applications*. CRC Press, New York, NY, pp. 165–222.

Labrie, S.J., Samson, J.E. and Moineau, S. (2010) Bacteriophage resistance mechanisms. *Nature Reviews Microbiology* 8, 317–327.

Lakshmanan, R.S., Guntupalli, R., Hu, J., Kim, D.J., Petrenko, V.A., Barbaree, J.M. and Chin, B.A. (2007) Phage immobilized magnetoelastic sensor for the detection of *Salmonella typhimurium*. *Journal of Microbiological Methods* 71, 55–60.

Letellier, L., Plancon, L., Bonhivers, M. and Boulanger, P. (1999) Phage DNA transport across membranes. *Research in Microbiology* 150, 499–505.

Leuvering, J.H., Thal, P.J., van der Waart, M. and Schuurs, A.H. (1980) Sol particle immunoassay (SPIA). *Journal of Immunoassay* 1, 77–91.

Li, L., Zhou, L., Yu, Y., Zhu, Z., Lin, C., Lu, C. and Yang, R. (2009) Development of up-converting phosphor technology-based lateral-flow assay for rapidly quantitative detection of hepatitis B surface antibody. *Diagnostic Microbiology and Infectious Disease* 63, 165–172.

Li, Z.B., Cai, W. and Chen, X. (2007) Semiconductor quantum dots for *in vivo* imaging. *Journal of Nanoscience and Nanotechnology* 7, 2567–2581.

Loeffler, J.M., Nelson, D. and Fischetti, V.A. (2001) Rapid killing of *Streptococcus pneumoniae* with a bacteriophage cell wall hydrolase. *Science* 294, 2170–2172.

Loessner, M.J. and Scherer, S. (1995) Organization and transcriptional analysis of the *Listeria* phage A511 late gene region comprising the major capsid and tail sheath protein genes *cps* and *tsh*. *Journal of Bacteriology* 177, 6601–6609.

Loessner, M.J., Rees, C.E., Stewart, G.S. and Scherer, S. (1996) Construction of luciferase reporter bacteriophage A511::*luxAB* for rapid and sensitive detection of viable *Listeria* cells. *Applied and Environmental Microbiology* 62, 1133–1140.

Loessner, M.J., Rudolf, M. and Scherer, S. (1997) Evaluation of luciferase reporter bacteriophage A511::*luxAB* for detection of *Listeria monocytogenes* in contaminated foods. *Applied and Environmental Microbiology* 63, 2961–2965.

Lyoo, Y.S., Kleiboeker, S.B., Jang, K.Y., Shin, N.K., Kang, J.M., Kim, C.H., Lee, S.J. and Sur, J.H. (2005) A simple and rapid chromatographic strip test for detection of antibody to porcine reproductive and respiratory syndrome virus. *Journal of Veterinary Diagnostic Investigation* 17, 469–473.

Madonna, A.J., Van Cuyk, S. and Voorhees, K.J. (2003) Detection of *Escherichia coli* using immunomagnetic separation and bacteriophage amplification coupled with matrix-assisted laser desorption/ionization time-of-flight mass spectrometry. *Rapid Communication in Mass Spectrometry* 17, 257–263.

Madonna, A.J., Voorhees, K.J. and Rees, J.C. (2007) Method for detection of low concentrations of a target bacterium that uses phages to infect target bacterial cells. US Patents and Trademark Office, Colorado School of Mines, Golden, CO.

Maiden, M.C., Bygraves, J.A., Feil, E., Morelli, G., Russell, J.E., Urwin, R., Zhang, Q., Zhou, J., Zurth, K., Caugant, D.A., Feavers, I.M., Achtman, M. and Spratt, B.G. (1998) Multilocus sequence typing: a portable approach to the identification of clones within populations of pathogenic microorganisms. *Proceedings of the National Academy of Sciences USA* 95, 3140–3145.

Maquelin, K., Dijkshoorn, L., van der Reijden, T.J. and Puppels, G.J. (2006) Rapid epidemiological analysis of *Acinetobacter* strains by Raman spectroscopy. *Journal of Microbiological Methods* 64, 126–131.

McAlpin, C., Cox, C.R., Matyi, S. and Voorhees, K.J. (2010) Enhanced MALDI-TOF MS analysis of bacteriophage major capsid proteins with b-mercaptoethanol pretreatment. *Rapid Communications in Mass Spectrometry* 24, 11–14.

McElroy, W.D. and DeLuca, M.A. (1983) Firefly and bacterial luminescence: basic science and applications. *Journal of Applied Biochemistry* 5, 197–209.

Meighen, E.A. (1994) Genetics of bacterial bioluminescence. *Annual Review of Genetics* 28, 117–139.

Moak, M. and Molineux, I.J. (2004) Peptidoglycan hydrolytic activities associated with bacteriophage virions. *Molecular Microbiology* 51, 1169–1183.

Murray, C.B., Kagan, C.R. and Bawendi, M.G. (2000) Synthesis and characterization of monodisperse nanocrystals and close-packed nanocrystal assemblies. *Annual Review of Materials* 30, 545–610.

Nelson, D., Loomis, L. and Fischetti, V.A. (2001) Prevention and elimination of upper respiratory colonization of mice by group A streptococci by using a bacteriophage lytic enzyme. *Proceedings of the National Academy of Sciences USA* 98, 4107–4112.

Nicolle, P., Le Minor, L., Buttiaux, R. and Ducrest, P. (1952) Phage typing of *Escherichia coli* isolated from cases of infantile gastroenteritis. II. Relative frequency of types in different areas and the epidemiological value of the method. *Bulletin de l'Académie Nationale de Médecine* 136, 483–485.

Oda, M., Morita, M., Unno, H. and Tanji, Y. (2004) Rapid detection of *Escherichia coli* O157:H7 by using green fluorescent protein-labeled PP01 bacteriophage. *Applied and Environmental Microbiology* 70, 527–534.

Oku, Y., Kamiya, K., Kamiya, H., Shibahara, Y., Ii, T. and Uesaka, Y. (2001) Development of oligonucleotide lateral-flow immunoassay for multi-parameter detection. *Journal of Immunological Methods* 258, 73–84.

Olson, K.R., McIntosh, J.R. and Olmsted, J.B. (1995) Analysis of MAP 4 function in living cells using green fluorescent protein (GFP) chimeras. *Journal of Cellular Biology* 130, 639–650.

Patel, I.S., Premasiri, W.R., Moir, D.T. and Ziegler, L.D. (2008) Barcoding bacterial cells: a SERS based methodology for pathogen identification. *Journal of Raman Spectroscopy* 39, 1660–1672.

Perez, G.L., Huynh, B., Slater, M. and Maloy, S. (2009) Transport of phage P22 DNA across the cytoplasmic membrane. *Journal of Bacteriology* 191, 135–140.

Piuri, M., Jacobs, W.R. Jr and Hatfull, G.F. (2009) Fluoromycobacteriophages for rapid, specific, and sensitive antibiotic susceptibility testing of *Mycobacterium tuberculosis*. *PLoS One* 4, e4870.

Pleceas, P. and Brandis, H. (1974) Rapid group and species identification of enterococci by means of tests with pooled phages. *Journal of Medical Microbiology* 7, 529–533.

Postic, B. and Finland, M. (1961) Observations on bacteriophage typing of *Pseudomonas aeruginosa*. *Journal of Clinical Investigation* 40, 2064–2075.

Rees, J.C. (2005). Detection of bacteria using matrix-assisted laser desorption-ionization mass spectrometry and immunodiagnostics with an emphasis on bacteriophage amplification detection. Thesis, Colorado School of Mines, Golden, CO.

Rees, J.C. and Voorhees, K.J. (2005) Simultaneous detection of two bacterial pathogens using bacteriophage amplification coupled with matrix-assisted laser desorption/ionization time-of-flight mass spectrometry. *Rapid Communications in Mass Spectrometry* 19, 2757–2761.

Ripp, S., Jegier, P., Birmele, M., Johnson, C.M., Daumer, K.A., Garland, J.L. and Sayler, G.S. (2006) Linking bacteriophage infection to quorum sensing signalling and bioluminescent bioreporter monitoring for direct detection of bacterial agents. *Journal of Applied Microbiology* 100, 488–499.

Rondon, L., Piuri, M., Jacobs, W.R. Jr, de Waard, J., Hatfull, G.F. and Takiff, H.E. (2011) Evaluation of fluoromycobacteriophages for detecting drug resistance in *Mycobacterium tuberculosis*. *Journal of Clinical Microbiology* 49, 1838–1842.

Ross, I.L., Davos, D.E., Mwanri, L., Raupach, J. and Heuzenroeder, M.W. (2011) MLVA and phage typing as complementary tools in the epidemiological investigation of *Salmonella enterica* serovar *typhimurium* clusters. *Current Microbiology* 62, 1034–1038.

Sarkis, G.J., Jacobs, W.R. Jr and Hatfull, G.F. (1995) L5 luciferase reporter mycobacteriophages: a sensitive tool for the detection and assay of live mycobacteria. *Molecular Microbiology* 15, 1055–1067.

Schmelcher, M., Shabarova, T., Eugster, M.R., Eichenseher, F., Tchang, V.S., Banz, M. and Loessner, M.J. (2010) Rapid multiplex detection and differentiation of *Listeria* cells by use of fluorescent phage endolysin cell wall binding domains. *Applied and Environmental Microbiology* 76, 5745–5756.

Schuch, R., Nelson, D. and Fischetti, V.A. (2002) A bacteriolytic agent that detects and kills *Bacillus anthracis*. *Nature* 418, 884–889.

Selander, R.K., Caugant, D.A., Ochman, H., Musser, J.M., Gilmour, M.N. and Whittam, T.S. (1986) Methods of multilocus enzyme electrophoresis for bacterial population genetics and systematics. *Applied and Environmental Microbiology* 51, 873–884.

Shcheglova, M.K. and Neidbailik, I.N. (1968) Experience in phage typing of *Listeria*. *Veterinariia* 45, 102–103.

Shimomura, O., Johnson, F.H. and Saiga, Y. (1962) Extraction, purification and properties of aequorin, a bioluminescent protein from the luminous hydromedusan, *Aequorea*. *Journal of Cellular and Comparative Physiology* 59, 223–239.

Sithigorngul, W., Rukpratanporn, S., Sittidilokratna, N., Pecharaburanin, N., Longyant, S., Chaivisuthangkura, P. and Sithigorngul, P. (2007) A convenient immunochromatographic test strip for rapid diagnosis of yellow head virus infection in shrimp. *Journal of Virology Methods* 140, 193–199.

Stanley, E.C., Mole, R.J., Smith, R.J., Glenn, S.M., Barer, M.R., McGowan, M. and Rees, C.E. (2007) Development of a new, combined rapid method using phage and PCR for detection and identification of viable *Mycobacterium paratuberculosis* bacteria within 48 hours. *Applied and Environmental Microbiology* 73, 1851–1857.

Stern, A. and Sorek, R. (2011) The phage–host arms race: shaping the evolution of microbes. *Bioessays* 33, 43–51.

Stewart, G.S., Jassim, S.A., Denyer, S.P., Newby, P., Linley, K. and Dhir, V.K. (1998) The specific and sensitive detection of bacterial pathogens within 4 h using bacteriophage amplification. *Journal of Applied Microbiology* 84, 777–783.

Sun, W., Brovko, L. and Griffiths, M. (2001) Use of bioluminescent *Salmonella* for assessing the efficiency of constructed phage-based biosorbent. *Journal of Industrial Microbiology and Biotechnology* 27, 126–128.

Tanaka, K., Waki, H., Ido, Y., Akita, S., Yoshida, Y. and Yoshida, T. (1988) Protein and polymer analyses up to *m/z* 100,000 by laser ionization time-of-flight mass spectrometry. *Rapid Communications in Mass Spectrometry* 2, 151–153.

Tanji, Y., Furukawa, C., Na, S.H., Hijikata, T., Miyanaga, K. and Unno, H. (2004) *Escherichia coli* detection by GFP-labeled lysozyme-inactivated T4 bacteriophage. *Journal of Biotechnology* 114, 11–20.

Thal, E. and Nordberg, B.K. (1968) On the diagnostic of *Bacillus anthracis* with bacteriophages. *Berliner und Münchener Tierärztliche Wochenschrift* 81, 11–13.

Thouand, G., Vachon, P., Liu, S., Dayre, M. and Griffiths, M.W. (2008) Optimization and validation of a simple method using P22::*luxAB* bacteriophage for rapid detection of *Salmonella enterica* serotypes A, B, and D in poultry samples. *Journal of Food Protection* 71, 380–385.

Tolba, M., Minikh, O., Brovko, L.Y., Evoy, S. and Griffiths, M.W. (2010) Oriented immobilization of bacteriophages for biosensor applications. *Applied and Environmental Microbiology* 76, 528–535.

Ulitzur, S. and Kuhn, J., (1987) Introduction of *lux* genes into bacteria: a new approach for specific determination of bacteria and their antibiotic susceptibility. In: Slomerich, R., Andreesen, R., Kapp, A., Ernst, M. and Woods, W.G. (eds) *Bioluminescence and Chemiluminescence: New Perspectives*. John Wiley & Sons, New York, NY, pp. 463–472.

Ulitzur, S. and Kuhn, J. (1989) Detection and/or identificaiton of microorganisms in a test sample using bioluminescence or other exogenous genetically introduced marker. US Patent no. 4,861,709.

Ulitzur, S. and Kuhn, J. (2000) Construction of *lux* bacteriophages and the determination of specific bacteria and their antibiotic sensitivities. *Methods in Enzymology* 305, 543–557.

Valyasevi, R., Sandine, W.E. and Geller, B.L. (1990) The bacteriophage kh receptor of *Lactococcus lactis* subsp. *cremoris* KH is the rhamnose of the extracellular wall polysaccharide. *Applied and Environmental Microbiology* 56, 1882–1889.

Viviani, V.R. (2002) The origin, diversity, and structure function relationships of insect luciferases. *Cellular and Molecular Life Sciences* 59, 1833–1850.

Voorhees, K.J., Madonna, A.J. and Rees, J.C. (2003). Method for detecting low concentrations of a target bacterium that uses phages to infect target bacterial cells. US Patent no. 7,166,425.

Waddell, T.E. and Poppe, C. (2000) Construction of mini-Tn*10luxABcam/Ptac*-ATS and its use for developing a bacteriophage that transduces bioluminescence to *Escherichia coli* O157:H7. *FEMS Microbiology Letters* 182, 285–289.

Wallmark, G. and Laurell, G. (1952) Phage typing of *Staphylococcus aureus* some bacteriological and clinical observations. *Acta Pathologica et Microbiologica Scandinavica* 30, 109–114.

Wang, S. and Hazelrigg, T. (1994) Implications for *bcd* mRNA localization from spatial distribution of *exu* protein in *Drosophila* oogenesis. *Nature* 369, 400–403.

Yim, P.B., Clarke, M.L., McKinstry, M., De Paoli Lacerda, S.H., Pease, L.F. III, Dobrovolskaia, M.A., Kang, H., Read, T.D., Sozhamannan, S. and Hwang, J. (2009) Quantitative characterization of quantum dot-labeled lambda phage for *Escherichia coli* detection. *Biotechnology and Bioengineering* 104, 1059–1067.

Zuiderwijk, M., Tanke, H.J., Sam Niedbala, R. and Corstjens, P.L. (2003) An amplification-free hybridization-based DNA assay to detect *Streptococcus pneumoniae* utilizing the up-converting phosphor technology. *Clinical Biochemistry* 36, 401–403.

11 Phage Detection as an Indication of Faecal Contamination

Lawrence D. Goodridge[1] and Travis Steiner[1]
[1]Department of Animal Sciences, Colorado State University.

The importance of water quality to public health has been demonstrated for thousands of years, as archaeological evidence of safe water usage and sanitary practices have been found from the Incas to the Romans (Rosen, 1993). Most current issues with water quality primarily impact on developing countries where there is a lack of adequate sanitation and treatment facilities (Carr, 2001). For example, it is estimated that there are approximately 4 billion global cases of diarrhoea annually (Carr, 2001), and inadequate water supply, sanitation and domestic hygiene account for 2.2 million deaths annually (Prüss and Havelaar, 2001).

While inadequate water quality is a major problem in developing countries, incidences of water contamination still remain a problem in many developed countries. For example, an outbreak in 1988 in Sweden affected 11,000 people when a chlorination failure occurred at a water-treatment plant (Andersson, 1991), and 2300 individuals in Walkerton, Ontario, Canada, were affected by a waterborne outbreak of *Escherichia coli* O157 when exposed to contaminated drinking water from a well that had most likely been contaminated by surface runoff water (Hrudey *et al.*, 2003). Out of the 2300 affected persons in the Walkerton outbreak, 65 people were hospitalized, with 27 of these developing haemolytic–uraemic syndrome and seven confirmed deaths. In the USA, a major outbreak of cryptosporidiosis in Milwaukee, WI, caused an estimated 403,000 people to become ill with 4400 hospitalizations (Corso *et al.*, 2003). For more on *E. coli* O157:H7, see Kuhl *et al.*, Christie *et al.*, Williams and LeJeune, Goodridge and Steiner, and Niu *et al.* (Chapters 3, 4, 6, 11 and 16, this volume).

Such outbreaks not only affect the health of the citizens but also have a severe economic impact within the community. It is estimated that the 1993 cryptosporidiosis outbreak had a financial cost of nearly US$100 million when taking into account medical costs and productivity losses (Corso *et al.*, 2003). These examples of waterborne illnesses highlight the need for rapid, sensitive and inexpensive methods to assess water quality to mitigate such costly outbreaks of waterborne disease in developing and developed countries. The quality of drinking water, wastewater treatment and recreational-use waters must constantly be assessed due to their potential to serve as a vehicle for spreading disease. In some instances, untreated wastewater is used for the irrigation of crops, creating a potential for widespread illness (Ensink *et al.*, 2002; Sears *et al.*, 1984; Srikanth and Naik, 2004). Water used in agriculture and food processing

(eds P. Hyman and S.T. Abedon)

can thus serve as a source for transmission of pathogens, becoming important from a food-safety standpoint (see Niu *et al.*, Chapter 16, this volume).

Assessing water quality as a consequence of these various issues can be crucial from a public health perspective. A number of methods for such assessment exist and can be differentiated into chemical, physical or biological approaches (as well as their use in combination). The purpose of this chapter, however, is to describe the scientific literature surrounding the use of coliphages and phages that infect other bacterial species as indicators of water quality. For additional reading, there are a number of earlier reviews of phage use as faecal indicators of water quality (Leclerc *et al.*, 2000; Ashbolt *et al.*, 2001; Gerba, 2006; Pillai, 2006; see also Cox, Chapter 10, this volume). We begin with an overview of the concept of the use of indicator organisms in general for assessment of water quality.

Assessing Water Quality Using Indicator Organisms

Microbial water quality currently is determined by testing for the presence of indicator organisms (Yates, 2007). The detection, isolation and identification of waterborne pathogens can be expensive, difficult and labour-intensive (Scott *et al.*, 2002). To alleviate these issues with waterborne pathogen testing, indicator microorganisms (organisms used to determine the presence of faecal pollution) and index microorganisms (organisms used to indicate the presence of specific groups of pathogenic microorganisms) are commonly used to determine the possible presence of faecal pollution and thereby the relative risk, as well as, in some cases, the actual pathogenic microorganisms found in a sample. As most of the microbial pathogens present in water are of faecal origin, the detection of faecal contamination has been the main aim of testing methodologies. Historically, coliforms, the thermotolerant coliform group, enterococci and *Clostridium perfringens* have been the bacterial indicators used to detect faecal contamination (Scott *et al.*, 2002; Bitton, 2005; Savichtcheva and Okabe, 2006; Yates, 2007), based on the rationale that these indicator organisms are indigenous to faeces and that their presence in the environment is therefore indicative of faecal pollution. Testing for these microorganisms in particular is necessary because testing for the numerous bacterial, viral and parasitical pathogens that can cause waterborne illness is generally considered to be impractical, particularly as doing so can require extensive laboratory knowledge, specialized equipment and a large quantity of time and other resources (Havelaar and Hogeboom, 1983; Yates, 2007).

In addition to testing for the presence of gross faecal pollution, it is also useful to develop methodology that can evaluate the possible presence of the pathogenic microorganisms themselves. Cabelli (1977) defined the basic criteria for a good index organism of water quality as the following: (i) the organism should be present when the pathogen is present and absent when the pathogen is not present; (ii) the organism must not proliferate in the environment; (iii) the organism should be present in numbers equal to or greater than the pathogens it indicates; (iv) the organism should be as resistant to environmental decay and water-treatment strategies as the pathogens it indicates; and (v) the organism should be both easily assayed and non-pathogenic. In addition to these criteria, the ideal indicator/index organism would also be able to discriminate between contamination sources (animal or human), also known as microbial source tracking, and would have the ability to predict the presence of waterborne viral pathogens (Scott *et al.*, 2002; Long *et al.*, 2005; Savichtcheva and Okabe, 2006).

The US Environmental Protection Agency (EPA) establishes maximum contaminant levels (MCL) for total coliforms in drinking water based on the number of samples taken per month, which corresponds to the size of the population for which the water source serves (EPA, 1989). If samples are positive for total coliforms, then further analyses for faecal coliforms or *E. coli* are required. The MCL for drinking water, water that has been treated and will be released into a distribution system, and water that is currently in the distribution system are all set at no detectable thermotolerant coliform or *E. coli* bacteria (EPA, 1989; World Health Organization, 2006). The MCL standard for recreational water is much more liberal and utilizes *E. coli* and enterococci as indicators. The MCL standards for recreational waters are set at 126 colony forming units (CFU) per 100 ml for *E. coli* and 35 CFU for enterococci (EPA, 1986). These criteria are based on the rationale that the absence of faecal coliforms, *E. coli* and enterococci indicate that the waters sampled are at a low risk of harbouring pathogens originating from faecal contamination (EPA, 1989). Many of these bacteria, however, are routinely isolated from soil and water environments that have not been impacted by faecal pollution. For example, faecal coliforms and *E. coli* have been known to survive and propagate within pristine and tropical environments, meaning that assessment of their initial numbers can be exaggerated (Hurst and Crawford, 2002). Additionally, the microbiological relationship between index microorganisms and the full range of pathogens they are intended to represent is not clearly understood, nor are the dynamics of microbial ecology expected to be globally homogeneous (National Research Council, 2004).

There are other difficulties in using these bacteria as markers of faecal contamination. The bacteria are able to grow in biofilms within drinking-water distribution systems. In addition, they are occasionally absent in water supplies during outbreaks of waterborne disease. Also, while the persistence of these bacteria in water-distribution systems is comparable to that of some bacterial pathogens, the relationship between bacterial index organisms and the presence of enteric viruses and protozoa is poor, which is important, as viruses and protozoa account for approximately 44% of waterborne outbreaks in the USA where the aetiological agent has been identified (Blackburn *et al.*, 2004). Finally, the methods used to detect the indicators/index bacteria are problematic. While there are established culture and molecular methods for the detection of most microbial pathogens, most of these methods have important limitations, including the length of time required for the test result (1–5 days) and the specificity and sensitivity of detection (Scott *et al.*, 2002). Another issue to consider is the fact that the bacterial indicators described above are not suited to tracking the source of faecal pollution when a contamination event is discovered. Microbial source tracking is extremely important as it identifies the source of the pollution, which enables containment and a decrease in the chance of waterborne disease outbreaks.

Due to the limitations of the bacterial indicators, including problems with their rapid detection, it is clear that there remains an acute need to identify better indicators of microbial quality that would determine faecal contamination in a rapid manner and also determine the source of that pollution so that corrective actions could be initiated. In an effort to find indicators and index organisms that fit the criteria previously described, novel alternatives to total coliform and *E. coli* as well as *Enterococcus* spp. have been explored. Bacteriophages as indicators of water quality, and as index organisms for enteric viruses, have been proposed as a substitute for bacteria.

Bacteriophages as Indicator and Index Microorganisms

Bacteriophages were recognized as being present in the intestinal tract of humans in the early 1900s (d'Hérelle, 1926). The use of phages to indicate the possible presence of pathogenic enteric bacteria was an idea that subsequently developed in the 1930s, and direct correlations between the presence of certain types of bacteriophages and the presence of faecal contamination were reported (Scarpino, 1978).

More recently, the tendency has been to use phages as indexes to demonstrate the presence of and model the survival of enteric viruses as a group. As with bacterial pathogens, the approach to monitoring for enteric viruses is based on the idea that it is better to monitor for the presence of faecal pollution than for specific pathogens, due to the fact that, as with bacteria, there are still unknown (and simply too many) enteric viral pathogens to detect individually. In addition, enteric viruses are often present in low concentrations, requiring the analysis of at least 10 litres of water, and while diagnostic methods have been proposed for many of the known enteric viruses, such methodology is labour-intensive, expensive and time-consuming. Due to their similarities to enteric viruses, certain phages are attractive candidates as indexes of enteric viruses in water.

The similarities between phages and enteric viruses begin with the simple fact that phages are viruses themselves. Bacteriophage assay conditions, however, are much simpler and cheaper than any of the enteric virus detection methods. For example, phages can be identified and quantified using several methods ranging from standard microbiological methods such as plating to more complex molecular methods based on genus- or species-specific antigenic properties and genetic identification (Metcalf *et al.*, 1995; Vinje *et al.*, 2004; Kirs and Smith, 2007). All these methods of detection can suffer the same limitations as bacterial detection, in some cases requiring even more laboratory expertise. Some phages are so similar in genetic structure that identification by antigenic activity against their capsid protein is the most effective method of discrimination. These antigenic tests can be performed using phage inactivation, lateral flow testing or latex agglutination (Love and Sobsey, 2007; see Cox, Chapter 10, this volume, for additional discussion of phage-detection methodologies).

As with bacterial indexes of faecal-borne bacterial pathogens, to be suitable sentinels of enteric viruses, phages should occur consistently and exclusively in human and animal faeces and sewage, should not multiply in the environment and should be present in greater numbers than the enteric viruses. Additionally, they should be at least as long-lived as the enteric viruses present in the environment, and their survival kinetics throughout the water-treatment process should be similar to that of enteric viruses (Keswick *et al.*, 1984; Payment and Franco, 1993; Hurst *et al.*, 1994).

Three groups of phages have been proposed as indicators of faecal pollution, and candidate index microorganisms for the enteric viruses including somatic coliphages (Kott, 1966; Hilton and Stotzky, 1973; Kott *et al.*, 1974; IAWPRC, 1991), male-specific coliphages (Havelaar and Hogeboom, 1984; IAWPRC, 1991) and phages infecting *Bacteroides fragilis* (Jofre *et al.*, 1986; Tartera and Jofre, 1987; IAWPRC, 1991; Armon and Kott, 1993; Grabow *et al.*, 1995; Gantzer *et al.*, 1998).

Somatic Coliphages

The somatic coliphage group includes all phages that require the presence of a receptor-binding protein for infection of their host *E. coli* and comprise bacteriophages from the families *Myoviridae, Siphoviridae, Podoviridae and Microviridae* (Muniesa *et al.*, 2003). These phages recognize outer-membrane proteins (OMPs) such as OMP C, OMP F and OMP K, or select lipopolysaccharide and sites within the O side chain as receptors for infection (Linberg, 1973). Somatic coliphages have been proposed as indicators of faecal contamination and guidelines currently exist for their detection and analysis via enrichment and plating methodologies (USEPA, 2001). The degradation of somatic coliphage genomes has been reported to be similar to that of enteric viruses, demonstrating that these phages can be used as reliable indexes of enteric virus presence when molecular methods are used to detect them (Skraber *et al.*, 2004). For example, Hot *et al.* (2003) studied whether the concentrations of somatic coliphages and infectious enteric viruses or the detection of enteric virus genomes were associated with the detection of human pathogenic viruses in surface water. The researchers tested water samples for the presence of somatic coliphages, and any water samples that contained phages were subsequently tested for the enteric viruses. Of the 68 surface-water samples positive for somatic coliphages, only two were positive for enteric viruses when detected by culturable methods, while 60 samples were positive for the enteric viruses when tested using RT-PCR.

Kott *et al.* (1969, 1974, 1978) were among the first researchers to investigate whether coliphages were valid indexes of enteric viral pollution and showed that the phages were present in wastewater and other faecally contaminated waters in numbers at least equal to the enteric viruses. Somatic phages also persist in wastewater and surface waters for longer time periods that enteroviruses (Kott *et al.*, 1969, 1974, 1978) and are detectable using routine methodologies (Grabow *et al.*, 1978). Somatic coliphages have been detected in sewage-contaminated waters but were not found in pristine waters (Toranzos *et al.*, 1988), and Suan and colleagues (1988) observed somatic coliphages to be highly correlated with faecal coliforms in tropical waters.

Justification for the use of coliphages as indicators in wastewater and other faecally contaminated waters has been reviewed by Gerba (1987), the International Association on Water Pollution Research and Control (IAWPRC) Study Group on Health Related Water Microbiology (IAWPRC, 1991) and Limsawat and Ohgaki (1997). Nevertheless, detractors of the use of somatic coliphages as indicators of faecal contamination cite data showing that propagation within the environment could occur whenever a host bacterium is present, causing their numbers to be over-represented, and somatic coliphages have been reported to replicate at temperatures as low as 15°C (Seeley and Primrose, 1980). Further research, however, has shown that optimal conditions required for replication of somatic coliphages in the environment are unlikely to occur, as considered by Muniesa and Jofre (2004).

Muniesa and Jofre and colleagues studied environmental bacterial host strains and found that, although environmental somatic coliphages could be propagated on laboratory *E. coli* strains WG5 and CN13, bacterial strains isolated from the environment were not susceptible to these coliphages (Muniesa *et al.*, 2003; Muniesa and Jofre, 2004). These results suggested that environmental replication of somatic coliphages is not a significant problem and therefore would not cause over-representation of somatic coliphages.

Other potential limitations exist that are relevant to the use of somatic coliphages as faecal indicators, including that they are not specific to *E. coli*. For instance, there is evidence that somatic coliphages may multiply in other species of Enterobacteriaceae, which are part of the total coliform group and often found associated with vegetation and biofilms. Of these, the two most common species are *Klebsiella pneumoniae* and *Enterobacter cloacae*. It is therefore possible that some coliphages might be produced that are not only unrelated to faecal contamination but, indeed, are unrelated to any health risk.

An additional issue is that there are many non-faecal sources of coliform bacteria. Accordingly, it is difficult to ascertain specifically whether an isolate of a somatic coliphage arose from any one point of contamination. Coliforms that can colonize biofilms may be present anywhere in the water collection treatment or distribution system. Furthermore, studies by Vaughn and Metcalf (1975) on coliphages and enteric viruses in sewage effluents, shellfish and shellfish-growing waters showed that coliphages are not adequate index microorganisms of enteric viruses. This was due to the replication of coliphages, the presence of more than one dominant phage type in estuarine and fresh waters making the results of a test for any one phage type not definitive (Vaughn and Metcalf, 1975; Seeley and Primrose, 1980; Parry *et al.*, 1981; Borrego *et al.*, 1990) and the lack of any correlation between the densities of coliphages and enteroviruses in raw sewage and farm pond water (Joyce and Weiser, 1967; Safferman and Morris, 1976; Havelaar, 1987; Nieuwstadt *et al.*, 1991; Wommack *et al.*, 1996). In addition, inconsistent occurrence of coliphages in raw sewage samples with the simultaneous isolation of enteroviruses has been observed, while other studies have reported a high number of enteroviruses isolated in many treated effluents with no coliphages detected (Vaughn and Metcalf, 1975).

Male-specific Coliphages

Male-specific coliphages from the families *Inoviridae* and *Leviviridae* are defined by their requirement for the expression of F pili on their host bacteria for successful infection (Long and Sobsey, 2004). F pili are typically expressed from a plasmid and are used by bacteria for the exchange of genetic material from one F^+ or male/donor bacterium to an F^- or female/recipient bacterium during conjugation (Novotny *et al.*, 1969). The expression of this appendage is regulated by several factors including existence of the F plasmid and temperature. F pili have been shown to be expressed at the highest numbers on the cell surface within a temperature range of 37–42°C, with expression non-existent at 25°C and below (Novotny and Lavin, 1971). The specific temperature range at which the F pili are expressed is one of the arguments for the use of these phages as indicators, because normal environmental temperatures do not support the expression of pili, thereby limiting phage propagation (Havelaar *et al.*, 1986).

Male-specific DNA coliphages

Male-specific coliphages are classified according to their nucleic acid type and divided into subgroups according to their morphology and serological properties (Long and Sobsey, 2004). Male-specific DNA (FDNA) coliphages from the family *Inoviridae* are unique in their morphology among phages in that they do not have a head capsid but rather are thin, filamentous structures that contain single-stranded DNA (Rasched and Oberer, 1986). The filamentous structures range from approximately 760 to 1950 nm in length and are approximately 6 nm wide. Their use as indicators of faecal contamination has been widely disputed as they are not as well characterized as the male-specific RNA (FRNA) coliphages and do not have morphologies or environmental stability similar to the pathogens they are intended to indicate. In addition, a link between FDNA phages and their sanitary significance has not been shown (Scott *et al.*, 2002; Sinton *et al.*, 1996). A study by Long (1998), however, indicated that these phages were present during times when FRNA phages were absent, and the author proposed the use of both FRNA and FDNA phages as possible water-quality indicators during different seasons of the year.

The predominating FDNA coliphages have received less attention as index organisms of enteric viruses because they are generally less plentiful than FRNA coliphages (Leclerc *et al.*, 2000), they do not resemble human enteric viruses morphologically and their ecology is poorly understood (Leclerc *et al.*, 2000). FDNA phages have been assayed utilizing plaque assays and membrane filtration and elution methods (Sinton *et al.*, 1996). As FDNA and FRNA utilize the same pili-expressing host, the standard method for identifying FDNA phages involves using medium that contains RNase to inactivate FRNA coliphages (Cole *et al.*, 2003).

Male-specific RNA coliphages

FRNA phages have received the most attention as indicators of faecal contamination and especially as index microorganisms. This prominence is due to the fact that many studies have confirmed that, for monitoring purposes, FRNA phages are reliable determinants of the possible presence of human enteric viruses, as they behave like waterborne viruses (Havelaar *et al.*, 1993). As noted, FRNA phages enter the host cell by adsorption to F pili and are a more homogeneous group than other microorganisms (Seeley and Primrose, 1980). Also as noted, F pili are only produced by *E. coli* cells at temperatures above 25°C, meaning that the phages would be unable to multiply in most environments. Therefore, detection of these phages indicates a recent contamination event.

FRNA phages have also attracted interest as useful alternatives to bacterial indexes because their morphology and survival characteristics closely resemble the human enteric gastrointestinal viruses (Scott *et al.*, 2002), meaning that, in addition to the usefulness of these phages as indicators of the presence of their host bacteria (*E. coli*) and therefore the presence of faecal pollution, they could also be used as surrogate markers for the presence of enteric viruses such as noroviruses or rotaviruses in water. They are also similar to enteric viruses with respect to inactivation kinetics following exposure to environmental factors and water treatments (Allwood *et al.*, 2003; Grabow, 2001). FRNA coliphage virions are approximately 26 nm in diameter and their capsids exhibit an icosahedral symmetry that consists of 180 copies of a dimer coat protein containing one copy of positive-sense, single-stranded RNA of between 3500 and 4200 nt (Bollback and Huelsenbeck, 2001; Stewart *et al.*, 2006). FRNA phages are grouped into two distinct genera that are defined by their genomic organization. FRNA phages from the genus *Levivirus* contain a gene locus encoding a lysis protein. Phages from the genus *Allolevivirus* contain a gene with a leaky stop codon that becomes a read-through approximately 5% of the time (Bollback and Huelsenbeck, 2001; Stewart *et al.*, 2006), encoding an extended coat protein. Unlike phages from the genus *Levivirus, Allolevivirus* phages do not produce a protein dedicated to host-cell lysis (Bollback and Huelsenbeck, 2001; Stewart *et al.*, 2006).

Both FRNA phage genera are further classified into four individual serogroups (I–IV) based on serological as well as genetic and physiochemical properties (Havelaar and Hogeboom, 1984). They can be identified via inactivation by antiserum, RT-PCR, latex agglutination and nucleic acid hybridization (Furuse *et al.*, 1978; Vinje *et al.*, 2004; Kirs & Smith, 2007; Love and Sobsey, 2007; Friedman *et al.*, 2009). Of these four groups, groups I and IV have been associated primarily with animal faecal sources, while groups II and III have been demonstrated to be associated primarily with faecal material from human faecal sources (Havelaar *et al.*, 1990; Fig. 11.1). This trait makes the FRNA coliphages attractive as indicators of faecal pollution as well as useful for microbial source tracking analysis.

Serotyping of FRNA phages has been suggested as a method to determine the origin of faecal pollution (Furuse *et al.*, 1978). In this work, the authors produced antisera against type phages from each of the four groups, including MS2 (group I), JP34 (group II), GA (group II), Q/β (group III), VK (group III) and SP (group IV). Unknown phages were incubated with antisera from each group and then grown on their respective *E. coli* host strain. If the phage was neutralized, then it belonged to the same group as the antisera and the phage failed to form plaques. In this way, the authors were able to determine that, of 52 phages that were serotyped, 46 belonged to group III (Furuse *et al.*, 1978). The serotyping method can be used in a source tracking scheme to discover the origin of faecal pollution. The serotyping method as described by Furuse *et al.* (1978), however, is

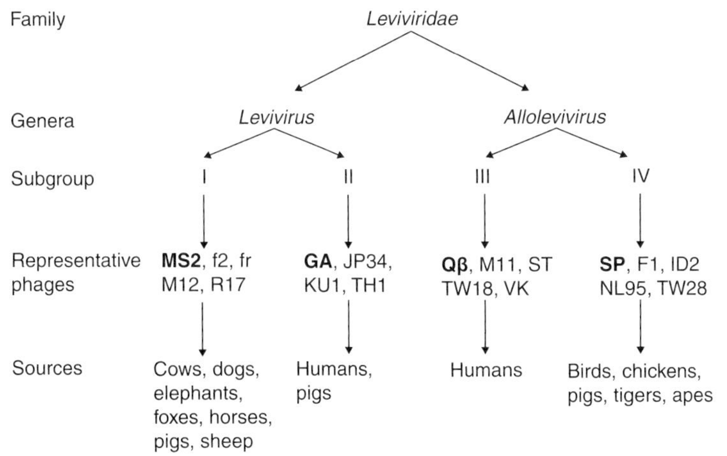

Fig. 11.1. Taxonomy of the FRNA phages. Phages in bold represent the type phage for each particular serotype or genogroup. Adapted from Smith (2006).

tedious, requires standard laboratory equipment (e.g. Petri plates, incubators) and takes at least 24 h to complete (Hsu *et al.*, 1995).

Hsu *et al.* (1995) showed that highly specific nucleic acid probes could be used to genotype FRNA phages as an alternative to serotyping. In this work, the authors developed oligonucleotide probes that were specific to each of the four serogroups and used these probes to genotype unknown phages. The genotyping method showed excellent agreement with the serotyping method ($P<0.05$) and the authors concluded that genotyping FRNA coliphages appears to be practical and reliable for typing unknown phage isolates in field samples. As with the serotyping method, genotyping has been demonstrated as a useful approach to source track the origin of faecal pollution in water samples. Using the genotyping method, Schaper *et al.* (2002) confirmed that genotypes II and III predominated in municipal sewage, illustrating that these genotypes were of human origin. Nevertheless, the genotyping method is laborious, time-consuming and requires extensive operator training. For example, the genotyping method as described by Hsu *et al.* (1995) includes isolating the phage from a water sample, conducting a plaque assay, obtaining a well-isolated plaque and transferring it to a membrane. Next, the membrane-bound phage are denatured, followed by fixation of the released nucleic acid on to the membrane with UV light. Finally, the genotype-specific probes are added, allowed to hybridize with the phage nucleic acid and detected colorimetrically (Hsu *et al.*, 1995).

Doré *et al.* (2000) investigated using FRNA phage to indicate the likely presence of enteric viruses in shellfish sold for consumption. FRNA phage and *E. coli* levels were determined over a 2-year period for oysters harvested from four commercial sites chosen to represent various degrees of sewage pollution. Three sites required cleaning (depuration) of the oysters before sale. Oysters from the fourth site could be sold directly without further processing. Samples were tested at the point of sale following commercial processing and packaging. All of the shellfish complied with the mandatory European Community *E. coli* standard (less than 230 CFU per 100 g of shellfish flesh), and the levels of contamination for more than 90% of the shellfish were at or below the level of sensitivity of the assay (20 *E. coli* CFU per 100 g), which indicated good quality based on this criterion. In contrast, FRNA phage were frequently detected at levels that exceeded 10^3 plaque-forming units (PFU) per 100 g. High levels of FRNA phage contamination were strongly associated with harvest area faecal pollution and with shellfish-associated disease outbreaks. Interestingly, FRNA phage contamination exhibited a marked seasonal trend that was consistent with the trend of oyster-associated gastroenteritis in the UK. The correlation between FRNA phage contamination and health risk was investigated further using a RT-PCR assay for Norwalk-like virus. Norwalk-like virus contamination of oysters was detected only at the most polluted site and also exhibited a seasonal trend that was consistent with the trend of FRNA phage contamination and with the incidence of disease. The authors concluded that the results of this study suggested that FRNA phages could be used as viral index organisms for market-ready oysters (Doré *et al.*, 2000).

Other studies have corroborated the finding that seasonal fluctuations in FRNA concentration mirror that of the Norwalk-like viruses. Doré *et al.* (2003) collected a total of 608 shellfish samples from 49 shellfish-harvesting areas and tested them for the presence of *E. coli* and FRNA phage. The results indicated that FRNA phage concentration in all samples was more than three times greater than that of *E. coli* (geometric mean counts of 1800 and 538 per 100 g for FRNA phage and *E. coli*, respectively). In addition, FRNA phage concentrations were strongly influenced by season, with a geometric mean count of 4503 PFU per 100 g in the winter (October–March) compared with 910 PFU per 100 g in the summer (April–September). The elevated levels of FRNA phages observed in the winter concur with the known increased virus risk associated with shellfish harvested at this time of year in the UK. The authors suggested that data from this study could be employed to introduce

FRNA phages as an index of the virus risk associated with shellfish (Doré *et al.*, 2003).

The concentrations at which these indicators exist in the environment are fairly low at about <10^4 PFU ml^{-1} (Furuse *et al.*, 1983; Havelaar *et al.*, 1986). To remedy this problem, the phages can be concentrated using charged membranes and elution. Alternatively, enrichment is possible for certain phages using the desired host bacterium. The latter is typically required to achieve numbers high enough for detection (Love and Sobsey, 2007). Furthermore, it has been shown that FRNA phages are not present in the faeces of all animals and have been reported to be isolated in the faeces of only about 3% of the human population (Havelaar *et al.*, 1990). This would suggest that the use of these coliphages as useful indicators of faecal contamination may be limited to water contamination from pooled sources such as sewage, septic tanks or agricultural runoff water.

Assaying for male-specific coliphages

One problem encountered when assaying for male-specific coliphages is the possibility of also detecting somatic coliphages. This is due to the existence of somatic receptors on the cellular surface of some of the host strains used to enumerate male-specific coliphages in water samples (Havelaar and Hogeboom, 1983; Hsu *et al.*, 1996; Rhodes and Kator, 1991). The male-specific coliphage host *E. coli* F_{amp} is susceptible to infection by FDNA phages, but this may not prove to be a detriment until the sanitary significance of FDNA phages is better understood (Hsu *et al.*, 1996; Sinton *et al.*, 1996). One route towards addressing these concerns is to use host strains besides *E. coli* F_{amp} that contain the F plasmid. *Salmonella enterica* subsp. *enterica* serovar Typhimurium WG49 and *E. coli* C3000 have both been proposed as candidates for detection of FRNA phages (Leclerc *et al.*, 2000). The use of these bacterial host strains is accompanied by its own set of problems, however. For example, while *E. coli* C3000 produces the required F pili, the bacterium is not antibiotic resistant and is susceptible to infection by both somatic coliphages and male-specific DNA phages (Hsu *et al.*, 1996). Furthermore, while the use of *Salmonella* Typhimurium WG49 solves the problem of attack by somatic coliphages, because, as a *Salmonella* species, it lacks the receptors required for somatic coliphage infection (Havelaar and Hogeboom, 1983), the strain is still susceptible to attack by somatic *Salmonella* phages (Stetler and Williams, 1996).

Real-time PCR is a sensitive, rapid method for identifying the presence of nucleic acids. FRNA bacteriophages, however, harbour their genetic information in the form of single-stranded positive-sense RNA which must first be transcribed into cDNA before a detectable amplicon can be produced. Despite this hurdle, multiplex PCR assays and real-time RT-PCR assays have been developed that have shown specificity for each subgroup of FRNA phages (Vinje *et al.*, 2004; Kirs & Smith, 2007; Friedman *et al.*, 2009). For example, Kirs and Smith (2007) developed a culture-independent multiplex real-time RT-PCR assay to effect simultaneous detection and quantification of the four FRNA groups. The assay detected as few as ten copies of isolated phage RNA and was able to quantify FRNA phages in seawater when culture-based methods (the double agar layer assay) failed. The authors concluded that accurate identification of the *in situ* concentration of FRNA phages using this method will facilitate more effective remediation strategies for impacted environments.

Friedman *et al.* (2009) developed a real-time RT-PCR for FRNA phages by first designing genogroup-specific primer sets based on a minimum of five and a maximum of ten complete phage genome sequences from strains in each FRNA genogroup. When a heat-release protocol that eliminated the need for RNA purification and the real-time RT-PCR method for genotype identification of FRNA phages were combined, the assay could detect multiple environmental FRNA phage strains. Following up on this work, Friedman *et al.* (2011) designed a real-time, quantitative RT-PCR assay to better differentiate the four genogroups of FRNA phages. In this work, primers and probes were designed using complete genomic sequences from 29 FRNA phages to develop a set of primer/probe sets that were based on: (i) ability to amplify a single, specific product; (ii) genogroup specificity; (iii) lack of cross-reactivity; and (iv) experimental reproducibility and sensitivity over a range of target concentrations. A custom RNA molecule was employed as an internal, non-competitive control. The assay was tested on a total of 49 FRNA phages isolated from various warm-blooded animals, sewage and combined sewage overflow. FRNA phages from animal wastes were genotyped as 86% group I, 4% group III Q-like and 9% group IV. Two sewage isolates typed to genogroup I and combined sewage overflow isolates genotyped as 40% group II and 52% group III.

Goodridge and Du Preez (2010) developed and evaluated an integrated concentration and detection method to rapidly assay for FRNA bacteriophages as indicators of faecal contamination in river water. Sampling for the study took place over a 3-week period during January 2010 in Phola Township (immediately adjacent to a sewage treatment plant) and Pretoria, South Africa. Following tropical rain events, 4 l of river water were collected from rivers at both sampling sites and transported back to the laboratory on ice. FRNA phages were concentrated from the water by addition of 1.5 g of anionic exchange resin (Amberlite® IRA 900; Sigma-Aldrich) to each water sample, followed by continuous agitation to keep the resin in suspension. After 4 h of concentration, the resin was recovered from the water samples and virions bound to the resin beads were detected by isolating viral RNA directly from the beads. The latter was accomplished with real-time RT-PCR using a SYBR Green assay and primers designed to individually detect the four FRNA phage genogroups. With an enrichment step, the real-time RT-PCR assay detected FRNA phages belonging to genogroups II and III from the Phola Township samples, clearly indicating the presence of human faecal pollution. When real-time RT-PCR was performed directly from the resin beads (without enrichment), FRNA phages belonging to genogroups II, III and IV (Phola Township) and II and IV (Pretoria) indicated the presence of animal faecal pollution in addition to human pollution. The entire concentration and detection assay (without enrichment) was completed within 8 h. These results suggest that anion exchange capture and real-time RT-PCR can be used as a rapid and sensitive assay for detection of FRNA bacteriophages as indicators of human and animal faecal pollution. Moreover, this method would be especially useful as a rapid presumptive assay to test compromised water supplies following hurricanes, monsoons and other similar weather events.

PCR methods have been criticized for potentially detecting non-viable organisms as well as free nucleic acids (Leclerc *et al.*, 2000; Gassilloud *et al.*, 2003). Research has shown, however, that free RNA from FRNA phage Qβ is degraded and undetectable by PCR within 1 h following exposure to wastewater (Limsawat and Ohgaki, 1997), which would suggest that if FRNA bacteriophage nucleic acid is detected, it would be from a viable organism or an extremely recent RNA contamination event.

B. fragilis Bacteriophages

B. fragilis has been suggested previously as an alternative bacterial indicator to standard indicators such as *E. coli* and total coliforms (Fiksdal *et al.*, 1985; Bitton, 2005). Phages that have specificity for *B. fragilis* infect one of the most abundant bacteria in the gastrointestinal tract and belong to the family *Siphoviridae*. Phages that infect *B. fragilis* strain RYC2056 are not human specific, while phages that infect host-strain *B. fragilis* HSP40, while being less numerous than those infecting strain RYC2056, are considered to be human specific (Puig *et al.*, 2000). Therefore, phages that infect *B. fragilis* HSP40 have been proposed as indicators of human faecal contamination because of their specificity along with their inability to propagate in the environment. A study by Tartera *et al.* (1989) replicated optimal environmental conditions for the propagation of phages infecting *B. fragilis*, including temperature, host-cell concentration and anaerobic conditions but excluding nutrients for host growth. Despite the combination of these conditions, which themselves are unlikely to be found in the environment, researchers did not observe replication of *Bacteroides*-specific bacteriophages in significant numbers in slaughterhouse wastewaters. They were also not present in faecally polluted waters containing faecal contamination from wildlife only. In contrast, phages active against *B. fragilis* HSP40 were detected in faeces (found in 10% of human faecal samples but not in animal faeces), sewage and other polluted aquatic environments (river water, seawater, groundwater and sediments) and were absent in non-polluted sites (Tartera and Jofre, 1987; Cornax *et al.*, 1990). In addition, Gantzer *et al.* (1998) demonstrated that *B. fragilis* phages were found to be reliable indexes of enterovirus contamination in the three different types of wastewater that were tested. These characteristics make *B. fragilis* HSP40 phages good candidates for microbe source tracking of a contamination event of human origin.

Standard methods for identification and quantification of phages that infect *B. fragilis* include double-layer agar most-probable-number methods (Tartera and Jofre, 1987). Several molecular and serological methods of detection have also been explored, including DNA hybridization and PCR. Plating-based methods, however, remain the standard for water assessment (Puig *et al.*, 2000), and certain hurdles exist that complicate the use of *B. fragilis* phages as indicators of faecal contamination, not least of which is that this host is an obligate anaerobe, requiring very specific growth conditions and laboratory skills to conduct phage assays (Leclerc *et al.*, 2000). In addition, studies conducted by Gantzer *et al.* (2002) confirmed previously reported data indicating that phages infecting *B. fragilis* are not found homogenously within the faeces of humans worldwide (Tartera and Jofre, 1987; Havelaar *et al.*, 1993; Grabow *et al.*, 1995). Finally, *B. fragilis* phages are detected in lower concentrations than other phages in fresh water (1–15 PFU per 100 ml; Araujo *et al.*, 1997), meaning that the use of concentration methods to increase the numbers of these phages prior to detection is needed, which may complicate downstream identification.

Conclusion

Bacteriophages continue to emerge as alternative indicators of faecal contamination and as index organisms identifying the presence of enteric viruses. While many peer-reviewed studies demonstrating their usefulness as indicators of water quality abound, a lack of standardized methods used for analysis of the presence of phages in various water types makes most of the data difficult to compare. The fact that there are many variables that affect the incidence, survival and behaviour of phages in different water environments, including the densities of both host bacteria and phages, temperature and pH, further complicates comparisons. Finally, the myriad of bacterial hosts used to recover phages during water-quality experiments has led to different results reported with respect to recovery efficiencies of phages isolated from the same environmental samples. Neverthelss, it is clear that phages are viable candidates as water-quality sentinels and, as more methods are developed and standardized for their efficient detection, it is expected that phages will find increased usefulness in assessments of microbial water quality, particularly as indexes of the presence of enteric viruses, the methods for detection of which are still in their infancy.

References

Allwood, P.B., Malik, Y.S., Hedberg, C.W. and Goyal, S.M. (2003) Survival of F-specific RNA coliphage, feline calicivirus, and *Escherichia coli* in water: a comparative study. *Applied and Environmental Microbiology* 69, 5707–5710.

Andersson, Y. (1991) A waterborne disease outbreak. *Water Science and Technology* 24, 13–15.

Araujo, R.M., Puig, A., Lasobras, J., Lucena, F. and Jofre, J. (1997) Phages of enteric bacteria in fresh water with different levels of faecal pollution. *Journal of Applied Bacteriology* 82, 281–286.

Armon, R. and Kott, Y. (1993) A simple, rapid, and sensitive presence/absence detection test for bacteriophage in drinking water. *Journal of Applied Bacteriology* 74, 490–496.

Ashbolt, N.J., Grabow, W.O.K. and Snozzi, M.

(2001) Indicators of microbial water quality. In: Fewtrell, L. and Bartram, J. (eds) *Water Quality: Guidelines, Standards and Health.* IWA Publishing, London, UK, pp. 289–315.

Bitton, G. (2005) Microbial indicators of fecal contamination: application to microbial source tracking. Report submitted to the Florida Stormwater Association <www.florida-stormwater.org/Files/FSA%20Educational%20Foundation/Research/Pathogens/FSAMicrobialSourceTrackingReport.pdf>.

Blackburn, B.G., Craun, G.F., Yoder, J.S., Hill, V., Calderon, R.L., Chen, N., Lee, S.H., Levy, D.A. and Beach, M.J. (2004) Surveillance for waterborne-disease outbreaks associated with drinking water – United States, 2001–2002. *Morbidity and Mortality Weekly Report Surveillance Summaries* 53, 23–45.

Bollback, J.P. and Huelsenbeck, J.P. (2001) Phylogeny, genome evolution, and host specificity of single-stranded RNA bacteriophage (family *Leviviridae*). *Journal of Molecular Evolution* 52, 117–128.

Borrego, J.J., Cornax, R., Morinigo, M.A., Martinez-Manzanores C. and Romero, P. (1990) Coliphage as an indicator of faecal pollution in water: their survival and productive infectivity in natural aquatic environments. *Water Research* 24, 111–116.

Cabelli, V.J. (1977) *Clostridium perfringens* as a water quality indicator. In: Hoadley, A.W. and Dutka. B.J. (eds) *Bacterial Indicator/Health Hazards Associated with Water.* American Society for Testing and Materials, Philadelphia, PA.

Carr, R. (2001) Excreta-related infections and the role of sanitation in the control of transmission. In: Fewtrell, L. and Bartram, J. (eds) *Water Quality: Guidelines, Standards and Health.* IWA Publishing, London, UK, pp. 89–113.

Cole, D., Long, S.C. and Sobsey, M.D. (2003) Evaluation of F^+ RNA and DNA coliphages as source-specific indicators of fecal contamination in surface waters. *Applied and Environmental Microbiology* 69, 6507–6514.

Cornax, R., Morinigo, M.A., Paez, I.G., Munoz, M.A. and Borrego, J.J. (1990) Application of direct plaque assay for detection and enumeration of bacteriophages of *Bacteroides fragilis* from contaminated water samples. *Applied and Environmental Microbiology* 56, 3170–3173.

Corso, P.S., Kramer, M.H., Blair, K.A., Addiss, D.G., Davis, J.P. and Haddix, A.C. (2003) Cost of illness in the 1993 waterborne *Cryptosporidium* outbreak, Milwaukee, Wisconsin. *Emerging Infectious Diseases* 9, 426–431.

D'Hérelle, F. (1926) *The Bacteriophage and Its Behavior* [English translation by G.H. Smith]. Williams and Wilkins, Baltimore, MD.

Doré, W.J., Henshilwood, K. and Lees, D.N. (2000) Evaluation of F-specific RNA bacteriophage as a candidate human enteric virus indicator for bivalve molluscan shellfish *Applied and Environmental Microbiology* 66, 1280–1285.

Doré, W.J., Mackie, M. and Lees, D.N. (2003) Levels of male-specific RNA bacteriophage and *Escherichia coli* in molluscan bivalve shellfish from commercial harvesting areas. *Letters in Applied Microbiology* 36, 92–96.

Ensink, J.H.J., van der Hoek, W., Matsuno, Y., Munir, S. and Aslam, M.R. (2002) *Use of Untreated Wastewater in Peri-urban Agriculture in Pakistan: Risks and Opportunities*. Research Report 64, IWMI Books, Reports H030848, International Water Management Institute.

EPA (1986) *Guide Standard and Protocol for Testing Microbiological Water Purifiers.* US Environmental Protection Agency, CI.

EPA (1989) *Safe Drinking Water.* Federal Register, 29 June 1989 (or *Code of Federal Regulations*, Title 40, Parts 141 and 142). US Environmental Protection Agency, CI.

Fiksdal, L., Maki, J.S., LaCroix, S.J. and Staley, J.T. (1985) Survival and detection of *Bacteroides* spp., prospective indicator bacteria. *Applied and Environmental Microbiology* 49, 148–150.

Friedman, S.D., Cooper, E.M., Calci, K.R. and Genthner, F.J. (2011) Design and assessment of a real time reverse transcription-PCR method to genotype single-stranded RNA male-specific coliphages (Family *Leviviridae*). *Journal of Virological Methods* 173, 196–202.

Friedman, S.D., Cooper, E.M., Casanova, L., Sobsey, M.D. and Genthner, F.J. (2009) A reverse transcription-PCR assay to distinguish the four genogroups of male-specific (F^+) RNA coliphages. *Journal of Virological Methods* 159, 47–52.

Furuse, K., Sakurai, T., Hirashima, A., Katsuki, M., Ando, A. and Watanabe, I. (1978) Distribution of ribonucleic acid coliphages in south and east Asia. *Applied and Environmental Microbiology* 35, 995–1002.

Furuse, K., Sakurai, T., Inokuchi, Y., Inoko, H., Ando, A. and Watanabe, I. (1983) Distribution of RNA coliphages in Senegal, Ghana, and Madagascar. *Microbiological Immunology* 27, 347–358.

Gantzer, C., Maul, A., Audic, J.M. and Schwartzbrod, L. (1998) Detection of infectious enteroviruses, enterovirus genomes, somatic coliphages, and *Bacteroides fragilis* phages in treated wastewater. *Applied and Environmental Microbiology* 64, 4307–4312.

Gantzer, C., Henny, J. and Schwartzbrod, L. (2002) *Bacteroides fragilis* and *Escherichia coli* bacteriophages in human faeces. *International Journal of Hygiene and Environmental Health* 205, 325–328.

Gassilloud, B., Schwartzbrod, L. and Gantzer, C. (2003) Presence of viral genomes in mineral water: a sufficient condition to assume infectious risk? *Applied and Environmental Microbiology* 69, 3965–3969.

Gerba, C.P. (1987) Phage as indicators of faecal pollution. In: Goyal, S.M., Gerba, C.M. and Bitton, G. (eds) *Phage Ecology*. J. Wiley & Sons, New York, NY, pp. 197–209.

Gerba, C.P. (2006) Bacteriophages as pollution indicators. In: Calendar, R. and Abedon, S.T. (eds) *The Bacteriophages*, 2nd edn. Oxford University Press, Oxford, UK, pp. 695–701.

Goodridge, L.D. and du Preez, M. (2010) Evaluation of a rapid assay for concentration and detection of the FRNA bacteriophages as microbial indicators of fecal pollution. In: *Society for Applied Microbiology Summer Meeting*, Brighton, UK.

Grabow, W.O.K. (2001) Bacteriophages: update on application as models for viruses in water. *Water SA* 27, 251–268.

Grabow, W.O.K., Neubrech, T.E., Holtzhausen, C.S. and Jofre, J. (1995) *Bacteroides fragilis* and *Escherichia coli* bacteriophages: excretion by humans and animals. *Water Science and Technology* 31, 223–230.

Grabow, W.O.K., Middendorff, I.G. and Basson, N.C. (1978) Role of lime treatment in the removal of bacteria, enteric viruses, and coliphages in wastewater reclamation plant. *Applied and Environmental Microbiology* 35, 663–669.

Havelaar, A.H. (1987) Virus, bacteriophages and water purification. *Veterinary Quarterly* 9, 356–360.

Havelaar, A.H. and Hogeboom, W.M. (1983) Factors affecting the enumeration of coliphages in sewage and sewage-polluted waters. *Antonie Van Leeuwenhoek* 49, 387–397.

Havelaar, A.H. and Hogeboom, W.M. (1984) A method for the enumeration of male-specific bacteriophages in sewage. *Journal of Applied Bacteriology* 56, 439–447.

Havelaar, A.H., Furuse, K. and Hogeboom, W.M. (1986) Bacteriophages and indicator bacteria in human and animal faeces. *Journal of Applied Bacteriology* 60, 255–262.

Havelaar, A.H., Pot-Hogeboom, W.M., Furuse, K., Pot, R. and Hormann, M.P. (1990) F-specific RNA bacteriophages and sensitive host strains in faeces and wastewater of human and animal origin. *Journal of Applied Bacteriology* 69, 30–37.

Havelaar, A.H., van Olphen, M. and Drost, Y.C. (1993) F-specific RNA bacteriophages are adequate model organisms for enteric viruses in fresh water. *Applied and Environmental Microbiology* 59, 2956–2962.

Hilton, M.C. and Stotzky, G. (1973) Use of coliphages as indicators of water pollution. *Canadian Journal of Microbiology* 19, 747–751.

Hot, D., Legeay, O., Jacques, J., Gantzer, C., Caudrelier, Y., Guyard, K., Lange, M. and Andreoletti, L. (2003) Detection of somatic phages, infectious enteroviruses and enterovirus genomes as indicators of human enteric viral pollution in surface water. *Water Research* 37, 4703–4710.

Hrudey, S.E., Payment, P., Huck, P.M., Gillham, R.W. and Hrudey, E.J. (2003) A fatal waterborne disease epidemic in Walkerton, Ontario: comparison with other waterborne outbreaks in the developed world. *Water Science and Technology* 47, 7–14.

Hsu, F.C., Shieh, Y.S., van Duin, J., Beekwilder, M.J. and Sobsey, M.D. (1995) Genotyping male-specific RNA coliphages by hybridization with oligonucleotide probes. *Applied and Environmental Microbiology* 61, 3960–3966.

Hsu, F.C., Chung, A., Amante, A., Shieh, Y.-S.C., Wait, D. and Sobsey, M.D. (1996) Distinguishing human from animal faecal contamination in water by typing male-specific RNA coliphages. In: *Proceedings of the AWWA Water Quality Technology Conference*, Boston, MA.

Hurst, C.J. and Crawford, R.L. (2002) *Manual of Environmental Microbiology*. ASM Press, Washington, DC.

Hurst, C.J., Blannon, J.C., Hardaway, R. and Jackson, W.C. (1994) Differential effect of tetrazolium dyes upon bacteriophage plaque assay titers. *Applied and Environmental Microbiology* 60, 3462–3465.

IAWPRC (1991) Bacteriophages as model viruses in water quality control. *Water Research* 25, 529–545.

Jofre, J., Bosch, A., Lucena, F., Girones, R. and Tartera, C. (1986) Evaluation of *Bacteroides fragilis* bacteriophages as indicators of the virological quality of water. *Water Science and Technology* 18, 167–177.

Joyce, G. and Weiser, H.H. (1967) Survival of enteroviruses and bacteriophage in farm pond waters. *Journal of the American Water Works Association* 59, 491–501.

Keswick, B.H., Gerba, C.P., Dupont, H.L. and Rose, J.B. (1984) Detection of enteric viruses in

treated drinking water. *Applied and Environmental Microbiology* 47, 1290–1294.
Kirs, M. and Smith, D.C. (2007) Multiplex quantitative real-time reverse transcriptase PCR for F^+-specific RNA coliphages: a method for use in microbial source tracking. *Applied and Environmental Microbiology* 73, 808–814.
Kott, Y. (1966) Estimation of low number of *Escherichia coli* bacteriophage by use of the most probable number method. *Applied Microbiology* 14, 141.
Kott, Y., Ben-Ari, H. and Buras, N. (1969) The fate of viruses in a marine environment. In: Jenkins, S.H. (ed.) *Proceedings of the 4th International Conference*. Pergamon Press, Oxford, UK, pp. 823–829.
Kott, Y., Ben-Ari, H. and Vinokur, L. (1978) Coliphages survival as viral indicator in various wastewater quality effluents. *Progress in Water Technology* 10, 337–346.
Kott, Y., Roze, N., Sperber, S. and Betzer, N. (1974) Bacteriophages as viral pollution indicators. *Water Research* 8, 165–171.
Leclerc, H., Edberg, S., Pierzo, V. and Delattre, J.M. (2000) Bacteriophages as indicators of enteric viruses and public health risk in groundwaters. *Journal of Applied Microbiology* 88, 5–21.
Limsawat, S. and Ohgaki, S. (1997) Fate of liberated viral RNA in wastewater determined by PCR. *Applied and Environmental Microbiology* 63, 2932–2933.
Linberg, A.A. (1973) Bacteriophage receptors. *Annual Review of Microbiology* 27, 205–241.
Long, S.C. (1998) Project report: development of methods to differentiate microorganisms in MDC reservoir watersheds. University of Massachusetts, Amherst, MA.
Long, S.C. and Sobsey, M.D. (2004) A comparison of the survival of F^+RNA and F^+DNA coliphages in lake water microcosms. *Journal of Water and Health* 2, 15–22.
Long, S.C., El-Khoury, S.S., Oudejans, S.J., Sobsey, M.D. and Vinje, J. (2005) Assessment of sources and diversity of male-specific coliphages for source tracking. *Environmental Engineering Science* 22, 367–377.
Love, D.C. and Sobsey, M.D. (2007) Simple and rapid F+ coliphage culture, latex agglutination, and typing assay to detect and source track fecal contamination. *Applied and Environmental Microbiology* 73, 4110–4118.
Metcalf, T.G., Melnick, J.L. and Estes, M.K. (1995) Environmental virology: from detection of virus in sewage and water by isolation to identification by molecular biology – a trip of over 50 years. *Annual Review of Microbiology* 49, 461–487.
Muniesa, M. and Jofre, J. (2004) Factors influencing the replication of somatic coliphages in the water environment. *Antonie Van Leeuwenhoek* 86, 65–76.
Muniesa, M., Moce-Llivina, L., Katayama, H. and Jofre, J. (2003) Bacterial host strains that support replication of somatic coliphages. *Antonie Van Leeuwenhoek* 83, 305–315.
National Research Council (2004) *Indicators for Waterborne Pathogens*. National Research Council (US) Committee on Indicators for Waterborne Pathogens. National Academies Press, Washington, DC.
Nieuwstadt, T.J., Havelaar, A.H. and van Olphen, M. (1991) Hydraulic and microbiological characterization of reactors for ultraviolet disinfection of secondary wastewater effluent. *Water Research* 25, 775–783.
Novotny, C.P. and Lavin, K. (1971) Some effects of temperature on the growth of F pili. *Journal of Bacteriology* 107, 671–682.
Novotny, C., Raizen, E., Knight, W.S. and Brinton, C.C. Jr (1969) Functions of F pili in mating-pair formation and male bacteriophage infection studies by blending spectra and reappearance kinetics. *Journal of Bacteriology* 98, 1307–1319.
Parry, O.T., Whitehead, J.A. and Dowling, L.T. (1981) Temperature sensitive coliphage in the environment. In: Goddard, M. and Butler, M. (eds) *Viruses and Wastewater Treatment*. Pergamon Press, Oxford, UK, pp. 150–154.
Payment, P. and Franco, E. (1993) *Clostridium perfringens* and somatic coliphages as indicators of the efficiency of drinking water treatment for viruses and protozoan cysts. *Applied and Environmental Microbiology* 59, 2418–2424.
Pillai, S.D. (2006) Bacteriophages as Fecal Indicator Organisms. In: Goyal, S. (ed.) *Food Virology*. Kluwer Academic/Plenum Publishers, New York, pp. 205–222.
Pruss, A. and Havelaar, A.H. (2001) The Global Burden of Disease study and applications in water, sanitation and hygiene. In: Fewtrell, L. and Bartram, J. (eds) *Water Quality: Guidelines, Standards and Health*. IWA Publishing, London, UK, pp. 43–59.
Puig, M., Jofre, J. and Girones, R. (2000) Detection of phages infecting *Bacteroides fragilis* HSP40 using a specific DNA probe. *Journal of Virological Methods* 88, 163–173.
Rasched, I. and Oberer, E. (1986) Ff coliphages: structural and functional relationships. *Microbiological Reviews* 50, 401–427.
Rhodes, M.W. and Kator, H.I. (1991) Use of *Salmonella typhimurium* WG49 to enumerate male-specific coliphages in an estuary and watershed subject to nonpoint pollution. *Water Research* 25, 1315–1323.

Rosen, G (1993) *A History of Public Health*. The Johns Hopkins University Press, Baltimore, MD.
Safferman, R.S. and Morris, M. (1976) Assessment of virus removal by a multi-stage activated sludge process. *Water Research* 10, 413–420.
Savichtcheva, O. and Okabe, S. (2006) Alternative indicators of fecal pollution: relations with pathogens and conventional indicators, current methodologies for direct pathogen monitoring and future application perspectives. *Water Research* 40, 2463–2476.
Scarpino, P.V. (1978) Bacteriophage indicators. In: Berg, G. (ed.) *Indicators of Viruses in Water and Food*. Ann Arbor Science Publishers, Ann Arbor, MI, pp. 201–208.
Schaper, M., Jofre, J., Uys, M. and Grabow, W.O.K. (2002) Distribution of genotypes of F-specific RNA bacteriophages in human and non-human sources of faecal pollution in South Africa and Spain. *Journal of Applied Microbiology* 92, 657–667.
Scott, T.M., Rose, J.B., Jenkins, T.M., Farrah, S.R. and Lukasik, J. (2002) Microbial source tracking: current methodology and future directions. *Applied and Environmental Microbiology* 68, 5796–5803.
Sears, S.D., Ferreccio, C., Levine, M.M., Cordano, A.M., Monreal, J., Black, R.E., D'Ottone, K. and Rowe, B. (1984) The use of Moore swabs for isolation of *Salmonella typhi* from irrigation water in Santiago, Chile. *Journal of Infectious Diseases* 149, 640–642.
Seeley, N.D. and Primrose, S.B. (1980) The effect of temperature on the ecology of aquatic bacteriophages. *Journal of General Virology* 46, 87–95.
Sinton, L.W., Finlay, R.K. and Reid, A.J. (1996) A simple membrane filtration-elution method for the enumeration of F-RNA, F-DNA and somatic coliphages in 100-ml water samples. *Journal of Microbiological Methods* 25, 257–269.
Skraber, S., Gassilloud, B. and Gantzer, C. (2004) Comparison of coliforms and coliphages as tools for assessment of viral contamination in river water. *Applied and Environmental Microbiology* 70, 3644–3649.
Smith, D.C. (2006) Microbial source tracking using F-specific coliphages and quantitative PCR. A Final Report submitted to the NOAA/UNH Cooperative Institute for Coastal and Estuarine Environmental Technology (CICEET).
Srikanth, R. and Naik, D. (2004) Health effects of wastewater reuse for agriculture in the suburbs of Asmara city, Eritrea. *International Journal of Occupational and Environmental Health* 10, 284–288.
Stetler, R.E. and Williams, F.P. Jr (1996) Pretreatment to reduce somatic *Salmonella* phage interference with FRNA coliphage assays: successful use in a one-year survey of vulnerable groundwaters. *Letters in Applied Microbiology* 23, 49–54.
Stewart, J.R., Vinje, J., Oudejans, S.J.G., Scott and Sobsey, M.D. (2006) Sequence variation among group III F-specific RNA coliphages from water samples and swine lagoons. *Applied and Environmental Microbiology* 72, 1226–1230.
Suan, S.T., Chuen, H.Y. and Sivaborvorn, K. (1988) Southeast Asian experiences with the coliphage test. *Environmental Toxicology* 3, 551–564.
Tartera, C. and Jofre, J. (1987) Bacteriophages active against *Bacteroides fragilis* in sewage-polluted waters. *Applied and Environmental Microbiology* 53, 1632–1637.
Tartera, C., Lucena, F. and Jofre, J. (1989) Human origin of *Bacteroides fragilis* bacteriophages present in the environment. *Applied and Environmental Microbiology* 55, 2696–2701.
Toranzos, G.A., Gerba, C.P. and Hanssen, H. (1988) Enteric viruses and coliphages in Latin America. *Environmental Toxicology* 3, 491–510.
USEPA (2001) Method 1601: male-specific (F^+) and somatic coliphage in water by two-step enrichment procedure. EPA Number: 821-R-01-030, April 2001, Washington, DC.
Vaughn, J.M. and Metcalf, T.G. (1975) Coliphages as indicators of enteric viruses in shellfish raising estuarine waters. *Water Research* 9, 613–616.
Vinje, J., Oudejans, S.J., Stewart, J.R., Sobsey, M.D. and Long, S.C. (2004) Molecular detection and genotyping of male-specific coliphages by reverse transcription-PCR and reverse line blot hybridization. *Applied and Environmental Microbiology* 70, 5996–6004.
Wommack, K.E., Hill, R.T., Muller, T.A. and Colwell, R.R. (1996) Effects of sunlight on bacteriophage viability and structure. *Applied and Environmental Microbiology* 62, 1336–1341.
World Health Organization (2006) Guidelines for drinking-water quality. <http://www.who.int/water_sanitation_health/dwq/gdwq0506.pdf>.
Yates, M.V. (2007) Classical indicators in the 21st century – far and beyond the coliform. *Water Environment Research* 79, 279–286.

12 Phage Translocation, Safety and Immunomodulation

Natasza Olszowska-Zaremba[1], Jan Borysowski[1], Krystyna Dąbrowska[2] and Andrzej Górski[2]
[1]*Department of Clinical Immunology, Medical University of Warsaw;* [2]*Institute of Immunology and Experimental Therapy, Polish Academy of Science.*

A growing body of data shows that bacteriophages are omnipresent in the environment. For example, phages are present in high numbers in different ecosystems including water ecosystems, the rhizosphere and soil. Moreover, phages have been detected in some foodstuffs and drinks; they also constitute a significant component of the natural microbiota of both humans and animals (see Letarov, Chapter 2, this volume). Interestingly, there are some data to suggest that phages, like bacteria, might undergo translocation from the gut to mesenteric lymph nodes and then to the extranodular space.

A major application of bacteriophages is the treatment of antibiotic-resistant bacterial infections (phage therapy; see Loc-Carrillo *et al.*, Burrowes and Harper, and Abedon, Chapters 13, 14 and 17, this volume). Several important issues need to be taken into consideration when discussing the safety of phage therapy. First, only virulent phages should be used for therapeutic purposes; temperate phages should be excluded from therapy (see Christie *et al.* and Abedon, Chapters 4 and 17, this volume). Secondly, the antibacterial range of phages is very narrow, so they are not likely to disturb the balance of the natural microflora. Thirdly, different components of bacterial cells may be present in phage preparations; these components may also be released from bacteria *in vivo* following phage-induced lysis of bacterial cells. A thorough analysis of the literature on phage therapy, however, clearly shows that phages are safe antibacterial agents.

Although the natural hosts of bacteriophages are bacterial cells, a number of studies have shown that phages can interact with some populations of eukaryotic cells, especially with immune cells. In fact, phages can affect a wide range of functions of different populations of immune cells involved in both innate and adaptive immunity, including the production of antibodies, the proliferation of T and B cells, phagocytosis and the respiratory burst of phagocytic cells, and the production of cytokines.

Background

With the discovery of penicillin, a period of intensive development of antibiotic therapy began, resulting in a belief that the problem of bacterial infections had been eliminated. A number of factors, including overuse of antibiotics in clinical medicine, animal production and agriculture, however, have resulted in the emergence and spread of multidrug-resistant bacterial strains. In view of the inadequate number of new antibiotics (in 2003 only five out of 400 drugs under clinical trials in the USA were antibiotics), the re-emergence of the pre-antibiotic era has become a real prospect (Nelson, 2003).

One of the classes of antibacterial agents that could be used to combat multidrug-resistant bacteria is bacteriophages. A large number of studies have been conducted on the use of phages in humans, and of particular importance are studies that were carried out in Georgia and Poland. Bacteriophages have several advantages over antibiotics, including high specificity of their antibacterial activity, a capacity to kill antibiotic-resistant bacteria, a lack of any significant side effects and the low cost of phage preparations (Górski *et al.*, 2009; Housby and Mann, 2009). However, it is essential to conduct formal clinical trials to confirm the efficacy and safety of phage therapy before bacteriophages become widely available as therapeutic agents (see Burrowes and Harper, Chapter 14, this volume).

An understudied aspect of phage biology is their potential effects on the immune system. It is known that infections with pathogenic viruses – a category that does not include bacteriophages – are associated with activation of coordinated mechanisms of immune responses (Horst *et al.*, 2011). At first, non-specific immune responses are induced following the recognition by immune cells of pathogen-associated molecular patterns in virions. Primary immune responses to viral infections are mediated by macrophages, natural killer cells, granulocytes, complement and interferons. Viral antigens are presented by class I and class II major histocompatibility complex (MHC) molecules to T cells, which results in the activation of these cells. Humoral and cellular specific immune responses are induced, resulting in the neutralization of viruses by antibodies and killing of the infected host cells by mechanisms of cellular cytotoxicity. In the course of evolution, however, many pathogenic viruses have developed mechanisms to avoid elimination by the immune system. Pathogenic viruses can avoid recognition by the host immune system as a result of high antigenic diversity, such as human immunodeficiency virus (HIV), or a decrease in the expression of class I and class II MHC molecules on the infected cells, as occurs in infection with cytomegalovirus, Epstein–Barr virus and herpes simplex virus type 1 (HSV-1) and HSV-2 (Ploegh, 1998; Olszewska and Radkowski, 2005). Moreover, pathogenic viruses can substantially affect antiviral immune responses by interfering with the activity of cytokines, such as the production by adenoviruses and Epstein–Barr virus of proteins with activity antagonistic to cytokines, and by interfering with the production of antibodies and components of the complement system, as occurs with HSV infection (Ploegh, 1998; Olszewska and Radkowski, 2005). So far, over 50 viral genes whose products exert immunomodulatory activity have been identified (Ploegh, 1998). Thus, it is not unreasonable that phages infecting pathogenic and normal flora bacteria would also be able to evolve immunomodulating proteins to aid in their retention in the body.

Prevalence of Bacteriophages and the Safety of Phage Therapy

Prevalence of bacteriophages in the environment

There is a surprising abundance and diversity of bacteriophages in the environment. It is estimated that there are between 10^4 and 10^8 phages ml^{-1} in water ecosystems (Weinbauer, 2004). Bacteriophages have also been isolated from sewage (mean concentration 10^4–10^8 ml^{-1}), the rhizosphere and soil (1.5×10^8 g^{-1}; Weinbauer, 2004). It was also reported that bacteriophages are present in yogurt (*Lactobacillus* phages) and sauerkraut, where as many as 26 different phages were found (Kiliç *et al.*, 1996; Lu *et al.*, 2003). It is known that phages are also prevalent in the bodies of humans and animals and in their secretions and excrements. Phages are particularly abundant in the intestines of mammals in view of the presence of abundant bacterial microflora. This issue is discussed in detail by Letarov (Chapter 2, this volume).

Safety of phage therapy – general considerations

The omnipresence of bacteriophages in the environment implies constant exposure of humans to phages. Thus, not surprisingly, bacteriophages administered for therapeutic purposes are well tolerated by patients. Side effects of phage therapy are observed very rarely. In a study by Ślopek *et al.* (1983), side effects included the intolerance of phage preparations, and allergic symptoms were observed in only three out of 138 patients (2.2%). Between days 3 and 5 of the treatment, a transient liver area discomfort was noted that could have been caused by massive endotoxin release following phage-mediated lysis of bacterial cells. As reported by Morello *et al.* (2011), however, components of bacterial cells released from the cells following phage-induced lysis do not exert pro-inflammatory activity. They showed that the administration of phages to mice infected by *Pseudomonas aeruginosa* is not associated with an increase in the production of cytokines.

Similar conclusions were drawn by Międzybrodzki *et al.* (2009), who examined the level of selected inflammatory markers – C-reactive protein (CRP), white blood cells (WBCs) and erythrocyte sedimentation rate (ESR) – in a group of 37 patients subjected to phage therapy. A retrospective analysis of the results showed no significant changes in CRP, WBCs or ESR values between days 5 and 8 of the treatment compared with the corresponding values found in the patients prior to phage administration. Moreover, between days 9 and 32 of the treatment, significant decreases in the levels of CRP and WBCs were observed. These results suggest that lysis of bacteria by phages does not exert any pro-inflammatory effects; in fact, they indicate that phages could exert anti-inflammatory activity (Międzybrodzki *et al.*, 2009).

A strong argument for the safety of phage therapy is the high specificity of the antibacterial activity of phages. Phages kill only bacteria from certain strains or subspecies. Therefore, unlike antibiotics, they are less likely to disturb the balance of the bacterial microflora. A lack of any deleterious effects of phages on the microflora was found in a study by Bruttin and Brüssow (2005). They reported that oral administration of T4 phage at a titre of 10^3 or 10^5 plaque-forming units (PFU) ml^{-1} to healthy individuals did not result in a decrease in *Escherichia coli* counts in faeces. Furthermore, no side effects attributable to the administration of phage preparations were observed in any subject.

An important topic relevant to the problem of the safety of phage therapy is the possibility of phage-mediated transfer of detrimental genes (e.g. those encoding virulence factors and toxins) between bacterial cells. This problem is discussed by Christie *et al.* (Chapter 4, this volume).

An additional potential source of side effects during phage therapy can be different components of bacterial cells, especially endotoxins contaminating bacteriophage preparations, which are released from the bacteria in which phages are propagated. In the early stages of the development of phage therapy (1930s and 1940s), it was observed that intravenous administration of preparations containing a large amount of endotoxin causes a number of side effects, including an elevation of temperature, shivers and headache (Chanishvili *et al.*, 2001). Effective methods have since been developed to purify phage preparations from different components of bacterial cells, including endotoxins. However, so far, phage lysates have been used for most treatment of humans, as purification of phage preparations from endotoxins is associated with a loss of active phage particles. To minimize side effects, such preparations are administered orally or topically (Sulakvelidze and Kutter, 2005). Studies on using phage lysates in the treatment of chronic bacterial infections in humans have been conducted since 1952 at the Hirszfeld Institute of Immunology and Experimental Therapy PAS, Wrocław, Poland. No significant side effects of phage therapy have been observed so far (Weber-Dąbrowska *et al.*, 2000a; Górski *et al.*, 2009; and unpublished results).

Preparations of phages specific to Gram-positive bacteria can also contain some components of bacterial cells. There are data in the literature, however, to show that it is possible to obtain safe preparations of these phages. For example, an apyrogenic preparation of a staphylococcal phage intended for intravenous administration was developed in Georgia. A study performed on 900 patients, including 494 subjects who were administered antibiotics with phages or phages only, and 406 subjects who were administered antibiotics only, showed that intravenous administration of this phage preparation did not cause any significant side effects (Chanishvili *et al.*, 2001).

Phages and oxidative stress

An essential component of non-specific immune responses associated with phagocytosis of microbes is the respiratory burst, in which reactive oxygen species (ROS) are generated. These species, including superoxide, hydrogen peroxide, hydroxyl radicals and singlet oxygen, exert bactericidal activity. When produced in excess, however, they can also induce oxidative stress and exert cytotoxic activity, resulting in damage of tissues (Gulam and Ahsan, 2006). Oxidative stress resulting from excessive production of ROS is implicated in the pathogenesis of cancer, sepsis, multiple organ dysfunction syndrome, atherosclerosis and neurodegenerative diseases, especially Alzheimer's disease and Parkinson's disease (Knight, 1995; Sikora, 2002; Gulam and Ahsan, 2006). Therefore, the evaluation of the effects of phages on ROS generation is essential.

It was shown that phages can decrease both *E. coli* and lipopolysaccharide (LPS)-induced production of ROS by neutrophils (Międzybrodzki *et al.*, 2008). Furthermore, the incubation of *E. coli* with homologous phages resulted in a decrease in the intensity of the respiratory burst in both neutrophils and monocytes, which could result from the lysis of bacteria by phages (Przerwa *et al.*, 2006). This result suggests that lysis of bacteria by phages is not likely to induce oxidative stress in cells, arguing for the safety of phage therapy. Another study revealed that both lysates and purified preparations of T4 phage and staphylococcal phage A3/R induced only a weak respiratory burst in human monocytes and neutrophils *in vitro* (Borysowski *et al.*, 2010). T4 and A3/R phage-induced production of ROS was so low that it probably would not induce oxidative stress *in vivo* (Przerwa *et al.*, 2006; Międzybrodzki *et al.*, 2008). The capacity of phages to decrease the production of ROS by phagocytic cells can explain the efficacy of phages in the treatment of sepsis (Weber-Dąbrowska *et al.*, 2003; Przerwa *et al.*, 2006). Our hypothesis to explain the effects of phages on phagocytic cells is based on the assumption that LPS is bound by phage virions, which could prevent the binding of LPS to its receptor that triggers the respiratory burst (Przerwa *et al.*, 2006).

Immune Responses to Exogenous Bacteriophages

Inactivation of phages by mechanisms of the innate immunity

Studies of the distribution of bacteriophages in mammals suggest that they can penetrate into the circulation following their administration by practically any standard route (Dąbrowska *et al.*, 2005; Sulakvelidze and Kutter, 2005). When there are no host bacteria in which phages could replicate, they are fairly rapidly removed from the blood and the majority of internal organs. As shown by Dubos *et al.* (1943), *Shigella dysenteriae* phages in healthy mice are cleared from the circulation within a few hours of their intraperitoneal administration, while in infected mice they are detectable in blood, even at 18 h post-administration (Sulakvelidze and Kutter, 2005). Phages are internalized and eliminated by cells of the reticuloendothelial system of the liver and spleen. Inchley (1969) noted that, in mice, over 99% of T4 phage virions were phagocytosed by Kupffer cells within 30 min of their intravenous administration. As well as Kupffer cells, spleen macrophages can also eliminate phages, but their activity in this regard is fourfold lower (Dąbrowska *et al.*,

2005). Consequently, phage titre in the liver decreases rapidly, while in the spleen phages remain at high titre for up to 5–7 days after their administration (Geier *et al.*, 1973). It has been suggested that phages entrapped in the spleen can be a source of antigen necessary for antibody generation (Geier *et al.*, 1973).

Anti-phage antibodies

General considerations

Exogenous phages can induce not only innate immune responses but also the production of neutralizing antibodies. Concentration of these antibodies depends, among other things, on the route of phage administration (topical and oral administration result in only a slight increase in the generation of antibodies) and dosage protocol (the level of antibodies is low following a single phage administration and when phages are administered several times with small intervals between doses; Sulakvelidze and Kutter, 2005). Phage immunogenicity also depends on phage strain. For example, T1 and T5 phages are less immunogenic than T4 phage (Adams, 1959).

Neutralizing antibodies can limit the efficacy of phage therapy. This was shown by Srivastava *et al.* (2004), who showed that the clearance of T7 phage from the circulation was slower in B cell-deficient mice compared with wild-type mice. Pagava *et al.* (2011) reported that 14 out of 31 blood samples from patients who were administered phages orally contained neutralizing antibodies that inactivated 52.5–97.3% of bacteriophages. Antibodies against *Staphylococcus aureus* phages were detected in 12 out of 57 patients with staphylococcal infections (Kucharewicz-Krukowska and Ślopek, 1987). During phage therapy, these antibodies were detected in 54% of patients. Further analysis of the results obtained in 30 patients revealed that phage therapy was ineffective in two out of five patients in whom anti-phage antibodies were detected prior to therapy (Kucharewicz-Krukowska and Ślopek, 1987). Interestingly, the clinical state of the remaining three patients improved following the treatment, which suggests that the presence of neutralizing antibodies does not always result in a lower therapeutic efficacy.

Immunogenic properties of ϕX174 phage

The immunogenic properties of phages have found use in the diagnosis of immune system diseases and monitoring of humoral immune responses. For example, ϕX174 phage has been used to evaluate immune responses in patients with primary and secondary immunodeficiencies, bone-marrow recipients and HIV patients (Bearden *et al.*, 2005). Pescovitz *et al.* (2011) used this phage to evaluate the effects of rituximab, a B cell-depleting antibody, on the production of antibodies in patients with type I diabetes. It was found that this assay is a sensitive indicator of antibody production that enables the evaluation of humoral immune responses. This study shows the utility of measuring anti-ϕX174 phage antibodies both when monitoring the efficacy of immunosuppressive therapy and when evaluating immunocompetence in patients.

Ochs *et al.* (1971) showed that this phage can also be used in differential diagnosis of immunodeficiencies. In healthy individuals, intravenous administration of ϕX174 results in the generation of IgM antibodies that neutralize the phage within 3–4 days (Bearden *et al.*, 2005). It was observed that a long presence of phages in the circulation (11–42 days) along with a lack of humoral immune response is typical of X-linked agammaglobulinaemia (in patients in whom a severe immunodeficiency was excluded; Ochs *et al.*, 1971). Phage ϕX174 was selected for diagnostic use because it is a strong antigen and does not cause any side effects in humans. Furthermore, the evaluation of its clearance from blood is relatively simple (Ochs *et al.*, 1971).

Anti-phage cellular immunity

Apart from non-specific immune responses and humoral immunity, cellular immunity also plays an important role in combating viral infections. Some of the first authors to show that phages can induce cellular immune responses were Langbeheim *et al.* (1978), who examined delayed-type hypersensitivity reactions to MS-2 phage and a synthetic fragment of one of its capsid proteins in guinea pigs. They found that subcutaneous injection of phages resulted in a strong hypersensitivity reaction in all animals. Cellular responses to MS-2 phage were also observed *in vitro*, as revealed by an intense proliferation of lymphocytes from phage-sensitized guinea pigs (Langbeheim *et al.*, 1978). On the other hand, the results of Srivastava *et al.* (2004) suggest that cellular immune responses play only a minor role in inactivation of phages. They showed that the clearance of T7 phage in T cell-deficient mice is similar to that found in T cell-proficient mice. Thus, the data are conflicting, and further studies are needed in this regard.

Immunomodulatory Effects of Bacteriophage Preparations

Effects of phages on granulocytes

Non-specific immune response efficacies are so high that they are often sufficient to clear microbes before the activation of specific immune responses. The first cells to migrate to the site of infection are macrophages and granulocytes, especially neutrophils. It has been shown that, in patients subjected to phage therapy, clearance of infection was accompanied by a decrease in the number of mature neutrophils and an increase in the number of neutrophil precursors in the peripheral blood (Weber-Dąbrowska *et al.*, 2002). Differentiation of neutrophils can be mediated by both phage virions and some components of bacterial cells present in phage lysates. Stimulatory effects of phage preparations on the generation of neutrophil precursors appear to be beneficial because they result in the enhancement of non-specific immune responses.

It was also shown that bacteriophages can affect the migration of phagocytic cells. Pre-incubation of human granulocytes with T4 phage results in a slight stimulation of migration *in vitro* (Kurzępa *et al.*, 2009). Moreover, phages were found to relieve the inhibitory effect of LPS on granulocyte migration (Kurzępa *et al.*, 2009). The results of other studies showed that phages had only a slight effect on the migration of human granulocytes and monocytes and had no effect on the intracellular killing of bacteria (unpublished results). Furthermore, T4 phage can decrease the migration of mononuclear cells, and to some extent of neutrophils, to a skin graft in mice, resulting in a marked decrease in cellular infiltration of the graft at days 7 and 8 post-transplantation (Górski *et al.*, 2003a).

Influence of phages on phagocytosis

Our studies have shown that phages also can affect phagocytosis (Przerwa *et al.*, 2006; Kurzępa *et al.*, 2009). Pre-incubation of T4 phage and *P. aeruginosa* F8 phage with human phagocytic cells *in vitro* resulted in a decrease in the phagocytosis of *E. coli* cells. A similar effect was noted when phagocytic cells were incubated with *E. coli* and T4 phage at a high titre (10^{10} PFU ml^{-1}).

This effect was weaker for F8 phage. Incubation of *E. coli* cells with homologous phage (T4), but not with heterologous phage (F8), resulted in an increase in the phagocytosis of bacteria by both neutrophils and monocytes (Przerwa *et al.*, 2006). It appears that homologous phages can coat bacteria ('phage opsonization'), thus facilitating their phagocytosis (Górski *et al.*, 2003a). T4 phage, however, had no effect on the *in vitro* phagocytosis of bacteria by monocytes and neutrophils isolated from either *E. coli*-infected or uninfected mice (Przerwa *et al.*, 2006).

Effects of phages on T cells and platelets – practical implications

Integrins comprise a family of cell-surface receptors that, by interacting with extracellular matrix proteins, are involved in various physiological processes including immune responses, tissue remodelling, haemostasis and angiogenesis. Moreover, they have been implicated in the pathogenesis of various diseases including infectious and non-infectious diseases (Perdih and Dolenc, 2010). The interactions of integrins with extracellular matrix proteins are mediated by their RGD (Arg-Gly-Asp) sequences. Several pathogenic viruses, including HIV, adenoviruses, picornaviruses and coxsackievirus A9, also contain this sequence for binding to target cells (Berinstein *et al.*, 1995; Triantafilou *et al.*, 2001; Lafrenie *et al.*, 2002).

In vitro experiments revealed that phages T4 and HAP1 (a T4 mutant strain; see below) can bind both platelets and T cells (Kniotek *et al.*, 2004a), possibly via integrin binding. In the case of T4 and HAP1 phages, this is very likely, because one of their capsid proteins, gp24, contains the KGD (Lys-Gly-Asp) sequence, which can be a ligand for β3-integrins that are present on both platelets and T cells (Górski *et al.*, 2003a; Wierzbicki *et al.*, 2006; Varon and Shai, 2009). An additional argument for β3-integrins mediating the interactions between phages and immune cells is the capacity of eptifibatide (an inhibitor of $\alpha_{IIb}\beta_3$ integrins) to block binding of phages to T cells and platelets (Kniotek *et al.*, 2004a). We also showed that phages can reduce the adhesion of platelets and, to a lesser extent, of T cells, to fibrinogen (Kniotek *et al.*, 2004a).

Our studies have shown that phages can exert immunosuppressive activity. In a study on the role of bacteriophages in the development of transplantation tolerance, Górski *et al.* (2006a) observed that bacteriophages can inhibit CD3-triggered activation of T cells. Intraperitoneal administration of purified preparations of phages T4, HAP1 and F8 at the dose 10^6 PFU per mouse resulted in significant inhibition of specific humoral responses *in vitro* (Kniotek *et al.*, 2004b). It was also found that a purified T4 phage preparation significantly decreased alloantigen-driven *in vitro* humoral responses (Kniotek *et al.*, 2004b). Furthermore, intraperitoneal administration of T4 phage extended allograft survival and reduced the inflammatory infiltration within the graft, suggestive of phage-mediated inhibition of anti-transplant immune responses (Górski *et al.*, 2003b, 2006a). According to our hypothesis, phages can affect graft rejection by inhibiting the interactions between T cells, fibrinogen and platelets (Kniotek *et al.*, 2004a).

Effects of phages on the production of cytokines

The results of some studies show that phages can affect the production of different cytokines (Weber-Dąbrowska *et al.*, 2000b; Kumari *et al.*, 2010). One of the mechanisms that could mediate these effects is the inhibition of nuclear factor (NF)-κB activity. NF-κB is a transcription factor that plays an important role in the expression of genes encoding cytokines, cellular surface receptors, acute phase proteins and adhesion molecules (Hiscott *et al.*, 1997). We have shown that T4 phage inhibited HSV-1-induced activity of NF-κB in mononuclear cells *in vitro* (Gorczyca *et al.*, 2007). The influence of phages on the activation of NF-κB may be important in view of the potential use of phages in the regulation of immune responses to allografts, infections and inflammation.

It was found that bacteriophages can decrease the production of cytokines induced by bacterial infections in animals. For example, in a murine model of lung infection caused by *P. aeruginosa*, Morello *et al.* (2011) determined the concentration of pro-inflammatory cytokines in bronchoalveolar lavage samples collected at 20 h after intranasal administration of bacteria. The authors noted a significant reduction in the concentration of interleukin (IL)-6 and keratinocyte chemoattractant in the bronchoalveolar lavage samples following the administration of a single dose of P3-CHA phage both 4 days before and 1 day after bacterial inoculation. Moreover, histopathological examination revealed that damage of pulmonary tissue was more pronounced in untreated mice compared with phage-treated animals. As suggested by Kumari *et al.* (2010), it is likely that phages, by decreasing the production of cytokines, limit the inflammatory reaction, thus reducing damage of tissues. It was also shown that treating burn wound infections caused by *Klebsiella pneumoniae* with Kpn5 phage resulted in a significant decrease in the level of both pro-inflammatory (IL-1β and tumour necrosis factor (TNF)-α) and anti-inflammatory (IL-10) cytokines in sera and lungs of the treated mice. It was suggested that a decrease in the production of cytokines during phage therapy can result from a reduction in the number of pathogenic bacteria that are eliminated by phages (Kumari *et al.*, 2010).

Our observations of patients subjected to experimental phage therapy has revealed that phages can substantially affect the production of IL-6 and TNF-α, two important pro-inflammatory cytokines (Weber-Dąbrowska *et al.*, 2000b). These results suggest that a positive outcome of phage therapy might, to some extent, be mediated by the regulation of TNF-α production. Interestingly, the effect of phages on the production of TNF-α depends on the level of this cytokine before the administration of phage preparations. In patients with a low or moderate serum level of TNF-α, phages increased the production of this cytokine, whereas in patients with a high level of TNF-α, phages decreased its production. In a similar (regulatory) way, phage therapy can affect spontaneous production of IL-6 and TNF-α by cultured mononuclear cells, as well as the LPS-induced production of TNF-α by mononuclear cells. A decrease in the LPS-induced production of TNF-α by mononuclear cells was observed in patients in whom therapeutic effects were partial or who did not respond to therapy. The regulatory effects of phage therapy on the production of cytokines can be attributed to one (or a combination) of three factors: (i) immunomodulatory activity of phages; (ii) natural recovery of the patient; and (iii) a combination of the effects of phages and bacterial components (Weber-Dąbrowska *et al.*, 2000b).

Bacteriophages can also affect the production of other cytokines. Purified T4 phage preparation significantly inhibited the production of IL-2 and, to a lesser extent, gamma interferon (IFN-γ) by phyto-haemagglutinin-activated human leukocytes (Przerwa *et al.*, 2005; Kurzępa *et al.*, 2009). On the other hand, Kleinschmidt *et al.* (1970) showed that the administration of a purified T4 phage preparation can result in an increase in the concentration of IFN-γ in the serum of mice. Moreover, DNA from T4 phages can induce the production of IFN in mice.

Immunomodulatory activity of staphage lysate

It appears that the lysate of a *S. aureus* phage, known as staphage lysate (SPL), could be used for immunodeficiency diagnosis, as well as in treating staphylococcal infections. It was reported that SPL induced proliferation in over 95% of lymphocytes isolated from healthy individuals (Dean *et al.*, 1975). Maximal lymphoproliferative responses were observed on day 6 of culture. Furthermore, unlike phytohaemagglutinin, SPL activates both T and B cells (Dean *et al.*, 1975). The authors suggested that SPL could be used as an adjunct to mitogens and antigens routinely used to evaluate cellular immune responses. In line with these data are results showing that SPL-induced immune responses are weaker or undetectable in cancer patients and in those with immunodeficiencies (Dean *et al.*, 1975).

SPL was shown to increase the production of immunoglobulins in human lymphocyte culture and to induce the production of polyclonal immunoglobulins by murine splenocytes (Sulakvelidze and Kutter, 2005). One dose of SPL administered to mice induced the production of IgG and, to a lesser extent, IgA (Esber *et al.*, 1985). According to Esber *et al.* (1985), SPL can also stimulate a specific humoral response in mice. Thus, an additional beneficial effect of SPL in the treatment of chronic staphylococcal infections may be the enhancement of antibacterial immune responses. Furthermore, SPL, as a preparation stimulating both humoral and cellular immunity, could be particularly useful in the treatment of infections in immunocompromised patients, including those receiving cytostatic drugs and individuals with immunodeficiencies (Esber *et al.*, 1985). The immunostimulatory effects of phage preparations are probably mediated by both phage virions and some components of bacterial cells present in preparations (Sulakvelidze and Kutter, 2005).

In view of the mitogenic properties of SPL and its capacity to induce IFN production, studies were conducted to evaluate its potential anticancer effects (Mathur *et al.*, 1988). In a rat model of breast cancer, the administration of SPL preceded by immunization with dead *S. aureus* cells resulted in a decrease in tumour diameter (Mathur *et al.*, 1988). A study performed on a murine sarcoma model, however, did not confirm the anticancer effects of SPL (Esber *et al.*, 1981). It was also shown that SPL reduces the migration of murine B16 melanoma cells *in vitro*. However, this effect may be mediated by some components of staphylococcal cells that are present in the lysate, as bacterial lysates without any phage particles also significantly decreased the migration. Furthermore, neither SPL nor control bacterial preparations inhibited the proliferation of cancer cells and did not stimulated their migration. This is an important result in regard to the potential therapeutic use of SPL (Dąbrowska *et al.*, 2010).

Phage capsid elements inducing immunomodulatory effects in mammals

Mammals represent an 'environment' for phages parasitizing mammalian symbiotic or pathogenic bacteria (see Letarov, Chapter 2, this volume). In this environment, phages are affected by a potent pressure of mammalian immunity that presumably drives selection and evolutionary adaptation. Thus, immunomodulatory effects of bacteriophages could result from adaptations to persistence in the mammalian environment. In general, both nucleic acids and phage capsid proteins may contribute to phage effects on the immunological system. The effects can be anticipated and unspecific, such as the general effect of large antigenic objects introduced into a system, or specific, such as directed interactions of active elements. Although data on specific means of interactions are poor – that is, which genes, polypeptides or proteins are involved – some possible mediators have been proposed.

In T4 phage, two proteins were proposed as active in interactions with mammalian immunity: gp24 and gpHoc (Górski *et al.*, 2003a; Dąbrowska *et al.*, 2006). These are both located on the phage head. Gp24 is an essential protein that builds head corners, while gpHoc is a dispensable external element of the capsid (Ishii and Yanagida, 1977; Fokine *et al.*, 2004). The full name of Hoc protein, 'highly antigenic outer capsid protein' (Ishii and Yanagida, 1975), indicates its effects on mammalian immunity. Studies of Hoc structure furthermore have revealed the presence of immunoglobulin-like folds in this protein (Bateman *et al.*, 1997; and unpublished results). These folds characterize cell-attachment molecules, often regulating immune systems, and are quite often present in phage genomes (Fraser *et al.*, 2006). Similarities of the phage protein to mammalian proteins engaged in interactions and regulation of immunity make Hoc a probable modulator of immune functions. Surprisingly, in light of its originally identified highly antigenic nature, data support the hypothesis that Hoc protein could help the phage to evade the immune system by causing Hoc-mediated suppression of immunity (Dąbrowska *et al.*, 2006).

Hoc was also shown to regulate the adhesion properties of the T4 phage particle (Dąbrowska *et al.*, 2004; Sathaliyawala *et al.*, 2010), which may contribute additionally to phage interactions with immune cells. Importantly, a nonsense mutant of *hoc* that lacks gpHoc in its capsid (HAP1 phage (high-affinity phage), selected for its affinity in binding to B16 murine melanoma cells) presented unique and unexpected properties toward mammalian cells. Although the mutation did not result in changes in the mutant's general morphology (as determined by electron microscopy), it resulted in substantial changes in bacteriophage metabolism and clearance, with clearance of HAP1 being significantly more rapid. The comparison of concentrations of active wild phages to concentrations of the Hoc^{-} mutant in murine blood and liver revealed a preponderance of the wild phage, by approximately one to three orders of magnitude (Dąbrowska *et al.*, 2007). It is probable that the degradation of HAP1 is more efficient than that of T4 phage, as gpHoc (the external head protein) covers other capsid elements that are more 'visible' and stimulatory to the degradation factors. Hoc protein seems to be a kind of protection for T4 phage, and this protection prevents rapid degradation of T4 in mammalian organisms (Dąbrowska *et al.*, 2007).

Phage clearance by reticuloendothelial system filtration is usually proposed as the main path of phage removal from non-immunized mammals (Merril, 1974). Thus, the innate immunity response to a phage can be modified according to its virion protein composition. Probably the very first virion protein whose properties influenced elements of innate immunity was identified by Merril *et al.* (1996) in phage λ. A substitution of the amino acid lysine for glutamic acid in the major head protein E caused a long-circulating phenotype, that is, λ-derived mutants that circulated for longer in murine blood in comparison with the wild-type phages. The mutated phages were also more effective in combating lethal bacterial infections in mice (Merril *et al.*, 1996).

Additional phage effects

T4 and HAP1 were investigated in *in vivo* cancer models with regard to their ability to interfere with cancer processes. This work was inspired by the work of Bloch (1940), who showed that phages can accumulate in cancer tissue and inhibit the growth of tumours. Next, it was demonstrated that phages bind cancer cells *in vitro* and *in vivo* and attach to the plasma membrane of lymphocytes (Kańtoch, 1958; Kańtoch and Mordarski, 1958; Wenger *et al.*, 1978). As described above, SPL can interfere with melanoma cell migration. T4 phage and HAP1 phage (Hoc^-) were investigated and compared in a murine model of intravenously injected melanoma. The HAP1 preparation was significantly more active in diminution of melanoma colonies in murine lungs (Dąbrowska *et al.*, 2004, 2007). This effect was correlated with the elevated adhesive potential of the Hoc^- phage on murine melanoma cells. Its adhesive potential towards mouse melanoma cells was estimated to be about eight times higher than that of T4. Elevated affinity to human melanoma was also observed. Adhesion was decreased by typical β3-integrin inhibitors such as eptifibatide or antibodies (Dąbrowska *et al.*, 2004).

Increased activity of phage lacking gpHoc supports the hypothesis that other capsid elements can interact and affect mammalian cells, as Hoc protein is located in the phage head. These differences in T4 and HAP1 inspired a hypothesis (Górski *et al.*, 2003a) on the potential activity of the KGD motif (a homologue of RGD) that is present in gp24 (head vertex protein) of T4. The motif is well exposed at the phage head, as revealed by the gp24 structure data (Protein Data Bank ID: 1YUE). The head vertex protein was postulated to interact with mammalian-specific receptors via its KGD motif and thus to affect cells or cancer processes through β3-integrins. β3-Integrins are a class of adhesion receptors that mediate cell–cell and cell–extracellular matrix interactions. Thus, they play a role in, for example, tissue integrity, cellular migration, cell survival, adhesion and differentiation. Disintegrins able to bind β3-integrins were reported to interfere with tumour growth, progression and metastasis (Giancotti and Ruoslahti, 1999; Adair and Yeager, 2002). Thus, gp24 was proposed as the protein active in phage adhesion to mammalian cells and in cancer processes (Dąbrowska *et al.*, 2004).

As adhesion processes as well as metastasis are related to cell migration, T4 and HAP1 phages were also investigated regarding their ability to interfere with cancer-cell migration *in vitro*. Migration assays were carried out on Matrigel^{TM} (an extracellular matrix substitution) or fibronectin. In both cases, phages were able to diminish cell migration, but no differences between T4 and HAP1 were observed (Dąbrowska *et al.*, 2009). HAP1 and T4 were also shown to decrease immunoglobulin production by human lymphocytes *in vitro* in response to allogenic cells, as well as anti-sheep red blood cells antibody production in mice. Both phages induced comparable effects, with no significant differences in their intensity (Kniotek *et al.*, 2004b).

Mammalian immunity can be also influenced by phage DNA, especially using phage DNA vaccines. This line of research is discussed in detail by Clark *et al.* (Chapter 7, this volume).

Bacteriophage Translocation

According to the hypothesis of Górski *et al.* (2006b), bacteriophages or their products, like bacteria, can undergo translocation migration from the gut lumen to mesenteric lymph nodes and subsequently to other extra-intestinal organs and sites (O'Boyle *et al.*, 1998; Wiest and Garcia-Tsao, 2005). Translocation of a small amount of bacteria can occur in 5–10% of healthy individuals and is considered a normal physiological phenomenon (Balzan *et al.*, 2007). If the function of the immune system is normal, then bacteria are rapidly cleared from the extra-intestinal space and are not deleterious. Intensive penetration of bacteria from within the gut, however, has been implicated in the pathogenesis of many severe diseases, including acute pancreatitis, sepsis and multiple organ dysfunction syndrome (O'Boyle *et al.*, 1998; Balzan *et al.*, 2007).

Reports on the abundance of phages in the gut microbiota raise the question of the possibility of phages undergoing translocation similarly to bacteria and some pathogenic human viruses. For example, coxsackievirus B3 could be detected in the hearts of mice as early as 24 h after its intragastric administration, which suggests a high efficiency of translocation (Harrath *et al.*, 2004). Viral translocation is mediated by M cells and dendritic cells, which bind viruses and transport them across the epithelium to Peyer's patches (Morin and Warner, 1994; Mossel and Ramig, 2003; Smith *et al.*, 2003). From Peyer's patches, viruses relocate to mesenteric lymph nodes and then to peripheral tissues, blood and internal organs (the spleen, liver and lungs) (Morin and Warner, 1994; Mossel and Ramig, 2003). Thus, if phages can undergo translocation, their presence can be expected in the peripheral blood of humans and animals. The presence of phages in the blood of healthy individuals, however, was reported in one paper only (Parent and Wilson, 1971). This study revealed the presence of mycobacteriophages not only in the sera of patients with Crohn's disease but also in four out of 18 blood samples taken from healthy individuals from the control group (Parent and Wilson, 1971). A larger body of data has been derived from animals. For example, Merril *et al.* (1972) detected phages in the blood of calves, sheep and hens. Orr *et al.* (1975) reported the presence of phages specific to *E. coli* in as many as 90% of blood samples from calves. It needs to be stressed, however, that in the majority of studies performed in the 1970s, only *E. coli* was used to detect the presence of phages in the sera of animals. It is impossible, therefore, to draw any general conclusions regarding the real scale of the presence of phages in the blood of animals.

Phage translocation was also demonstrated in studies conducted to evaluate the distribution of phages following their oral administration to humans and animals. For example, Keller and Engley (1958) reported that *Bacillus megaterium* phages could be detected in blood 5 min after intragastric administration to mice. Likewise, Bradley *et al.* (1963) showed that phages can translocate fairly rapidly across the gut wall following oral administration to mice. The pace of penetration of phages from the gut to the blood suggests that it can occur by diffusion (Dąbrowska *et al.*, 2005). Duerr *et al.* (2004), however, suggested that translocation of phages across the intestinal barrier can be a more complex process involving M cells, dendritic cells and enterocytes. By using the phage-display technique, it was shown that translocation of phages requires certain amino acid sequences in capsid proteins. These are probable ligands for receptors present on cells that mediate transport across the intestinal barrier and M13 phage lacking such sequences did not undergo translocation.

Little is known about phage translocation in humans. Weber-Dąbrowska *et al.* (1987) evaluated the distribution of phages in patients who were administered phages orally to treat bacterial infections. The presence of phages was detected in 47 out of 56 blood samples taken from patients on day 10 of treatment. A recent study by Pagava *et al.* (2011) performed on 102 children with bacterial infections confirmed that phages can be absorbed from the digestive tract. Therapeutic phages were detected in six out of seven blood samples and in 48 out of 55 urine samples collected from patients between days 3 and 5 of the treatment. The translocation of orally administered phages through the gut wall enables phage therapy to be used also in the treatment of systemic infections and infections of the urinary tract.

In conflict with these results are those of Bruttin and Brüssow (2005), who evaluated the safety of oral administration of T4 phage in 15 healthy volunteers. Phages were not detected in blood samples taken from patients either at the beginning or at the end of the study. There are two potential reasons for the lack of phage translocation in this study. First, these investigators did not neutralize the gastric juice prior to the administration of bacteriophages, which could result in the inactivation of virions due to a strongly acidic pH in the digestive tract. Secondly, the titre of phage preparations used by these authors was about 1000 times lower than the titre of phage lysates employed by Weber-Dąbrowska *et al.* (1987) for therapeutic purposes. Moreover, it needs to be taken into account that, in patients with chronic bacterial infections, the gut wall can be penetrated by microorganisms more easily than in healthy individuals (Górski *et al.*, 2006b).

A growing body of data shows that the bacterial gut microflora is very important for the proper function of the human organism (Blaser, 2011; see also Letarov, Chapter 2, this volume). Consistent with this, it is likely that bacteriophages present in the intestines, like bacteria, play an important role in maintaining the physiological balance of the host organism. According to the hypothesis of Górski *et al.* (2006b), endogenous phages can affect the function of the immune system in both the gut and the whole body (as a result of their translocation). Bacteriophages present in the gut can also limit bacterial translocation, thus reducing the inflammatory reaction caused by the migration of bacteria through the wall of the digestive tract (Górski *et al.*, 2006b). Another important function of phages present in the gut could be inhibition of the activity of dendritic cells, which are essential for inducing immune responses in the digestive tract. In this way, phages could control inflammatory reactions in the gut (Górski and Weber-Dąbrowska, 2005; Górski *et al.*, 2006b).

Concluding Remarks

Theoretically, side effects of phage therapy might be caused by phage virions themselves, some components of bacterial cells present in phage preparations and/or the products of phage-mediated lysis of bacteria. Both experimental and clinical studies, however, show that the administration of phage preparations does not result in any serious side effects. Moreover, in view of the high specificity of antibacterial activity, phages are less likely to disturb the balance of the normal microflora compared with traditional antibiotics. In general, then, bacteriophages appear to be safe antibacterial agents.

The interactions between phages and the immune system are quite complex. First, phage antigens can induce specific humoral and cellular immune responses. From the point of view of phage therapy, the former seem to be more important because the production of neutralizing antibodies is one of the main factors potentially reducing the therapeutic effectiveness of phages. Moreover, a number of studies have shown that phages can affect different functions of major populations of immune cells including phagocytosis and the respiratory burst of phagocytic cells and the production of cytokines. It appears that at least some of the interactions between phages and immune cells are mediated by capsid proteins.

An interesting question is whether phages, being a significant component of the gut microbiota, can undergo translocation across the wall of the digestive tract. Although the translocation of bacteria is a relatively well-documented phenomenon and one that occurs both in normal physiological settings and in different diseases, the knowledge about the translocation of phages is scanty. There are some data in the literature, however, to suggest that bacteriophages, like bacteria, could translocate across the wall of the gut. It is possible that phage translocation could play a role in the development and/or function of the immune system in both the gut mucosa and the whole body.

Acknowledgements

This work was supported by funds from the Operational Program 'Innovative Economy, 2007–2013' (Priority axis 1. Research and Development of Modern Technologies, Measure 1.3 Support for R&D projects for entrepreneurs carried out by scientific entities, Submeasure 1.3.1, Development project no. POIG 01.03.01-02-003/08 entitled 'Optimization of the production characterization of bacteriophage preparations for therapeutic use') and an intramural grant from Warsaw Medical University (1MG/W1/09). We are grateful to R. Ashcroft for correcting the English version of the manuscript.

References

Adair, B.D. and Yeager, M. (2002) Three-dimensional model of the human platelet integrin $\alpha_{IIb}\beta_3$ based on electron cryomicroscopy and X-ray crystallography. *Proceedings of the National Academy of Sciences USA* 99, 14059–14064.

Adams, M.H. (1959) Antigenic properties. In: Adams, M.H. (ed.) *Bacteriophages.* Interscience Publishers, New York, NY, pp. 97–119.

Balzan, S., Quadros, C.A., Cleva, R., Zilberstein, B. and Cecconello, I. (2007) Bacterial translocation: overview of mechanisms and clinical impact. *Journal of Gastroenterology and Hepatology* 22, 464–471.

Bateman, A., Eddy, S.R. and Mesyanzhinov, V.V. (1997) A member of the immunoglobulin superfamily in bacteriophage T4. *Virus Genes* 14, 163–165.

Bearden, C.M., Agarwal, A., Book, B.K., Vieira, C.A., Sidner, R.A., Ochs, H.D., Young, M. and Pescovitz, M.D. (2005) Rituximab inhibits the *in vivo* primary and secondary antibody response to a neoantigen, bacteriophage ϕX174. *American Journal of Transplantation* 5, 50–57.

Berinstein, A., Roivainen, M., Hovi, T., Mason, P.W. and Baxt, B. (1995) Antibodies to the vitronectin receptor (integrin $\alpha_V\beta_3$) inhibit binding and infection of foot-and-mouth disease virus to cultured cells. *Journal of Virology* 69, 2664–2666.

Blaser, M. (2011) Stop the killing of beneficial bacteria. *Nature* 476, 393–394.

Bloch, H. (1940) Experimental investigation on the relationships between bacteriophages and malignant tumors. *Archives in Virology* 1, 481–496 (in German).

Borysowski, J., Wierzbicki, P., Kłosowska, D., Korczak-Kowalska, G., Weber-Dąbrowska, B. and Górski, A. (2010) The effects of T4 and A3/R phage preparations on whole-blood monocyte and neutrophil respiratory burst. *Viral Immunology* 23, 1–4.

Bradley, S.G., Kim, Y.B. and Watso, D.W. (1963) Immune response by the mouse to orally administered actinophage. *Proceedings of the Society for Experimental Biology and Medicine* 113, 686–688.

Bruttin, A. and Brüssow, H. (2005) Human volunteers receiving *Escherichia coli* phage T4 orally: a safety test of phage therapy. *Antimicrobial Agents and Chemotherapy* 49, 2874–2878.

Chanishvili, N., Chanishvili, T., Tediashvili, M. and Barrow, P.A. (2001) Phages and their application against drug-resistant bacteria. *Journal of Chemical Technology and Biotechnology* 76, 689–699.

Dąbrowska, K., Opolski, A., Wietrzyk, J., Świtała-Jeleń, K., Boratyński, J., Nasulewicz, A., Lipińska, L., Chybicka, A., Kujawa, M., Zabel, M., Dolińska-Krajewska, B., Piasecki, E., Weber-Dąbrowska, B., Rybka, J., Salwa, J., Wojdat, E., Nowaczyk, M. and Górski, A. (2004) Antitumor activity of bacteriophages in murine experimental cancer models caused possibly by inhibition of β3 integrin signaling pathway. *Acta Viriologica* 48, 241–248.

Dąbrowska, K., Świtała-Jeleń, K., Opolski, A., Weber-Dąbrowska, B. and Górski, A. (2005) Bacteriophage penetration in vertebrates. *Journal of Applied Microbiology* 98, 7–13.

Dąbrowska, K., Świtała-Jeleń, K., Opolski, A. and Górski, A. (2006) Possible association between phages, Hoc protein and the immune system. *Archives of Virology* 151, 209–215.

Dąbrowska, K., Zembala, M., Boratyński, J., Świtała-Jeleń, K., Wietrzyk, J., Opolski, A., Szczaurska, K., Kujawa, M., Godlewska, J. and Górski, A. (2007) Hoc protein regulates the biological effects of T4 phage in mammals. *Archives of Microbiology* 187, 489–498.

Dąbrowska, K., Skaradziński, G., Jończyk, P., Kurzępa, A., Wietrzyk, J., Owczarek, B., Żaczek, M., Świtała-Jeleń, K., Boratyński, J., Poźniak,

G., Maciejewska, M. and Górski, A. (2009) The effect of bacteriophages T4 and HAP1 on *in vitro* melanoma migration. *BMC Microbiology* 9, 13

Dąbrowska, K., Skaradziński, G., Kurzępa, A., Owczarek, B., Żaczek, M., Weber-Dąbrowska, B., Wietrzyk, J., Maciejewska, M., Budynek, P. and Górski, A. (2010) The effects of staphylococcal bacteriophage lysates on cancer cells *in vitro*. *Clinical and Experimental Medicine* 10, 81–85.

Dean, J.H., Silva, J.S., McCoy, J.L., Chan, S.P., Baker, J.J., Leonard, C. and Herberman, R.B. (1975) *In vitro* human reactivity to staphylococcal phage lysate. *Journal of Immunology* 115, 1060–1064.

Dubos, R.J., Straus, J.H. and Pierce, C. (1943) The multiplication of bacteriophage *in vivo* and its protective effect against an experimental infection with *Shigella dysenteriae*. *Journal of Experimental Medicine* 20, 161–168.

Duerr, D.M., White, S.J. and Schluesener, H.J. (2004) Identification of peptide sequences that induce the transport of phage across the gastrointestinal mucosal barrier. *Journal of Virological Methods* 116, 177–180.

Esber, H.J., DeCourcy, J. and Bogden, A.E. (1981) Specific and nonspecific immune resistance enhancing activity of staphage lysate. *Journal of Immunopharmacology* 3, 79–92.

Esber, H.J., Ganfield, D. and Rosenkrantz, H. (1985) Staphage lysate: an immunomodulator of the primary immune response in mice. *Immunopharmacology* 10, 77–82.

Fokine, A., Chipman, P.R., Leiman, P.G., Mesyanzhinov, V.V., Rao, V.B. and Rossmann, M.G. (2004) Molecular architecture of the prolate head of bacteriophage T4. *Proceedings of the National Academy of Sciences USA* 101, 6003–6008.

Fraser, J.S., Yu, Z., Maxwell, K.L. and Davidson, A.R. (2006) Ig-like domains on bacteriophages: a tale of promiscuity and deceit. *Journal of Molecular Biology* 359, 496–507.

Geier, M.R., Trigg, M.E. and Merril, C.R. (1973) Fate of bacteriophage lambda in non-immune germ-free mice. *Nature* 246, 221–223.

Giancotti, F.G. and Ruoslahti, E. (1999) Integrin signaling. *Science* 285, 1028–1032.

Gorczyca, W.A., Mitkiewicz, M., Siednienko, J., Kurowska, E., Piasecki, E., Weber-Dąbrowska, B. and Górski, A. (2007) Bacteriophages decrease activity of NF-κb induced in human mononuclear cells by Human Herpesvirus-1. In: Kalil, J., Neto-Cunha, E. and Rizzo, L.V. (eds) *Proceedings of 13th International Congress of Immunology.* MEDIMOND S.r.l., Rio de Janeiro, Brazil, pp. 73–77.

Górski, A. and Weber-Dąbrowska, B. (2005) The potential role of endogenous bacteriophages in controlling invading pathogens. *Cellular and Molecular Life Sciences* 62, 511–519.

Górski, A., Dąbrowska, K., Świtała-Jeleń, K., Nowaczyk, M., Weber-Dąbrowska, B., Boratyński, J., Wietrzyk, J. and Opolski, A. (2003a) New insights into the possible role of bacteriophages in host defense and disease. *Medical Immunology* 2, 2.

Górski, A., Nowaczyk, M., Weber-Dąbrowska, B., Kniotek, M., Boratyński, J., Achmed, A., Dąbrowska, K., Wierzbicki, P., Świtała-Jeleń, K. and Opolski, A. (2003b) New insights into the possible role of bacteriophages in transplantation. *Transplantation Proceedings* 35, 2372–2373.

Górski, A., Kniotek, M., Perkowska-Ptasińska, A., Mróz, A., Przerwa, A., Gorczyca, W., Dąbrowska, K., Weber-Dąbrowska, B. and Nowaczyk, M. (2006a) Bacteriophages and transplantation tolerance. *Transplantation Proceedings* 38, 331–333.

Górski, A., Ważna, E., Weber-Dąbrowska, B., Dąbrowska, K., Świtała-Jeleń, K. and Międzybrodzki, R. (2006b) Bacteriophage translocation. *FEMS Immunology and Medical Microbiology* 46, 313–319.

Górski, A., Międzybrodzki, R., Borysowski, J., Weber-Dąbrowska, B., Łobocka, M., Fortuna, W., Letkiewicz, S., Zimecki, M. and Filby, G. (2009) Bacteriophage therapy for the treatment of infections. *Current Opinion in Investigational Drugs* 10, 766–774.

Gulam, W. and Ahsan, H. (2006) Reactive oxygen species: role in the development of cancer and various chronic conditions. *Journal of Carcinogenesis* 5, 14.

Harrath, R., Bourlet, T., Delézay, O., Douche-Aourik, F., Omar, S., Aouni, M. and Pozzetto, B. (2004) Coxsackievirus B3 replication and persistence in intestinal cells from mice infected orally and the human CaCo-2 cell line. *Journal of Medical Virology* 74, 283–290.

Hiscott, J., Beauparlant, P., Crepieux, P., DeLuca, C., Kwon, H., Lin, R. and Petropoulos, L. (1997) Cellular and viral protein interactions regulating IκBα activity during human retrovirus infection. *Journal of Leukocyte Biology* 62, 82–92.

Horst, D., Verweij, M.C., Davison, A.J., Ressing, M.E. and Wiertz, E.J.H.J. (2011) Viral evasion of T cell immunity: ancient mechanisms offering new applications. *Current Opinion in Immunology* 23, 96–103.

Housby, J.N. and Mann, N.H. (2009) Phage therapy. *Drug Discovery Today* 14, 536–540.

Inchley, C.J. (1969) The activity of mouse Kupffer

cells following intravenous injection of T4 bacteriophage. *Clinical and Experimental Immunology* 5, 173–187.

Ishii, T. and Yanagida, M. (1975) Molecular organization of the shell of the T-even bacteriophage head. *Journal of Molecular Biology* 97, 655–660.

Ishii, T. and Yanagida, M. (1977) The two dispensable structural proteins (*soc* and *hoc*) of the T4 phage capsid; their purification and properties, isolation and characterization of the defective mutants, and their binding with the defective heads *in vitro*. *Journal of Molecular Biology* 109, 487–514.

Kańtoch, M. (1958) Studies on phagocytosis of bacterial viruses. *Archivum Immunologiae et Therapiae Experimentalis* 6, 63–84 (in Polish).

Kańtoch, M. and Mordarski, M. (1958) Binding of bacterial viruses by cancer cells *in vitro*. *Postępy Higieny i Medycyny Doświadczalnej* 12, 191–192 (in Polish).

Keller, R. and Engley, F.B. (1958) Fate of bacteriophage particles introduced into mice by various routes. *Proceedings of the Society for Experimental Biology and Medicine* 98, 557–580.

Kiliç, A.O., Pavlova, S.I., Ma, W. and Tao, L. (1996) Analysis of *Lactobacillus* phages and bacteriocins in American dairy products and characterization of a phage isolated from yogurt. *Applied and Environmental Microbiology* 62, 2111–2116.

Kleinschmidt, W.J., Douthardt, R.J. and Murphy, E.B. (1970) Interferon production by T4 coliphage. *Nature* 228, 27–30.

Knight, J.A. (1995) Diseases related to oxygen-derived free radicals. *Annals of Clinical and Laboratory Science* 25, 111–121.

Kniotek, M., Ahmed, A.M.A., Dąbrowska, K., Świtała-Jeleń, K., Weber-Dąbrowska, B., Boratyński, J., Nowaczyk, M., Opolski, A. and Górski, A. (2004a) Bacteriophage interactions with T cells and platelets. In: Monduzzi Editore, *Genomic Issues, Immune System Activation and Allergy (Immunology 2004)*. Medimond International Proceedings, Montreal, Canada, pp. 189–193.

Kniotek, M., Weber-Dąbrowska, B., Dąbrowska, K., Świtała-Jeleń, K., Boratyński, J., Wiszniowski, M., Glinkowski, W., Babiak, I., Górecki, A., Nowaczyk, M. and Górski, A. (2004b) Phages as immunomodulators of antibody production. In: Monduzzi Editore, *Genomic Issues, Immune System Activation and Allergy (Immunology 2004)*. Medimond International Proceedings Montreal, Canada, pp. 33–37.

Kucharewicz-Krukowska, A. and Ślopek, S. (1987) Immunogenic effect of bacteriophage in patients subjected to phage therapy. *Archivum Immunologiae et Therapiae Experimentalis* 35, 553–561.

Kumari, S., Harjai, K. and Chhibber, S. (2010) Evidence to support the therapeutic potential of bacteriophage Kpn5 in burn wound infection caused by *Klebsiella pneumoniae* in BALB/c mice. *Journal of Microbiology and Biotechnology* 20, 935–941.

Kurzępa, A., Dąbrowska, K., Skaradziński, G. and Górski, A. (2009) Bacteriophage interactions with phagocytes and their potential significance in experimental therapy. *Clinical and Experimental Medicine* 9, 93–100.

Lafrenie, R.M., Lee, S.F., Hewlett, I.K., Yamada, K.M. and Dhawan, S. (2002) Involvement of integrin $\alpha_v\beta_3$ in the pathogenesis of human immunodeficiency virus type 1 infection in monocytes. *Virology* 297, 31–38.

Langbeheim, H., Teitelbaum, D. and Arnon, R. (1978) Cellular immune response toward MS-2 phage and a synthetic fragment of its coat protein. *Cellular Immunology* 38, 193–197.

Lu, Z., Breidt, F., Plengvidhya, V. and Fleming, H.P. (2003) Bacteriophage ecology in commercial sauerkraut fermentations. *Applied and Environmental Microbiology* 79, 3192–3202.

Mathur, A., Narayanan, K., Zerbe, A., Ganfield, D., Ramasastry, S.S. and Futrell, J.W. (1988) Immunomodulation of intradermal mammary carcinoma using staphage lysate in a rat model. *Journal of Investigative Surgery* 1, 117–123.

Merril, C.R. (1974) Bacteriophage interactions with higher organisms. *Transactions of the New York Academy of Sciences* 36, 265–272.

Merril, C.R., Friedman, T.B., Attallah, A.F.M., Geier, M.R., Krell, K. and Yarkin, R. (1972) Isolation of bacteriophages from commercial sera. *In Vitro* 8, 91–93.

Merril, C.R., Biswas, B., Carlton, R., Jensen, N.C., Creed, G.J., Zullo, S. and Adhya, S. (1996) Long-circulating bacteriophage as antibacterial agents. *Proceedings of the National Academy of Science USA* 93, 3188–3192.

Międzybrodzki, R., Świtała-Jeleń, K., Fortuna, W., Weber-Dąbrowska, B., Przerwa, A., Łusiak-Szelachowska, M., Dąbrowska, K., Kurzypa, A., Boratyński, J., Syper, D., Poźniak, G., Ługowski, C. and Górski, A. (2008) Bacteriophage preparation inhibition of reactive oxygen species generation by endotoxin-stimulated polymorphonuclear leukocytes. *Virus Research* 131, 233–242.

Międzybrodzki, R., Fortuna, W., Weber-Dąbrowska,

B. and Górski, A. (2009) A retrospective analysis of changes in inflammatory markers in patients treated with bacterial viruses. *Clinical and Experimental Medicine* 9, 303–312.

Morello, E., Saussereau, E., Maura, D., Huerre, M., Touqui, L. and Debarbieux, L. (2011) Pulmonary bacteriophage therapy on *Pseudomonas aeruginosa* cystic fibrosis strains: first steps towards treatment and prevention. *PLoS ONE* 6, 1–9.

Morin, M.J. and Warner, A. (1994) A pathway for entry of reoviruses into the host through M cells of the respiratory tract. *Journal of Experimental Medicine* 180, 1523–1527.

Mossel, E.C. and Ramig, R.F. (2003) A lymphatic mechanism of rotavirus extraintestinal spread in the neonatal mouse. *Journal of Virology* 77, 12352–12356.

Nelson, R. (2003) Antibiotic development pipeline runs dry. *Lancet* 362, 1726–1727.

O'Boyle, C., MacFie, J., Mitchell, C., Johnstone, D., Sagar, P. and Sedman, P. (1998) Microbiology of bacterial translocation in humans. *Gut* 42, 29–35.

Ochs, H.D., Davis, S.D. and Wedgwood, R.J. (1971) Immunologic responses to bacteriophage ϕX174 in immunodeficiency diseases. *Journal of Clinical Investigation* 50, 2559–2568.

Olszewska, D. and Radkowski, H. (2005) Odporność przeciwzakaźna. In: Gołąb, J., Jakóbisiak, M. and Lasek, W. (eds) *Immunologia.* Wydawnictwo Naukowe PWN, Warszawa, pp. 337–355.

Orr, H.C., Sibinovic, K.H., Probst, P.G., Hochstein, D. and Littlejohn, D.C. (1975) Bacteriological activity in unfiltered calf sera collected for tissue culture use. *In Vitro* 11, 230–233.

Pagava, K., Gachechiladze, K., Korinteli, I., Dzuliashvili, M., Alavidze, Z., Hoyle, N. and Metskhvarishvili, G. (2011) What happens when the child gets bacteriophage per os? *Georgian Medical News* 196–197, 101–105.

Parent, K. and Wilson, I.D. (1971) Mycobacteriophage in Crohn's disease. *Gut* 12, 1019–1020.

Perdih, A. and Dolenc, M.S. (2010) Small molecule antagonists of integrin receptors. *Current Medical Chemistry* 17, 2371–2392.

Pescovitz, M.D., Torgerson, T.R., Ochs, H.D., Ocheltree, E., McGee, P., Krause-Steinrauf, H., Lachin, J.M., Canniff, J., Greenbaum, C., Herold, K.C., Skyler, J.S., Weinberg, A. and the Type 1 Diabetes TrialNet Study Group (2011) Effect of rituximab on human *in vivo* antibody immune responses. *Journal of Allergy and Clinical Immunology* 128, 1295–1302.e5.

Ploegh, H.L. (1998) Viral strategies of immune evasion. *Science* 280, 248–253.

Przerwa, A., Kniotek, M., Nowaczyk, M., Weber-Dąbrowska, B., Świtała-Jeleń, K., Dąbrowska, K. and Górski, A. (2005) Bacteriophages inhibit interleukin-2 production by human T lymphocytes. In: *12th Congress of the European Society for Organ Transplantation*, Geneva, Switzerland.

Przerwa, A., Zimecki, M., Świtała-Jeleń, K., Dąbrowska, K., Krawczyk, E., Łuczak, M., Weber-Dąbrowska, B., Syper, D., Międzybrodzki, R. and Górski, A. (2006) Effects of bacteriophages on free radical production and phagocytic functions. *Medical Microbiology and Immunology* 195, 143–150.

Sathaliyawala, T., Islam, M.Z., Li, Q., Fokine, A., Rossmann, M.G. and Rao, V.B. (2010) Functional analysis of the highly antigenic outer capsid protein, Hoc, a virus decoration protein from T4-like bacteriophages. *Molecular Microbiology* 77, 444–455.

Sikora, J.P. (2002) Immunotherapy in the management of sepsis. *Archivum Immunologiae et Therapiae Experimentalis* 50, 317–324.

Ślopek, S., Durlakowa, I., Weber-Dąbrowska, B., Kucharewicz-Krukowska, A., Dąbrowski, M. and Bisikiewicz, R. (1983) Results of bacteriophage treatment of suppurative bacterial infections. *Archivum Immunologiae et Therapiae Experimentalis* 31, 267–291.

Smith, P.D., Meng, G., Salazar-Gonzalez, J.C. and Shaw, G.M. (2003) Macrophage HIV-1 infection and the gastrointestinal tract reservoir. *Journal of Leukocyte Biology* 74, 642–649.

Srivastava, A.S., Kaido, T. and Carrier, E. (2004) Immunological factors that affect the *in vivo* fate of T7 phage in the mouse. *Journal of Virological Methods* 115, 99–104.

Sulakvelidze, A. and Kutter, E. (2005) Bacteriophage therapy in humans. In: Sulakvelidze, A. and Kutter, E. (eds) *Bacteriophages: Biology and Application.* CRC Press, Boca Raton, FL, pp. 381–436.

Triantafilou, K., Takada, Y. and Traintafilou, M. (2001) Mechanisms of integrin-mediated virus attachment and internalization process. *Critical Reviews in Immunology* 21, 311–322.

Varon, D. and Shai, E. (2009) Role of platelet-derived microparticles in angiogenesis and tumor progression. *Discovery Medicine* 8, 237–241.

Weber-Dąbrowska, B., Dąbrowska, M. and Ślopek, S. (1987) Studies on bacteriophage penetration in patients subjected to phage therapy. *Archivum Immunologiae et Therapiae Experimentalis* 35, 563–568.

Weber-Dąbrowska, B., Mulczyk, M. and Górski, A. (2000a) Bacteriophage therapy of bacterial infections: an update of our institute's experience. *Archivum Immunologiae et Therapiae Experimentalis* 48, 547–551.

Weber-Dąbrowska, B., Zimecki, M. and Mulczyk, M. (2000b) Effective phage therapy is associated with normalization of cytokine production by blood cell cultures. *Archivum Immunologiae et Therapiae Experimentalis* 48, 31–37.

Weber-Dąbrowska, B., Zimecki, M., Muczyk, M. and Górski, A. (2002) Effect of phage therapy on the turnover and function of peripheral neutrophils. *FEMS Immunology and Medical Microbiology* 34, 135–138.

Weber-Dąbrowska, B., Mulczyk, M. and Górski, A. (2003) Bacteriophages as an efficient therapy for antibiotic-resistant septicemia in man. *Transplantation Proceedings* 35, 1385–1386.

Weinbauer, M.G. (2004) Ecology of prokaryotic viruses. *FEMS Microbiology Reviews* 28, 127–181.

Wenger, S.L., Turner, J.H. and Petricciani, J.C. (1978) The cytogenic, proliferative and viability effects of four bacteriophages on human lymphocytes. *In Vitro* 14, 543–549.

Wierzbicki, P., Kłosowska, D., Wyzgał, J., Nowaczyk, M., Przerwa, A., Kniotek, M. and Górski, A. (2006) β3 integrin expression on T cells from renal allograft recipients. *Transplantation Proceedings* 38, 338–339.

Wiest, R. and Garcia-Tsao, G. (2005) Bacterial translocation (BT) in cirrhosis. *Hepatology* 41, 422–430.

13 Phage Therapy of Wounds and Related Purulent Infections

Catherine Loc-Carrillo[1], Sijia Wu[1] and James Peter Beck[1]
[1]*Department of Orthopaedics, The University of Utah.*

'Every convenience brings its own inconvenience.' – Roman proverb.

The widespread use of antibiotics for over half a century has contributed to the emergence of antibiotic-resistant bacteria. The ever-increasing problem of treating infections caused by these 'superbugs' is pushing the healthcare system and research scientists to look at alternative antimicrobial agents that can be harnessed to aid in the fight against pathogens while leaving our normal microbial flora untouched. It is well established that for every prokaryotic cell in the human body there are approximately ten bacterial cells, although we are only able to culture 10–50% of them (Teibelbaum and Walker, 2002; Gill *et al.*, 2006; Morowitz *et al.*, 2011). Bacteria inhabit the human body as symbionts, contributing to the structure and function of tissues around them and playing an important part in the balance between health and disease by promoting the development and function of our adaptive immunity (Lai *et al.*, 2009; Lee and Mazmanian, 2010). However, once a physiological barrier has been breached – that is, the occurrence of a wound – there is an increased risk of colonization of the deep tissues by opportunistic pathogens, including those with antibiotic-resistant properties, and subsequent infection (Hermans and Treadwell, 2010; Morowitz *et al.*, 2011). Phage therapy represents an important strategy by which otherwise untreatable antibiotic-resistant infections may be brought under control or cured.

This chapter provides a general definition of what wounds are and describes specific wound infections that have been investigated for their receptiveness to phage therapy. We highlight some of the early clinical work, from the pre-antibiotic era, when phage therapy was first used to prevent and treat wound infections, as well as other localized infections. We also provide insights into some of the reasons for failure of these approaches as judged by today's standards. Finally, we discuss the strengths and weaknesses in experimental design of various animal models used to determine the efficacy of bacteriophages as antimicrobials, as it is only by following a methodical scientific approach that the potential use of this 'unconventional' treatment of wound infections will be accepted by 21st-century scientists and clinicians alike. For more on other aspects of phage therapy, see Olszowska-Zaremba *et al.*, Burrowes and Harper, and Abedon, Chapters 12, 14 and 17, this volume, as well as reviews by Abedon *et al.* (2011), Maura and Debarbieux (2011) and Kutateladze and Adamia (2010).

 Bacteriophages in Health and Disease (eds P. Hyman and S.T. Abedon)

Sharing our Bodies with Microbes

The surface of the skin is colonized by microorganisms, the ecology of which is dependent on the 'topographical location, endogenous host factors and exogenous environmental factors' (Grice and Segre, 2011). In humans, the acidic surface of the skin (~pH 5) provides favourable conditions for bacteria such as coagulase-negative staphylococci and corynebacteria. Certain skin locations, where large quantities of triglyceride-containing sebum are found, encourage the presence of *Propionibacterium acnes*, while other regions with conditions of higher temperature and humidity support the growth of Gram-negative bacilli and *Staphylococcus aureus*. Host factors such as age and sex can also have a great effect on the colonizing microbiota. A disinfected skin surface will be recolonized within hours by microorganisms that reside in deeper undisinfected locations such as hair follicles, glands, the oral cavity and urogenital openings.

Although the skin is estimated to have ~10^{12} bacterial inhabitants, the gut holds the vast majority of microorganisms, containing an impressive 10^{14} bacteria, the majority of which are composed of just four phyla: Actinobacteria (Gram-positive bacteria that include the genera *Corynebacterium*, *Mycobacterium* and *Bifidobacterium*), Firmicutes (Gram-positive bacteria that include clostridia and bacilli), Bacteroidetes (Gram-negative bacteria including members of the genus *Bacteroides*) and Proteobacteria (Gram-negative bacteria including the genera *Escherichia*, *Salmonella* and *Helicobacter*), although at vastly different frequencies. Unlike the skin, however, the gut is permeable, allowing the efficient absorption of nutrients while possessing innate defences that can help prevent bacteria from also crossing the mucosal barrier (Hopper and Gordon, 2001; Grice and Segre, 2011).

Defining Wounds and Related Purulent Infections

For our purposes, we define the term 'wound' as any penetration of the protective integumentary organ of the body that we commonly call the skin. This penetration could be from external mechanical forces such as surgery or trauma, or from infectious breakdown of the skin from an underlying abscess. The outcome of such penetration is open communication between the exterior and the underlying deep tissues of the body. All of these tissues, under healthy, normal, uninjured, homeostatic circumstances, are in a sterile environment that is kept free of bacteria by the innate and adaptive immune systems. Swab samples taken from an incision during elective surgery, following proper surgical prepping and draping procedures, thus should always be sterile.

The main types of wounds encountered clinically include: surgical wounds, thermal wounds (involving partial- or full-thickness burns), chemical injuries, wounds due to trauma and chronic wounds (such as venous, diabetic and pressure ulcers) (Hermans and Treadwell, 2010). During the treatment of any type of wound or wound infection, proper wound care is a race to obtain infection-free closure before nosocomial infection occurs or superinfection with antibiotic-resistant bacteria takes over the infection process. Table 13.1 summarizes historical studies investigating the efficacy of phage therapy of wounds.

Acute wounds

Uninfected surgical wounds or incisions tend to heal within 3 weeks. However, the rate of wound healing depends on a healthy blood supply and varies according to the tissue's blood supply, which is dependent on its distance from the heart. The highly vascular tissues of the face thus allow suture removal within days of 'clean' (or sterile) incision, while sutures in the lower extremities remain for 2–3 weeks until primary uncomplicated wound healing is achieved.

Burn wounds

Partial superficial burns, also known as 'first-degree burns', tend to blister but heal without scarring. Infections are uncommon and healing occurs within 2 weeks (Hermans,

Table 13.1. Summary of historical studies investigating the efficacy of phage therapy of wounds.

Reference	Affiliation	Bacteria	Wound type (*n*)	Route of administration	Outcome	Details
McKinley, 1923	Baylor University College of Medicine	*Staphylococcus* spp.	Open wounds (4)	Injected directly into wound or subcutaneous, or both	100% success rate	Resulted in decreased discharge and eventual improvement of patient condition
Larkum, 1929	Michigan Department of Health	*Staphylococcus* spp.	Furunculosis (208)	Subcutaneous injection in combination with local application	78% successfully treated; 19% had recurrence; 3% showed no improvement	A 2-year study: phage preparations were provided to participating physicians in return for clinical outcome of patients;
			Osteomyelitis (6)		50% had rapid and complete recovery; 50% had marked improvement	149 patients had records of whether a reaction to the inoculation was seen: 42% showed no reaction; 47% had mild, local erythema and soreness; 10% had a rise in temperature and malaise; and 1% had severe reactions
Rice, 1930	Indiana University School of Medicine	*Staphylococcus* spp. and *E. coli*	Boils and carbuncles (66)	Applied locally as a wet dressing; injected into deep lesions; some abscesses were injected into the area as a 1% agar jelly preparation	83% had excellent results; 8% had intermediate results; 8% failure rate	Most of the failures for boils and carbuncles were due to the isolate being resistant to the phage strains available for treatment
			Abscesses (27)		89% had excellent results; 4% had intermediate results; 4% had no improvement	Excellent results were obtained in cases where the abscess had not been opened prior to treatment
			Infected wounds (44)		91% had excellent results; 7% had good outcomes; 2% had no effect seen	An increase in pus production of infected wounds was commonly seen after the first application for the first 24–48 h, but wounds rapidly healed after cleaning out the pus
			Osteomyelitis (11)		36% had excellent results; 27% had fair outcomes; 36% had no effect seen	If there was dead bone in the lesion, the phage treatment had no value; however, if all necrotic bone was removed, good results were obtained

Continued

Table 13.1. Continued

Reference	Affiliation	Bacteria	Wound type (*n*)	Route of administration	Outcome	Details
Walker, 1931	Research Laboratories of E.R. Squibb & Sons	*Staphylococcus* spp.	Abscess (15 animals; 40 experiments)	0.5 ml subcutaneous injection into abscess or 5 ml intravenous injections	100% failure	Rabbits were infected experimentally; treatment was administered either 2 min after the bacterial inoculum or once lesions had developed; each route of administration was set up as a separate study
Shultz, 1932	Department of Bacteriology & Experimental Pathology, Stanford University	*Staphylococcus* spp.	Furunculosis (63)	Subcutaneous injections; used to irrigate wounds	70% success rate	A 2-year study: appropriate phage stocks were supplied based on bacterial cultures isolated from infected patients
			Carbuncles (16)		88% success rate	
Albee and Patterson, 1930	Department of Orthopaedic Surgery, New York Medical School, NY	*Staphylococcus* spp.	Osteomyelitis	Gauzes used to dress the wounds were soaked in phage filtrate and Vaseline®	Not determined	Clinical isolates were tested for their susceptibility to phage filtrate prior to treatment; seven cases that resulted in successful outcome were showcases in this study; in three of these cases, 'native' (temperate) phage were induced from the bacteria found infecting the wound

2010). Major burn wounds, involving >15% of the total body surface area, may look superficial but tend to develop into deeper, full-thickness wounds, where excision of necrotic tissue and closure of the wound by grafting skin over the area is necessary. Factors that increase the risk of infection in these wounds include the prior use of irradiation and some medications such as corticosteroids and other immunosuppressive drugs, stress-induced hyperglycaemia, low serum cortisol levels, anaemia and the length of hospital stay. Individuals with burns covering >30% of the total body surface area are at risk of rapid bacterial colonization due to the immunosuppressive effect of the injury (Rowley-Conwy, 2010). The prolonged presence of an open wound or delayed initial burn wound care leads to high incidence rates of morbidity and mortality.

Typically, within the first week, endogenous Gram-positive bacteria initially colonize burns but are then rapidly replaced by antibiotic-susceptible Gram-negative bacteria. The use of broad-spectrum antibiotics for treating these infections tends to favour subsequent colonization by exogenous nosocomial organisms, which are generally antibiotic-resistant bacteria such as methicillin-resistant *S. aureus* (MRSA), vancomycin-resistant enterococci and extended-spectrum β-lactamase (ESBL)-producing Gram-negative rods (Church *et al.*, 2006; Rafla and Tredget, 2011; see Kuhl *et al.*, Chapter 3, this volume, for additional discussion of some of these organisms).

Chronic wounds

Any impediment to the wound-healing process affects the timely physiological repair needed for anatomical and functional integrity. The resulting chronic wounds are generally defined as taking longer than 3 weeks to heal (Hermans and Treadwell, 2010). The chronic wound state may occur as part of disorders such as diabetes or pressure regions (e.g. bed sores) and is not always initiated by infections.

Once infection occurs, it is imperative to intervene by undertaking an effective wound-care strategy that encourages deep tissue homeostasis, allows the recolonization of commensals and ultimately achieves wound healing. In all cases, repeated irrigation of the wound that removes foreign material and infectious exudate, and thorough debridement of all potentially devitalized and necrotic tissue, is necessary. For superficial local infections, the appropriate topical antimicrobial(s) and wound dressings may be all that is required, but for deep wound and systemic infections, systemic antimicrobials are necessary (Percival and Dowd, 2010). Co-morbidities and complications that modify the strategy needed to treat a particular wound infection include advanced age, obesity, poor nutritional status, dysvascular diseases such as those seen with smokers and diabetic patients, or circumstances where poor blood circulation occurs around infection sites such as in osteomyelitis and amputations.

Diabetic wounds

Acute as well as chronic wounds are common complications of diabetes mellitus, which is one of the most common causes of chronic wound states, with or without infection. Most diabetic wounds are on the feet (far away from the heart and with the poorest blood supply). Acute infections tend to be monomicrobial and due to aerobic Gram-positive cocci, such as *S. aureus* and β-haemolytic streptococci (i.e. groups A, C and G but especially B), while chronic infections tend towards more complex mixtures of organisms (polymicrobial), including enterococci, various Enterobacteriaceae, obligate anaerobes, *Pseudomonas aeruginosa* and occasionally other non-fermentative Gram-negative rods (Lipsky *et al.*, 2004; see Kuhl *et al.*, Chapter 3, this volume).

Osteomyelitis

Osteomyelitis is an acute or chronic infection of bone. It is most commonly caused by bacteria but may also be due to fungi. It can occur as a secondary complication of trauma, surgery or the insertion of a foreign device such as a joint prosthesis. It may also occur by contiguous spread in regions of poor vascular function, such as diabetic foot infections or from a haematogenous dissemination from infected skin sores or insect bites, urinary tract infections or dental manipulation. Haematogenous spread is commonly seen in children (Lew and Waldvogel, 1997). Various types of osteomyelitis require different medical and surgical strategies, ranging from short-term antibiotic therapy and limited surgical incision and drainage for acute cases to more extensive surgical debridement of dead bone and long-term antibiotic treatment (e.g. 6–8 weeks of intravenous administration) in chronic cases. The infecting organism may vary with age and underlying disease. Children suffering from acute haematogenous osteomyelitis may tend to be infected by *Streptococcus pyogenes*, whereas adults suffering from chronic open wounds to the bone and soft tissue may have *P. aeruginosa*. Overall, 80% of osteomyelitis cases are due to *S. aureus* (Howell and Goulston, 2011; see also Kuhl *et al.*, Chapter 3, this volume).

Early Phage Therapy of Wounds

This section highlights an early phage-therapy study by MacNeal *et al.* (1942); we direct the reader to Table 13.1 for an overview of earlier studies. We also consider the Soviet Union phage-therapy experience.

A 500-case study treating staphylococcaemia

The last large phage-therapy study carried out in the West, following the phage-therapy-critical Eaton and Bayne-Jones' reviews (1934a,b,c), was completed by MacNeal *et al.* (1942). They reported on the use of phage therapy in 500 patients suffering from staphylococcaemia in the years 1931–1940, collected from 169 hospitals in various parts of the USA. The systemic infection in most of these cases had resulted from local infection such as abscesses, carbuncles and osteomyelitis. Patient inclusion in this series of clinical trials required a request from the attending physician to use phage therapy on their patient, and a confirmation of staphylococci being present in their bloodstream. It was noted that a larger number of patients were male, possibly due to a greater exposure to external wounds and having 'less meticulous care of the skin'. *In vitro* testing of 416 cases that produced staphylococci isolates from patients found that 89% of these isolates were susceptible to the phage stock preparation employed in this study.

Thirty patients did not receive the phage therapy either due to death occurring prior to commencing treatment or not wanting treatment despite being initially included in the trial – these were counted as the negative-control group. Twenty-five patients received phage therapy despite their blood cultures becoming negative prior to commencing treatment, and seven of them died. This latter group was considered the positive-control group (see Abedon, Chapter 17, this volume, for discussion of the use of controls during phage therapy). Table 13.2 provides a summary of the results.

From their findings, MacNeal *et al.* (1942) concluded that early recognition of symptoms, a willingness to execute the phage-therapy programme as prescribed and the good judgement to use complementary therapeutic measures would all improve the rate of recovery. Unfortunately, the combination of other measures such as using sulfathiazole and fractional transfusions with phage therapy, which they stated offered a more favourable outlook in the severe forms of infection, make it difficult to determine whether bacteriophages played a part in treating the infections.

This study, as well as those outlined in Table 13.1, used phage filtrate preparations that were only void of bacterial cells through filtration. Although some researchers speculated on the presence of bacterial artefacts in these preparations, they were unaware of bacterial toxins (i.e. endotoxins) that we now recognize as responsible for some of the side effects observed from phage treatments (see Olszowska-Zaremba *et al.*, Chapter 12, this volume, for additional discussion of these issues).

Table 13.2. A summary showing staphylococcaemia patients included in the phage therapy study by MacNeal *et al.* (1942)

	No phage therapy – positive blood culture		Phage therapy despite negative blood culture		Phage therapy following positive blood culture		
Patients	Died	Survived	Died	Survived	Died within 3 days	Died after 3 days	Survived
Number of patients	25	5	7	18	137	166	142
Percentage of patients	83.3	16.7	28.0	72.0	30.8	37.3	31.9

The Soviet-era experience

Chanishvili *et al.* (2009) recently collected and collated Soviet reports conducted during the 1930s to 1970s and translated them into English. The results are described in this section. When we consider that this therapy was all that was available in the pre-antibiotic era and later that the economics of medical care in the Soviet Union only allowed limited access to the variety of antibiotics available in the West, it can be understood that the need to care for patients, by whatever means available, took precedence over rigorous scientific methodology. Viewed in this context, it is possible to glean some useful clinical information from their data.

Early phage-therapy trials established methods for the intramuscular and intravenous use of phage preparations. Phages were most commonly applied topically at the infection site and also mixed with local anaesthetics and injected around the wound. The results of these early trials indicated a success rate of 80% in 114 surgical cases infected with *staphylococcus* and a 72% positive outcome in 1888 cases, in which the 28% of failed patient treatments was attributed to mixed bacterial infections. These reports prompted the Soviet Military to institute the universal use of phage wound therapy during the Finnish Campaign and the Second World War. At the end of the Soviet Era, the Eliava Institute, in the Republic of Georgia, was producing 5 t of phage products each week primarily for the military. See Burrowes and Harper (Chapter 14, this volume) for additional discussion of the Eliava Institute.

Phage therapy in the Soviet military

In the Finland Campaign, between 1938 and 1939, phage cocktails containing bacteriophages infecting strict anaerobes, staphylococci and streptococci were used to treat patients with gas gangrene. In one study, the mixture was applied to 767 patients with only a 19% death rate compared with the control group, which was treated with 'other methods' and resulted in 42% deaths. Another series, using the same phage mixture produced at the Eliava Institute, had 19% deaths compared with 54% using 'other medications'. Surgical debridement and amputation presumably were carried out as required. The death/survival rates appeared to be the end points, with no mention of limb salvage. These studies have been reviewed by Chanishvili *et al.* (2009).

Subsequent to the initial therapeutic use of phage therapy, its prophylactic capacity against gas gangrene was tested experimentally 'in the field'. Three 'mobile brigades', each composed of six persons (three surgeons and three bacteriologists), were to test bacteriophage cocktails specific for staphylococci, streptococci and *Clostridium perfringens*. The first brigade treated 2500 wounded soldiers, with the result of only 1.4% displaying symptoms of gas gangrene. In comparison, 4.3% of soldiers in the control group (n = 7918) became infected. Data from the other two brigades showed that, of 941 phage-treated soldiers, only 1.4% developed gas gangrene compared with 6.8% of the untreated controls (n is unknown). In another set of 2584 phage-treated soldiers, only 0.7%

developed gas gangrene compared with 2.3% of the untreated controls (*n* is unknown). The combined data indicated a 30% decrease in the incidence of gas gangrene as a direct consequence of prophylactic phage therapy. Time-dependent results of prophylactic therapy were revealed and it was concluded that 'in those that were treated earlier, the wounds were clear of infection sooner and granulation appeared rapidly, temperature was normalized in a shorter period of time and unpleasant odours did not develop or were insignificant' (Chanishvili *et al.*, 2009).

In 1940, one mobile brigade was moved into more permanent hospital facilities. This move allowed a more controlled scientific environment and more standardized techniques of blood analysis prior to phage therapy. The times and descriptions of surgical wound manipulation and phage application were systematically recorded and patients were grouped according to wound infection type and anatomical location. This work was published in 1941 in a book written by Professor Tsulukidze entitled *Experience of the Use of Bacteriophages in Conditions of War Trauma*. Chanishvili *et al.* (2009) described Tsulukidze's work. In summary, he found that, on arrival from the front, mixed bacterial infections were the general rule and were unlike infections seen after long hospital stays and prolonged manipulations, which produced more complicated mixed infections and 'therapeutic' difficulties including nosocomial infection with *Proteus* species (against which bacterium they had no phage preparations). To avoid the in-hospital evolution of wound infections to those with complex polymicrobial communities and *Proteus* superinfection (which seemed to dominate other infections), bandaging protocols were modified to prevent cross-contamination between patients. In an attempt to prevent the development of anti-phage antibodies, subcutaneous injection of phages was done three to four times each second day rather than once daily. Apparently, their assumption was that daily phage injection was more immunogenic than their regimen (see Olszowska-Zaremba *et al.*, Chapter 12, this volume, for further discussion of phage–immune system interactions). Phages were also sprayed on to the wound with each dressing change.

Soft-tissue wounds treated with 'ordinary therapy' had treatment with 'chloramines, rivanol and Vishnevsky's salve' and were dressed with tampons. Many such wounds had abundant pus and complicated infections with surrounding inflammation as well as necrotic foci, and retained bullet and wood fragments. To control serious intoxication due to sepsis and toxaemia, as well as high fever and gangrenous inflammation, these wounds required extensive debridement and 'wound purification' with iodine and alcohol followed by wound irrigation with a 2% sodium chloride solution. The wounds were then sprayed with phages and dressed with phage-soaked gauze but not packed with tampons. A volume of 5–10 ml of phage was injected simultaneously into the abdominal wall, shoulder or hip. 'After the first one or two phage applications the temperature was normalized. Only three or four such treatments were normally required to achieve a complete cure and the blood composition was also improved. The wounds were stitched after phage therapy on the 6th-8th day of treatment, so that further infection was unlikely.' (Chanishvilli *et al.*, 2009). Treatment with phage therapy took a number of days whereas 'ordinary therapy' took several weeks. No phage-therapy cases required additional incisions or any other surgical intervention.

Bone injuries were treated with the same phage protocol including debridement, phage wound dressing and phage injections remote from the regions of the wounds. Fractures were immobilized in casts or traction. Phage-treated open-bone injuries did not smell under a closed cast and the cast could remain unchanged for 60 days allowing fracture healing. In such wounds, without phage therapy, it would be anticipated that the casts would rapidly become saturated with purulent wound drainage and have a putrid smell, requiring frequent cast changes and disturbing fracture fixation and healing. In cases that were less successfully treated with phages, the reasons given included: the presence of *Proteus* as the cause of the infection (which the phages were not active against), the occurrence of frostbite injuries, long-term hospitalization (21–40 days) and accompanying diseases such as pneumonia, influenza and throat infections.

Topical phage applications were found to cause few or no side effects. Injecting near the wound was found to be painful for the patient, while injecting remotely from the wound was less painful but apparently similarly effective. Sometimes redness occurred at the injection site 'caused by the meat bullion' used to prepare the phages. A slight rise in temperature (0.5–1.0°C) was also noted. No reaction was seen following the second or third injections, however (see Olszowska-Zaremba *et al.*, Chapter 12, this volume, for additional discussion). Tsulukidze had noted that phage therapy shortened hospital stays and that, by avoiding large surgical interventions, he could decrease the number of new infections.

Phage therapy trials on civilians

The compiled data of Chanishvili *et al.* (2009) in general reported comparable responses using similar techniques when applied in civilian wound care and infections. Several additional noteworthy observations are highlighted here. First, when treating furunculosis, subcutaneous phage injections up to three times in increasing doses (e.g. 1 ml, 2 ml and 3 ml over 3 days), with lancing and drainage of the abscess and additional phage application into the abscess along with phage-soaked dressings, resulted in more successful outcomes than extensive incision and drainage while otherwise following the same protocol. In summary, Chanishvili *et al.* (2009) concluded that phage therapy could shorten the duration of treatments, although with an increase in duration if combined with invasive methods such as lancing, incisions or surgery, the latter resulting in the longest increase.

Secondly, persistent soft-tissue infections (i.e. lasting 1–3 months), chronic osteomyelitis and chronic amputation stump wounds could be treated with topical liquid phage cocktails using phage-soaked tampons or bathing in phages. Such means of phage administration were less traumatic to the patient than subcutaneous or intramuscular injections. Mixtures of phages active against aerobes and anaerobes were used to soak bandages. Dressings were changed every 2–3 days. In one group of 15 patients treated in this way, 13 patients were completely cured in 2–3 weeks, with the remaining two having improved outcomes.

Lastly, to avoid disturbing open surgical wounds following debridement of osteomyelitis, phages were mixed with 0.7% agar and used to fill the wound. This allowed phages to be released slowly and to linger in the wound, promoting phage persistence for 7–8 days.

In 1974, Krasnoshekova and Soboleva reported having used a standard piobacteriophage cocktail – produced by the Eliava Institute (containing phages against staphylococci and haemolytic *Streptococcus* species, *E. coli*, *P. aeruginosa* and *Proteus vulgaris*, also translated as 'pyo' bacteriophage) and adapted further to strains endemic to the Kazan region – for treating 85 patients with chronic wound infections. They found the pathogens to consist of *Staphylococcus* species in 93% of the wounds and selected specific phages against those pathogens. They reported 98% effective outcomes when phage therapy was based on the *in vitro* activity of the phages previously adapted to the bacterial host infecting these patients (Chanishvili *et al.*, 2009).

Also in 1974, Matusis and co-workers reported treating 94 patients with non-unions (where a broken bone is not able to heal), pseudoarthroses (false joint or established non-union) of the tibia and amputation stumps complicated by osteomyelitis. In 93% of cases, osteomyelitis was caused by staphylococci and 72% of these isolated strains were susceptible to the corresponding phage preparation (Chanishvili *et al.*, 2009). They had three treatment groups. For group 1, antibiotic therapy alone was employed ($n = 24$). Healing of stumps took 1–12 months (13 cases). Union (i.e. healing) of fractures, took 6–13 months and of these, four occurred within a 4-month period. Six osteomyelitic stumps failed to heal, as did three fracture non-unions. With group 2, a combination of topical phages along with topical antibiotics were used (n = 12). Healing of stumps took 0.5–3 months (12 cases). Fracture healing occurred in 4–9 months. Two cases 'relapsed'. In group 3, treatments consisted solely of topical phage therapy and/or intramuscular phage injections (n = 53). Forty-seven of the stump-wound infections healed in 0.3–2 months, with six cases taking 7–9 months. There were no relapses. In this series, phage therapy proved superior to both combined phage and antibiotic therapy as well as antibiotic therapy alone.

Human Wound Studies in the Late 20th Century

Human treatment of wounds using phages has been ongoing in Eastern Europe, most notably in Poland and the former Soviet Union republic of Georgia.

Polish studies

In addition to the studies highlighted by Chanishvili *et al.* (2009), there have been more recent phage-therapy studies from Eastern Europe that have been published in English language journals. Slopek *et al.* (1987), for example, reported on 550 hospitalized patients treated with phage therapy between 1981 and 1986, with most treated within the surgery departments, in Wrocław, Poland. Of the 40 cases of osteomyelitis seen, 29 cases were *Staphylococcus* infections, three cases were *Pseudomonas* infections and one was a *Klebsiella* infection. Polymicrobial infections were found in 11 cases, and 38 cases had been resistant to antibiotics. Also summarized was the treatment of cases of furunculosis, decubitus ulcers (bed sores) and open wounds (including burns). Table 13.3 summarizes the outcomes of this study. See also

Table 13.3. Summary of the various types of wound infections treated with phage therapy by Slopek *et al.* (1987)

Wound	Infection type	Result of treatment[a]					Total number of cases	Number of cases with positive outcome
		0	+1	+2	+3	+4		
Osteomyelitis	Single-microbe infection	0	2	3	19	5	40	36 (90%)
	Mixed-microbial infection	0	0	0	10	1		
Furunculosis	Single-microbe infection	0	0	2	19	29	55	55 (100%)
	Mixed-microbial infection	0	0	0	3	2		
Decubitus ulcer	Single-microbe infection	0	0	0	2	0	16	13 (81%)
	Mixed-microbial infection	0	3	0	9	2		
Open wound (including burns)	Single-microbe infection	1	1	0	13	7	49	42 (86%)
	Mixed-microbial infection	0	5	4	12	6		

[a]Scores represent the following results: 0, no effect; +1, transient improvement; +2, marked improvement with tendency to healing; +3, elimination of pus and healing of wounds; +4, complete recovery. Positive outcomes were classed as wounds that displayed scores of +2 or higher.

Olszowska-Zaremba *et al.* (Chapter 12, this volume) for additional consideration of the Poland experience.

Cislo *et al.* (1987) applied phage therapy to 31 selected patients. Patients suffering from suppurative skin infections were chosen based on 'no improvements seen' after treatment with routine methods including antibiotics, or the bacteria isolated from the lesions displaying resistance to antibiotics. The same phage administration strategy as used by Slopek *et al.* (1987) was followed. No other antimicrobial treatment (including local disinfectants) was used at the same time as phage therapy was administered. They reported 16 cases resulting in outstanding therapeutic effects, seven cases with marked improvement and negative cultures, and two cases with transient improvement of the wound. Treatment was abandoned in six cases due to 'side effects', although what these side effects were was not specified in the study. One case did not result in any improvement.

PhagoBioDerm™

PhagoBioDerm™ is a wound-dressing product co-developed by Georgian scientists Zemphira Alavidze and Ramaz Katsarava, and supported by an American company (Intralytix Inc.) along with the University of Maryland, MD, USA. It is a slow-releasing biodegradable polymer film impregnated with the antibiotic ciprofloxacin (0.6 mg cm^{-2}) and a mixture of bacteriophage (PyoPhage, 1×10^6 plaque-forming units (PFU) cm^{-2}) active against *S. aureus*, *P. aeruginosa*, *E. coli*, *Streptococcus* and *Proteus*. To date, the clinical experience of PhagoBioDerm has been published in two uncontrolled studies.

Markoishvili *et al.* (2002) reported on a series of clinical cases over a 1-year period, presenting wounds that were healing poorly due to inadequate vascularization, including in patients with diabetes, that were refractory to standard infection treatments. The wound dressing was used on 96 patients aged between 31 and 101 years. Treatment involved irrigating wounds, applying PhagoBioDerm film and dressing with sterile bandages to immobilize the film. The wounds were examined daily for the first 5 days and once every 2–4 days thereafter. Films were replaced any time they became fragmented or had completely degraded (i.e. every 3–7 days). They were left untouched if the films had 'tightly attached to the wound' or the wound had healed. Completely healed wounds were seen in 67 patients (70%) and the time required for infection-free treated sites ranged from 6 days to 15 months. The limitations to their studies were discussed and included the lack of a placebo group, which they had avoided because of concerns that continued application of conventional treatments would remain ineffective.

Jikia *et al.* (2005) reported on treating three patients who were exposed to strontium-90 and had developed radiation burns, which had subsequently become infected with MRSA. After being treated with a variety of antibiotics and topical ointments, their infections persisted. Admission to the hospital and failure to cure the infection by standard hospital procedures prompted the topical application of PhagoBioDerm. Only the two patients who had persistent infections were treated. Two days after the films were placed on the skin ulcers, both patients showed a dramatic decrease in purulent drainage as well as wound pain. On day 7, the ulcers were culture negative for *S. aureus*. The authors concluded that the elimination of the infecting bacteria and possibly the 'marked improvement in wound healing observed' were due to PhagoBioDerm.

Animal Models

Western interest in phage therapy was renewed by a series of animal studies in the 1980s investigated by Smith and colleagues in the UK (Smith and Huggins, 1982, 1983; Smith *et al.*, 1987a,b; Sulakvelidze *et al.*, 2001; Abedon *et al.*, 2011). They found that phage therapy could be used to treat mice infected with *E. coli*, and that it was even more effective than multiple doses of an array of antibiotics (Smith and Huggins, 1982). A decade later, Soothill (1994) published one of the first *in vivo* animal experiments related to phage therapy of wounds. A more detailed review of several of these studies is given by Burrowes and Harper (Chapter 14, this volume). In this section, we review animal models that have been used to investigate the efficacy of phage therapy in specific types of wounds and discuss some of the noteworthy experimental designs and potential drawbacks.

Burn wound models

Most of the published literature investigating the efficacy of phage therapy on wounds has described infected burn wounds. Many studies have shown phage therapy to be successful at protecting animals from succumbing to infections that would otherwise lead to death if untreated. Not all phage products/treatments have been shown to be equally potent, however.

Soothill (1994) demonstrated that phages could be used prophylactically to prevent destruction of skin grafts. To represent an excised burn, he removed a full-thickness skin section from the back of guinea pigs. The wound was then inoculated with 100 μl of bacterial suspension, along with phage or control suspension. The 0.2 mm thick graft area that was initially removed was placed back on to the defect area and covered by a compression pack followed by a large dressing. Skin grafts were blindly assessed after 5 days. Grafts were considered to be successfully attached if the skin was pink and blanched when pressure was applied. Six of seven animals with grafts infected with *P. aeruginosa* (6.0×10^5 colony-forming units (CFU)) and treated with a purified single phage, BS24 (1.2×10^7 PFU), had successful outcomes. This compared with all seven negative phage-treatment control animals that had infected grafts but remained untreated, resulting in grafts that failed.

McVay *et al.* (2007) conducted a study to determine the protective ability of a phage cocktail containing three phages on thermally injured mice infected with a *P. aeruginosa* strain ($\sim 3 \times 10^2$ CFU) that was directly injected under the anterior end of the burn. The cocktail inoculums ($\sim 3 \times 10^8$ PFU) were administered intraperitoneally (IP), intramuscularly (IM) or subcutaneously (SC) immediately after the bacterial challenge. Mice treated with IP injection of purified phage showed a mortality rate of 12% compared with 72% for IM, 78% for SC and 94% for untreated mice. The investigators suggested that the reason the IP route was most effective was due to an increased rate of phage dissemination and therefore a higher dose of phages delivery to their targets, which they demonstrate from the detectable levels of phage concentrated in the liver, spleen and blood of the treated mice over a 48 h period.

Kumari *et al.* (2009a) evaluated five 'purified' and well-characterized bacteriophages individually and as a cocktail. *Klebsiella pneumoniae* (100% lethal dose (LD_{100}) = 10^6 CFU) was injected subcutaneously directly under the anterior end of the burn. Treated mice ($n = 12$) were immediately administered phages (multiplicity of infection (MOI) of 1.0) via the IP route to test their protective capacity. After 72 h, there was only an ~6% survival rate for untreated mice versus 81–97% with the different individual phages and 94% with the five-phage cocktail treatment. Full recovery and complete regeneration of skin layers was seen on day 20 after phage treatment. Continuing on from this study, Kumari *et al.* (2010a) extended their experiments using one of the phages in the cocktail, Kpn5, to determine whether the protective property of this phage was dose dependent. Following SC injection of the same previously used lethal dose of *K. pneumoniae*, mice ($n = 10$) were injected IP with 100 μl of the phage preparation at various doses. Survival rates of 96% were seen with higher phage doses. At 10-, 100- and 1000-fold lower doses, survival rates were 53, 13 and 0%, respectively. Delayed phage treatment appeared to rescue 73, 47, 27 and 7% of the animals, if administered at higher doses with delays of 6, 12, 18 and 24 h, respectively.

Similar results were reported by Malik and Chhibber (2009) for a mouse third-degree burn wound model for which a subcutaneous challenge with ten times the LD_{100} of *K. pneumoniae* suspension (i.e. 10^7 CFU) was treated with IP or SC injection of pure bacteriophage KØ1 (10^8 PFU) at 30 min and 6 h post-infection. While phage therapy appeared to have protective effects against mice with experimental burn wounds infected with *K. pneumoniae,* similar burn wound infections caused by *P. aeruginosa* (10^7 CFU ml^{-1}) could not be treated and resulted in 100% mortality (Kumari *et al.*, 2009b). Kumari *et al.* (2010b) compared the use of other alternative antimicrobial agents (i.e. honey and aloe vera gel) against phage therapy. Topical phage treatment using a single phage isolate suspended in 3% hydrogel provided a high level of protection (67% survival) at day 7 compared with 27–33% with non-phage treatment and 0% survival without treatment. In a similar study, Kumari *et al.* (2011) compared the daily topical application of silver nitrate (5%) or gentamicin (1000 mg l^{-1}) with the same burn wound model and found these to have survival rates of 57 and 53%, respectively, on day 7, whereas a single phage treatment in 3% hydrogel resulted in 63% survivors. These differences were not found to be statistically significant.

Some concerns regarding the studies carried out by Kumari and colleagues include their only stating the concentration and not the volume of bacteria used to inoculate the wound. In addition, one must assume that the phage treatment was administered immediately after the bacterial application, as with their previous studies, although this is not stated in some of their papers. In addition, although the MOIs are stated in their articles, it is difficult to determine the actual dose of phage used for their therapeutic work. Their explanations of the phage therapy failing to treat their *P. aeruginosa* infected mice include the possibility of biofilm formation inhibiting phage action, although this bacterial process would seem unlikely to have occurred by the time the phage were administered, that is, if phage had been used as a protective treatment. The development of phage-resistant bacteria was another reason given as to why this therapy did not work. However, they had tested bacteria isolated from different organs of the dead mice and found them to remain sensitive to the phages, which they actually state would indicate 'absence of such mutants'. See Abedon (Chapter 17, this volume) for a more general discussion of phage-therapy experimental design including the various issues applicable – commendably as well as critically – to the studies of Kumari and co-workers. Despite not specifying some of the technical parameters used in their studies, Chhibber's group have still carried out phage-therapy research using methodological scientific standards and are producing very promising data.

Abscess models

While most modern *in vivo* phage-therapy investigations of animal wound models have been of infected burn wounds, there have been a few papers considering animal models of *S. aureus* abscesses. Wills *et al.* (2005), for example, tested the efficacy of treatment using a single *Staphylococcus* phage (LS2a) in three rabbit studies. In the first study, rabbits ($n = 8$) were inoculated with SC injection of *S. aureus* at 8×10^7 CFU and immediately treated with SC administration of phage at 2×10^9 PFU (i.e. an MOI of 25) at the same site. After 4 days, only one rabbit developed an abscess (area = 64 mm^2). This abscess was smaller than those found in the untreated rabbits (median area = 106 mm^2). In the second study, infected rabbits ($n = 4$ per treated group) received various doses of phage treatment of 6×10^7, 6×10^6 or 6×10^5 PFU. On day 4, the results indicated that the phages had some dose-dependent prophylactic effect. The median area of the abscesses in particular decreased as the dose of phage administered increased. The third study tested the implications of delayed treatment. Rabbits ($n = 4$) were infected with 5×10^7 CFU of *S. aureus*, and 3×10^9 PFU of a *Staphylococcus* phage at 6, 12 or 24 h after bacterial injection. There was no significant decrease found in the median area of the abscesses between the treated and untreated groups ($n = 5$). Overall, Wills and colleagues concluded that, although some protection was achieved against

S. aureus infection, the results were not as notable as those seen previously with Gram-negative bacteria.

Capparelli *et al.* (2007) published a thorough investigation on the efficacy of phage therapy to treat *S. aureus*-infected mice. Part of their investigation involved testing the ability of their *Staphylococcus* phage (M^{Sa}, a long-circulating mutant) to be used as a prophylactic agent to inhibit formation of abscesses. Abscesses were induced in mice ($n = 5$) through SC injections to the abdomen using 10^8 CFU of a *S. aureus* strain. Phage treatment (10^9 PFU) was administered immediately after, via the same route as the bacteria. Additional experiments were also used to determine whether their phages could help infected mice recover from established abscesses. A single phage dose, administered 4 days after the bacterial inoculum, was compared with four daily doses of phage treatment. They found that the formation of abscesses could be inhibited if the single phage treatment had been given concurrently with the bacteria. If phage treatment as single or multiple doses was provided 4 days after infection, however, abscesses were not inhibited from forming, although the bacterial load present in the abscesses was significantly reduced.

Interestingly, the multiple-dose treatment was more effective at reducing the overall weight of the abscesses than that of the single dose. This observation was interpreted as being due to the need for the bacterial density to pass a required threshold in order for 'active therapy' to occur with the single dose (as seen with their systemic infection model), whereas the multiple-dose treatment of the abscesses mimicked 'passive therapy', allowing the phages to kill bacterial cells through contact with large loads of phage particles but without phage multiplication. This speculation seems unlikely for two reasons: (i) the original bacterial dosage was sufficient to support active treatment; and (ii) the original phage dosage, at least potentially, was sufficient to supply a good approximation of a passive treatment. To date, only a few published animal studies have used multiple phage dosages in their treatment, and not enough effort has been made to understand the kinetics of the phage and target bacteria. We may now be at a good point, with our increasing understanding of phage biology, to stop just thinking of using phage therapy as a 'one-time shot' and instead start studying how phages can be utilized in their 'best-case scenario' such as with multiple dosing (see Abedon, Chapter 17, this volume, for additional consideration of this point).

Experimental design limitations

It is worth mentioning that some investigators, including McVay and Kumari, used blood collection tubes containing EDTA. EDTA is a metal chelator and from our experience it has been shown to affect phage attachment to their bacterial hosts, resulting in a reduction in the number of infective phage detected in a sample. When 1 mM EDTA was present in the growth medium, phage propagation was inhibited and resulted in a 4 log difference in phage amplification when compared with the control flask, over a 2 h period (unpublished data). Citrate, another chelating agent, has also been used in certain phage propagation media to inhibit adsorption by those phages that employ divalent cations as adsorption cofactors (Hyman and Abedon, 2009). Therefore, when collecting samples for enumerating infective phage present, it may be wise to take this into consideration and use collection tubes without EDTA or other chelating agents such as citrate ion present.

When designing the infected-animal models needed to test the efficacy of phage therapy, one should also consider the pathogenicity of the bacterial strain to the animal species chosen to study a particular infection. In Walker's study (1931), the bacterial strain used to infect his rabbits was found to be avirulent, and therefore 'large doses were required to produce any sort of lesion'. One must bear in mind, however, that, by using large doses for bacterial challenge, the conditions produced may not be an accurate representation of infection as it occurs in humans. Thus, the general practice of using an inoculum to ensure death as an end point (if the infection remains untreated) might induce a vastly different host response than if the infection had originated from a low inoculum, as with most naturally occurring infections (Zak and O'Reilly, 1991). That said, investigators must also understand the bacteria and infections they intend to treat. Using *S. aureus* as an example, it is known that this organism's virulence in humans is in part a consequence of its ability to extract iron from human red blood cells. One virulent strain of the human pathogen was found to be less effective at extracting iron from mouse red blood cells and this impedes infection in mice (Lowry, 2011). It is therefore important to determine which animal species and bacterial strains are suitable to represent the infection that the bacteriophages are being tested to treat. See Zak and O'Reilly (1991) for a review on using animal models in the evaluation of antimicrobial agents.

Modern-day Clinical Phage-therapy Trials

It is only through evidence-based science that the appropriate application of phage therapy can be determined. Phages are not truly 'magic bullets' capable of curing all bacterial infections but they do have the potential, teamed with the larger arsenal of anti-bacterials and other techniques, to combat the increasing problem of antibiotic-resistant bacteria. Below are some examples of clinical studies that have tried to follow the modern standards required by the regulatory agencies to support the development of therapeutics that can be used as adjuncts or alternatives to the currently available antibiotic regimens.

In 2006, Marza *et al.* published a single case report involving a 27-year-old man with a 50% total body surface area burn wound. The burn wound developed an infection with *P. aeruginosa* and, despite antibiotic treatment, grafted areas required regrafting a number of times due to rapid bacterial enzymatic graft breakdown. After obtaining ethical approval and informed consent, phage therapy was attempted based on previous work by Soothill (1994) showing its potential. The purified phage chosen for the treatment was shown to cause lysis of the *P. aeruginosa* isolated from the patient and not to be toxic to human keratinocytes. Phage suspension (10^3 PFU) was applied to paper discs (25 mm diameter) and placed on the infected areas. After 48 h, phage counts in the discs had apparently amplified by up to 1200-fold. Three days after phage treatment, no *P. aeruginosa* was cultured from swabs. Marza *et al.* (2006) carefully remarked that the improvement in the infection could also have been due to intravenous ceftazidime the patient was receiving at the same time as the phage treatment.

Rhoads *et al.* (2009) reported results from a phase I clinical safety trial. The safety of an eight-phage cocktail suspension (WPP-201; containing 1×10^9 PFU ml^{-1} of each phage) against *S. aureus*, *P. aeruginosa* and *E. coli* was tested through a randomized, controlled, double-blind study. However, the report did not specifically mention any improvements in the infected patients. This may be because the study was biased towards not harming the patients and there was no mention about ensuring that the phages used were appropriate for the pathogens present in the infected wounds, which may explain the poor efficacy seen with this phage therapy.

Merabishvili *et al.* (2009) produced a phage cocktail that met the regulatory standards needed for a therapeutic product to be used in human clinical trials. Their cocktail (BFC-1 at 3×10^9 PFU ml^{-1}) consisted of two *Pseudomonas* phages (14/1 and PNM) and one *Staphylococcus* phage (ISP), all found to exhibit a large host range specifically against burn-wound bacterial isolates. Eight patients suffering from infected burn wounds were treated without any adverse effects being seen. However, no efficacy results were published in their paper.

Future endeavours

We are currently investigating the use of bacteriophages as prophylactic and therapeutic agents in a rat model of osteomyelitis. If proved efficacious, we would anticipate that this therapy could also be used as a post-operative prophylactic treatment for orthopaedic device implantations or as infection prevention sprays for wounded soldiers on the battlefront. In relation to orthopaedic devices, one particular interest in our laboratory is the use of phages in infection prevention sprays or washes for patients with percutaneous osseointegrated devices, which are implanted through the skin into the bone of an amputated limb. Bone integrates with these implants and the protruding stem provides a robust skeletal docking system for artificial limbs. This attachment is especially suited to short-stump amputees and patients with multiple limb loss, in whom conventional socket attachment technology fails, which is a population that has grown as a result of the Iraq and Afghanistan wars. The significant advantages of this system have been proven in human volunteers in Europe (Tillander *et al.* 2010), although infection remains a problem in 10–30% of these implants.

With regard to battlefield wounds, despite current infection intervention strategies implemented to treat combat-related injuries in soldiers, including antibiotics, one-third of such injuries have been complicated by infections (Murray, 2008; Brown *et al.* 2010). Therefore, we must endeavour to find alternative treatments to current practices, which could even mean revisiting the early practices of using phages in the battlefield, duplicating the experience of Soviet military medicine.

Closing Remarks

Animal models seem to indicate that the application of bacteriophages, as prophylactic agents, has certain advantages over currently approved therapies and other alternative approaches. The use of phage as therapeutic agents for treating acute or chronic infections, however, still remains outside the standards of Western medical practice. Furthermore, the data from Eastern Europe has not been felt worthy of scrutiny by mainstream Western medicine, the impression being that these studies lack the prospective, double-blind evidence required to meet current regulatory standards for the approved use of a therapeutic agent. It is therefore imperative that investigators follow good phage-therapy practices when designing and conducting their animal studies (see Abedon, Chapter 17, this volume), and most importantly when conducting human clinical trials. From the growing number of phage-therapy studies published more recently, and with the particular emphasis on developing phage preparations that meet the standards of regulatory agencies, one could speculate that a small yet momentous wave of clinical phage-therapy trials will begin to be performed within the next 5 years (as declared by Rhoads *et al.*, 2009 and Merabishvili *et al.*, 2009; see also Burrowes and Harper, Chapter 14, this volume). The results should help determine whether phage therapy can truly become a standard of care in Western medicine.

References

Abedon, S.T., Kuhl, S.J., Blasdel, B.G. and Kutter, E.M. (2011) Phage treatment of human infections. *Bacteriophage* 1, 1–20.

Albee, F.H. and Patterson, M.B. (1930) The bacteriophage in surgery. *Annals of Surgery* 91, 855–874.

Brown, K.V., Murray, C.K. and Clasper, J.C. (2010) Infectious complications of combat-related mangled extremity injuries in the British military. *Journal of Trauma* 69, S109–S115.

Capparelli, R., Parlato, M., Borriello, G., Salvatore, P. and Iannelli, D. (2007) Experimental phage therapy against *Staphylococcus aureus* in mice. *Antimicrobial Agents and Chemotherapy* 51, 2765–2773.

Chanishvili, N., Khurtsia, N. and Malkhazova, I. (2009) Phage therapy in surgery and wound treatment. In: Chanishvili, N. and Sharp, R. (eds) *A Literature Review of the Practical Application of Bacteriophage Research*. Eliava Institute of Bacteriophage, Microbiology and Virology, Tbilisi, Georgia. pp. 21–32.

Church, D., Elsayed, S., Reid, O., Winston, B. and Lindsay, R. (2006) Burn wound infections. *Clinical Microbiology Reviews* 19, 403–434.

Cislo, M., Dabrowki, M., Weber-Dabrowska, B. and Woyton, A. (1987) Bacteriophage treatment of suppurative skin infections. *Archivum Immunologiae et Therapiae Experimentalis* 35, 175–183.

Eaton, M.D. and Bayne-Jones, S. (1934a) Bacteriophage therapy: review of the principles and results of the use of bacteriophage in the treatment of infections. *Journal of the American Medical Association* 103, 1769–1776.

Eaton, MD. and Bayne-Jones, S. (1934b) Bacteriophage therapy: review of the principles and results of the use of bacteriophage in the treatment of infections. *Journal of the American Medical Association* 103, 1847–1853.

Eaton, MD. and Bayne-Jones, S. (1934c) Bacteriophage therapy: review of the principles and results of the use of bacteriophage in the treatment of infections. *Journal of the American Medical Association* 103, 1934–1939.

Gill, S.R., Pop, M., DeBoy, R.T., Eckburg, P.B., Turnbaugh, P.J., Samuel, B.S., Gordon, J.I., Relman, D.A., Fraser-Liggett, C.M. and Nelson, K.E. (2006) Metagenomic analysis of the human distal gut microbiome. *Science* 312, 1355–1359.

Grice, E.A. and Segre, J.A. (2011) The skin microbiome. *Nature Reviews Microbiology* 9, 244–253.

Hermans, M.H.E. (2010) Burn wound management. In: Percival, P. and Cutting, K. (eds) *Microbiology of Wounds*. CRC Press, Boca Raton, FL, pp. 135–149.

Hermans, M.H.E. and Treadwell, T. (2010) An introduction to wounds. In: Percival, P. and Cutting, K. (eds) *Microbiology of Wounds*. CRC Press, Boca Raton, FL, pp. 83–134.

Hopper, L.V. and Gordon, J.I. (2001) Commensal host–bacterial relationships in the gut. *Science* 292, 1115–1118.

Howell, W.R. and Goulston, C. (2011) Osteomyelitis: an update for hospitalists. *Hospital Practice* 39, 153–160.

Hyman, P. and Abedon, S.T. (2009) Practical methods for determining phage growth parameters. In: Clokie, M.R.J. and Kropinski, A.M. (eds) *Bacteriophage: Methods and Protocols*. Humana Press, New York, NY, pp. 175–202.

Jikia, D., Chkhaide, N., Imedashvili, E., Mgaloblishvili, I., Tsitlanadze, G., Katsarava, R., Glenn Morris, J. Jr and Sulakvelidze, A. (2005) The use of a novel biodegradable preparation capable of the sustained release of bacteriophages and ciprofloxacin, in the complex treatment of multidrug-resistant *Staphylococcus aureus*-infected local radiation injuries caused by exposure to Sr90. *Clinical Dermatology* 30, 23–26.

Kumari, S., Harjai, K., and Chhibber, S. (2009a) Efficacy of bacteriophage treatment in murine burn wound infection induced by *Klebsiella pneumoniae*. *Journal of Microbiology and Biotechnology* 19, 622–628.

Kumari, S., Harjai, K., and Chhibber, S. (2009b) Bacteriophage treatment of burn wound infection caused *by Pseudomonas aeruginosa* PAO in BALB/c mice. *American Journal of Biomedical Sciences* 1, 385–394.

Kumari, S., Harjai, K. and Chhibber, S. (2010a) Evidence to support the therapeutic potential of bacteriophage Kpn5 in burn wound infection caused by *Klebsiella pneumoniae* in BALB/c mice. *Journal of Microbiology and Biotechnology* 20, 935–941.

Kumari, S., Harjai, K. and Chhibber, S. (2010b) Topical treatment of *Klebsiella pneumoniae* B5055 induced burn wound infection in mice using natural products. *Journal of Infection in Developing Countries* 4, 367–377.

Kumari, S., Harjai, K. and Chhibber, S. (2011) Bacteriophage versus antimicrobial agents for the treatment of murine burn wound infection caused by *Klebsiella pneumoniae* B5055. *Journal of Medical Microbiology* 60, 205–210.

Kutateladze, M. and Adamia, R. (2010) Bacteriophages as potential new therapeutics to replace or supplement antibiotics. *Trends in Biotechnology* 28, 591–595.

Lai, Y., Di Nardos, A., Nakatsuji, T., Leichtle, A., Yang, Y., Cogen, A.L., Wu, Z.R., Hooper, L.V., von Aulock, S., Radek, K.A., Huang, C.M., Ryan, A.F. and Gallo, R.L. (2009) Commensal bacteria regulate TLR3-dependent inflammation following skin injury. *Nature Medicine* 15, 1377–1382.

Larkum, N.W. (1929) Bacteriophage treatment of *Staphylococcus* infections. *Journal of Infectious Diseases* 45, 34–41.

Lee, Y.K. and Mazmanian, S.K. (2010) Has the microbiota played a critical role in the evolution of the adaptive immune system? *Science* 330, 1768–1773.

Lew, D.P. and Waldvogel, F.A. (1997) Osteomyelitis. *New England Journal of Medicine* 336, 999–1007.

Lipsky, B.A., Berendt, A.R., Deery, H.G., Embil, J.M., Joseph, W.S., Karchmer, A.W., LeFrock, J.L., Lew, D.P., Mader, J.T., Norden, C. and Tan, J.S. (2004) Diagnosis and treatment of diabetic foot infections. *Clinical Infectious Diseases* 39, 885–910.

Lowry, F.D. (2011) How *Staphylococcus aureus* adapts to its host. *New England Journal of Medicine* 364, 1987–1990.

MacNeal, W.J., Frisbee, F.C. and McRae, M.A. (1942) Staphylococcemia 1931–1940: five hundred patients. *American Journal of Clinical Pathology* 12, 281–294.

Malik, R and Chhibber, S. (2009) Protection with bacteriophage KØ1 against fatal *Klebsiella pneumoniae* – induced burn wound infection in mice. *Journal of Microbiology, Immunology and Infection* 42, 134–140.

Markoishvili, K., Tsitlanadze, G., Katsarava, R., Morris, J.G. and Sulakvelidze, A. (2002) A novel sustained-release matrix based on biodegradable poly(ester amide)s and impregnated with bacteriophages and an antibiotic shows promise in management of infected venous stasis ulcers and other poorly healing wounds. *International Journal of Dermatology* 41, 453–458.

Marza, J.A., Soothhill, J.S., Boydell, P. and Collyns, T.A. (2006) Multiplication of therapeutically administered bacteriophages in *Pseudomonas aeruginosa* infected patients. *Burns* 32, 644–646.

Maura, D. and Debarbieux, L. (2011) Bacteriophages as twenty-first century antibacterial tools for food and medicine. *Applied Microbiology and Biotechnology* 90, 851–859.

McKinley, E.B. (1923) The bacteriophage in the treatment of infections. *Archives of Internal Medicine* 32, 899–910.

McVay, C.S., Velasquez, M. and Fralick, J.A. (2007) Phage therapy of *Pseudomonas aeruginosa* infection in a mouse burn wound model. *Antimicrobial Agents and Chemotherapy* 51, 1934–1938.

Merabishvili, M., Pirnay, J.P., Verbeken, G., Chanishvili, N., Tediashvili, M., Lashkhi, N., Glonti, T., Krylov, V., Mast, J., van Parys, L., Lavigne, R., Volckaert, G., Mattheus, W., Verween, G., de Corte, P., Rose, T., Jennes, S., Zizi, M., de Vos, D. and Veneechoutte, M. (2009) Quality-controlled small-scale production of a well-defined bacteriophage cocktail for use in human clinical trials. *PLoS One* 4, e4944.

Morowitz, M.J., Babrowski, T., Carlisle, E.M., Olivas, A., Romanowski, K.S., Seal, J.B., Liu, D.C. and Alverdy, J.C. (2011) The human microbiome and surgical disease. *Annals of Surgery* 253, 1094–1101.

Murray, C.K. (2008) Epidemiology of infections associated with combat-related injuries in Iraq and Afghanistan. *Journal of Trauma* 64, S232–S238.

Percival, S.L. and Dowd, S.E. (2010) The microbiology of wounds. In: Percival, P. and Cutting, K (eds) *Microbiology of Wounds*. CRC Press, Boca Raton, FL, pp. 187–217.

Rafla, K. and Tredget, E.E. (2011) Infection control in the burn unit. *Burns* 37, 5–15.

Rhoads, D.D., Wolcott, R.D., Kuskowski, M.A., Wolcott, B.M., Ward, L.S. and Sulakvelidze, A. (2009) Bacteriophage therapy of venous leg ulcers in humans: results of a phase I safety trial. *Journal of Wound Care* 16, 237–243.

Rice, T.B. (1930) Use of bacteriophage filtrates in treatment of suppurative conditions: report of 300 cases. *American Journal of the Medical Sciences* 179, 345–360.

Rowley-Conwy, G. (2010) Infection prevention and treatment in patients with major burn injuries. *Nursing Standard* 25, 51–60.

Shultz, E.W. (1932) Bacteriophage: a possible therapeutic aid in dental infections. *Journal of Dental Research* 12, 295–310.

Slopek, S., Weber-Dabrowska, B., Dabrowski, M. and Kucharewicz-Krukowska, A. (1987) Results of bacteriophage treatment of suppurative bacterial infections in the years 1981–1986. *Archivum Immunologiae et Therapiae Experimentalis* 35, 569–583.

Smith, H.W. and Huggins, M.B. (1982) Successful treatment of experimental *Escherichia coli* infections in mice using phages: its general superiority over antibiotics. *Journal of General Microbiology* 128, 307–318.

Smith, H.W. and Huggins, M.B. (1983) Effectiveness of phage in treating experimental *Escherichia coli* diarrhoea in calves, piglets and lambs. *Journal of General Microbiology* 129, 2659–2675.

Smith, H.W., Huggins, M.B. and Shaw, K.M. (1987a) The control of experimental *Escherichia coli* diarrhoea in calves by means of bacteriophages. *Journal of General Microbiology* 133, 1111–1126.

Smith, H.W., Huggins, M.B. and Shaw, K.M. (1987b) Factors influencing the survival and multiplication of bacteriophages in calves and in their environment. *Journal of General Microbiology* 133, 1127–1135.

Soothill, J.S. (1994) Bacteriophage prevents destruction of skin grafts by *Pseudomonas aeruginosa*. *Burns* 20, 209–211.

Sulakvelidze, A., Alavidze, Z. and Morris, J.G. (2001) Bacteriophage therapy. *Antimicrobial Agents and Chemotherapy* 45, 649–659.

Teibelbaum, J.E. and Walker, W.A. (2002) Nutritional impact of pre- and probiotics as protective gastro-intestinal organisms. *Annual Reviews in Nutrition* 22, 107–138.

Tillander, J., Hagberg, K., Harberg, L. and Branemark, R. (2010) Osseointegrated titanium implants for limb prostheses attachments. *Clinical Orthopaedics and Related Research* 468, 2781–2788.

Walker, J.E. (1931) The effect of bacteriophage in experimental *Staphylococcus* and *Streptococcus* skin infections. *Southern Medical Journal* 24, 1087–1089.

Wills, Q.F., Kerrigan, C. and Soothill, J.S. (2005) Experimental bacteriophage protection against *Staphylococcus aureus* abscesses in a rabbit model. *Antimicrobial Agents and Chemotherapy* 49, 1220–1221.

Zak, O. and O'Reilly, T. (1991) Animal models in the evaluation of antimicrobial agents. *Antimicrobial Agents and Chemotherapy* 35, 1527–1531.

14 Phage Therapy of Non-wound Infections

Ben Burrowes[1] and David R. Harper[1]
[1]*Ampliphi Biosciences Corporation.*

Antibiotic resistance had been observed – both clinically and experimentally – among bacterial isolates even before the widespread clinical use of antibiotics (Waksman *et al.*, 1945; Wainwright and Swan, 1986). At first, such resistance was of little concern, as new antibiotics were routinely becoming available. Bacteria, however, have continued to become ever more drug resistant and, with regard to the treatment of certain resistant pathogens, entire classes of antibiotics have become ineffective (Boucher *et al.*, 2009). The relentless evolutionary pressure imposed by decades of both antibiotic use and close proximity of infected or susceptible patients (e.g. in hospitals) has diminished the clinical value of most if not all antibiotics. Pathogens and opportunists such as vancomycin-resistant *Enterococcus* (VRE), methicillin-resistant and vancomycin-resistant *Staphylococcus aureus* (MRSA and VRSA, respectively), *Pseudomonas aeruginosa* and *Acinetobacter baumannii* are extremely difficult to treat using conventional approaches and have required concerted efforts to prevent colonization of new patients – principally by following strict barrier nursing procedures (Boyce *et al.*, 1994; Jarvis, 2010).

Phage therapy is an old idea but one that is gaining renewed interest in the face of the clinical challenges being brought by bacterial drug resistance. As such, phage therapy and biology have been reviewed extensively in journals (Sulakvelidze *et al.*, 2001; Summers, 2001; Thiel, 2004; Abedon, 2010; Kutter *et al.*, 2010) and monographs (Summers, 1999; Kutter and Sulakvelidze, 2005; Häusler, 2006; Clokie and Kropinski, 2009; Abedon, 2011b), all of which offer valuable insights into various aspects of phage therapy and biology.

For the purposes of this book, bacterial infections have been divided into two arbitrary groups: wound infections (see Loc-Carrillo *et al.*, Chapter 13, this volume), in which the bacteria infect damaged or broken tissues, and non-wound infections, in which bacteria infect previously undamaged tissue. It is possible, however, that one type of infection can lead to the other, for example, bacteraemias. In this chapter, we briefly consider the history of phage therapy as it pertains to non-wound infections, before detailing more up-to-date experimental models and results.

Discovery and history

Very soon after their discovery (d'Hérelle, 2007), phages were put to use in treating bacterial diseases (d'Hérelle, 1922). Diarrhoeal diseases were among the first to be treated, and the positive results led to work on other human and animal diseases.

Numerous scientists and clinicians were sufficiently convinced by the results to devote time and resources to studying or administering phage therapy (d'Hérelle, 1921, 1929; Smith, 1924; d'Hérelle *et al.*, 1928; Asheshov *et al.*, 1930, 1931; Morison, 1932). Of particular note was the so-called 'Bacteriophage Enquiry' in which phages were used experimentally in India to treat or prevent cholera, with apparent success (d'Hérelle, 1929; Morison, 1932; Summers, 1993). Subsequent mixed results and the rise of antibiotic use, however, led to substantial reductions in the use of phage therapy worldwide (Eaton and Bayne-Jones, 1934; Krueger and Scribner, 1941). Broader consideration of the history of phage therapy can be found elsewhere (Sulakvelidze *et al.*, 2001; Summers, 1999, 2001; Thiel, 2004; Häusler, 2006; Chanishvili and Sharp, 2009). For the remainder of this section, we briefly provide an overview of work that went on, particularly in the former-Soviet republic of Georgia and Poland and which continues to the present. Details of work elsewhere, such as that in France and Switzerland, are discussed by Abedon *et al.* (2011).

Georgia

Following positive outcomes of early phage-therapy trials conducted in various countries, the Soviet Union commissioned the creation of a bacteriophage research institution in Tbilisi, Georgia – now called the Eliava Institute of Bacteriophage, Microbiology and Virology (IBMV), and this institute is actively isolating, characterizing and producing therapeutic phages to this day. Phage therapy is in fact widely used in Georgia, so much so that it is essentially considered a standard of care among local practitioners (Kutateladze and Adamia, 2008; Chanishvili and Sharp, 2009).

In appreciating the issues of phage host range and bacterial resistance, the IBMV has essentially always employed therapeutic phage cocktails, which are mixtures of phages that can be active against a wide range of bacterial strains and species associated with different disease types (Chan and Abedon, 2012). Cocktails were generated and maintained for pyogenic, intestinal, urological and gynaecological diseases, as well as developing monophage preparations for individual patients or bacterial strains, and generation of bespoke cocktails as required. Two phage preparations in particular are produced in large quantities for general use and are even available over the counter at Georgian pharmacies. Both are derivations of cocktails originally developed by d'Hérelle and are known, variously, as Intestiphage and Pyophage.

Intestiphage is one of the longest standing phage preparations (Chanishvili and Sharp, 2009). The preparation is made empirically using both environmentally isolated and laboratory-developed phages that are only tested for their plaque morphology and host range. The phages are then propagated in bulk by growing multiple phages on a small number of hosts (Dr Z. Alavidze, IBMV, personal communication). Intestiphage therefore consists of an essentially unknown number of poorly characterized phages active against common gut pathogens. The preparation is still available over the counter at pharmacies across the former Soviet Union. Phages active against *Shigella* spp., *Staphylococcus*, *Salmonella* spp., *Escherichia coli*, *Proteus* spp. and *P. aeruginosa* isolates were pooled and sold as a filter-sterilized crude lysate to be taken orally, either as a treatment or prophylactically. Human studies of this preparation, in its myriad forms, were carried out and usually reported to be successful. Many of these studies have been thoroughly reviewed by Chanishvili and Sharp (2009). In one study, more than 15,000 children received anti-dysentery phages prophylactically with significant success compared with a no-phage control group of nearly 3000 children (Babalova *et al.*, 1968).

Another study, reviewed by Kutateladze and Adamia (2008), investigated phage treatment of children using an Intestiphage preparation. The study divided the children into three treatment groups: phage treatment alone, phages plus antibiotics, and antibiotics alone. The result was a reported superiority of phage treatment alone compared with treatment with antibiotics, with or without co-treatment with phage. Recovery time fell from 29 days with antibiotics alone to 9 days in the group treated only with phage. In yet another study, discussed in the same review, more than 18,000 children were enrolled in a study to assess prophylactic use of anti-cholera phages. Nearly 16,000 of these children were administered phages (each dose equating to approximately 25 ml of phage culture – perhaps 10^8–10^{10} plaque forming units (PFU)) resulting in as much as a fivefold reduction in the number of cholera cases (Kutateladze and Adamia, 2008).

The Pyophage preparation, commonly used to treat wound infections in particular, includes phages active against the common pathogens *E. coli, S. aureus, P. aeruginosa, Streptococcus* and *Proteus* strains. These organisms overlap with common vaginal and urinary tract infections, and Pyophage has therefore been used to treat non-wound infections as well. Where Pyophage was compared with antibiotic treatment, it was found that combined treatment (phages plus antibiotics) was preferable (Kutateladze and Adamia, 2008). Other reports from the Institute suggest that Pyophage is also beneficial for treating mastitis (Chanishvili and Sharp, 2009). See Loc-Carrillo *et al.* (Chapter 13, this volume) for a more extensive consideration of the phage therapy of wound infections.

Poland

In 1952, the Institute of Immunology and Experimental Therapy was founded in Wrocław, Poland, by the Polish Academy of Sciences. The institute is now named the Hirszfeld Institute after its first director and founder, Ludwig Hirszfeld. The Hirszfeld Institute has carried out many studies of phage therapy (as summarized by Slopek *et al.*, 1987; see also Olszowska-Zaremba *et al.* and Loc-Carrillo *et al.*, Chapters 12 and 13, this volume), typically with phage treatment attempted only after conventional therapies have failed. Recently, the Institute has also become involved in phage therapy unrelated to infection control, such as the potential of phages, as immunologically active agents, to treat cancer. For a more in-depth discussion of this work, see Olszowska-Zaremba *et al.* (Chapter 12, this volume).

The 1980s 'rediscovery' and 1990s renaissance

Western research into phage therapy was rekindled when Smith and Huggins began to carry out well-designed, thorough and scientifically rigorous animal trials of phage therapy (Smith and Huggins, 1982, 1983; Smith *et al.*, 1987a,b). Central to their design was the use of either naturally occurring infections or animal models that closely paralleled human infections, much the same approach that d'Hérelle had practised (Summers, 1999).

Of mice and ruminants

Smith and Huggins began by using a human *E. coli* isolate that was known to be virulent in both humans and mice (Smith and Huggins, 1982). They then isolated a phage that infected these cells via the bacterial K1 antigen, an *E. coli* outer-membrane antigen that is important in the virulence of this organism. Mice were infected intramuscularly (IM) with 3×10^7 colony forming units (CFU; equivalent to around 100 times the 50% lethal dose (LD_{50})) or intracerebrally with 500 CFU (around ten times the LD_{50}). A single dose of phages containing 3×10^8 PFU was then inoculated IM – into the opposite flank for those mice that received an IM bacterial inoculation – and compared with mice who received antibiotics. Both single and repeat doses of streptomycin, tetracycline, ampicillin, chloramphenicol and trimethoprim/sulfafurazole were used as comparison groups. The only antibiotic regimen that rescued as many mice as a single phage dose was eight treatments with streptomycin (two and three mice died, respectively, from groups of 30), leading the authors to conclude that phages were 'generally superior' to currently available treatments (Smith and Huggins, 1982).

Smith and Huggins then treated juvenile pigs, sheep and cows with phages as therapy against experimental infections with enteropathogenic *E. coli* strains of demonstrated pathogenicity in these animals (Smith and Huggins, 1983). Experimental animals were infected orally with sufficient bacterial cells to induce diarrhoea and were then administered orally with between 10^9 and 10^{11} PFU of a mixture of two phages, depending on the exact protocol. Phages were given either at 1 or 8 h post-infection, or at the onset of diarrhoea. Although earlier phage treatment led to higher survival rates and faster recovery, even application of phages at the onset of diarrhoea resulted in clearly demonstrated phage-mediated control of diarrhoeal disease. They also suggested that phages in the faeces of the treated animals could have a protective effect for other animals sharing the same pen. No direct comparison was made with antibiotic treatment, but the use of a single dose of phage and the protective effect of phage treatment to other animals were convincing arguments for the efficacy of phage therapy.

Prevention of *E. coli* infection was later demonstrated in calf pens by spraying phages in pens, or even just by housing calves in pens previously inhabited by phage-shedding animals (Smith *et al.*, 1987a). In these experiments, K1-tropic phages were also used and the resultant $K1^-$ *E. coli* mutants (i.e. those bacterial mutants that had lost their K1 capsule and were therefore selected for in the presence of the K1-tropic phage) were less virulent than the $K1^+$ parent – again highlighting advantageous properties of phage use over antibiotic treatment (this issue is discussed at length by Levin and Bull, 1996). Perhaps the most important aspect of Smith and Huggins' work was that their experiments were rigorously designed, using appropriate controls and large enough groups of animals to confirm their central conclusions. Significantly, their work also was conducted in the West (the UK, in particular) and was published in English language journals.

At the same time as Smith and Huggins' work was being published, the Hirszfeld Institute in Poland also began publishing in English (see discussion of Polish work above; see also Olszowska-Zaremba *et al.* (Chapter 12, this volume). Their trials were not as rigorous as those of Smith and co-workers but did involve human subjects, and they reported no serious side effects and provided general evidence of efficacy.

Going beyond Smith *et al.*

During the 1980s and into the 1990s, antibiotic resistance was starting to become a significant clinical concern. Alternatives were needed and both researchers and clinicians were more open to ideas that previously had been rejected out of hand. Phage therapy, as a consequence, gained momentum once more. This time, however, both the scientific and regulatory frameworks were far more stringent than they had been during d'Hérelle's day, meaning that much of the work of the Eliava and Hirszfeld Institutes had to be built on to provide clinical data acceptable to the regulators.

Addressing this need, the British clinician James Soothill developed multiple experimental approaches to demonstrate the efficacy of phage therapy for pathogens other than *E. coli* (Soothill *et al.*, 1988; Soothill, 1992, 1994). Phages were used successfully, *in vitro*, to prevent destruction of pig skin samples by *P. aeruginosa* (Soothill *et al.*, 1988). Importantly, Soothill developed a simple mouse model to assess phage treatment of experimental *A. baumannii*, *P. aeruginosa* and *S. aureus* infections (Soothill, 1992), thereby enabling phage therapy of multiple diseases to be assessed in a single animal model. The results showed efficacy of phages in treating *Pseudomonas* and *Acinetobacter* infections, although they did not show efficacy in treating *S. aureus*. This latter result may have been associated with the poor *in vitro* activity of the phages they used, as generally, and as seen similarly with antibiotics, poor antibacterial performance in the 'test tube' is typically expected to be predictive of poor performance *in vivo* (see Abedon, Chapter 17, this volume, for additional discussion along these lines). As will be discussed below, however, this 'rule of thumb' may not always hold.

Recent animal models

Current regulatory frameworks demand stringent testing of disease treatment protocols in animals before testing in humans is considered appropriate. In spite of its history, this applies as much to phage therapy as to other medicines. Many *in vitro* models of phage activity and therapy exist, but they are limited in their usefulness for studying the complex pharmacology of the large, self-replicating nucleoprotein complexes that are phages (Payne and Jansen, 2003; Levin and Bull, 2004; Abedon and Thomas-Abedon, 2010; see also Abedon, Chapter 17, this volume, for additional discussion of phage-therapy pharmacology). In order to understand the pharmacology of phage therapy, we must determine the pharmacokinetics (the fate of the administered agent *in vivo*) and pharmacodynamics (the physiological and therapeutic activity of the administered agent) of phage therapeutics, which, due to both the complexity of phages and their ability to self-replicate, are fundamentally different in phage therapy compared with chemotherapeutic approaches, a theme that has been explored at length by Curtright and Abedon (2011). It is therefore extremely valuable to develop relevant, high-quality animal disease models to study the application and outcomes of phage therapy. We will now briefly discuss recent animal models of particular relevance to clinical applications.

Septicaemia

The ease with which blood-borne bacterial infections can be induced in mouse models has been exploited for many bacterial pathogens, and phage treatment is usually successful in these models. One of the most common models involves intraperitoneal (IP) injection of bacteria to induce bacteraemia in mice, which is challenged with phages either prior to infection, simultaneously with infection or post-infection. Such models have been used for the treatment of bacterial infections including *E. coli* (Merril *et al.*, 1996), *Salmonella typhimurium* (Merril *et al.*, 1996), *S. aureus* (Matsuzaki *et al.*, 2003), VRE (Biswas *et al.*, 2002; Wang *et al.*, 2006b), *Klebsiella pneumoniae* (Vinodkumar *et al.*, 2005) and *P. aeruginosa* (Wang *et al.*, 2006a; Vinodkumar *et al.*, 2008; Heo *et al.*, 2009). These animal models typically report efficacy of phage treatment, although with varying degrees of therapeutic success. In general, it appears that factors such as the half-life of phages *in vivo* (a pharmacokinetic property), virulence of the phage(s) chosen both *in vitro* and *in vivo* (pharmacodynamic properties) and other components of the animal/phage/host system influence the success of phage treatment or prophylaxis. The numbers of phages required to effectively rescue mice from infection depends on the ability of the phages to kill their target cells *in situ*: phages that are less efficient at killing their target cells will require larger numbers to sufficiently lyse host bacteria than would be needed when using more efficient phages. Dosing parameters are particularly significant when considering 'active' versus 'passive' phage therapy. Active phage therapy requires phage propagation *in situ* to bring about a therapeutic effect, whereas passive phage therapy uses higher doses of phages and further phage propagation is not required to generate therapeutic efficacy. This is discussed further by Abedon (Chapter 17, this volume).

Phage pharmacology is in its infancy but has been discussed in several studies and theoretical papers (Geier *et al.*, 1973; Payne and Jansen, 2003; Levin and Bull, 2004; Dabrowska *et al.*, 2005; Gill *et al.*, 2006; McVay *et al.*, 2007; Abedon and Thomas-Abedon, 2010; Abedon, 2011a). Perhaps most importantly, no correlation between antibiotic resistance and phage resistance has been reported so far, suggesting that phage therapy could prove especially useful in treating drug-resistant bacteria.

Lung infections

Chronic bacterial colonization of the lung, especially in immunocompromised patients, represents a significant challenge clinically because it can be difficult to maintain sufficient chemotherapeutic concentrations at the site of infection. Phage treatment of such infections may be desirable, as phage propagation may lead to sufficient phage numbers *in situ* to control infections. Conversely, it may also be problematic to deliver sufficient phage numbers into the deep lung tissue to bring about clearing of the bacteria. Animal models therefore are highly useful to ascertain such limitations, as well as the potential of phage treatment of lung infections.

Intranasal (IN) administration of lung pathogens has been used by several groups to assess phage treatment of murine pulmonary infections, and phages can then be applied by any route to determine treatment or prophylactic efficacy. This model has been used to study phage treatment of *K. pneumoniae* (Chhibber *et al.*, 2008), which causes bacterial pneumonia and is associated with significant mortality, especially resulting from nosocomial spread (Podschun and Ullmann, 1998). A similar model of *Burkholderia cenocepacia* infection also demonstrated efficacy of phage treatment (Carmody *et al.*, 2010). In both the *K. pneumoniae* and *B. cenocepacia* models, however, IP administration of phages proved efficacious whereas IN administration did not. This could be problematic, as IP administration is a time-consuming, invasive and potentially risky procedure that is rarely used clinically. When essentially the same model was used for *P. aeruginosa* infection, IN administration was shown to be effective (Debarbieux *et al.*, 2010; Morello *et al.*, 2011), raising the question of why IN administration is effective in this model compared with others. More work is required to ascertain the factors involved in phage treatment success or failure.

Gastrointestinal disease

Despite the long history of phage treatment of alimentary diseases, few animal studies of phage therapy of this class of disease are found in the modern literature. Such models do not appear to be difficult to work with, as the bacteria can be applied orally, as well as the phage, although intragastric administration of phages is also used. Studies with *E. coli* (Chibani-Chennoufi *et al.*, 2004; Denou *et al.*, 2009) and *P. aeruginosa* (Watanabe *et al.*, 2007) have all shown phage therapy to be effective in reducing pathogen load and rescuing mice from infection, albeit to varying extents. A rabbit model has also been used to assess phage treatment of cholera, again with apparent success (Bhowmick *et al.*, 2009). It is important to note that prophylactic use of phages will require phage numbers to be maintained at sufficient levels at the site of future infection to prevent bacterial colonization. In some cases, therefore, non-pathogenic phage hosts present in the normal flora could sustain phage numbers prior to pathogen challenge (see Letarov, Chapter 2, this volume). *E. coli*, for example, is a major bacterial component of the normal gut flora and as such may potentially maintain phage numbers in the gut prior to infection, whereas *Vibrio cholerae* is not present in the gut of healthy animals, implying that prophylaxis may be less useful for this type of infection.

While temperate phages are usually avoided for use in phage therapy, temperate phages were used for treatment of *Clostridium difficile* ileocecitis in a hamster model system (Ramesh *et al.*, 1999). Currently, no published reports exist of the isolation of virulent *C. difficile* phages, so a temperate phage was the only option in this case. Phages were administered orally following neutralization of gastric acids to minimize phage degradation in the stomach. Both phages and bacterial cells were administered intragastrically under anaesthesia, and phage-treated animals were significantly rescued from death compared with antibiotic-treated or untreated animals. Temperate phages of *C. difficile* are known to encode or regulate toxin production in this species (Govind *et al.*, 2009), so it would be of great value (see Abedon, Chapter 17, this volume) to ascertain whether phage therapy using such phages is potentially harmful in this model by assessing toxin production and lysogeny in faecal *C. difficile* isolates after phage treatment. It is also interesting to note that gastric neutralization may not always be necessary, as some phages are able to survive the gastric pH and reach the bowel in significant numbers without this step (Chibani-Chennoufi *et al.*, 2004; Bruttin and Brussow, 2005), although this characteristic would still need to be demonstrated on a phage-by-phage basis.

Although temperate phages are usually considered unfit agents for phage therapy, it appears that they may nevertheless reduce bacterial load and symptoms in various bacterial diseases. Merril *et al.* (1996) used a derivative of the temperate bacteriophage λ, which had been serially passaged to increase its half-life in the murine bloodstream, to treat bacteraemic mice. This work demonstrated that even temperate phages are capable of mitigating bacteraemia symptoms in their model, especially when the phage had been evolved to perform better *in vivo* by serial passage. While temperate phages are generally proscribed for therapy due to their potential to augment the pathogenicity of their bacterial hosts, nevertheless they possibly may still be used where no alternatives exist, although with due consideration to their enhanced potential for genetic transfer (see Christie *et al.* and Hendrickson, Chapters 4 and 5, this volume).

Canine otitis

While not all animal diseases are relevant to human disease, veterinary trials can still be used to pave the way for human clinical trials where there is overlap between the animal and human disease, such as in diseases of similar bacterial aetiology. Natural occurrences of canine otitis were used for a field trial as a veterinary precursor to regulatory-approved human clinical trials in the UK (Wright *et al.*, 2009; Hawkins *et al.*, 2010). Pet dogs were selected that demonstrated chronic, refractory otitis caused by *P. aeruginosa* and were treated with a cocktail of phages by placing the purified phage preparation into the ear canal. Ten dogs were treated once with 0.2 ml of a cocktail containing six phage strains (1×10^5 PFU per phage strain, equivalent to a total of 2.4 ng of therapeutic agent). In all treated animals, the mean bacterial count fell by 67% at 48 h after phage inoculation, while phage numbers had increased approximately 100-fold. Clinical observations showed that otitis symptoms had been reduced in all treated dogs. The positive results of this work were sufficient to allow this phage preparation to progress to UK Medicines and Healthcare Products Regulatory Agency-regulated phase I/II clinical trials, which will be discussed below (see also Abedon, Chapter 17, this volume).

Modern clinical trials

A brief list of human uses of phage therapy is given in Table 14.1. Ultimately, phage therapy is unlikely to be accepted in the West until it satisfies the regulatory authorities by demonstrating efficacy and safety in fully regulated trials. While it has been suggested that phage therapy warrants a loosening of the regulatory standards (Verbeken *et al.*, 2007), it must none the less prove itself to the same standards as all modern drugs, no matter how distinct it is from current therapies. Meanwhile, the resurgence of phage therapy in the West has led to the creation of several small companies. In addition, some large, multinational companies have expressed an interest in developing and, ultimately, commercializing phage products for human use. In some cases, the clinical research has focused on wound-care applications, which are discussed by Loc-Carrillo *et al.* (Chapter 13, this volume).

The first modern Western trial of phage therapy was carried out by the Port Washington, New York-based company Exponential Biotherapies Inc. (Thiel, 2004). The trial was an approved phase I assessment of toxicity of phages active against strains of VRE but which unfortunately has yet to be reported in the literature. Exponential Biotherapies Inc. is currently trying to develop small-molecule therapeutics (see http://www.expobio.com/company-profile/index.php), and the phage-therapy arm of the company was spun off as EBI Food Safety (now called Micreos Food Safety), which is active in control of food-borne bacterial pathogens (see http://www.micreosfoodsafety.com) (see also Niu *et al.*, Chapter 16, this volume).

Table 14.1. A brief list of human uses of potential applications for phage therapy.

Infection site	Target organism(s)	References
Blood (septicemia)	*S. aureus*, *Klebsiella* spp., *E. coli*, *P. aeruginosa*, *Proteus* spp.	Slopek *et al.*, 1987; Weber-Dabrowska *et al.*, 2003; Häusler, 2006; Kutateladze and Adamia, 2008; Chanishvili and Sharp, 2009
Eye	*Staphylococcus* spp., *Klebsiella* spp., *P. aeruginosa*, *Proteus* spp., *Neisseria* spp., *Streptococcus pyogenes*	Slopek *et al.*, 1987; Chanishvili and Sharp, 2009
Ear	*P. aeruginosa*	Weber-Dabrowska *et al.*, 2000; **Wright *et al.*, 2009**
Respiratory	*S. aureus*, *E. coli*, *Klebsiella* spp., *Proteus* spp., *P. aeruginosa*, *Streptococcus viridans*	Slopek *et al.*, 1987; Kutateladze and Adamia, 2008
Alimentary	*Salmonella* spp., *E. coli*, *Shigella* spp., *Enterococcus* spp., *Staphylococcus* spp., *P. aeruginosa*, *Proteus* spp., *Klebsiella* spp.	Knouf *et al.*, 1946; Slopek *et al.*, 1987; **Bruttin and Brussow, 2005**; Chanishvili and Sharp, 2009
Genitourinary	*P. aeruginosa*, *Staphylococcus* spp., *E. coli*, *Proteus* spp.	Slopek *et al.*, 1987; Chanishvili and Sharp, 2009
Skeletal	*Staphylococcus* spp., *P. aeruginosa*, *Proteus* spp., *Klebsiella* spp., *E. coli*	Slopek *et al.*, 1987; Weber-Dabrowska *et al.*, 2000

The dearth of Western human trials to date means that the work referenced here is primarily from the Eliava (Kutateladze and Adamia, 2008; Chanishvili and Sharp, 2009) and Hirszfeld (Slopek *et al.*, 1987; Weber-Dabrowska *et al.*, 2000) Institutes. Western trial references are shown in bold. Further details of Hirszfeld Institute human trials can be found in Slopek *et al.* (1983a,b, 1984, 1985a,b,c). Some details of other human applications, research and clinical trials and other studies are given by Häusler (2006) and Abedon (2011a,b). This table, however, illustrates some of the wide range of potential applications for phage therapy.

A recent report described small-scale production of a purified phage preparation for use in an as yet unreported clinical trial to be carried out with the Belgian military (Merabishvili *et al.*, 2009). While the trial was for burn patients (a wound), the techniques and regulatory processes are of great importance in paving the way for production of phage preparations for potentially any application, although more stringent requirements may be imposed for parenteral phage use. Conversely, a relatively crude preparation was used by Nestlé for oral administration in a test of phage safety (Bruttin and Brussow, 2005). T4 phage was prepared simply by amplification, filtration and density-gradient ultracentrifugation before being given to volunteers orally. No adverse effects were reported and the company is now moving ahead with a large trial of phage treatment of *E. coli* diarrhoea in children in Bangladesh using an in-house cocktail comprised of T4-like phages as well as a commercially supplied phage cocktail and standard-of-care treatments (http://clinicaltrials.gov/ct2/show/NCT00937274).

Another phase I clinical trial of phage therapy has been carried out, using a cocktail of phages active against several bacterial species, in this case of chronically infected wounds (Rhoads *et al.*, 2009). No adverse effects due to phage treatment were reported after topical administration of the cocktail (see also Loc-Carrillo *et al.*, Chapter 13, this volume).

As mentioned above, a canine otitis trial in the UK (Hawkins *et al.*, 2010) was carried out as a precursor to clinical trials regulated by the Medicines and Healthcare Products Regulatory Agency in the UK (Wright *et al.*, 2009). This is the first and currently the only phase II trial of phage therapy to have been carried out under full Western regulation. As a placebo-controlled, double-blinded, randomized trial, this is the first work to truly bring phage therapy to the levels of scrutiny required for clinical acceptance in the West. A total of 24 patients were recruited with chronic otitis of *Pseudomonas* aetiology. The infecting bacteria were demonstrated to be susceptible to the cocktail prior to the beginning of the trial. As in the canine otitis trial, a single dose of 600,000 phages was administered into the ear canal, and clinical assessments made on days 7, 21 and 42, including microbiological analyses and physician and patient assessments. Both the test and control groups received ear cleaning at every visit and the results were the first human clinical data to demonstrate to Western regulators that phage therapy could be superior to existing therapies in reducing symptoms and bacterial load. In two patients, phage application led to bacterial titres falling below the detection limit within 7 days and remained undetectable up to 42 days post-treatment. In three patients, including the two whose bacterial infection had resolved, phage treatment led to complete resolution of the infection. Physician- and patient-assessed scores, as well as microbiological analysis, demonstrated the efficacy of phage treatment compared with the placebo. The results of this trial were sufficient to allow an extension of this work into phase III trials in Europe. If successful, then this trial should pave the way for full regulatory approval.

Additional issues

Should phage therapy successfully complete clinical trials and become a regular treatment for bacterial disease, then the issue of phage resistance and maintenance of phage preparations will come to the fore. It is likely that bacteria will eventually become resistant to any phage preparation, as they have to antibiotics. Do we therefore attempt to create phage preparations that will be broadly effective both geographically and temporally, or instead should the focus be on development of bespoke approaches that tailor phages for individuals or small groups of patients? These issues are of importance and are already being discussed by phage-therapy researchers and companies (Pirnay *et al.*, 2011). However, within the existing Western regulatory framework, it is unlikely that bespoke approaches will be acceptable, at least in the near term. As mentioned above, it has been suggested that *in vitro* phage activities may be poorly predictive of *in vivo* activity (D.R. Harper, AmpliPhi Biosciences, UK, J.A. Fralick, Texas Tech University Health Sciences Center, USA and L. Debarbieux, Institut Pasteur, France, personal communication) and some experimental work is emerging to corroborate these observations (Bull *et al.*, 2010; see also Abedon, 2012). If this is the case, we will need superior tools to predict the pharmacodynamic properties of therapeutic phages.

The route of administration is also important to consider. The applications discussed above used multiple ways to apply phages, and phages are partitioned differently depending on how they are applied (McVay *et al.*, 2007). Debarbieux reported the efficacy of phage treatment of *P. aeruginosa* lung infections in mice, as discussed above (Debarbieux *et al.*, 2010; Morello *et al.*, 2011). This model, however, is probably not sufficient to truly mimic phage treatment of the cystic fibrosis lung, which is severely colonized by mature, polymicrobial biofilm (Harrison, 2007; Hogardt and Heesemann, 2010; Høiby *et al.*, 2010; see also Shen *et al.*, Chapter 15, this volume), as opposed to recently applied bacterial cells of a single species. In response to this potential need, research has been carried out to find mechanisms to aerosolize phages as dried particulates for inhalational delivery into the deep lung space (Golshahi *et al.*, 2008). Furthermore, what degree of phage purification will be required for a given application? Some authors have purified their phages by the gold standard method of caesium chloride density-gradient ultracentrifugation (Biswas *et al.*, 2002), while others have used alternative methods of purification (e.g. Merabishvili *et al.*, 2009; Wright *et al.*, 2009). The degree of purification may depend on the route of administration, presumably with parenteral administration requiring far more highly purified phages than oral applications, for example (see also Gill and Hyman, 2010).

Is it possible to use temperate phages for therapy? If no virulent phages can be found for a given bacterial pathogen, as is currently the case for *C. difficile*, then it could be possible to use temperate phages that have been modified to block their ability to lysogenize their hosts. One company, the UK-based Novolytics, has already applied for a patent for such an approach (Rapson *et al.*, 2008), although trials have not been reported to date. Considering that so many temperate phages augment the pathogenicity of their hosts (see Christie *et al.*, Chapter 4, this volume), it is unlikely that this approach would be used for all but the most extreme circumstances (see Abedon, Chapter 17, this volume).

One of the most valued traits of phages is the ease with which they can be manipulated *in vitro*. As mentioned above, directed evolution has been used successfully to alter the pharmacokinetic properties of bacteriophage λ and P22 such that they remain longer in the bloodstream of mice (Merril *et al.*, 1996). The huge global diversity of phages, combined with phage-display technology (see Siegel, Chapter 8, this volume) and other molecular approaches raises the possibility that phages that target specific tissues or body compartments could be selected for. Molecular techniques have also been used to augment phage antibiofilm properties (Lu and Collins, 2007) or alter host-cell antibiotic sensitivity (Lu and Collins, 2009), which could extend the efficacy of phage therapy as either a stand-alone or a combined therapy, albeit at the cost of complicating the regulatory approval pathway (see also Goodridge, 2010, and Shen *et al.*, Chapter 15, this volume).

Phages can have unpredicted effects on antibiotic sensitivity (reviewed briefly by Heinemann *et al.*, 2000) and antibiotics can cause changes in how phages propagate (Comeau *et al.*, 2007), so it is too early to predict exactly how phage therapy will work in combination with antibiotics. It is hoped that reducing antibiotic use, and thereby reducing the selective pressure on bacteria to maintain antibiotic resistance, will reinstate a certain degree of antibiotic susceptibility, and there may be evidence to support this (Kutateladze and Adamia, 2008). Alternatively, one group has developed a combinatorial approach to greatly expand the host range of a single phage by creating a highly diverse library of phages with modified host ranges (Pouillot *et al.*, 2010). Currently, Western regulators require that any phage formulation should be trialled as it is intended to be used, so maintenance of the phage preparation will require dialogue with the regulators and perhaps using regulatory pathways such as those used for vaccines that permit some modification of the formulation without the need for clinical trials. Otherwise, addition of new phages to a formula to maintain efficacy if bacteria become resistant could mean that the new formulation will have to be tested again through clinical trials. If a small number of phages that have already been shown to be safe for use can be modified combinatorially to expand their host range, then such requirements might be avoided.

Conclusions

Phage therapy still has a long way to go before it becomes a common or preferred treatment modality for a large range of bacterial diseases. It is likely that, even after Western regulatory approval, phage therapy will be used initially for refractory diseases. Antibiotic resistance is still on the rise, however, and the emerging paradigm of bacterial biofilms also represents a major issue for antibiotics. Ultimately, this may lead clinicians to prefer phages to current conventional treatments, even for some antibiotic-susceptible bacterial isolates (e.g. where it is difficult to maintain adequate antibiotic concentrations, as is the case with many eye infections). The long history of phage therapy may finally result, given regulatory approval, in a treatment regimen of worldwide appeal, with proven efficacy and the ability to combat bacterial diseases for at least another century.

References

Abedon, S.T. (2010) The 'nuts and bolts' of phage therapy. *Current Pharmaceutical Biotechnology* 11, 1.

Abedon, S. (2011a) Phage therapy pharmacology: calculating phage dosing. *Advances in Applied Microbiology* 77, 1–40.

Abedon, S. (2011b) *Bacteriophages and Biofilms: Ecology, Phage Therapy, Plaques.* Nova Science Publishers, New York.

Abedon, S.T. (2012) Bacteriophages as drugs: the pharmacology of phage therapy. In: Borysowski, J., Miêdzybrodzki, R. and Górski, A. (eds) *Phage Therapy: Current Research and Applications.* Caister Academic Press, UK.

Abedon, S.T. and Thomas-Abedon, C. (2010) Phage therapy pharmacology. *Current Pharmaceutical Biotechnology* 11, 28–47.

Abedon, S.T., Kuhl, S.J., Blasdel, R.G. and Kutter, E.M. (2011) Phage treatment of human infections. *Bacteriophage* 1, 66–85.

Asheshov, I., Khan, S. and Lahiri, M. (1931) The treatment of cholera with bacteriophage. *Indian Medical Gazette* 66, 179–184.

Asheshov, I.N., Asheshov, I., Khan, S. and Lahiri, M. (1930) Bacteriophage inquiry. Report on the work during the period from 1st January to 1st September 1929. *Indian Journal of Medical Research* 17, 971–984.

Babalova, E., Katsitadze, K., Sakvarelidze, L., Imnaishvili, N.S., Sharashidze, T., Badashvili, V., Kiknadze, G., MeÄpariani, A., Gendzekhadze, N. and Machavariani, E. (1968) Preventive value of dried dysentery bacteriophage. *Zhurnal Mikrobiologii, Epidemiologii, i Immunobiologii* 45, 143–145.

Bhowmick, T.S., Koley, H., Das, M., Saha, D.R. and Sarkar, B. (2009) Pathogenic potential of vibriophages against an experimental infection with *Vibrio cholerae* O1 in the RITARD model. *International Journal of Antimicrobial Agents* 33, 569–573.

Biswas, B., Adhya, S., Washart, P., Paul, B., Trostel, A.N., Powell, B., Carlton, R. and Merril, C.R. (2002) Bacteriophage therapy rescues mice bacteremic from a clinical isolate of vancomycin-resistant *Enterococcus faecium. Infection and Immunity* 70, 204–210.

Boucher, H.W., Talbot, G.H., Bradley, J.S., Edwards, J.E., Gilbert, D., Rice, L.B., Scheld, M., Spellberg, B. and Bartlett, J. (2009) Bad bugs, no drugs: no ESKAPE! An update from the Infectious Diseases Society of America. *Clinical Infectious Diseases* 48, 1–12.

Boyce, J.M., Jackson, M.M., Pugliese, G., Batt, M.D., Fleming, D., Garner, J.S., Hartstein, A.I., Kauffman, C.A., Simmons, M. and Weinstein, R. (1994) Methicillin-resistant *Staphylococcus aureus* (MRSA): a briefing for acute care hospitals and nursing facilities. *Infection Control and Hospital Epidemiology* 15, 105–115.

Bruttin, A. and Brussow, H. (2005) Human volunteers receiving *Escherichia coli* phage T4 orally: a safety test of phage therapy. *Antimicrobial Agents and Chemotherapy* 49, 2874–2878.

Bull, J., Vimr, E. and Molineux, I. (2010) A tale of tails: sialidase is key to success in a model of phage therapy against K1-capsulated *Escherichia coli. Virology* 398, 79–86.

Carmody, L.A., Gill, J.J., Summer, E.J., Sajjan, U.S., Gonzalez, C.F., Young, R.F. and LiPuma, J.J. (2010) Efficacy of bacteriophage therapy in a model of *Burkholderia cenocepacia* pulmonary infection. *Journal of Infectious Diseases* 201, 264–271.

Chan, B. and Abedon, S. (2012) Phage therapy pharmacology: phage cocktails. *Advances in Applied Microbiology* 78 (in press)

Chanishvili, N. and Sharp, R. (2009) *A Literature Review of the Practical Application of Bacteriophage Research.* Eliava Institute of Bacteriophages, Microbiology and Virology, Tbilisi, Georgia.

Chhibber, S., Kaur, S. and Kumari, S. (2008) Therapeutic potential of bacteriophage in treating *Klebsiella pneumoniae* B5055-mediated lobar pneumonia in mice. *Journal of Medical Microbiology* 57, 1508–1513.

Chibani-Chennoufi, S., Sidoti, J., Bruttin, A., Kutter, E., Sarker, S. and Brussow, H. (2004) In vitro and in vivo bacteriolytic activities of *Escherichia coli* phages: implications for phage therapy. *Antimicrobial Agents and Chemotherapy* 48, 2558–2569.

Clokie, M.R.J. and Kropinski, A.M.B. (2009) *Bacteriophages: Methods and Protocols: Molecular and Applied Aspects.* Humana Press, Totowa, NJ.

Comeau, A.M., Tétart, F., Trojet, S.N., Prère, M.F. and Krisch, H. (2007) Phage–antibiotic synergy (PAS): β-lactam and quinolone antibiotics stimulate virulent phage growth. *PLoS One* 2, e799.

Curtright, A. and Abedon, S. (2011) Phage therapy: emergent property pharmacology. *Journal of Bioanalysis and Biomedicine* S6, doi: 10.4172/1948-593X.S6-002 (Epub ahead of print).

Dabrowska, K., Switała-Jelen, K., Opolski, A., Weber-Dabrowska, B. and Górski, A. (2005) Bacteriophage penetration in vertebrates. *Journal of Applied Microbiology* 98, 7–13.

Debarbieux, L., Leduc, D., Maura, D., Morello, E., Criscuolo, A., Grossi, O., Balloy, V. and Touqui, L. (2010) Bacteriophages can treat and prevent *Pseudomonas aeruginosa* lung infections. *Journal of Infectious Diseases* 201, 1096–1104.

Denou, E., Bruttin, A., Barretto, C., Ngom-Bru, C.,

Brüssow, H. and Zuber, S. (2009) T4 phages against *Escherichia coli* diarrhea: potential and problems. *Virology* 388, 21–30.
d'Hérelle, F. (1921) Le microbe bactériophage, agent d'immunité dans la peste et la barbone. *Comptes Rendus de l'Académie des Sciences* 172, 99–100.
d'Hérelle, F. (1922) *The Bacteriophage, its Role in Immunity*. Williams & Wilkins, Baltimore, MD.
d'Hérelle, F. (1929) Studies upon asiatic cholera. *Yale Journal of Biology and Medicine* 1, 195–219.
d'Hérelle, F. (2007) On an invisible microbe antagonistic toward dysenteric bacilli: brief note by Mr. F. D'Hérelle, presented by Mr. Roux. 1917. *Research in Microbiology* 158, 553–554.
d'Hérelle, F., Malone, R. and Lahiri, M. (1928) The treatment and prophylaxis of infectious diseases of the intestinal tract and of cholera in particular. In: *Transaction of the 7th Congress held in British India, December 1927, Far Eastern Association of Tropical Medicine*, Vol. II. Thacker, Calcutta, pp. 284–287.
Eaton, M.D. and Bayne-Jones, S. (1934) Bacteriophage therapy. *Journal of the American Medical Association* 103, 1847–1853.
Geier, M.R., Trigg, M.E. and Merril, C.R. (1973) Fate of bacteriophage lambda in non-immune germ-free mice. *Nature* 246, 221–222.
Gill, J., Pacan, J., Carson, M., Leslie, K., Griffiths, M. and Sabour, P. (2006) Efficacy and pharmacokinetics of bacteriophage therapy in treatment of subclinical *Staphylococcus aureus* mastitis in lactating dairy cattle. *Antimicrobial Agents and Chemotherapy* 50, 2912–2918.
Gill, J.J. and Hyman, P. (2010) Phage choice, isolation and preparation for phage therapy. *Current Pharmaceutical Biotechnology* 11, 2–14.
Golshahi, L., Seed, K.D., Dennis, J.J. and Finlay, W.H. (2008) Toward modern inhalational bacteriophage therapy: nebulization of bacteriophages of *Burkholderia cepacia* complex. *Journal of Aerosol Medicine and Pulmonary Drug Delivery* 21, 351–360.
Goodridge, L.D. (2010) Designing phage therapeutics. *Current Pharmaceutical Biotechnology* 11, 15–27.
Govind, R., Vediyappan, G., Rolfe, R.D., Dupuy, B. and Fralick, J.A. (2009) Bacteriophage-mediated toxin gene regulation in *Clostridium difficile*. *Journal of Virology* 83, 12037–12045.
Harrison, F. (2007) Microbial ecology of the cystic fibrosis lung. *Microbiology* 153, 917–923.
Häusler, T. (2006) *Viruses vs. Superbugs: a Solution to the Antibiotics Crisis?* Palgrave Macmillan, Hampshire, UK.
Hawkins, C., Harper, D., Burch, D., Anggard, E. and Soothill, J. (2010) Topical treatment of *Pseudomonas aeruginosa* otitis of dogs with a bacteriophage mixture: a before/after clinical trial. *Veterinary Microbiology* 146, 309–313.
Heinemann, J.A., Ankenbauer, R.G. and Amabile-Cuevas, C.F. (2000) Do antibiotics maintain antibiotic resistance? *Drug Discovery Today* 5, 195–204.
Heo, Y.J., Lee, Y.R., Jung, H.H., Lee, J.E., Ko, G.P. and Cho, Y.H. (2009) Antibacterial efficacy of phages against *Pseudomonas aeruginosa* infections in mice and *Drosophila melanogaster*. *Antimicrobial Agents and Chemotherapy* 53, 2469–2474.
Hogardt, M. and Heesemann, J. (2010) Adaptation of *Pseudomonas aeruginosa* during persistence in the cystic fibrosis lung. *International Journal of Medical Microbiology* 300, 557–562.
Høiby, N., Ciofu, O. and Bjarnsholt, T. (2010) *Pseudomonas aeruginosa* biofilms in cystic fibrosis. *Future Microbiology* 5, 1663–1674.
Jarvis, W.R. (2010) Prevention and control of methicillin-resistant *Staphylococcus aureus*: dealing with reality, resistance and resistance to reality. *Clinical Infectious Diseases* 50, 218–220.
Knouf, E.G., Ward, W.E., Reichle, P.A., Bower, A. and Hamilton, P.M. (1946) Treatment of typhoid fever with type specific bacteriophage. *Journal of the American Medical Association* 132, 134–138.
Krueger, A. and Scribner, E. (1941) The bacteriophage: its nature and its therapeutic use. *Journal of the American Medical Association* 116, 2269–2277.
Kutateladze, M. and Adamia, R. (2008) Phage therapy experience at the Eliava Institute. *Medecine et Maladies Infectieuses* 38, 426–430.
Kutter, E. and Sulakvelidze, A. (2005) *Bacteriophages: Biology and Applications*. CRC Press, Boca Raton, FL.
Kutter, E., de Vos, D., Gvasalia, G., Alavidze, Z., Gogokhia, L., Kuhl, S. and Abedon, S.T. (2010) Phage therapy in clinical practice: treatment of human infections. *Current Pharmaceutical Biotechnology* 11, 69–86.
Levin, B.R. and Bull, J. (1996) Phage therapy revisited: the population biology of a bacterial infection and its treatment with bacteriophage and antibiotics. *American Naturalist* 881–898.
Levin, B.R. and Bull, J.J. (2004) Population and evolutionary dynamics of phage therapy. *Nature Reviews Microbiology* 2, 166–173.
Lu, T.K. and Collins, J.J. (2007) Dispersing biofilms with engineered enzymatic bacteriophage.

Proceedings of the National Academy of Sciences USA 104, 11197–11202.

Lu, T.K. and Collins, J.J. (2009) Engineered bacteriophage targeting gene networks as adjuvants for antibiotic therapy. *Proceedings of the National Academy of Sciences USA.* 106, 4629–4634.

Matsuzaki, S., Yasuda, M., Nishikawa, H., Kuroda, M., Ujihara, T., Shuin, T., Shen, Y., Jin, Z., Fujimoto, S., Nasimuzzaman, M.D., Wakiguchi, H., Sugihara, S., Sugiura, T., Koda, S., Muraoka, A. and Imai, S. (2003) Experimental protection of mice against lethal *Staphylococcus aureus* infection by novel bacteriophage φMR11. *Journal of Infectious Diseases* 187, 613–624.

McVay, C.S., Velasquez, M. and Fralick, J.A. (2007) Phage therapy of *Pseudomonas aeruginosa* infection in a mouse burn wound model. *Antimicrobial Agents and Chemotherapy* 51, 1934–1938.

Merabishvili, M., Pirnay, J.P., Verbeken, G., Chanishvili, N., Tediashvili, M., Lashkhi, N., Glonti, T., Krylov, V., Mast, J. and Van Parys, L. (2009) Quality-controlled small-scale production of a well-defined bacteriophage cocktail for use in human clinical trials. *PLoS One* 4, e4944.

Merril, C.R., Biswas, B., Carlton, R., Jensen, N.C., Creed, G.J., Zullo, S. and Adhya, S. (1996) Long-circulating bacteriophage as antibacterial agents. *Proceedings of the National Academy of Sciences USA* 93, 3188–3192.

Morello, E., Saussereau, E., Maura, D., Huerre, M., Touqui, L., Debarbieux, L. and Aziz, R. (2011) Pulmonary bacteriophage therapy on cystic fibrosis strains: first steps towards treatment and prevention. *PLoS One* 6, e16963.

Morison, J. (1932) *Bacteriophage in the Treatment and Prevention of Cholera.* H.K. Lewis, London, UK.

Payne, R.J.H. and Jansen, V.A.A. (2003) Pharmacokinetic principles of bacteriophage therapy. *Clinical Pharmacokinetics* 42, 315–325.

Pirnay, J.P., de Vos, D., Verbeken, G., Merabishvili, M., Chanishvili, N., Vaneechoutte, M., Zizi, M., Laire, G., Lavigne, R. and Huys, I. (2011) The phage therapy paradigm: Prêt-à-Porter or Sur-mesure? *Pharmaceutical Research* 28, 934–937.

Podschun, R. and Ullmann, U. (1998) *Klebsiella* spp. as nosocomial pathogens: epidemiology, taxonomy, typing methods and pathogenicity factors. *Clinical Microbiology Reviews* 11, 589–603.

Pouillot, F., Blois, H. and Iris, F. (2010) Genetically engineered virulent phage banks in the detection and control of emergent pathogenic bacteria. *Biosecurity and Bioterrorism: Biodefense Strategy, Practice, and Science* 8, 155–169.

Ramesh, V., Fralick, J.A. and Rolfe, R.D. (1999) Prevention of *Clostridium difficile*-induced ileocecitis with bacteriophage. *Anaerobe* 5, 69–78.

Rapson, M.E., Burden, F.A., Glancey, L.P., Hodgson, D.A. and Mann, N.H. (2008) US Patent Application 20080317715.

Rhoads, D.D., Wolcott, R.D., Kuskowski, M.A., Wolcott, B.M., Ward, L.S. and Sulakvelidze, A. (2009) Bacteriophage therapy of venous leg ulcers in humans: results of a phase I safety trial. *Journal of Wound Care* 18, 237–238.

Slopek, S., Durlakowa, I., Weber-Dabrowska, B., Kucharewicz-Krukowska, A., Dabrowski, M. and Bisikiewicz, R. (1983a) Results of bacteriophage treatment of suppurative bacterial infections. I. General evaluation of the results. *Archivum Immunologiae et Therapiae Experimentalis* 31, 267–291.

Slopek, S., Durlakowa, I., Weber-Dabrowska, B., Kucharewicz-Krukowska, A., Dabrowski, M. and Bisikiewicz, R. (1983b) Results of bacteriophage treatment of suppurative bacterial infections. II. Detailed evaluation of the results. *Archivum Immunologiae et Therapiae Experimentalis* 31, 293–327.

Slopek, S., Durlakowa, I., Weber-Dabrowska, B., Dabrowski, M. and Kucharewicz-Krukowska, A. (1984) Results of bacteriophage treatment of suppurative bacterial infections. III. Detailed evaluation of the results obtained in further 150 cases. *Archivum Immunologiae et Therapiae Experimentalis* 32, 317–335.

Slopek, S., Kucharewicz-Krukowska, A., Weber-Dabrowska, B. and Dabrowski, M. (1985a) Results of bacteriophage treatment of suppurative bacterial infections. VI. Analysis of treatment of suppurative staphylococcal infections. *Archivum Immunologiae et Therapiae Experimentalis* 33, 261–273.

Slopek, S., Kucharewicz-Krukowska, A., Weber-Dabrowska, B. and Dabrowski, M. (1985b) Results of bacteriophage treatment of suppurative bacterial infections. V. Evaluation of the results obtained in children. *Archivum Immunologiae et Therapiae Experimentalis* 33, 241–259.

Slopek, S., Kucharewicz-Krukowska, A., Weber-Dabrowska, B. and Dabrowski, M. (1985c) Results of bacteriophage treatment of suppurative bacterial infections. IV. Evaluation of the results obtained in 370 cases. *Archivum Immunologiae et Therapiae Experimentalis* 33, 219–240.

Slopek, S., Weber-Dabrowska, B., Dabrowski, M.

and Kucharewicz-Krukowska, A. (1987) Results of bacteriophage treatment of suppurative bacterial infections in the years 1981–1986. *Archivum Immunologiae et Therapiae Experimentalis* 35, 569–583.

Smith, H.W. and Huggins, M. (1982) Successful treatment of experimental *Escherichia coli* infections in mice using phage: its general superiority over antibiotics. *Journal of General Microbiology* 128, 307–318.

Smith, H.W. and Huggins, M. (1983) Effectiveness of phages in treating experimental *Escherichia coli* diarrhoea in calves, piglets and lambs. *Journal of General Microbiology* 129, 2659–2675.

Smith, H.W., Huggins, M.B. and Shaw, K.M. (1987a) The control of experimental *Escherichia coli* diarrhoea in calves by means of bacteriophages. *Journal of General Microbiology* 133, 1111–1126.

Smith, H.W., Huggins, M.B. and Shaw, K.M. (1987b) Factors influencing the survival and multiplication of bacteriophages in calves and in their environment. *Journal of General Microbiology* 133, 1127–1135.

Smith, J. (1924) The bacteriophage in the treatment of typhoid fever. *British Medical Journal* 2, 47–49.

Soothill, J. (1992) Treatment of experimental infections of mice with bacteriophages. *Journal of Medical Microbiology* 37, 258–261.

Soothill, J. (1994) Bacteriophage prevents destruction of skin grafts by *Pseudomonas aeruginosa*. *Burns* 20, 209–211.

Soothill, J., Lawrence, J. and Ayliffe, G. (1988) The efficacy of phages in the prevention of the destruction of pig skin *in vitro* by *Pseudomonas aeruginosa*. *Medical Science Research* 16, 1287–1288.

Sulakvelidze, A., Alavidze, Z. and Morris, J.G. Jr (2001) Bacteriophage therapy. *Antimicrobial Agents and Chemotherapy* 45, 649–659.

Summers, W.C. (1993) Cholera and plague in India: the bacteriophage inquiry of 1927–1936. *Journal of the History of Medicine and Allied Sciences* 48, 275–301.

Summers, W.C. (1999) *Félix d'Herelle and the Origins of Molecular Biology*. Yale University Press, New Haven, CT.

Summers, W.C. (2001) Bacteriophage therapy. *Annual Reviews in Microbiology* 55, 437–451.

Thiel, K. (2004) Old dogma, new tricks – 21st century phage therapy. *Nature Biotechnology* 22, 31–36.

Verbeken, G., de Vos, D., Vaneechoutte, M., Merabishvili, M., Zizi, M. and Pirnay, J.P. (2007) European regulatory conundrum of phage therapy. *Future Microbiology* 2, 485–491.

Vinodkumar, C., Neelagund, Y. and Kalsurmath, S. (2005) Bacteriophage in the treatment of experimental septicemic mice from a clinical isolate of multidrug resistant *Klebsiella pneumoniae*. *Journal of Communicable Diseases* 37, 18–29.

Vinodkumar, C., Kalsurmath, S. and Neelagund, Y. (2008) Utility of lytic bacteriophage in the treatment of multidrug-resistant *Pseudomonas aeruginosa* septicemia in mice. *Indian Journal of Pathology and Microbiology* 51, 360–366.

Wainwright, M. and Swan, H.T. (1986) C.G. Paine and the earliest surviving clinical records of penicillin therapy. *Medical History* 30, 42–56.

Waksman, S.A., Reilly, H.C. and Schatz, A. (1945) Strain specificity and production of antibiotic substances: V. strain resistance of bacteria to antibiotic substances, especially to streptomycin. *Proceedings of the National Academy of Sciences USA* 31, 157–164.

Wang, J., Hu, B., Xu, M., Yan, Q., Liu, S., Zhu, X., Sun, Z., Reed, E., Ding, L. and Gong, J. (2006a) Use of bacteriophage in the treatment of experimental animal bacteremia from imipenem-resistant *Pseudomonas aeruginosa*. *International Journal of Molecular Medicine* 17, 309–317.

Wang, J., Hu, B., Xu, M., Yan, Q., Liu, S., Zhu, X., Sun, Z., Tao, D., Ding, L. and Reed, E. (2006b) Therapeutic effectiveness of bacteriophages in the rescue of mice with extended spectrum β-lactamase-producing *Escherichia coli* bacteremia. *International Journal of Molecular Medicine* 17, 347–355.

Watanabe, R., Matsumoto, T., Sano, G., Ishii, Y., Tateda, K., Sumiyama, Y., Uchiyama, J., Sakurai, S., Matsuzaki, S. and Imai, S. (2007) Efficacy of bacteriophage therapy against gut-derived sepsis caused by *Pseudomonas aeruginosa* in mice. *Antimicrobial Agents and Chemotherapy* 51, 446–452.

Weber-Dabrowska, B., Mulczyk, M. and Górski, A. (2000) Bacteriophage therapy of bacterial infections: an update of our institute's experience. *Archivum Immunologiae et Therapiae Experimentalis* 48, 547–551.

Weber-Dabrowska, B., Mulczyk, M. and Górski, A. (2003) Bacteriophages as an efficient therapy for antibiotic-resistant septicemia in man. *Transplantation Proceedings* 35, 1385–1386.

Wright, A., Hawkins, C., Änggård, E. and Harper, D. (2009) A controlled clinical trial of a therapeutic bacteriophage preparation in chronic otitis due to antibiotic resistant *Pseudomonas aeruginosa*; a preliminary report of efficacy. *Clinical Otolaryngology* 34, 349–357.

15 Phage-based Enzybiotics

Yang Shen[1], Michael S. Mitchell[1], David M. Donovan[2] and Daniel C. Nelson[1]
[1]Institute for Bioscience and Biotechnology Research, University of Maryland;
[2]Animal Biosciences and Biotechnology Lab, ANRI, ARS, USDA.

The term 'enzybiotic' was coined by Vincent Fischetti's group in 2001 to describe both the enzymatic and antibiotic properties of bacteriophage-encoded endolysins (Nelson *et al.*, 2001). Endolysins are peptidoglycan hydrolases that function to lyse the bacterial cell wall for release of progeny virions during the phage lytic cycle. Timing of lysis is initiated by holins, which permeabilize the bacterial membrane and allow the cytoplasmically accumulated endolysins access to the peptidoglycan (Young, 1992; Wang *et al.*, 2000). Significantly, in the absence of holins or parental bacteriophage, these enzymes can be used exogenously to lyse the peptidoglycan of susceptible organisms. Once the structural peptidoglycan is compromised, internal turgor pressure, measured at 20–50 atmospheres for Gram-positive organisms (Whatmore and Reed, 1990; Doyle and Marquis, 1994; Arnoldi *et al.*, 2000), causes a rapid osmotic lysis of the bacterial membrane resulting in cell death. Non-endolysin enzybiotics, such as virion-associated enzymes, have also been identified (Takac and Blasi, 2005; Rodriguez *et al.*, 2011). With few exceptions, this enzyme-mediated 'lysis from without' phenomenon – a term that has been used to describe a variety of phenomena in which an extracellular agent destroys a bacterial cell envelope (Abedon, 2011a) – is restricted to Gram-positive species, as the Gram-negative peptidoglycan is covered by a protective outer membrane that is not permeable to an exogenous enzyme under normal conditions. None the less, enzybiotics encoded by Gram-positive phage represent an attractive therapeutic option (Loessner, 2005; Borysowski *et al.*, 2006; Donovan, 2007; Hermoso *et al.*, 2007; Fischetti, 2010).

In recent years, the enzybiotic moniker has been extended by some authors to describe any enzyme that displays general antibacterial or antifungal properties, whether phage derived or not (Veiga-Crespo *et al.*, 2007). These include a wide range of enzymes such as autolysins and exolysins. Autolysins are peptidoglycan hydrolases encoded by bacteria for growth, division and repair of the bacterial peptidoglycan. Like their phage-encoded endolysin counterparts, some autolysins can produce lysis from without when administered in sufficient quantities. For example, the major pneumococcal autolysin, LytA, was shown to effect a 5 log drop in bacterial counts 4 h after intraperitoneal injection in a mouse model of pneumococcal septicaemia (Rodriguez-Cerrato *et al.*, 2007). Likewise, fungal glucanases and chitinases could be considered autolysins with a potential for therapeutic use (Veiga-Crespo and Villa, 2010). In contrast to endolysins and autolysins, an exolysin is an enzyme secreted by a bacterial cell to lyse the peptidoglycan of a different strain or species. One of the most-studied bacterial exolysins is lysostaphin, a peptidoglycan hydrolase secreted by *Staphylococcus simulans* that cleaves the *Staphylococcus aureus* cell wall but does not hydrolyse the *S. simulans* cell wall (Schindler and Schuhardt, 1964). This enzyme has been investigated for therapeutic efficacy against *S. aureus* in such diverse animal models as burn infections (Cui *et al.*, 2011), ocular infections (Dajcs *et al.*, 2001, 2002), systemic infections (Kokai-Kun *et al.*, 2007), keratitis models (Dajcs *et al.*, 2000), nasal colonization (Kokai-Kun *et al.*, 2003) and aortic valve endocarditis (Climo *et al.*, 1998; Patron *et al.*, 1999). Similarly, zoocin A is an exolysin secreted by *Streptococcus equi* subs. *zooepidemicus* that hydrolyses the peptidoglycan of *Streptococcus pyogenes* (Akesson *et al.*, 2007). In addition to bacterial exolysins, eukaryotic cells can secrete their own exolysins. For example, lysozymes found in human tears and saliva are eukaryotic exolysins that are part of the innate immune system providing protection against bacterial invasion.

While the most comprehensive definition of 'enzybiotics' would include all endolysins, autolysins, exolysins and perhaps other enzymes, we will focus in this chapter on the phage-encoded enzymes, primarily the endolysins. Current knowledge about the mechanisms of action for these enzymes, *in vitro* and *in vivo* activity, synergy with other enzymes or antibiotics, immune responses and resistance to these enzymes will be discussed. In addition, at the end of the chapter, we will briefly review bacteriocins and depolymerases, two other classes of phage-encoded enzymes that may also be considered enzybiotics.

Endolysins

History

The history of endolysins dates back to the discovery of bacteriophages themselves. It has been well documented in the literature that Frederick Twort in 1915 and Felix d'Hérelle in 1917 are credited with independently discovering bacteriophages. What is rarely mentioned in historical reviews, however, is that Twort's seminal manuscript also contained perhaps the first evidence for phage endolysins. Specifically, he noted that, in addition to the transmissible properties of the ultramicroscopic agent (i.e. a staphylococcal phage), there also appeared to be a non-transmissible, heat-labile property responsible for producing transparent zones of lysis (Twort, 1915). In hindsight, it is likely that Twort was observing lysis from without, attributable to either the endolysin or a virion-associated enzyme of the staphylococcal phage. By 1925, he had theorized that, by definition, a 'transmissible virus' will only act on live bacteria whereas a non-transmissible 'bacterial lysin' secreted by the virus should also act on dead bacteria (Twort, 1925). He then demonstrated that dead staphylococci could not be lysed by phage unless a small amount of live staphylococci was added, which liberated some nascent agent.

In 1926, Reynals not only confirmed Twort's lytic findings on dead staphylococci but also showed that this phenomenon is restricted to Gram-positive organisms. When identical experiments were performed on *Escherichia* or *Shigella* species with their corresponding phage, no lysis of dead cells was shown, even in the presence of susceptible live cells (Reynals, 1926). By 1933, Rakieten reported that staphylococcal strains resistant to infection by a staphylococcal phage were efficiently lysed when a small amount of sensitive staphylococci was added (Rakieten, 1933). He concluded that this was due to a non-transmissible, 'bacteriophage lysin', which was presumably produced during the process of 'bacteriophagy'. By the end of the 1930s, Alice Evans had shown that lysis from without was not unique to staphylococci. In several elegant papers, she showed that phage B only infected group C streptococci but its 'potent principle' could lyse group A, C and E streptococci (Evans, 1934, 1936, 1940). More than 30 years later, Vincent Fischetti would purify Evans' potent principle (i.e. an endolysin) (Fischetti *et al.*, 1971) and, eventually, his laboratory would use it in the first *in vivo* therapeutic studies of a bacteriophage endolysin, henceforth known as an enzybiotic (Nelson *et al.*, 2001).

Basic cell-wall architecture and types of endolysin activities

The bacterial peptidoglycan is an essential scaffold found on the outside of the cytoplasmic membrane of most bacteria. It functions to maintain a rigid structure and preserve cell integrity. As its name implies, the main structural features of peptidoglycan are linear glycan strands cross-linked by short peptide stems, which in turn are cross-linked either directly together or through a peptide 'bridge' (Fig. 15.1). The glycan strand is a polymer of alternating *N*-acetylmuramic acid (MurNAc) and *N*-acetylglucosamine (GlcNAc) residues coupled by $\beta(1\rightarrow4)$ linkages. The peptide stem is covalently attached to the glycan polymer by an amide bond between each MurNAc and an L-alanine residue. The remainder of the peptide stem contains alternating L- and D-form amino acids. For many Gram-positive organisms, the third residue of the stem peptide is L-lysine, which is cross-linked to an opposing stem peptide on a separate glycan polymer through an interpeptide bridge, the composition of which varies among species. For example, the interpeptide bridge of *S. aureus* is pentaglycine whereas di-alanine is present in *S. pyogenes*. In Gram-negative organisms and some genera of Gram-positive bacteria (i.e. *Bacillus* and *Listeria*), a meso-diaminopimelic acid (mDAP) residue is present at position number three of the stem peptide instead of L-lysine. In these organisms, mDAP directly cross-links to the terminal D-alanine of the opposite stem peptide (i.e. there is no interpeptide bridge). Whether an interpeptide bridge is present or not, the joining of opposing stem peptides gives rise to the three-dimensional lattice that is the defining characteristic of the bacterial peptidoglycan.

There are in fact only a limited number of covalent bonds that are available for

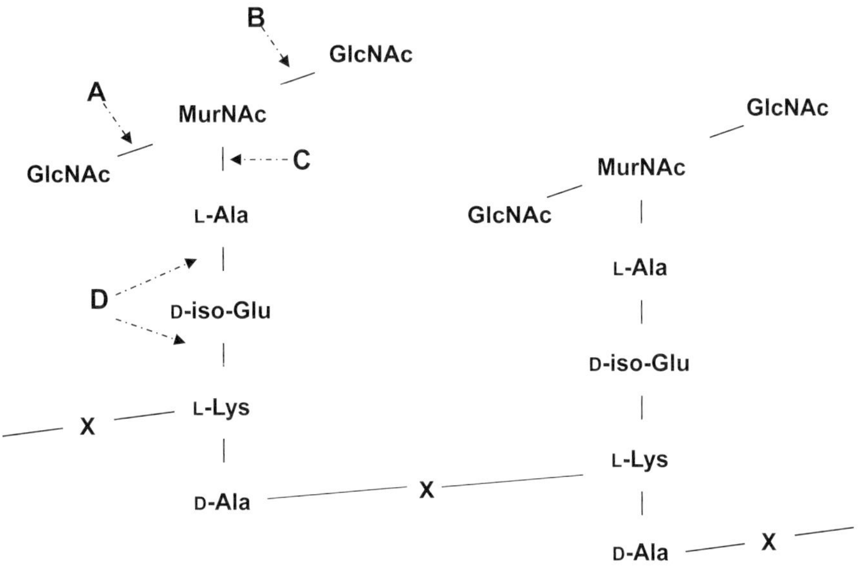

Fig. 15.1. Structure of the peptidoglycan and cleavage sites by generic endolysins. Glycosidases cleave the carbohydrate backbone. (A) *N*-Acetylglucosaminidase cleaves on the reducing side of *N*-acetylglucosamine (GlcNAc). (B) Lysozymes, also called muramidases, are *N*-acetylmuramidases that cleave on the reducing side of *N*-acetylmuramic acid (MurNAc). Another glycosidase-like enzyme that cleaves this same bond but does not require hydrolysis of water is the lytic transglycosylase. (C) The *N*-acetylmuramoyl-L-alanine amidase cleaves the first amide bond between the peptide moiety (L -alanine) and the glycan moiety (MurNAc) (D). Numerous endopeptidase cleavage sites are present on the peptidoglycan. Note, X is the transpeptide bridge, the length and composition of which varies depending on species.

cleavage by endolysins due to the reasonably conserved overall structure of the bacterial peptidoglycan. In general, endolysins (as well as autolysins and exolysins) can be classified into three groups depending on their cleavage specificity: those that cleave between two sugar residues (i.e. glycosidases and lytic transglycosylases), those that cleave between a sugar residue and an amino acid (i.e. *N*-acetylmuramoyl-L-alanine amidases) and those that cleave between two amino acids (i.e. endopeptidases), each of which is summarized in Fig. 15.1 and select examples of which are presented in Table 15.1.

Sugar-cleaving enzymes include glycosidases (i.e. *N*-acetylglucosaminidases and

Table 15.1. Examples of *in vitro* studies with Gram-positive phage endolysins.

Endolysin	Target	Catalytic activities	*In vitro* activity	Reference(s)
λSa2	*Streptococcus agalactiae*	Glycosidase + endopeptidase	Activity on purified peptidoglycan as analysed by mass spectrometry	Pritchard *et al.* (2007)
Cpl-1	*Streptococcus pneumoniae*	*N*-acetylmuramidase	5 log decrease in CFU in 30 s with 100 μg	Garcia *et al.* (1987)
PAL	*Streptococcus pneumoniae*	*N*-acetylmuramoyl-L-alanine amidase	4 log decrease in CFU in 30 s with 100 U ml^{-1}	Loeffler *et al.* (2001)
PlyGBS (aka B30)	*Streptococcus agalactiae*	*N*-acetylmuramidase + endopeptidase	2 log decrease in CFU in 60 min with 40 U	Cheng *et al.* (2005); Pritchard *et al.* (2004)
PlyC	*Streptococcus pyogenes*	*N*-acetylmuramoyl-L-alanine amidase	7 log decrease in CFU in 5 s with 10 ng	Nelson *et al.* (2001)
Ply700	*Streptococcus uberis*	*N*-acetylmuramoyl-L-alanine amidase	0.5 log decrease in CFU in milk in 15 min with 50 μg ml^{-1}	Celia *et al.* (2008)
PlyG	*Bacillus anthracis/ cereus*	*N*-acetylmuramoyl-L-alanine amidase	6 log decrease in CFU in 15 min with 20 U	Schuch *et al.* (2002)
PlyB	*Bacillus cereus*	*N*-acetylmuramidase	OD_{600} decrease from 0.45 to 0.05 with 2.5 μM; PlyG also had similar results	Porter *et al.* (2007)
Φ11	*Staphylococcus aureus*	*N*-acetylmuramoyl-L-alanine amidase + endopeptidase	OD_{600} decrease from 0.3 to 0.15 in 20 min with 20 μg ml^{-1}	Navarre *et al.* (1999)
LysK	*Staphylococcus aureus*	*N*-acetylmuramoyl-L-alanine amidases + endopeptidase	3 log decrease in CFU in 1 h with crude lysate	O'Flaherty *et al.* (2005)
ClyS	*Staphylococcus aureus*	*N*-acetylmuramoyl-L-alanine amidase	3 log decrease in CFU in 30 min with 250 μg	Daniel *et al.* (2010)
Ply511	*Listeria monocytogenes*	*N*-acetylmuramoyl-L-alanine amidase	OD_{600} decrease from 1.6 to 0.25 in 20 min with 180 U ml^{-1}	Gaeng *et al.* (2000)
PlyPSA	*Listeria monocytogenes*	*N*-acetylmuramoyl-L-alanine amidase	OD_{600} decrease from 1.0 to 0.2 in 10 min with 2.8 nmol	Korndorfer *et al.* (2006)
Ply500	*Listeria monocytogenes*	L-alanyl-D-glutamate endopeptidase	OD_{600} decrease from 1.0 to 0.2 in 1 min at 4.8 μg/ml or 3 min at 1.6 μg ml^{-1}	Schmelcher *et al.* (2011)
Ply118	*Listeria monocytogenes*	L-alanyl-D-glutamate endopeptidase	OD_{600} decrease from 1.6 to 0.5 in 20 min with 60 U ml^{-1}	Gaeng *et al.* (2000)

N-acetylmuramidases) and lytic transglycosylases. *N*-Acetylglucosaminidases cleave the glycan component of the peptidoglycan on the reducing side of the GlcNAc residue ('A' in Fig. 15.1). This type of activity, which is commonly found in autolysins, does not appear to be prevalent in endolysins. None the less, an *N*-acetylglucosaminidase has been described as one of two catalytic activities associated with the streptococcal λSa2 endolysin (Pritchard *et al.*, 2007). In contrast, an *N*-acetylmuramidase, more commonly known as a lysozyme or muramidase, cleaves the glycan on the reducing side of the MurNac residue ('B' in Fig. 15.1) and is commonly associated with endolysins. Examples include the pneumococcal Cpl-1 endolysin (Garcia *et al.*, 1987) and the streptococcal B30 endolysin (Pritchard *et al.*, 2004). Lytic transglycosylases are very similar to *N*-acetylmuramidases in that they also cleave the β(1→4) linkages between MurNAc and GlcNAc. However, lytic transglycosylases do not require water to catalyse the reaction whereas *N*-acetylmuramidases are hydrolases (i.e. they require water for bond lysis). The λ endolysin (Taylor and Gorazdowska, 1974) and the gp144 endolysin from the ΦKZ bacteriophage (Paradis-Bleau *et al.*, 2007) are two examples of known phage lytic transglycosylases.

The second mechanistic class includes the *N*-acetylmuramoyl-L-alanine amidase. This enzyme is often referred to simply as an 'amidase', which has become confusing in the literature because the term amidase suggests any type of generalized protease activity. In reality, this enzyme is an amidohydrolase that only cleaves a specific amide bond between the glycan component (MurNac) and the peptide component (L-alanine) of the peptidoglycan ('C' in Fig. 15.1). The *N*-acetylmuramoyl-L-alanine amidase activity has been demonstrated for the staphylococcal endolysins Φ11 (Navarre *et al.*, 1999) and LysK (Becker *et al.*, 2009a; Donovan *et al.*, 2009), the *Listeria* endolysins Ply511 (Loessner *et al.*, 1995) and PlyPSA (Korndorfer *et al.*, 2006) and the streptococcal endolysin PlyC (Nelson *et al.*, 2006).

The third class is defined by a protease, or endopeptidase, activity ('D' in Fig. 15.1). Endopeptidases could also be described as amidases, but in contrast to the *N*-acetylmuramoyl-L-alanine amidase above, endopeptidases hydrolyse an amide bond between two amino acids, which defines a peptide bond. These cleavages often occur in the stem peptide, such as the *Listeria* Ply500 and Ply118 L-alanyl-D-glutamate endolysins (Loessner *et al.*, 1995), or in the interpeptide bridge, such as the staphylococcal Φ11 endolysin that cleaves an alanine–glycine bond (Navarre *et al.*, 1999).

Endolysin Modular Structure

Gram-negative endolysin structure

The Gram-negative peptidoglycan is contained within the periplasmic space between the inner and outer bacterial membranes. It is relatively thin (five to ten layers) compared with Gram-positive peptidoglycan (up to 40 layers) and lysis 'from within' is via a holin and endolysin during the phage lytic cycle. Accordingly, endolysins from phages that infect Gram-negative hosts are typically comprised of a single globular catalytic domain, which in most cases supplies one of the two glycosidase activities. The two notable exceptions include the *Pseudomonas* phage endolysins KZ144 and EL188, which have a distinct cell-wall-binding domain in addition to a catalytic domain (Briers *et al.*, 2007). These binding domains alone were shown to be sufficient to direct high-affinity binding to purified *Pseudomonas aeruginosa* cell walls (Briers *et al.*, 2009). Otherwise, the catalytic domains do not appear to require any specific binding domain to recognize and digest the Gram-negative peptidoglycan.

Gram-positive endolysin structure

The Gram-positive endolysins have a modular design with one or more catalytic domains and a cell-wall-binding domain (Garcia *et al.*, 1988, 1990). One of the best-characterized endolysin catalytic domains is the cysteine, histidine-dependent amidohydrolase/peptidase (CHAP) domain (Bateman and

Rawlings, 2003; Rigden *et al.*, 2003) with active site residues that are conserved across many species and have been shown to be essential for activity by site-directed mutagenesis (Pritchard *et al.*, 2004; Nelson *et al.*, 2006). Other catalytic domains exist and have been described more fully by Nelson *et al.* (2012).

In contrast to Gram-negative bacteria, Gram-positive organisms contain no outer membrane and their considerable peptidoglycan layer is highly cross-linked with surface carbohydrates and proteins. Consequently, endolysins from Gram-positive-infecting phages also possess a cell-wall-binding domain (CBD) that recognizes epitopes on the peptidoglycan surface. These epitopes can consist of carbohydrates, teichoic acids or peptide moieties of the peptidoglycan and can impart a species- or even strain-specific binding of the endolysin (Schmelcher *et al.*, 2011), although the necessity of the CBD for lysis is still debated. Frequently, the CBD has been found to be an absolute requirement for lysis-from-without activity of the endolysin, as the catalytic domain alone has highly diminished activity against the host organism, presumably as a consequence of catalysis occurring after only random encounters between the catalytic domain and the sessile bond of the peptidoglycan, or the enzyme is completely inactive (Porter *et al.*, 2007). By contrast, however, there are other reports where deleting the CBD does not affect the lytic activity (Low *et al.*, 2005) and even results in increased lysis from without (Cheng and Fischetti, 2007; Horgan *et al.*, 2009; Gerova *et al.* 2011).

In addition to a catalytic domain and a CBD, Gram-positive phage endolysins often have a flexible linker sequence of 10–20 amino acids that connect the globular domains. This inherent flexibility has led to considerable trouble crystallizing full-length endolysins. Many attempts have yielded only the structures of individual catalytic domains or isolated CBDs (Low *et al.*, 2005; Porter *et al.*, 2007; Korndorfer *et al.*, 2008; Silva-Martin *et al.*, 2010). None the less, recent efforts to produce full-length structures of PlyPSA, a listerial *N*-acetylmuramoyl-L-alanine amidase (Korndorfer *et al.*, 2006), and Cpl-1, a pneumococcal *N*-acetylmuramidase (Hermoso *et al.*, 2003), have been successful. These structures reveal extreme compartmentalization of the globular domains, which suggests that domains can be added, deleted or swapped to tailor activity (Hermoso *et al.*, 2007). Indeed, the specificity of clostridial, lactococcal, pneumococcal and listerial endolysins has already been modified or changed by insertion or deletion of CBDs (Croux *et al.*, 1993; Lopez *et al.*, 1997; Schmelcher *et al.*, 2011). Although the modular nature of phage endolysins was initially described in the early 1990s (Diaz *et al.*, 1990, 1991; Garcia *et al.*, 1990; Croux *et al.*, 1993), it was not until recently that exploiting the modular nature of endolysins has become a major biotechnological focus for the development of enzybiotics.

Multiple-domain endolysins

While endolysins from phages infecting Gram-positive bacteria usually contain both a catalytic domain and a CBD, as mentioned above, some endolysins contain multiple catalytic domains. This is common in those derived from staphylococcal or streptococcal phage. The presence of two catalytic domains, however, does not necessarily indicate that each contributes equally to the lytic activity. For example, the staphylococcal endolysin Φ11 has both *N*-acetylmuramoyl-L-alanine amidase and D-alanyl-glycyl endopeptidase catalytic domains (Navarre *et al.*, 1999). While the endopeptidase domain was active by itself in deletion constructs (Donovan *et al.*, 2006a; Sass and Bierbaum, 2007), the amidase domain was shown to be silent by deletion analysis (Sass and Bierbaum, 2007). Similarly, LysK, the endolysin from the staphylococcal phage K, has an active *N*-terminal endopeptidase domain and a virtually inactive amidase domain (Becker *et al.*, 2009a; Horgan *et al.*, 2009).

The same phenomenon is also prevalent in streptococcal endolysins. The streptococcal λSA2 endolysin, for instance, contains a D-glutaminyl-L-lysine endopeptidase activity and an *N*-acetylglucosaminidase activity (Pritchard *et al.*, 2007). Further analysis revealed, however, that the endopeptidase domain possessed nearly all of the catalytic activity and that the glucosaminidase was almost devoid of activity (Donovan and Foster-Frey, 2008). At present, it is not clear if these 'silent' domains are an artefact of the assay or substrate used for the assay, were once catalytically active but lost their activity through evolution, or function as a secondary binding domain by positioning the substrate for optimal cleavage by the 'non-silent' catalytic domain.

Most of these findings rely on gene deletion analysis that yields a truncated protein product. These strategies raise the potential for aberrant tertiary structure in the truncated protein that could be misinterpreted as diminished domain activity. To address this concern, the streptococcal B30 endolysin (also known as PlyGBS; Cheng *et al.*, 2005) was subjected to site-directed mutagenesis of key amino acid residues that resulted in eradication of a single domain's enzymatic activity in full-length protein constructs. Although this does not fully rule out protein misfolding, it certainly is much less likely. The result was that the *N*-acetylmuramidase domain was nearly silent, with virtually all lysis from without due to the D-alanyl-L-alanyl endopeptidase catalytic domain (Donovan *et al.*, 2006b). This is despite the fact that each lytic domain showed enzymatic activity in the native protein on purified streptococcal peptidoglycan (Pritchard *et al.*, 2004).

At present, it is not clear why an endolysin would evolve two disparate catalytic domains, particularly when one of the two has little to no activity. One of us (D.M.D.) has speculated that the observed differences in activity may be attributable to inherent variations in cleavage when done from within versus from without (Donovan and Foster-Frey, 2008). Alternatively, the 'silent' domain might aid in binding to the peptidoglycan and position the active domain in the proper orientation for optimal cleavage, or cleavage with the highly active domain might create a substrate that is more amenable to the second domain activity.

Although we can only speculate as to the advantages conferred by the dual domains, one putative mechanism for the modular nature and accumulation of two domains is likely to be due to horizontal gene transfer and a consequence of recombination. Indeed, group I introns are often found within endolysin genes from phage that infect *Streptococcus* (Foley *et al.*, 2000) and *Staphylococcus* species (O'Flaherty *et al.*, 2004; Kasparek *et al.*, 2007; Becker *et al.*, 2009b). Specifically, Foley and colleagues found group I introns interrupting the endolysin genes of half of the 62 *Streptococcus thermophilus* phage genomes studied (Foley *et al.*, 2000). Additionally, in at least one pair of more than 90% identical staphylococcal phage endolysin genes, the intron is positioned between the two lytic domains and thus would allow 'exonic shuffling' of the lytic domains (Becker *et al.*, 2009b). Clearly, identification of additional intron-containing multi-domain enzymes and further experimentation on the current enzymes is needed to understand more fully the role of introns in the evolution of these important enzymes.

Nearly all endolysins, whether derived from Gram-negative or Gram-positive phages, and whether they are single- or multi-domained, are products of individual genes. The only known exception is PlyC, an endolysin from the streptococcal C1 bacteriophage. Originally, this enzyme was believed to be a single protein like all other endolysins (Fischetti *et al.*, 1971; Nelson *et al.*, 2001). Later, the gene for this endolysin was thought to be interrupted when whole-genome sequencing of the C1 phage uncovered a putative group I intron between the two coding sequences known to be present in the endolysin (Nelson *et al.*, 2003). However, a subsequent investigation aimed at cloning the gene(s) and systematically investigating each open reading frame revealed that PlyC is composed of two dissimilar gene products: PlyCA, which contains the catalytically active L-alanine amidase domain, and PlyCB, which functions as the CBD (Nelson *et al.*, 2006). Moreover, PlyCB self-assembles into an octamer, whether in the presence of PlyCA or expressed alone. Thus, the mature PlyC holoenzyme is a nonamer composed of one PlyCA and eight PlyCB subunits. No other multi-subunit endolysins have been described and the implications for such a complex are not clear. None the less, nanogram quantities of PlyC can achieve approximately 7 log killing of streptococcal cells within seconds, which is several orders of magnitude more active than any other endolysin (Nelson *et al.*, 2001). Future structure/function studies may provide additional insight into this novel endolysin.

Endolysin *In Vitro* Activity

Gram-negative organisms

The use of endolysins as enzybiotics against Gram-negative pathogens has been very limited. The endolysin-susceptible peptidoglycan layer resides between an inner and outer membrane in Gram-negative organisms and as such is not directly exposed to the extracellular environment (i.e. lysis from without via endolysin digestion of peptidoglycan has not been reported in the literature for untreated Gram-negative organisms). In order to overcome the hurdle of the outer membrane, the use of peptides, detergents and chelators in combination with endolysins was first suggested by Vaara (1992). For example, 50 μg ml^{-1} of the EL188 endolysin in the presence of 10 mM EDTA (i.e. a metal chelator) can reduce *P. aeruginosa* titres by 3–4 logs in 30 min depending on the strain tested (Briers *et al.*, 2011). Additionally, there have been studies in which various chemical moieties conjugated to exolysins or hydrophobic peptides have been genetically fused to endolysins to alter the membrane permeability of these enzymes (Ito *et al.*, 1997; Masschalck and Michiels, 2003).

Sometimes naturally occurring membrane permeability domains are present on endolysins. This was shown recently for LysAB2, the endolysin from the *Acinetobacter baumannii* phage ΦAB2, which was reported to lyse live Gram-positive and Gram-negative bacteria (Lai *et al.*, 2011). Notably, this endolysin contains a C-terminal amphipathic region that was shown to be necessary for the observed antibacterial activity. Another example is the lys1521 endolysin from a *Bacillus amyloliquefaciens* phage. This enzyme possesses two cationic C-terminal regions that were shown to be able to permeabilize the outer membrane of *P. aeruginosa* (Muyombwe *et al.*, 1999). The wild-type enzyme, which contained an N-terminal catalytic domain as well as both C-terminal cationic domains, displayed antibacterial activity against live *P. aeruginosa* (Orito *et al.*, 2004). However, caution should be exercised when considering any of the approaches presented in this section for human therapeutics, as agents known to destabilize the Gram-negative outer membrane are also often toxic to eukaryotic cell membranes. To date, no therapeutic use of an enzybiotic against a Gram-negative organism has been demonstrated successfully in an animal model.

Gram-positive organisms

The vast majority of the literature on endolysins/enzybiotics is devoted to describing *in vitro* activity (i.e. lysis from without) of endolysins against Gram-positive pathogens. The 'substrate' for these enzymes is complex (often the three-dimensional superstructure of the peptidoglycan as well as secondary binding sites for the CBDs). Thus, synthetic small-molecular-weight chromogenic or fluorogenic substrates are rarely adequate as substrates for these enzymes. Instead, isolated cell walls or even whole cells remain the substrates of choice. As such, the most common assay involves the use of a spectrophotometer to measure the loss of turbidity of a suspension of live cells due to lysis from without (i.e. drop in optical density at 600 nm (OD_{600}) per unit time). This method has been used to describe a 'unit' of enzyme activity based on the dilution of an endolysin solution needed to reduce the optical density by half in a defined amount of time (Fischetti *et al.*, 1971; Loeffler *et al.*, 2001).

While approaches that rely on turbidometric analysis are simplistic and easy to use, they are an indirect measure of lysis (and cell death) and are altered by changes in numerous biological factors. First, the phase of growth affects the thickness of the peptidoglycan, which in turn affects the apparent rate of lysis. An enzyme tested on mid-exponential-phase cells may have several orders of magnitude greater activity than the same enzyme assayed on stationary-phase cells. Likewise, the turgor pressure of the cell is important, as is the osmolarity of the solution. Similarly, cells in medium lyse much more slowly than cells resuspended in distilled water. Also, as cells lyse, they release DNA and other cellular components, causing the viscosity of the reaction tube to rise, making further lysis less efficient. Finally, because endolysins are often species specific, it is difficult to compare the activity of one enzyme against another when the substrate is a different bacterial species. Despite these caveats, Mitchell and colleagues have developed a mathematical approach to estimate kinetic constants based on a turbidometric reduction assay (Mitchell *et al.*, 2010). In addition, autoclaved *S. aureus* cells have been used to reduce the day-to-day variability of cultured cells for the determination of kinetics for the LysK endolysin (Filatova *et al.*, 2010).

Another common type of *in vitro* endolysin assay is based on colony counts whereby a known amount of bacteria is exposed to an endolysin for a given amount of time, serially diluted and then plated. Activity is then reported as the log-fold decrease in colony-forming units (CFU) over a defined interval. While this method is often used to quantify cell death, the high affinity of the CBDs for the bacterial surface can allow a situation where the reported activity (i.e. rate of cell death) is overestimated. For example, an endolysin may bind the bacterial surface via the CBD in seconds, but catalytic lysis may not occur immediately. If the endolysin remains bound during the serial-dilution step, actual lysis and cell death may take place on the agar plate at a later time point. As such, reported activity, especially for short incubation times, can be overestimated.

Table 15.1 contains a list of *in vitro* results for many endolysins that produce lysis from without on Gram-positive organisms. This list is not intended to be comprehensive. Instead, its purpose is to provide activity data for many of the endolysins mentioned in this chapter and to also highlight the diversity of activity displayed by these enzymes in different assays. A thorough inspection of this table will reveal something already obvious to those in the field, namely that not all endolysins are created equal. Some kill/lyse very efficiently in seconds at microgram

Table 15.2. Summary of *in vivo* studies with phage endolysins.

Bacteria	Endolysin	Reference
Streptococcus pyogenes	C1[a]	Nelson *et al.* (2001)
Streptococcus agalactiae	PlyGBS	Cheng *et al.* (2005)
Bacillus anthracis	PlyG	Schuch *et al.* (2002)
	PlyPH	Yoong *et al.* (2006)
Streptococcus pneumoniae	Cpl-1	Loeffler *et al.* (2001, 2003); Loeffler and Fischetti (2003); Jado *et al.* (2003); Entenza *et al.* (2005); McCullers *et al.* (2007); Grandgirard *et al.* (2008); Witzenrath *et al.* (2009)
	PAL	Jado *et al.* (2003); Loeffler and Fischetti (2003)
Staphylococcus aureus	MV-L	Rashel *et al.* (2007)
	$CHAP_k$	Fenton *et al.* (2010)
	LysGH15	Gu *et al.* (2011)
	ClyS[b]	Daniel et al. (2010); Pastagia et al. (2011)

[a]Renamed PlyC according to Nelson *et al.* (2006).
[b]Chimaeric construct from the bacteriophage Twort and ΦNM3 endolysins.

or even nanogram quantities, whereas others need much higher doses and longer incubation times to show any lysis.

Endolysin *In Vivo* Studies

Although endolysins have been studied for their role in the bacteriophage replication cycle for over half a century, it has only been in the past 10 years that scientists have begun evaluating the use of endolysins in animal infection models of human disease. Table 15.2 shows a complete list to date of all reported *in vivo* therapeutic trials that utilize endolysins, which are further summarized below.

Fischetti and co-workers were the first to use a purified endolysin in an *in vivo* model of streptococcal infection (Nelson *et al.*, 2001). It was found that oral administration of an endolysin from the streptococcal C1 bacteriophage, named PlyC in a subsequent publication (Nelson *et al.*, 2006), provided protection from colonization in mice challenged with 10^7 *S. pyogenes* (group A streptococci) (28.5% infected for endolysin treatment versus 70.5% infected for PBS treatment). Furthermore, when this enzyme was administered orally to nine heavily colonized mice, no detectable streptococci were observed at 2 h after endolysin treatment and only one mouse had any measurable streptococcal counts 24 and 48 h later (which possibly were from intracellular streptococci that had repopulated the pharyngeal surface).

PlyGBS, also known as the (97% identical) B30 endolysin (Pritchard *et al.*, 2004), is another phage endolysin that is active against group A streptococci as well as against groups B, C, G and L streptococci. This enzyme was tested in a murine vaginal model of *Streptococcus agalactiae* (group B streptococci) colonization to see whether PlyGBS could be a potential therapeutic agent for pregnant women to prevent transmission of neonatal meningitis-causing streptococci to newborns (Cheng *et al.*, 2005). A single vaginal dose was shown to decrease colonization of pathogenic group B streptococci by approximately 3 logs. Notably, PlyGBS was found to have a pH optimum of around 5.0, which is similar to the range normally found within the human vaginal tract. Furthermore, this enzyme did not possess bacteriolytic activity against common vaginal microflora such as *Lactobacillus acidophilus*, suggesting a pathogen-specific therapeutic that, unlike broad-range antibiotics, would probably reduce the concern of resistance development in exposed commensal bacteria.

Several phage endolysins have also been used against germinating spores and vegetative cells of *Bacillus* species. PlyG, an endolysin isolated from the *Bacillus anthracis* γ phage, was shown to rescue 13 out of 19 mice in an intraperitoneal mouse model of septicaemia and extended the life of the remaining mice several fold over the controls (Schuch *et al.*, 2002). Significantly, this enzyme displayed a favourable temperature profile and was able to remain fully active after heating to 60°C for 1 h. A second *Bacillus* endolysin, PlyPH, also active against *B. anthracis*, is unique in that it has a high activity over a broad pH range, from pH 4.0 to 10.5. This enzyme also protected 40% of mice in an intraperitoneal *Bacillus* infection model compared with 100% death in control mice (Yoong *et al.*, 2006).

The most extensively studied endolysins in animal models are Cpl-1, an *N*-acetylmuramidase from the Cp-1 pneumococcal phage, and PAL, an *N*-acetylmuramoyl-L-alanine amidase from the Dp-1 pneumococcal phage. PAL was shown to need only 30 s to cause an approximate 4 log drop in viability of 15 different *Streptococcus pneumoniae* serotypes representing multidrug-resistant isolates as well as those that contained a thick polysaccharide capsule (Loeffler *et al.*, 2001). In a mouse model of nasopharyngeal carriage, PAL was shown to eliminate all pneumococci in a dose-responsive manner. In another study, Cpl-1 was shown to be effective in both a mucosal colonization model and in blood via a pneumococcal bacteraemia model (Loeffler *et al.*, 2003). These two enzymes were also shown to be synergistic in a mouse interperitoneal model (Jado *et al.*, 2003; see the following section for more information about synergy). Cpl-1 was also shown to work on established pneumococcal biofilms in a rat endocarditis model (Entenza *et al.*, 2005). In an infant rat model of pneumococcal meningitis, an intracisternal injection of Cpl-1 resulted in a 3 log decrease in pneumococci in the cerebrospinal fluid, and an intraperitoneal injection led to a decrease of two orders of magnitude (Grandgirard *et al.*, 2008). Additionally, Cp-1 was shown to save 100% of mice from fatal pneumonia when administered 24 h after infection and 42% of mice when administered 48 h after infection, a time point at which bacteraemia was fully established (Witzenrath *et al.*, 2009). Lastly, Cpl-1 treatment of mice colonized with *S. pneumoniae* in an otitis media model was shown to significantly reduce co-colonization following challenge with influenza virus (McCullers *et al.*, 2007). Given that pneumococci are early colonizers to which additional pathogens and viruses adhere, eliminating this population could have a multiplier effect on controlling infectious diseases.

S. aureus, particularly methicillin-resistant *S. aureus* (MRSA), is a major public health concern and a primary source of nosocomial and community-acquired infections (see Kuhl *et al.*, Chapter 3, this volume). There is considerable interest in identifying and evaluating highly active staphylococcal endolysins. The first staphylococcal-specific endolysin investigated *in vivo* was MV-L, which was cloned from the ΦMR11 bacteriophage (Rashel *et al.*, 2007). This enzyme rapidly lysed all tested staphylococcal strains *in vitro*, including MRSA and vancomycin-resistant clones. *In vivo*, this enzyme reduced MRSA nasal colonization by approximately 3 logs and provided complete protection in an intraperitoneal model of staphylococcal infection when administered at 30 min post-infection. At 60 min post-infection, the same amount of MV-L provided protection in 60% of mice compared with controls. More recently, an endolysin from the GH15 phage, LysGH15, showed 100% protection in a mouse intraperitoneal model of septicaemia (Gu *et al.*, 2011). Likewise, $CHAP_K$, a truncated version of the endolysin LysK, effected a 2 log drop in nasal colonization of mice when given 1 h post-challenge (Fenton *et al.*, 2010). Lastly, ClyS is the first engineered endolysin to be tested in an animal model (Daniel *et al.*, 2010). This enzyme is a chimaera between the N-terminal catalytic domain of the Twort phage endolysin (Loessner *et al.*, 1998) and the C-terminal CBD of the ΦNM3 phage endolysin. Like MV-L, ClyS displayed potent bacteriolytic properties against multidrug-resistant staphylococci *in vitro*. In a mouse MRSA decolonization model, 2 log reductions in viability were observed 1 h following a single treatment of ClyS. Similarly, a single dose of ClyS provided protection when administered 3 h after staphylococcal challenge in an intraperitoneal septicaemia model. Additionally, ClyS was shown to be effective at treating topical infections of *S. aureus* (Pastagia *et al.*, 2011).

Synergy

Antimicrobial synergy has been demonstrated for multiple endolysins, either with other endolysins or with antibiotics. Synergy studies typically use a checkerboard broth microdilution method that allows the concurrent determination of the minimal inhibitory concentration (MIC) of each agent (endolysin or antibiotic). The fractional MIC values of each agent are then put on an *x*/*y* plot, which is called an isobologram. A linear relationship corresponds to an 'additive' effect. For example, if 0.5 MIC of agent A + 0.5 MIC of agent B displays the same efficacy as 1.0 MIC of either agent A or B, the two agents are additive. However, if the relationship has an inverse, non-linear curve, the effect is said to be 'synergistic' (i.e. if 0.25 MIC each of agents A and B is equal to 1.0 MIC of either agent alone).

Using this method, the pneumococcal Cpl-1 endolysin, a muramidase, was shown to be synergistic with PAL, an L-alanine amidase (Jado *et al.*, 2003; Loeffler and Fischetti, 2003). Because these enzymes hydrolyse different bonds, synergy is believed to be due to a greater destabilization of the three-dimensional peptidoglycan matrix. Cpl-1 was also found to be synergistic with penicillin as well as with gentamicin, but not with levofloxacin or azithromycin (Djurkovic *et al.*, 2005). Interestingly, the synergy with penicillin was greatest on strains with the highest levels of penicillin resistance. In a similar fashion, the staphylococcal endolysin LysH5 was found to be synergistic with nisin, an antimicrobial peptide (Garcia *et al.*, 2010), and LysK was shown to be synergistic with lysostaphin, a staphylococcal exolysin (Becker *et al.*, 2008). Finally, ClyS, a fusion endolysin described above, was shown to be synergistic with oxacillin and vancomycin *in vitro* and with oxacillin *in vivo* in a mouse model of *S. aureus* septicaemia (Daniel *et al.*, 2010).

Immune responses

Because endolysins are globular proteins, they would be expected to elicit antibodies that may render them inactive and could hinder their future development as human or animal therapeutics. Towards this end, two decisive studies used *in vivo* models to address this issue. First, Loeffler *et al.* (2003) injected Cpl-1, a pneumococcal endolysin, intravenously into mice three times a week for 4 weeks resulting in positive IgG antibodies against the endolysin in five out of six mice. Next, these 'immunized' mice and naïve control mice were challenged with intravenous pneumococci followed by a 200 μg dose of Cpl-1 10 h post challenge. Surprisingly, pneumococcal titres were reduced to the same level in both groups of mice at 15 min. Moreover, the authors found that hyperimmune rabbit serum did not neutralize the bacteriolytic activity of Cpl-1 when tested against pneumococci *in vitro*. Similar *in vitro* results have been reported for endolysins active against *B. anthracis* and *S. pyogenes* (Fischetti, 2005).

In the second study, Jado *et al.* (2003) challenged mice with pneumococci followed by treatment with either of the pneumococcal endolysins, Cpl-1 or PAL. Ten days later they confirmed that the recovered mice had high IgG antibody titres to both enzymes, re-challenged the mice with pneumococci and retreated them with the enzybiotics. Significantly, pneumococcal titres fell by 2–3 logs upon administration of the enzymes and all mice survived with no signs of anaphylaxis or adverse side events. These studies suggest that, while antibodies can readily be raised to endolysins due to their proteinaceous nature, the antibodies do not effectively neutralize the bacteriolytic actions of these enzymes *in vitro* or *in vivo*.

Resistance Development

Because of the narrow specificity in which phage endolysins produce lysis from without, it can be difficult to determine whether a strain has developed true resistance to an endolysin or whether it was never susceptible in the first place. To date, there are no reports of strains sensitive to an endolysin developing resistance to the same endolysin, however. In fact, there have been a few reports where researchers have actively tried, but failed, to develop resistance to endolysins. In one study, *S. pneumonia, S. pyogenes,* and *B. anthracis* were exposed to sublethal doses of Cpl-1, PlyC and PlyG, respectively (Fischetti, 2005). Surviving colonies were grown and once again exposed to sublethal doses of the corresponding endolysin. In some cases, over 100 rounds of screening took place. At different cycles, surviving colonies were tested with lethal doses and in no instance was resistance observed.

In a separate study, resistance to an endolysin was investigated more formally (Schuch *et al.*, 2002). Here, *Bacillus* species were screened for spontaneous resistance to PlyG, a *Bacillus*-specific endolysin, as well as two antibiotics, streptomycin and novobiocin. Resistant colonies were readily isolated for both antibiotics at a frequency of approximately 1×10^{-9}, but no resistance was found for PlyG at the same screening frequency. Next, bacilli were exposed to a chemical mutagen known to induce random mutations and the screening was repeated. This time, spontaneous resistance to the antibiotics occurred at frequencies of around 1×10^{-6}, but no resistance was observed for PlyG, even at a frequency of more than 5×10^{-9}. Therefore, even under conditions that promoted spontaneous resistance to antibiotics by approximately 3 logs, no resistance could be detected for this endolysin.

An explanation for the lack of observed resistance to endolysins has been put forth by Fischetti (2005). In brief, he postulated that phage and their bacterial hosts have co-evolved over the millennia such that phages have evolved to target conserved bonds in the peptidoglycan in order to guarantee survival of the progeny phage. As such, resistance, if present, would be a rare event. None the less, resistance to these enzymes will always be a real concern. Most famously, resistance has been well documented for lysostaphin, an exolysin secreted by *S. simulans* (Thumm and Gotz, 1997). While this enzyme is not a phage endolysin, it is a peptidoglycan hydrolase with a similar catalytic and binding domain to many known phage endolysins (Schindler and Schuhardt, 1964). One must wonder if it is only a matter of time before endolysin resistance is identified.

Bacteriocins

Bacteriocins represent another class of enzybiotics. Bacteriocins are proteins and peptides that comprise a large and functionally diverse family of potent toxins capable of killing bacteria at nanomolar concentrations. First discovered in 1925 (Gratia, 1925), bacteriocin activity is usually directed against a narrow spectrum of bacteria taxonomically similar to the bacteriocin producer, although broader activity against more distantly related bacteria and even toxicity towards eukaryotic cells has been recorded (Gratia, 1989). It has been suggested that some classes of bacteriocins have a common origin with bacteriophage or may even be phage remnants. While this is still a hotly debated area, we provide a brief overview of the bacteriocin field below.

Initially, bacteriocins were classified either as low-molecular-weight peptides that were usually thermostable, non-sedimentable by ultracentrifugation and undetectable on electron micrographs, or as high-molecular-weight molecules that were easily sedimented, heat-sensitive, protease resistant and resolvable by electron microscopy (Bradley, 1967). Over time, it became apparent that the bacteriocins of Gram-negative and Gram-positive bacteria were distinct from one another and only shared their proteinaceous and antimicrobial natures in common (Daw and Falkiner, 1996).

Gram-positive bacteriocins

The small-peptide bacteriocins of Gram-positive bacteria are currently classified into class I lantibiotics and class II unmodified peptides (Rea *et al.*, 2011), which are each further divided into several subclasses. Bacteriolysins represent a third class of Gram-positive antimicrobial proteins that were formerly categorized as class III bacteriocins, with lysostaphin being the best known member of this class (Kumar, 2008; Bastos *et al.*, 2009). With the exception of the 'particulate bacteriocins' discussed below, most Gram-positive bacteriocins currently lack an obvious phage connection, although it was reported recently that production of the broad-spectrum lantibiotic sublancin 168 by *Bacillus subtilis* strain 168, along with its dedicated immunity system, is determined by the SP-β prophage (Dubois *et al.*, 2009).

Gram-negative bacteriocins

The bacteriocins of Gram-negative bacteria primarily belong to two families: the colicin and colicin-like high-molecular-weight proteins (30–80 kDa), and the low-molecular-weight peptides (<10 kDa) that are also referred to as microcins. *Escherichia coli* colicins were the first bacteriocins identified and, along with the pyocins of *P. aeruginosa* (Fyfe *et al.*, 1984), have served as a model system for understanding bacteriocin structure and function, genetic organization and the role of bacteriocins in microbial diversity and evolution. These bacteriocins are located in gene clusters that include a closely linked lysis gene involved in toxin release via lysis or pseudo-lysis (Pugsley and Schwartz, 1983a,b) and a dedicated immunity system that protects the producing strain from the lethal activity of the toxin (Braun *et al.*, 1994; Smarda and Smajs, 1998). The majority of colicin and colicin-like bacteriocins studied to date exhibit a similar three-domain structure, consisting of an *N*-terminal translocation domain, a central receptor-binding domain and a C-terminal 'killing activity' domain (Braun *et al.*, 1994; Lazdunski *et al.*, 1998). The killing activity can be a nuclease, an inhibitor of DNA replication, or one that blocks protein synthesis.

The Gram-negative bacteriocins have drawn many parallels to bacteriophages. The bacteriocin receptors on the target cell surface often double as bacteriophage receptors (Guterman *et al.*, 1975). Additionally, the host system confers immunity to the producer against the toxic effects of the bacteriocin, which has been compared to host immunity from lysogeny. Lastly, open reading frames with sequence similarity to phage genes are often found within or in proximity to bacteriocin gene clusters (Chavan *et al.*, 2005), implicating bacteriophages in bacteriocin acquisition and evolution.

Particulate bacteriocins

Defective bacteriophage and phage-like complexes have frequently been observed in the supernatants of bacteriocin-producing Gram-positive and Gram-negative cells (Bradley, 1967; Reanney and Ackermann, 1982). Some of these 'particulate' bacteriocins resemble bacteriophages, with small heads that may contain randomly packaged DNA (Hirokawa and Kadlubar, 1969). The PBSX-like defective prophages of *Bacillus* are one well-studied example of phage-like particles that are inhibitory on a Gram-positive host. The PBSX 'killer particles' appear to be complete phage-like particles that do not lack any components but package random pieces of genomic DNA into heads that are too small to package a full genome, and do not inject their contents into sensitive cells upon adsorbing to the surface (Okamoto *et al.*, 1968). Similar to temperate bacteriophage, these non-infectious particles are spontaneously produced at low levels and can be induced by chemical agents or UV light.

Other particulate bacteriocins resemble phage heads or headless tails. The phage tail-like complexes are often classified according to their resemblance to the R-type and F-type pyocins of *P. aeruginosa* (Fyfe *et al.*, 1984; Nakayama *et al.*, 2000). R-type pyocins appear as rigid contractile sheaths that resemble the tails of members of the *Myoviridae*, while the F-type pyocins are sheathless, flexible rod-like structures with similarity to the tails of members of the *Siphoviridae*. The killing activity of the phage tail-like bacteriocins occurs by forming pores in the cell membrane, leading to a rapid loss of ions and membrane potential (Nakayama *et al.*, 2000; Strauch *et al.*, 2001, 2003). This is visualized by noticeable shrinkage of the cytoplasmic membrane and cell contents (al-Jumaili, 1976). Significantly, tail-like bacteriocins have been shown to have therapeutic potential both *in vitro* and *in vivo*. For example, enterocoliticin, a phage tail-like bacteriocin produced by pathogenic *Yersinia enterocolitica* serotype O3, was shown to be therapeutic against *Y. enterocolitica* growth in challenged BALB/c mice (Damasko *et al.*, 2005). Similarly, serracin P from *Serratia plymithicum* J7, was active against strains of the fire blight pathogen *Erwinia amylovora*, suggesting that this bacteriocin would make a good biopesticide candidate (Jabrane *et al.*, 2002).

Depolymerases

'Depolymerase' is a generic term used to describe any number of glycanase-like activities associated with the phage particle that are used to degrade the bacterial capsule (Yurewicz *et al.*, 1971; Sutherland *et al.*, 2004). The capsule serves as a secondary receptor for the phage but as a primary receptor for the depolymerase (Hughes *et al.*, 1998a). After the depolymerase degrades the polymer of the capsule, the phage can bind the primary receptor on the bacterial surface allowing infection to proceed. Due to the diversity of bacterial carbohydrates/capsules, depolymerase activities could include glucosidases, endohexosaminidases, alginate lyases, exopolygalacturonic acid lyases, hyaluronate lyases, guluronan lyases, amylases, cellulases, dextranases, galactosidases and pullulanases, to name just a few.

There is considerable interest in using bacteriophage depolymerases therapeutically to degrade bacterial biofilms. Biofilms are formed when planktonic bacteria (i.e. free, individual cells) adsorb on to a surface (e.g. wound or medical implant) and form colonies (reviewed by Stoodley *et al.*, 2002; Hall-Stoodley *et al.*, 2004; Stewart and Franklin, 2008). Once the colonies have become established, phenotypic changes cause them to oversecrete the capsular carbohydrates (i.e. polysaccharides) that serve as the structural backbone for the biofilm. Non-cellular components and debris, including additional carbohydrates, proteins, lipids and nucleic acids, become entangled in the polysaccharide backbone and constitute the 'slime' layer of a biofilm, which is technically known as the extrapolymeric substance (EPS). Appreciably, the EPS superstructure of the biofilm is known to protect internal bacteria from antimicrobials, antibodies and circulating immune cells (Stoodley *et al.*, 2002; Folkesson *et al.*, 2008).

Several mechanisms are thought to contribute to the antimicrobial resistance associated with biofilms: (i) delayed or restricted penetration of antimicrobial agents through the biofilm EPS matrix; (ii) decreased metabolism and growth rate of biofilm organisms, which resist killing by compounds that only attack actively growing cells; (iii) increased accumulation of antimicrobial-degrading enzymes; (iv) enhanced exchange rates of drug resistance genes; and (v) increased antibiotic tolerance (as opposed to resistance) through expression of stress-response genes, phase variation and biofilm-specific phenotype development (Lewis, 2001; Fux *et al.*, 2003; Hall-Stoodley *et al.*, 2004; Keren *et al.*, 2004). Significantly, several groups have shown that whole-phage-based therapy can be used effectively to reduce biofilms (Bartell *et al.*, 1966; Bartell and Orr, 1969a; Yurewicz *et al.*, 1971; Sutherland *et al.*, 2004; reviewed by Abedon, 2011b). Unlike the use of whole-phage therapy, application of depolymerases to a biofilm would not directly kill bacterial pathogens. However, anything that digests the EPS layer would act to 're-sensitize' the underlying cells to antibiotics, complement, antibodies and immune cells. As such, we classify depolymerases as a phage-derived enzybiotic, along with the endolysins and bacteriocins.

Numerous depolymerase activities have been described in the literature, although the depolymerase nomenclature was more prevalent 30–40 years ago. More recently, these enzymes are referred to by the specific enzyme activity they possess (i.e. an endorhamnosidase) rather than the more generic depolymerase terminology. Due to the thick alginate layer, the earliest and most well-studied depolymerases were from phages that infect *Pseudomonas* species (Bartell *et al.*, 1966, 1968; Bartell and Orr, 1969a,b). In addition, depolymerases have been purified and characterized for phage that infect *E. amylovora* (Vandenbergh *et al.*, 1985), *Azotobacter vinelandii* (Davidson *et al.*, 1976, 1977) and *Klebsiella pneumoniae* (Bessler *et al.*, 1973; Thurow *et al.*, 1974, 1975). More recently, Hughes *et al.* (1998a,b) specifically showed that bacteriophage depolymerases could penetrate the EPS layers of *Enterobacter agglomerans* and *Serratia marcescens* biofilms. Finally, *in vivo* studies have shown that a recombinant depolymerase can reduce colonization of *E. coli* in rats, presumably through increased ingestion of the de-capsulated bacteria by macrophages (Mushtaq *et al.*, 2005).

Tailspike proteins, common to phage belonging to the family *Podoviridae*, may represent a well-conserved depolymerase family that offers untapped potential for development of therapeutic depolymerases. Tailspikes have long been known to degrade bacterial polysaccharides, but until recently, have only been investigated in the context of phage–host relationships and not as potential enzybiotics. Tailspikes are trimers, contain an N-terminal 'head-binding' domain that interacts with the base-plate structural proteins, and a C-terminal 'receptor binding and cleaving' domain (for an excellent review, see Casjens and Thuman-Commike, 2011). Significantly, the receptor-binding and cleaving domains can self-assemble into trimers and can cleave polysaccharides, even in the absence of the N-terminal head-binding domains. Much like the cell-wall-binding domains of the Gram-positive endolysins, tailspikes can display extreme species specificity due to the unique polysaccharide epitope displayed by susceptible organisms. Tailspike proteins are often resistant to proteolysis, are SDS stable and can tolerate high temperatures, which makes them well suited for therapeutic development. Indeed, due to the excellent thermostability properties of tailspike proteins, a heating step has been proposed as a purification tool for a *Klebsiella* depolymerase (Kassa and Chhibber, 2012).

Finally, purified tailspike proteins have been shown to degrade the capsule of *Pseudomonas putida in vitro* (Cornelissen *et al.*, 2011) and decrease the colonization of *Salmonella* species in chicken when administered orally (Waseh *et al.*, 2010).

Concluding Remarks

The spread of antibiotic-resistant pathogens has forced physicians and biotechnology and pharmaceutical companies to consider alternatives. One option is bacteriophage therapy, which is described in detail by Olszowska-Zaremba *et al.*, Loc-Carrillo *et al.*, Burrowes and Harper, and Abedon (Chapters 12–14 and 17, this volume). Exploiting phages for their enzymes (endolysins, bacteriocins or depolymerises) is another valid approach. Much like phage therapy, the use of these enzymes has benefited from billions of years of co-evolution between phages and their hosts to develop antimicrobial proteins that can specifically target pathogens with little to no expected adverse effects on other organisms, such as beneficial commensal species.

While endolysins are not self-replicating like phages, they are enzymatic in their actions and thus can act repeatedly to generate lysis from without. Moreover, because they are proteins and lack nucleic acids or genetic mechanisms to recombine or alter the host genome, endolysins should be safer and thus face fewer regulatory hurdles than whole phages for therapeutic use. Endolysins have been shown to be active *in vitro*, *in vivo* and on sessile bacteria such as those found in biofilms (Sass and Bierbaum, 2007). They also can act synergistically with each other as well as with conventional antibiotics. Although antigenic, antibodies to endolysins do not neutralize their activity, and as proteins they are more readily degraded and thus more eco-friendly than conventional antibiotics. Endolysins lyse their target organism on contact and only interact with the most external parts of the cell surface, thereby avoiding many common resistance mechanisms to antibiotics (e.g. modification of target, modification of the therapeutic agent and efflux pumps to eliminate the therapeutic). Indeed, resistance has never been reported in organisms that are originally sensitive to a particular endolysin. Finally, the modular nature of endolysins allows molecular biologists and protein biochemists an opportunity to engineer these enzymes for optimal activity and/or optimal host selection. Phage endolysins thus possess many qualities that make them of high interest as a novel class of future anti-microbials.

References

Abedon, S.T. (2011a) Lysis from without. *Bacteriophage* 1, 46–49.

Abedon, S.T. (2011b) *Bacteriophages and Biofilms: Ecology, Phage Therapy, Plaques.* Nova Science Publishers, Hauppauge, NY.

Akesson, M., Dufour, M., Sloan, G.L. and Simmonds, R.S. (2007) Targeting of streptococci by zoocin A. *FEMS Microbiology Letters* 270, 155–161.

al-Jumaili, I.J. (1976) Physical properties and the fine structure of proteocines. *Zentralblatt für Bakteriologie, Parasitenkunde, Infektion-skrankheiten und Hygiene. Erste Abteilung Originale. Reihe A: Medizinische Mikrobiologie und Parasitologie* 235, 421–432.

Arnoldi, M., Fritz, M., Bauerlein, E., Radmacher, M., Sackmann, E. and Boulbitch, A. (2000) Bacterial turgor pressure can be measured by atomic force microscopy. *Physical Review E Statistical Physics, Plasmas, Fluids, and Related Interdisciplinary Topics* 62, 1034–1044.

Bartell, P.F. and Orr, T.E. (1969a) Origin of polysaccharide depolymerase associated with bacteriophage infection. *Journal of Virology* 3, 290–296.

Bartell, P.F. and Orr, T.E. (1969b) Distinct slime polysaccharide depolymerases of bacteriophage-

infected *Pseudomonas aeruginosa*: evidence of close association with the structured bacteriophage particle. *Journal of Virology* 4, 580–584.
Bartell, P.F., Orr, T.E. and Lam, G.K. (1966) Polysaccharide depolymerase associated with bacteriophage infection. *Journal of Bacteriology* 92, 56–62.
Bartell, P.F., Lam, G.K. and Orr, T.E. (1968) Purification and properties of polysaccharide depolymerase associated with phage-infected *Pseudomonas aeruginosa*. *Journal of Biological Chemistry* 243, 2077–2080.
Bastos, M.C., Ceotto, H., Coelho, M.L. and Nascimento, J.S. (2009) Staphylococcal antimicrobial peptides: relevant properties and potential biotechnological applications. *Current Pharmaceutical Biotechnology* 10, 38–61.
Bateman, A. and Rawlings, N.D. (2003) The CHAP domain: a large family of amidases including GSP amidase and peptidoglycan hydrolases. *Trends in Biochemical Sciences* 28, 234–237.
Becker, S.C., Foster-Frey, J. and Donovan, D.M. (2008) The phage K lytic enzyme LysK and lysostaphin act synergistically to kill MRSA. *FEMS Microbiology Letters* 287, 185–191.
Becker, S.C., Dong, S., Baker, J.R., Foster-Frey, J., Pritchard, D.G. and Donovan, D.M. (2009a) LysK CHAP endopeptidase domain is required for lysis of live staphylococcal cells. *FEMS Microbiology Letters* 294, 52–60.
Becker, S.C., Foster-Frey, J., Stodola, A.J., Anacker, D. and Donovan, D.M. (2009b) Differentially conserved staphylococcal SH3b_5 cell wall binding domains confer increased staphylolytic and streptolytic activity to a streptococcal prophage endolysin domain. *Gene* 443, 32–41.
Bessler, W., Freund-Mölbert, E., Knufermann, H., Rudolph, C., Thurow, H. and Stirm, S. (1973) A bacteriophage-induced depolymerase active on *Klebsiella* K11 capsular polysaccharide. *Virology* 56, 134–151.
Borysowski, J., Weber-Dabrowska, B. and Górski, A. (2006) Bacteriophage endolysins as a novel class of antibacterial agents. *Experimental Biology and Medicine* 231, 366–377.
Bradley, D.E. (1967) Ultrastructure of bacteriophage and bacteriocins. *Bacteriological Reviews* 31, 230–314.
Braun, V., Pilsl, H. and Gross, P. (1994) Colicins: structures, modes of action, transfer through membranes, and evolution. *Archives of Microbiology* 161, 199–206.
Briers, Y., Volckaert, G., Cornelissen, A., Lagaert, S., Michiels, C.W., Hertveldt, K. and Lavigne, R. (2007) Muralytic activity and modular structure of the endolysins of *Pseudomonas aeruginosa* bacteriophages phiKZ and EL. *Molecular Microbiology* 65, 1334–1344.
Briers, Y., Schmelcher, M., Loessner, M.J., Hendrix, J., Engelborghs, Y., Volckaert, G. and Lavigne, R. (2009) The high-affinity peptidoglycan binding domain of *Pseudomonas* phage endolysin KZ144. *Biochemical and Biophysical Research Communications* 383, 187–191.
Briers, Y., Walmagh, M. and Lavigne, R. (2011) Use of bacteriophage endolysin EL188 and outer membrane permeabilizers against *Pseudomonas aeruginosa*. *Journal of Applied Microbiology* 110, 778–785.
Casjens, S.R. and Thuman-Commike, P.A. (2011) Evolution of mosaically related tailed bacteriophage genomes seen through the lens of phage P22 virion assembly. *Virology* 411, 393–415.
Celia, L.K., Nelson, D. and Kerr, D.E. (2008) Characterization of a bacteriophage lysin (Ply700) from *Streptococcus uberis*. *Veterinary Microbiology* 130, 107–117.
Chavan, M., Rafi, H., Wertz, J., Goldstone, C. and Riley, M.A. (2005) Phage associated bacteriocins reveal a novel mechanism for bacteriocin diversification in *Klebsiella*. *Journal of Molecular Evolution* 60, 546–556.
Cheng, Q. and Fischetti, V.A. (2007) Mutagenesis of a bacteriophage lytic enzyme PlyGBS significantly increases its antibacterial activity against group B streptococci. *Applied Microbiology and Biotechnology* 74, 1284–1291.
Cheng, Q., Nelson, D., Zhu, S. and Fischetti, V.A. (2005) Removal of group B streptococci colonizing the vagina and oropharynx of mice with a bacteriophage lytic enzyme. *Animicrobial Agents and Chemotherapy* 49, 111–117.
Climo, M.W., Patron, R.L., Goldstein, B.P. and Archer, G.L. (1998) Lysostaphin treatment of experimental methicillin-resistant *Staphylococcus aureus* aortic valve endocarditis. *Antimicrobial Agents and Chemotherapy* 42, 1355–1360.
Cornelissen, A., Ceyssens, P.J., T'Syen, J., Van Praet, H., Noben, J.P., Shaburova, O.V., Krylov, V.N., Volckaert, G. and Lavigne, R. (2011) The T7-related *Pseudomonas putida* phage φ15 displays virion-associated biofilm degradation properties. *PLoS One* 6, e18597.
Croux, C., Ronda, C., Lopez, R. and Garcia, J.L. (1993) Interchange of functional domains switches enzyme specificity: construction of a chimeric pneumococcal-clostridial cell wall lytic enzyme. *Molecular Microbiology* 9, 1019–1025.
Cui, F., Li, G., Huang, J., Zhang, J., Lu, M., Lu, W., Huan, J. and Huang, Q. (2011) Development of

chitosan-collagen hydrogel incorporated with lysostaphin (CCHL) burn dressing with anti-methicillin-resistant *Staphylococcus aureus* and promotion wound healing properties. *Drug Delivery* 18, 173–180.

Dajcs, J.J., Hume, E.B., Moreau, J.M., Caballero, A.R., Cannon, B.M. and O'Callaghan, R.J. (2000) Lysostaphin treatment of methicillin-resistant *Staphylococcus aureus* keratitis in the rabbit. *Investigative Ophthalmology and Visual Science* 41, 1432–1437.

Dajcs, J.J., Thibodeaux, B.A., Hume, E.B., Zheng, X., Sloop, G.D. and O'Callaghan, R.J. (2001) Lysostaphin is effective in treating methicillin-resistant *Staphylococcus aureus* endophthalmitis in the rabbit. *Current Eye Research* 22, 451–457.

Dajcs, J.J., Austin, M.S., Sloop, G.D., Moreau, J.M., Hume, E.B., Thompson, H.W., McAleese, F.M., Foster, T.J. and O'Callaghan, R.J. (2002) Corneal pathogenesis of *Staphylococcus aureus* strain Newman. *Investigative Ophthalmology and Visual Science* 43, 1109–1115.

Damasko, C., Konietzny, A., Kaspar, H., Appel, B., Dersch, P. and Strauch, E. (2005) Studies of the efficacy of Enterocoliticin, a phage-tail like bacteriocin, as antimicrobial agent against *Yersinia enterocolitica* serotype O3 in a cell culture system and in mice. *Journal of Veterinary Medicine B Infectious Diseases and Veterinary Public Health* 52, 171–179.

Daniel, A., Euler, C., Collin, M., Chahales, P., Gorelick, K. and Fischetti, V.A. (2010) Synergism between a novel chimeric lysin and oxacillin protects against infection by methicillin-resistant *Staphylococcus aureus*. *Antimicrobial Agents and Chemotherapy.*

Davidson, I.W., Sutherland, I.W. and Lawson, C.J. (1976) Purification and properties of an alginate lyase from a marine bacterium. *Biochemistry Journal* 159, 707–713.

Davidson, I.W., Lawson, C.J. and Sutherland, I.W. (1977) An alginate lysate from *Azotobacter vinelandii* phage. *Journal of General Microbiology* 98, 223–229.

Daw, M.A. and Falkiner, F.R. (1996) Bacteriocins: nature, function and structure. *Micron* 27, 467–479.

d'Herelle, F.H. (1917) Sur un microbe invisible antagoniste des bacilles dysenteriques. *Comptes Rendus de l'Académie des Sciences (Paris)* 165, 373–375.

Diaz, E., Lopez, R. and Garcia, J.L. (1990) Chimeric phage-bacterial enzymes: a clue to the modular evolution of genes. *Proceedings of the National Academy of Sciences USA* 87, 8125–8129.

Diaz, E., Lopez, R. and Garcia, J.L. (1991) Chimeric pneumococcal cell wall lytic enzymes reveal important physiological and evolutionary traits. *Journal of Biological Chemistry* 266, 5464–5471.

Djurkovic, S., Loeffler, J.M. and Fischetti, V.A. (2005) Synergistic killing of *Streptococcus pneumoniae* with the bacteriophage lytic enzyme Cpl-1 and penicillin or gentamicin depends on the level of penicillin resistance. *Antimicrobial Agents and Chemotherapy* 49, 1225–1228.

Donovan, D.M. (2007) Bacteriophage and peptidoglycan degrading enzymes with antimicrobial applications. *Recent Patents in Biotechnology* 1, 113–122.

Donovan, D.M. and Foster-Frey, J. (2008) LambdaSa2 prophage endolysin requires Cpl-7-binding domains and amidase-5 domain for antimicrobial lysis of streptococci. *FEMS Microbiology Letters* 287, 22–33.

Donovan, D.M., Lardeo, M. and Foster-Frey, J. (2006a) Lysis of staphylococcal mastitis pathogens by bacteriophage phi11 endolysin. *FEMS Microbiology Letters* 265, 133–139.

Donovan, D.M., Dong, S., Garrett, W., Rousseau, G.M., Moineau, S. and Pritchard, D.G. (2006b) Peptidoglycan hydrolase fusions maintain their parental specificities. *Applied and Environmental Microbiology* 72, 2988–2996.

Donovan, D.M., Becker, S.C., Dong, S., Baker, J.R., Foster-Frey, J. and Pritchard, D.G. (2009) Peptidoglycan hydrolase enzyme fusions for treating multi-drug resistant pathogens. *Biotech International* 21, 6–10.

Doyle, R.J. and Marquis, R.E. (1994) Elastic, flexible peptidoglycan and bacterial cell wall properties. *Trends in Microbiology* 2, 57–60.

Dubois, J.Y., Kouwen, T.R., Schurich, A.K., Reis, C.R., Ensing, H.T., Trip, E.N., Zweers, J.C. and van Dijl, J.M. (2009) Immunity to the bacteriocin sublancin 168 is determined by the SunI (YolF) protein of *Bacillus subtilis*. *Antimicrobial Agents and Chemotherapy* 53, 651–661.

Entenza, J.M., Loeffler, J.M., Grandgirard, D., Fischetti, V.A. and Moreillon, P. (2005) Therapeutic effects of bacteriophage Cpl-1 lysin against *Streptococcus pneumoniae* endocarditis in rats. *Antimicrobial Agents and Chemotherapy* 49, 4789–4792.

Evans, A.C. (1934) Streptococcus bacteriophage: a study of four serological types. *Public Health Reports* 49, 1386–1401.

Evans, A.C. (1936) Studies on hemolytic streptococci. I. Methods of classification. *Journal of Bacteriology* 31, 423–437.

Evans, A.C. (1940) The potency of nascent *Streptococcus* bacteriophage B. *Journal of Bacteriology* 39, 597–604.

Fenton, M., Casey, P.G., Hill, C., Gahan, C.G., Ross, R.P., McAuliffe, O., O'Mahony, J., Maher, F. and Coffey, A. (2010) The truncated phage lysin $CHAP_k$ eliminates *Staphylococcus aureus* in the nares of mice. *Bioengineered Bugs* 1, 404–407.

Filatova, L.Y., Becker, S.C., Donovan, D.M., Gladilin, A.K. and Klyachko, N.L. (2010) LysK, the enzyme lysing *Staphylococcus aureus* cells: specific kinetic features and approaches towards stabilization. *Biochimie* 92, 507–513.

Fischetti, V.A. (2005) Bacteriophage lytic enzymes: novel anti-infectives. *Trends in Microbiology* 13, 491–496.

Fischetti, V.A. (2010) Bacteriophage endolysins: a novel anti-infective to control Gram-positive pathogens. *International Journal of Medical Microbiology* 300, 357–362.

Fischetti, V.A., Gotschlich, E.C. and Bernheimer, A.W. (1971) Purification and physical properties of group C streptococcal phage-associated lysin. *Journal of Experimental Medicine* 133, 1105–1117.

Foley, S., Bruttin, A. and Brussow, H. (2000) Widespread distribution of a group I intron and its three deletion derivatives in the lysin gene of *Streptococcus thermophilus* bacteriophages. *Journal of Virology* 74, 611–618.

Folkesson, A., Haagensen, J.A., Zampaloni, C., Sternberg, C. and Molin, S. (2008) Biofilm induced tolerance towards antimicrobial peptides. *PLoS One* 3, e1891.

Fux, C.A., Stoodley, P., Hall-Stoodley, L. and Costerton, J.W. (2003) Bacterial biofilms: a diagnostic and therapeutic challenge. *Expert Review of Anti-infective Therapy* 1, 667–683.

Fyfe, J.A., Harris, G. and Govan, J.R. (1984) Revised pyocin typing method for *Pseudomonas aeruginosa*. *Journal of Clinical Microbiology* 20, 47–50.

Gaeng, S., Scherer, S., Neve, H. and Loessner, M.J. (2000) Gene cloning and expression and secretion of *Listeria monocytogenes* bacteriophage-lytic enzymes in *Lactococcus lactis*. *Applied and Environmental Microbiology* 66, 2951–2958.

Garcia, E., Garcia, J.L., Garcia, P., Arraras, A., Sanchez-Puelles, J.M. and Lopez, R. (1988) Molecular evolution of lytic enzymes of *Streptococcus pneumoniae* and its bacteriophages. *Proceedings of the National Academy of Science USA* 85, 914–918.

Garcia, J.L., Garcia, E., Arraras, A., Garcia, P., Ronda, C. and Lopez, R. (1987) Cloning, purification and biochemical characterization of the pneumococcal bacteriophage Cp-1 lysin. *Journal of Virology* 61, 2573–2580.

Garcia, P., Garcia, J.L., Garcia, E., Sanchez-Puelles, J.M. and Lopez, R. (1990) Modular organization of the lytic enzymes of *Streptococcus pneumoniae* and its bacteriophages. *Gene* 86, 81–88.

Garcia, P., Martinez, B., Rodriguez, L. and Rodriguez, A. (2010) Synergy between the phage endolysin LysH5 and nisin to kill *Staphylococcus aureus* in pasteurized milk. *International Journal of Food Microbiology* 141, 151–155.

Gerova, M., Halgasova, N., Ugorcakova, J. and Bukovska, G. (2011) Endolysin of bacteriophage BFK20: evidence of a catalytic and a cell wall binding domain. *FEMS Microbiology Letters* 321, 83–91.

Grandgirard, D., Loeffler, J.M., Fischetti, V.A. and Leib, S.L. (2008) Phage lytic enzyme Cpl-1 for antibacterial therapy in experimental pneumococcal meningitis. *Journal of Infectious Diseases* 197, 1519–1522.

Gratia, A. (1925) Sur un remarquable exemple d'antagonisme entre deux souches de *Colibacille*. *Comptes Rendus de la Société de Biologie* 93, 1040–1041.

Gratia, J.P. (1989) Products of defective lysogeny in *Serratia marcescens* SMG 38 and their activity against *Escherichia coli* and other *Enterobacteria*. *Journal of General Microbiology* 135, 25–35.

Gu, J., Xu, W., Lei, L., Huang, J., Feng, X., Sun, C., Du, C., Zuo, J., Li, Y., Du, T., Li, L. and Han, W. (2011) LysGH15, a novel bacteriophage lysin, protects a murine bacteremia model efficiently against lethal methicillin-resistant *Staphylococcus aureus* infection. *Journal of Clinical Microbiology* 49, 111–117.

Guterman, S.K., Wright, A. and Boyd, D.H. (1975) Genes affecting coliphage BF23 and E colicin sensitivity in *Salmonella typhimurium*. *Journal of Bacteriology* 124, 1351–1358.

Hall-Stoodley, L., Costerton, J.W. and Stoodley, P. (2004) Bacterial biofilms: from the natural environment to infectious diseases. *Nature Reviews Microbiology* 2, 95–108.

Hermoso, J.A., Monterroso, B., Albert, A., Galan, B., Ahrazem, O., Garcia, P., Martinez-Ripoli, M., Garcia, J.L. and Menendez, M. (2003) Structural basis for selective recognition of pneumococcal cell wall by modular endolysin from phage Cp-1. *Structure* 11, 1239–1249.

Hermoso, J.A., Garcia, J.L. and Garcia, P. (2007) Taking aim on bacterial pathogens: from phage therapy to enzybiotics. *Current Opinion on Microbiology* 10, 461–472.

Hirokawa, H. and Kadlubar, F. (1969) Length of deoxyribonucleic acid of PBSX-like particles of

Bacillus subtilis induced by 4-nitroquinoline-1-oxide. *Journal of Virology* 3, 205–209.
Horgan, M., O'Flynn, G., Garry, J., Cooney, J., Coffey, A., Fitzgerald, G.F., Ross, R.P. and McAuliffe, O. (2009) Phage lysin LysK can be truncated to its CHAP domain and retain lytic activity against live antibiotic-resistant staphylococci. *Applied and Environmental Microbiology* 75, 872–874.
Hughes, K.A., Sutherland, I.W. and Jones, M.V. (1998a) Biofilm susceptibility to bacteriophage attack: the role of phage-borne polysaccharide depolymerase. *Microbiology* 144, 3039–3047.
Hughes, K.A., Sutherland, I.W., Clark, J. and Jones, M.V. (1998b) Bacteriophage and associated polysaccharide depolymerases – novel tools for study of bacterial biofilms. *Journal of Applied Microbiology* 85, 583–590.
Ito, Y., Kwon, O.H., Ueda, M., Tanaka, A. and Imanishi, Y. (1997) Bactericidal activity of human lysozymes carrying various lengths of polyproline chain at the C-terminus. *FEBS Letters* 415, 285–288.
Jabrane, A., Sabri, A., Compere, P., Jacques, P., Vandenberghe, I., Van Beeumen, J. and Thonart, P. (2002) Characterization of serracin P, a phage-tail-like bacteriocin and its activity against *Erwinia amylovora*, the fire blight pathogen. *Applied and Environmental Microbiology* 68, 5704–5710.
Jado, I., Lopez, R., Garcia, E., Fenoll, A., Casal, J. and Garcia, P. (2003) Phage lytic enzymes as therapy for antibiotic-resistant *Streptococcus pneumoniae* infection in a murine sepsis model. *Journal of Antimicrobial Chemotherapy* 52, 967–973.
Kasparek, P., Pantucek, R., Kahankova, J., Ruzickova, V. and Doskar, J. (2007) Genome rearrangements in host-range mutants of the polyvalent staphylococcal bacteriophage 812. *Folia Microbiologica* 52, 331–338.
Kassa, T. and Chhibber, S. (2012) Thermal treatment of the bacteriophage lysate of *Klebsiella pneumoniae* B5055 as a step for the purification of capsular depolymerase enzyme. *Journal of Virological Methods* 179, 135–141.
Keren, I., Kaldalu, N., Spoering, A., Wang, Y. and Lewis, K. (2004) Persister cells and tolerance to antimicrobials. *FEMS Microbiology Letters* 230, 13–18.
Kokai-Kun, J.F., Walsh, S.M., Chanturiya, T. and Mond, J.J. (2003) Lysostaphin cream eradicates *Staphylococcus aureus* nasal colonization in a cotton rat model. *Antimicrobial Agents and Chemotherapy* 47, 1589–1597.
Kokai-Kun, J.F., Chanturiya, T. and Mond, J.J. (2007) Lysostaphin as a treatment for systemic *Staphylococcus aureus* infection in a mouse model. *Journal of Antimicrobial Chemotherapy* 60, 1051–1059.
Korndorfer, I.P., Danzer, J., Schmelcher, M., Zimmer, M., Skerra, A. and Loessner, M.J. (2006) The crystal structure of the bacteriophage PSA endolysin reveals a unique fold responsible for specific recognition of *Listeria* cell walls. *Journal of Molecular Biology* 364, 678–689.
Korndorfer, I.P., Kanitz, A., Danzer, J., Zimmer, M., Loessner, M.J. and Skerra, A. (2008) Structural analysis of the L-alanoyl-D-glutamate endopeptidase domain of *Listeria* bacteriophage endolysin Ply500 reveals a new member of the LAS peptidase family. *Acta Crystallographica D Biological Crystallography* 64, 644–650.
Kumar, J.K. (2008) Lysostaphin: an anti-staphylococcal agent. *Applied Microbiology and Biotechnology* 80, 555–561.
Lai, M.J., Lin, N.T., Hu, A., Soo, P.C., Chen, L.K., Chen, L.H. and Chang, K.C. (2011) Antibacterial activity of *Acinetobacter baumannii* phage varphiAB2 endolysin (LysAB2) against both Gram-positive and Gram-negative bacteria. *Applied Microbiology and Biotechnology* 90, 529–539.
Lazdunski, C.J., Bouveret, E., Rigal, A., Journet, L., Lloubes, R. and Benedetti, H. (1998) Colicin import into *Escherichia coli* cells. *Journal of Bacteriology* 180, 4993–5002.
Lewis, K. (2001) Riddle of biofilm resistance. *Antimicrobial Agents and Chemotherapy* 45, 999–1007.
Loeffler, J.M. and Fischetti, V.A. (2003) Synergistic lethal effect of a combination of phage lytic enzymes with different activities on penicillin-sensitive and -resistant *Streptococcus pneumoniae* strains. *Antimicrobial Agents and Chemotherapy* 47, 375–377.
Loeffler, J.M., Nelson, D. and Fischetti, V.A. (2001) Rapid killing of *Streptococcus pneumoniae* with a bacteriophage cell wall hydrolase. *Science* 294, 2170–2172.
Loeffler, J.M., Djurkovic, S. and Fischetti, V.A. (2003) Phage lytic enzyme Cpl-1 as a novel antimicrobial for pneumococcal bacteremia. *Infection and Immunity* 71, 6199–6204.
Loessner, M.J. (2005) Bacteriophage endolysins – current state of research and applications. *Current Opinion on Microbiology* 8, 480–487.
Loessner, M.J., Wendlinger, G. and Scherer, S. (1995) Heterogeneous endolysins in *Listeria monocytogenes* bacteriophages: a new class of enzymes and evidence for conserved holin genes within the siphoviral lysis cassettes. *Molecular Microbiology* 16, 1231–1241.
Loessner, M.J., Gaeng, S., Wendlinger, G., Maier,

S.K. and Scherer, S. (1998) The two-component lysis system of *Staphylococcus aureus* bacteriophage Twort: a large TTG-start holin and an associated amidase endolysin. *FEMS Microbiology Letters* 162, 265–274.

Lopez, R., Garcia, E., Garcia, P. and Garcia, J.L. (1997) The pneumococcal cell wall degrading enzymes: a modular design to create new lysins? *Microbial Drug Resistance* 3, 199–211.

Low, L.Y., Yang, C., Perego, M., Osterman, A. and Liddington, R.C. (2005) Structure and lytic activity of a *Bacillus anthracis* prophage endolysin. *Journal of Biological Chemistry* 280, 35433–35439.

Masschalck, B. and Michiels, C.W. (2003) Antimicrobial properties of lysozyme in relation to foodborne vegetative bacteria. *Critical Reviews of Microbiology* 29, 191–214.

McCullers, J.A., Karlstrom, A., Iverson, A.R., Loeffler, J.M. and Fischetti, V.A. (2007) Novel strategy to prevent otitis media caused by colonizing *Streptococcus pneumoniae. PLoS Pathogens* 3, e28.

Mitchell, G.J., Nelson, D.C. and Weitz, J.S. (2010) Quantifying enzymatic lysis: estimating the combined effects of chemistry, physiology and physics. *Physical Biology* 7, 046002.

Mushtaq, N., Redpath, M.B., Luzio, J.P. and Taylor, P.W. (2005) Treatment of experimental *Escherichia coli* infection with recombinant bacteriophage-derived capsule depolymerase. *Journal of Antimicrobial Chemotherapy* 56, 160–165.

Muyombwe, A., Tanji, Y. and Unno, H. (1999) Cloning and expression of a gene encoding the lytic functions of *Bacillus amyloliquefaciens* phage: evidence of an auxiliary lysis system. *Journal of Bioscience and Engineering* 88, 221–225.

Nakayama, K., Takashima, K., Ishihara, H., Shinomiya, T., Kageyama, M., Kanaya, S., Ohnishi, M., Murata, T., Mori, H. and Hayashi, T. (2000) The R-type pyocin of *Pseudomonas aeruginosa* is related to P2 phage, and the F-type is related to lambda phage. *Molecular Microbiology* 38, 213–231.

Navarre, W.W., Ton-That, H., Faull, K.F. and Schneewind, O. (1999) Multiple enzymatic activities of the murein hydrolase from staphylococcal phage phi11. Identification of a D-alanyl-glycine endopeptidase activity. *Journal of Biological Chemistry* 274, 15847–15856.

Nelson, D., Loomis, L. and Fischetti, V.A. (2001) Prevention and elimination of upper respiratory colonization of mice by group A streptococci using a bacteriophage lytic enzyme. *Proceedings of the National Academy of Science USA* 98, 4107–4112.

Nelson, D., Schuch, R., Zhu, S., Tscherne, D.M. and Fischetti, V.A. (2003) The genomic sequence of C_1: the first streptococcal phage. *Journal of Bacteriology* 185, 3325–3332.

Nelson, D., Schuch, R., Chahales, P., Zhu, S. and Fischetti, V.A. (2006) PlyC: a multimeric bacteriophage lysin. *Proceedings of the National Academy of Science USA* 103, 10765–10770.

Nelson, D.C., Rodriguez, L., Schmelcher, M., Klumpp, J. and Donovan, D.M. (2012) Endolysins as antimicrobials. *Advances in Virus Research* .

O'Flaherty, S., Coffey, A., Edwards, R., Meancy, W., Fitzgerald, G.F. and Ross, R.P. (2004) Genome of staphylococcal phage K: a new lineage of *Myoviridae* infecting Gram-positive bacteria with a low G+C content. *Journal of Bacteriology* 186, 2862–2871.

O'Flaherty, S., Coffey, A., Meaney, W., Fitzgerald, G.F. and Ross, R.P. (2005) The recombinant phage lysin LysK has a broad spectrum of lytic activity against clinically relevant staphylococci, including methicillin-resistant *Staphylococcus aureus. Journal of Bacteriology* 187, 7161–7164.

Okamoto, K., Mudd, J.A., Mangan, J., Huang, W.M., Subbaiah, T.V. and Marmur, J. (1968) Properties of the defective phage of *Bacillus subtilis. Journal of Molecular Biology* 34, 413–428.

Orito, Y., Morita, M., Hori, K., Unno, H. and Tanji, Y. (2004) *Bacillus amyloliquefaciens* phage endolysin can enhance permeability of *Pseudomonas aeruginosa* outer membrane and induce cell lysis. *Applied Microbiology and Biotechnology* 65, 105–109.

Paradis-Bleau, C., Cloutier, I., Lemieux, L., Sanschagrin, F., Laroche, J., Auger, M., Garnier, A. and Levesque, R.C. (2007) Peptidoglycan lytic activity of the *Pseudomonas aeruginosa* phage ϕKZ gp144 lytic transglycosylase. *FEMS Microbiology Letters* 266, 201–209.

Pastagia, M., Euler, C., Chahales, P., Fuentes-Duculan, J., Krueger, J.G. and Fischetti, V.A. (2011) A novel chimeric lysin shows superiority to mupirocin for skin decolonization of methicillin-resistant and -sensitive *Staphylococcus aureus* strains. *Antimicrobial Agents and Chemotherapy* 55, 738–744.

Patron, R.L., Climo, M.W., Goldstein, B.P. and Archer, G.L. (1999) Lysostaphin treatment of experimental aortic valve endocarditis caused by a *Staphylococcus aureus* isolate with reduced susceptibility to vancomycin. *Antimicrobial Agents and Chemotherapy* 43, 1754–1755.

Porter, C.J., Schuch, R., Pelzek, A.J., Buckle, A.M., McGowan, S., Wilce, M.C., Rossjohn, J., Russell, R., Nelson, D., Fischetti, V.A. and Whisstock, J.C. (2007) The 1.6 Å crystal structure of the catalytic domain of PlyB, a bacteriophage lysin active against *Bacillus anthracis*. *Journal of Molecular Biology* 366, 540–550.

Pritchard, D.G., Dong, S., Baker, J.R. and Engler, J.A. (2004) The bifunctional peptidoglycan lysin of *Streptococcus agalactiae* bacteriophage B30. *Microbiology*150, 2079–2087.

Pritchard, D.G., Dong, S., Kirk, M.C., Cartee, R.T. and Baker, J.R. (2007) LambdaSa1 and LambdaSa2 prophage lysins of *Streptococcus agalactiae*. *Applied and Environmental Microbiology* 73, 7150–7154.

Pugsley, A.P. and Schwartz, M. (1983a) Expression of a gene in a 400-base-pair fragment of colicin plasmid ColE2-P9 is sufficient to cause host cell lysis. *Journal of Bacteriology* 156, 109–114.

Pugsley, A.P. and Schwartz, M. (1983b) A genetic approach to the study of mitomycin-induced lysis of *Escherichia coli* K-12 strains which produce colicin E2. *Molecular and General Genetics* 190, 366–372.

Rakieten, M.L. (1933) The effect of staphylococcus bacteriophage lysin on resistant strains of staphylococci. *Journal of Immunology* 25, 127–137.

Rashel, M., Uchiyama, J., Ujihara, T., Uehara, Y., Kuramoto, S., Sugihara, S., Yagyu, K., Muraoka, A., Sugai, M., Hiramatsu, K., Honke, K. and Matsuzaki, S. (2007) Efficient elimination of multidrug-resistant *Staphylococcus aureus* by cloned lysin derived from bacteriophage phi MR11. *Journal of Infectious Diseases* 196, 1237–1247.

Rea, M.C., Ross, R.P., Cotter, P.D. and Hill, C. (2011) Classification of bacteriocins from Gram-positive bacteria. In: Drider, D. and Rebuffat, S. (eds) *Prokaryotic Antimicrobial Peptides: From Genes to Applications*. Springer, New York, NY, pp. 29–53.

Reanney, D.C. and Ackermann, H.W. (1982) Comparative biology and evolution of bacteriophages. *Advances in Virus Research* 27, 205–280.

Reynals, F.D. (1926) Bacteriophage et microbes tues. *Comptes Rendus de la Société de Biologie* 94, 242–243.

Rigden, D.J., Jedrzejas, M.J. and Galperin, M.Y. (2003) Amidase domains from bacterial and phage autolysins define a family of γ-D,L-glutamate-specific amidohydrolases. *Trends in Biochemical Sciences* 28, 230–234.

Rodriguez, L., Martinez, B., Zhou, Y., Rodriguez, A., Donovan, D.M. and Garcia, P. (2011) Lytic activity of the virion-associated peptidoglycan hydrolase HydH5 of *Staphylococcus aureus* bacteriophage vB_SauS-philPLA88. *BMC Microbiology* 11, 138.

Rodriguez-Cerrato, V., Garcia, P., Huelves, L., Garcia, E., Del Prado, G., Gracia, M., Ponte, C., Lopez, R. and Soriano, F. (2007) Pneumococcal LytA autolysin, a potent therapeutic agent in experimental peritonitis-sepsis caused by highly β-lactam-resistant *Streptococcus pneumoniae*. *Antimicrobial Agents and Chemotherapy* 51, 3371–3373.

Sass, P. and Bierbaum, G. (2007) Lytic activity of recombinant bacteriophage φ11 and φ12 endolysins on whole cells and biofilms of *Staphylococcus aureus*. *Applied and Environmental Microbiology* 73, 347–352.

Schindler, C.A. and Schuhardt, V.T. (1964) Lysostaphin: a new bacteriolytic agent for the *Staphylococcus*. *Proceedings of the National Academy of Science USA* 51, 414–421.

Schmelcher, M., Tchang, V.S. and Loessner, M.J. (2011) Domain shuffling and module engineering of *Listeria* phage endolysins for enhanced lytic activity and binding affinity. *Microbial Biotechnology* 5, 651–662.

Schuch, R., Nelson, D. and Fischetti, V.A. (2002) A bacteriolytic agent that detects and kills *Bacillus anthracis*. *Nature* 418, 884–889.

Silva-Martin, N., Molina, R., Angulo, I., Mancheno, J.M., Garcia, P. and Hermoso, J.A. (2010) Crystallization and preliminary crystallographic analysis of the catalytic module of endolysin from Cp-7, a phage infecting *Streptococcus pneumoniae*. *Acta Crystallographica F Structural Biology and Crystallization Communications* 66, 670–673.

Smarda, J. and Smajs, D. (1998) Colicins – exocellular lethal proteins of *Escherichia coli*. *Folia Microbiologica* 43, 563–582.

Stewart, P.S. and Franklin, M.J. (2008) Physiological heterogeneity in biofilms. *Nature Reviews Microbiology* 6, 199–210.

Stoodley, P., Sauer, K., Davies, D.G. and Costerton, J.W. (2002) Biofilms as complex differentiated communities. *Annual Review of Microbiology* 56, 187–209.

Strauch, E., Kaspar, H., Schaudinn, C., Dersch, P., Madela, K., Gewinner, C., Hertwig, S., Wecke, J. and Appel, B. (2001) Characterization of enterocoliticin, a phage tail-like bacteriocin and its effect on pathogenic *Yersinia enterocolitica* strains. *Applied and Environmental Microbiology* 67, 5634–5642.

Strauch, E., Kaspar, H., Schaudinn, C., Damasko, C., Konietzny, A., Dersch, P., Skurnik, M. and

Appel, B. (2003) Analysis of enterocoliticin, a phage tail-like bacteriocin. *Advances in Experimental Medicine and Biology* 529, 249–251.

Sutherland, I.W., Hughes, K.A., Skillman, L.C. and Tait, K. (2004) The interaction of phage and biofilms. *FEMS Microbiology Letters* 232, 1–6.

Takac, M. and Blasi, U. (2005) Phage P68 virion-associated protein 17 displays activity against clinical isolates of *Staphylococcus aureus*. *Antimicrobial Agents and Chemotherapy* 49, 2934–2940.

Taylor, A. and Gorazdowska, M. (1974) Conversion of murein to non-reducing fragments by enzymes from phage lambda and Vi II lysates. *Biochimica et Biophysica Acta* 342, 133–136.

Thumm, G. and Gotz, F. (1997) Studies on prolysostaphin processing and characterization of the lysostaphin immunity factor (Lif) of *Staphylococcus simulans* biovar *staphylolyticus*. *Molecular Microbiology* 23, 1251–1265.

Thurow, H., Niemann, H., Rudolph, C. and Stirm, S. (1974) Host capsule depolymerase activity of bacteriophage particles active on *Klebsiella* K20 and K24 strains. *Virology* 58, 306–309.

Thurow, H., Niemann, H. and Stirm, S. (1975) Bacteriophage-borne enzymes in carbohydrate chemistry. Part I. On the glycanase activity associated with particles of *Klebsiella* bacteriophage No. 11. *Carbohydrate Research* 41, 257–271.

Twort, F.W. (1915) An investigation on the nature of ultra-microscopic viruses. *Lancet* 186, 1241–1246.

Twort, F.W. (1925) The transmissible bacterial lysin and its action on dead bacteria. *Lancet* 206, 642–644.

Vaara, M. (1992) Agents that increase the permeability of the outer membrane. *Microbiological Reviews* 56, 395–411.

Vandenbergh, P.A., Wright, A.M. and Vidaver, A.K. (1985) Partial purification and characterization of a polysaccharide depolymerase associated with phage-infected *Erwinia amylovora*. *Applied and Environmental Microbiology* 49, 994–996.

Veiga-Crespo, P. and Villa, T.G. (2010) Advantages and disadvantages in the use of antibiotics or phages as therapeutic agents. In: Veiga-Crespo, P. and Villa, T.G. (eds) *Enzybiotics*. John Wiley & Sons, Hoboken, NJ, pp. 27–58.

Veiga-Crespo, P., Ageitos, J.M., Poza, M. and Villa, T.G. (2007) Enzybiotics: a look to the future, recalling the past. *Journal of Pharmaceutical Sciences* 96, 1917–1924.

Wang, I.-N., Smith, D.L. and Young, R. (2000) Holins: the protein clocks of bacteriophage infections. *Annual Review of Microbiology* 54, 799–825.

Waseh, S., Hanifi-Moghaddam, P., Coleman, R., Masotti, M., Ryan, S., Foss, M., MacKenzie, R., Henry, M., Szymanski, C.M. and Tanha, J. (2010) Orally administered P22 phage tailspike protein reduces *Salmonella* colonization in chickens: prospects of a novel therapy against bacterial infections. *PLoS One* 5, e13904.

Whatmore, A.M. and Reed, R.H. (1990) Determination of turgor pressure in *Bacillus subtilis*: a possible role for K^+ in turgor regulation. *Journal of General Microbiology* 136, 2521–2526.

Witzenrath, M., Schmeck, B., Doehn, J.M., Tschernig, T., Zahlten, J., Loeffler, J.M., Zemlin, M., Muller, H., Gutbier, B., Schütte, H., Hippenstiel, S., Fischetti, V.A., Suttorp, N. and Rosseau, S. (2009) Systemic use of the endolysin Cpl-1 rescues mice with fatal pneumococcal pneumonia. *Critical Care Medicine* 37, 642–649.

Yoong, P., Schuch, R., Nelson, D. and Fischetti, V.A. (2006) PlyPH, a bacteriolytic enzyme with a broad pH range of activity and lytic action against *Bacillus anthracis*. *Journal of Bacteriology* 188, 2711–2714.

Young, R. (1992) Bacteriophage lysis: mechanism and regulation. *Microbiological Reviews* 56, 430–481.

Yurewicz, E.C., Ghalambor, M.A., Duckworth, D.H. and Heath, E.C. (1971) Catalytic and molecular properties of a phage-induced capsular polysaccharide depolymerase. *Journal of Biological Chemistry* 246, 5607–5616.

16 Role of Phages in the Control of Bacterial Pathogens in Food

Yan D. Niu[1], Kim Stanford[2], Tim A. McAllister[1] and Todd R. Callaway[3]
[1]*Agriculture and Agri-Food Canada;* [2]*Alberta Agriculture and Rural Development;*
[3]*Food and Feed Safety Research Unit Agricultural Research Service, USDA.*

In spite of the numerous hurdles that have been implemented throughout the food production chain to ensure food safety, foodborne illnesses still affect more than 48 million Americans each year (Scallan *et al.*, 2011) at a total cost of more than US$150 billion (Scharff, 2010). Many of the human foodborne illnesses result from bacterial contamination of foods, both produce and meat-based (Scallan *et al.*, 2011). The indirect and direct cost each year of the five most common foodborne pathogenic bacteria in the US totals more than US$43 billion and cause more than 1600 deaths (ERS/USDA, 2001; Scharff, 2010; see Kuhl *et al.*, Chapter 3, this volume, for details of some specific diseases). To combat this immense drain on both health and finances, researchers in the US, the EU and around the world have developed pathogen reduction strategies for use in the food supply. While these strategies have typically improved human health and food safety, they have been far from perfect. Thus, additional strategies are currently being developed and evaluated for use to reduce foodborne pathogens in our food supply, with one of these strategies being the use of bacteriophages. Phages have been examined for use in food products, as well as on-farm interventions and environmental cleaning agents. In this chapter, we will review: (i) the nature of bacterial contamination of the food supply; (ii) the isolation of zoonotic-bacteria-specific phages; and (iii) the use of phages against foodborne bacterial pathogens, with emphasis on treatment at the farm, in food processing plants and in ready-to-eat foods. For more information on phage therapy, please see Olszowska-Zaremba *et al.*, Loc-Carrillo *et al.*, Burrowes and Harper, and Abedon (Chapters 12, 13, 14 and 17, this volume) and for more on bacterial detection and identification see Williams and LeJeune, Cox, and Goodridge and Steiner (Chapters 6, 10, and 11, this volume).

Bacterial Contamination of the Food Supply

Foods for human consumption naturally contain bacteria acquired during growth, harvest, preparation, processing and production. These bacteria are often reflective of environmental contamination throughout the food production and preparation chain (Doane *et al.*, 2007; Arthur *et al.*, 2008). Typically, this bacterial population is harmless to humans and provides a first immunological primer to the body to determine 'self' from 'non-self', and a lack of exposure to these bacteria has been linked to the rise in human allergies (Callaway *et al.*, 2006b).

When animal and vegetable products are not cooked or pasteurized, however, they can often be sources of human infections by pathogenic bacteria (e.g. Cody *et al.*, 1999; LeJeune and Rajala-Schultz, 2009).

Foodborne pathogenic bacteria can survive in a variety of environments, including animals (below), and have been isolated from soil. Pathogenic bacteria can colonize produce by being taken up by plant roots (Natvig *et al.*, 2002) or can be transferred during handling/processing (Prazak *et al.*, 2002; Ilic *et al.*, 2008). These bacteria can be transmitted to fruits and vegetables by direct contact with animals, as well as by contact with vectors such as insects, mice, birds and other mammals, as well as irrigation water (Manshadi *et al.*, 2001; Jay *et al.*, 2007; Talley *et al.*, 2009).

Foodborne pathogens are found on all types of animal production farms and in all stages of animal growth and production (Oliver *et al.*, 2005; Alali *et al.*, 2010; LeJeune and Kersting, 2010) and can be transmitted to meat products during processing (Mackey and Derrick, 1979). Foodborne pathogenic bacteria can also live asymptomatically in the gut or on the skin and hide of food animals (Porter *et al.*, 1997). Enterohaemorrhagic *Escherichia coli* (EHEC; including *E. coli* O157:H7), *Salmonella, Campylobacter* and *Listeria* are some of the most common foodborne pathogenic bacteria isolated from human outbreaks, and have all been isolated from cattle, swine and poultry (Borland, 1975; Oliver *et al.*, 2005).

The food production industry has focused for many years on the development of strategies that reduce foodborne pathogenic bacteria in the food supply (Koohmaraie *et al.*, 2005; Ilic *et al.*, 2008). While many of these have been focused on strategies within the processing plant, increasingly research has focused on reducing populations of pathogens in live animals to enhance the effectiveness of processing plant-level interventions (Sargeant *et al.*, 2007; Vandeplas *et al.*, 2010). Perhaps more significantly from a public health perspective, human foodborne illness outbreaks have been linked to indirect human contact with faeces via water supplies (both drinking and irrigation) as well as direct faecal contact (see Goodridge and Steiner, Chapter 11, this volume). Increasing incidences of human illnesses linked to direct animal contact via petting zoos and fairs/open farms emphasizes the need for some strategies to focus on reduction of pathogens on the farm (Chapman *et al.*, 2000; Durso *et al.*, 2005). While the point source behind many foodborne illness outbreaks often remains unknown, it is critical for public health nevertheless to reduce the prevalence of these pathogens generally on the farm (Steinmuller *et al.*, 2006). One of the environmental factors that has been suggested to play an important role in reducing bacterial distribution in various food-related environments is bacteriophages (Wetzel and LeJeune, 2007; Niu *et al.*, 2009b).

Natural Phages in Food-producing Environments

Phages are found naturally throughout the food chain (Jorgensen *et al.*, 2002; Atterbury *et al.*, 2003a; Tsuei *et al.*, 2007) and throughout animals in general (see Letarov, Chapter 2, this volume). As interest in using phage as anti-pathogen interventions in the food supply has grown, questions about their role in the environment and food supply have grown as well. The widespread nature of phages associated with crops as well as live animal (and human) intestinal environments strongly suggests that humans have and do consume phages regularly. Other studies have demonstrated the common presence of phages in many traditional fermented foods eaten around the world, such as yogurt and cheeses (Goodridge, 2008). Thus, the addition of phages to foods to reduce foodborne pathogenic bacteria is not significantly different from the phage intake that occurs with normal food consumption.

Phages against foodborne (zoonotic) pathogenic bacteria have been isolated from a wide variety of environmental sources (Ronner and Cliver, 1990; Atterbury *et al.*, 2003a; Kim *et al.*, 2008). In animals, researchers have found that 23–55% of feedlot cattle pens, 48% of swine faecal samples and 1–30% of chicken faecal samples contained phages (Connerton *et al.*, 2004; Callaway *et al.*, 2006a, 2010; Niu *et al.*, 2009b). Coliphages were isolated from 69–79% of grasses and silages fed to cattle, as well as from 5% of processed feeds (Hutchison *et al.*, 2006). Anti-*Salmonella* phage were isolated from swine manure slurries and lagoon effluents, indicating that they are common components of the swine faecal population (McLaughlin *et al.*, 2006; McLaughlin and King, 2008). Phages specifically active against *Listeria* species have been isolated from turkey processing plants (Kim *et al.*, 2008), and naturally occurring anti-*Campylobacter* phages were also widely isolated from retail poultry products in the UK (Atterbury *et al.*, 2003a). Collectively, these data illustrate the fact that humans already consume a great many phages via the food chain, and have probably done so for millennia.

Phages against Foodborne Bacterial Pathogens

In the West, the pursuit of phages and antibacterial therapeutics remained relatively dormant until the 1980s when Williams Smith and co-workers demonstrated superior efficacy with phage therapy as compared with antibiotics in a series of experiments with mice, pigs, calves and lambs (see Loc-Carrillo *et al.* and Burrowes and Harper, Chapters 13 and 14, this volume). In these early studies, phages were used against enteropathogenic *E. coli* (EPEC) induced diarrhoea and splenic enterotoxigenic *E. coli* (ETEC) colonization in calves (Barrow, 2001). Other research groups concluded that phages had both prophylactic and therapeutic effects on the reduction of experimentally induced diarrhoea in weaned pigs experimentally infected with *E. coli* (Jamalludeen *et al.*, 2009). Further studies have shown that phage treatment is capable of reducing necrotic enteritis caused by *Clostridium perfringens* in broiler chickens (Miller *et al.*, 2010).

Phages have also been used in attempts to treat other non-gastrointestinal veterinary diseases, such as bovine mastitis caused by *Staphylococcus aureus* (Gill *et al.*, 2006) and *E. coli* (Bicalho *et al.*, 2010). Other *in vitro* studies have demonstrated that field isolates of phages isolated from the uteri of postpartum Holstein dairy cows have activity against pathogenic *E. coli* (Bicalho *et al.*, 2010; Santos *et al.*, 2010). Researchers have isolated phages that are active against avian pathogenic *E. coli* (APEC) (Huff *et al.*, 2002, 2005; Oliveira *et al.*, 2010) that cause air sacculitis (a production disease) in chickens. Oliveira *et al.* (2010) found that a cocktail of three phages were able to protect contaminated birds and naturally infected flocks from severe colibacillosis APEC infections. These efforts are in addition to a large literature considering phage application to animals (and humans) more generally (see Olszowska-Zaremba *et al.*, Loc-Carrillo *et al.*, Burrowes and Harper, and Abedon, Chapters 12–14 and 17, this volume).

On-farm Use of Phages Against Zoonotic Bacteria

Of the 1415 species of infectious organisms recorded as pathogenic for humans, 868 (61%) are characterized as zoonotic (Cleaveland *et al.*, 2001; Taylor *et al.*, 2001). Of 28 pathogens causing clinically significant bacterial zoonoses (Christou, 2011), at least ten including *Helicobacter* sp., *Campylobacter* sp., *Listeria* sp., *Salmonella* sp., *Shigella* sp., *S. aureus*, *E. coli* O157:H7, *Streptococcus* sp., *Vibrio* sp. and *Yersinia* sp. have been demonstrated to be sensitive to phages in both *in vitro* and *in vivo* clinical trials (O'Flaherty *et al.*, 2005). One of the most effective ways to protect humans from these infections is to reduce the prevalence and incidence of the pathogens on farm, although total elimination is unlikely (Hynes and Wachsmuth, 2000). Thus, in recent years, efforts have focused on developing a continuum of non-antibiotic pathogen reduction strategies from farm to fork, which has included the development of pre-harvest pathogen interventions for use in and on live animals (Callaway *et al.*, 2004; Loneragan and Brashears, 2005; LeJeune and Wetzel, 2007; Sargeant *et al.*, 2007; Vandeplas *et al.*, 2010). In North America, while the use of phages in the treatment of bacterial diseases in humans and animals has received the 'most attention', the use of phages to combat pathogenic bacterial contamination of food as a public health measure is far more developed from both a regulatory and a commercial perspective. In this section we review the literature concerning phage use to control zoonotic bacterial pathogens in food. Additional, recent reviews of this subject include Greer (2005), Johnson *et al.* (2008), Goodridge (2010) and Hagens and Loessner (2010).

On-farm use of phages against foodborne pathogenic bacteria

Phage use is being explored as a control measure for a variety of infectious diseases in livestock, but is most advanced in poultry (Connerton *et al.*, 2004; Higgins *et al.*, 2005; Loc Carrillo *et al.*, 2005; Sillankorva *et al.*, 2010). The use of phage in broilers is most effective against *Salmonella* and *Campylobacter* in growing flocks (Sklar and Joerger, 2001; Loc Carrillo *et al.*, 2005; Sillankorva *et al.*, 2010). Poultry are often found to be positive for *Salmonella* and *Campylobacter* in studies (Jorgensen *et al.*, 2002), and poultry account for most of the food-associated salmonellosis in humans (Guo *et al.*, 2011). While natural anti-*Salmonella* phage populations in poultry occur at a low incidence (approximately 1%; T.R. Callaway, unpublished data), they have been isolated from a variety of sources and can be used to reduce populations of *Campylobacter* and *Salmonella* in the crop and intestine of chickens (Sklar and Joerger, 2001; Loc Carrillo *et al.*, 2005; Toro *et al.*, 2005; Wagenaar *et al.*, 2005; Sillankorva *et al.*, 2010). While these experiments have shown promise as proof of concept, to date phages have not been commercially introduced into the poultry industry because of expense and the need to apply treatments to entire poultry flocks or broiler houses simultaneously.

Swine are often associated with carriage of *Salmonella* spp. (Davies, 2011; Rostagno and Callaway, 2012), with prevalence in faeces ranging from 7 to 20% (Davies, 2011). Reports suggest that as many as 25% of US human cases of salmonellosis can be attributed to consumption of *Salmonella*-contaminated pork products (Guo *et al.*, 2011; Rostagno and Callaway, 2012). Naturally occurring phages that are active against *Salmonella* spp. have been isolated from swine faeces (Callaway *et al.*, 2010), as well as from swine wastewater lagoons (McLaughlin *et al.*, 2006; McLaughlin and King, 2008), which can harbour faecal *Salmonella* populations (Hill and Sobsey, 2003). Researchers found that, while 48% of the swine faecal samples contained phages, the occurrence of phages that specifically infected *Salmonella* Typhimurium was approximately 1% (Callaway *et al.*, 2010). They suggested that, while phages may be widespread in the national swine herd, finding a phage active against a specific *Salmonella* serotype is possibly linked to the presence of that specific serotype in the animal environment. When these researchers amplified the anti-*Salmonella* phages from commecial swine, and inoculated them into swine experimentally infected with *Salmonella* Typhimurium, populations in the cecum and rectum were reduced (Callaway *et al.*, 2011).

When other researchers used *Salmonella* phages isolated from municipal wastewater, they found that phage treatment reduced experimentally inoculated *Salmonella* populations in finished swine (Wall *et al.*, 2010). These researchers found that a phage cocktail reduced *Salmonella* populations in the ileum and caecum under conditions similar to those encountered in commercial production systems (e.g. in pens in an open barn) (Wall *et al.*, 2010; Zhang *et al.*, 2010). Results from a follow-up study from this research group also appeared to be promising in the development of a viable phage cocktail product for use in the swine industry (Zhang *et al.*, 2010). In both of these proof-of-concept studies, the number of pigs that were positive for *Salmonella* was reduced by phage treatment, but the pathogen was never completely eliminated from all phage-treated swine (Wall *et al.*, 2010; Zhang *et al.*, 2010; Callaway *et al.*, 2011). Despite the lack of elimination of *Salmonella* in these animals, some individual pigs did stop faecal *Salmonella* shedding (reduction in prevalence). Additionally, the degree of reduction of *Salmonella* populations (concentrations reduced from 1 to 3 $\log_{10}$ colony-forming units (CFU) g^{-1} digesta) associated with phage treatment was significant enough to cause a reduction in the total *Salmonella* pathogen load entering the food chain and should thus result in a significant reduction in human foodborne illnesses (Hynes and Wachsmuth, 2000). In an intriguing study, a cocktail of phages significantly reduced the severity of diarrhoea in pigs experimentally infected with ETEC O149 but did not affect total intestinal *E. coli* populations (Jamalludeen *et al.*, 2009). Phages were demonstrated to replicate in ETEC-infected pigs within 2 days (Jamalludeen *et al.*, 2009).

Unfortunately for swine and poultry producers, *Salmonella* intervention strategies present unique challenges. There are more than 2500 serotypes of *Salmonella* currently known, and thousands of these serotypes have been isolated from food animals in recent years (Popoff *et al.*, 2004; USDA/FSIS, 2006). Studies using phages to control *Salmonella* in live animals have been hampered by this broad diversity of serotypes, because typically phages have been selected against one or two common serotypes (Callaway *et al.*, 2010, 2011). The development of phage cocktails that are active against multiple serotypes is likely to alleviate this hurdle, but much research into the ecology and specificity of serotypes and their attendant phages is needed to be able to use phage effectively to control *Salmonella* in live animals on a large scale.

Phages in ruminant animals

Because of their unique physiology, ruminants pose more challenges than single-stomached animals with regard to the use of any intervention strategies against foodborne pathogenic bacteria (Loneragan and Brashears, 2005; LeJeune and Wetzel, 2007; Sargeant *et al.*, 2007; Oliver *et al.*, 2008). Ruminants also have relatively large caecum where a distinct microbial population performs a secondary fermentation. The relatively recent discovery of *E. coli* O157:H7 and other EHEC (also known as *Shiga* toxigenic *E. coli* or STEC) as natural residents of the ruminant gastrointestinal tract, especially at the recto-anal junction, has brought new focus to research into ways to reduce pathogens, especially EHEC, in the intestinal tract of live cattle (Loneragan and Brashears, 2005; LeJeune and Wetzel, 2007; Sargeant *et al.*, 2007). Because *E. coli* O157:H7 is classified as an adulterant in ground beef, there is zero tolerance for this organism, meaning that any positive lots of ground beef must be destroyed, which represents a significant cost to the cattle industry. Over the past 18 years, the beef industry has spent in excess of US$1 billion in combating *E. coli* O157:H7 at both the processing plant and live animal levels (Kay, 2003). A wide variety of potential *E. coli* O157:H7 intervention strategies have been proposed for use in cattle and are reviewed elsewhere (Loneragan and Brashears, 2005; LeJeune and Wetzel, 2007; Sargeant *et al.*, 2007). We will focus on the use of phages in this role, because the highly impactful nature of this pathogen has resulted in some elegant proofs of the ability of phage to target specific foodborne pathogenic bacteria in the gastrointestinal tract.

Phages have been isolated from ruminant animals for many years (Adams *et al.*, 1966; Callaway *et al.*, 2006a; Viazis *et al.*, 2011), and have been used in attempts to modify ruminal fermentation with variable success (Orpin and Munn, 1973; Klieve *et al.*, 1991). Phages that target *E. coli* O157:H7 were isolated from 10% of bovine and ovine faecal samples (Kudva *et al.*, 1999). In other studies, anti-*E. coli* O157:H7 phages were found in 15% of the individual faecal samples and in 55% of US commercial feedlot pens (Callaway *et al.*, 2006a) and from approximately 23% of the faecal samples in Canadian feedlot pens (Niu *et al.*, 2009b). Phages from these and other sources (Jamalludeen *et al.*, 2007; Viazis *et al.*, 2011) have been used in a variety of studies in ruminants to reduce *E. coli* O157:H7 populations in the gastrointestinal tract, albeit with variable success (Kudva *et al.*, 1999; Bach *et al.*, 2002; Bach *et al.*, 2003; Sheng *et al.*, 2006; Bach *et al.*, 2009; Rozema *et al.*, 2009). Phages have exhibited high virulence in killing STEC/EHEC O157:H7 and non-O157 STEC/EHEC in broth culture (Niu *et al.*, 2008, 2009a, 2010; Stanford *et al.*, 2010; Viazis *et al.*, 2011) as well as experimentally infected cattle and sheep (Callaway *et al.*, 2006a; Raya *et al.*, 2006, 2011; Sheng *et al.*, 2006). Other studies have had less definitive results, although this was dependent on the method used to treat the cattle with phage (i.e. oral versus rectal administration, or feed versus waterborne dosing) or on the phages selected for use (Bach *et al.*, 2003, 2009; Rozema *et al.*, 2009).

Early studies examining the effect of phages on *E. coli* O157:H7 *in vitro* involved isolation of phages from ovine and bovine sources. These studies found that the phages were only highly active against *E. coli* O157:H7 when aerated, which mitigated their utility in the anaerobic intestinal tract (Kudva *et al.*, 1999). In one of the first studies applying phages orally to ruminants, a single phage type that had worked well against *E. coli* O157:H7 *in vitro* (Bach *et al.*, 2002) did not reduce experimentally inoculated *E. coli* O157:H7 populations in sheep, but this was hypothesized to be related to adsorption or inactivation of the phage (Bach *et al.*, 2003). In the first successful proof-of-concept study, oral addition of a single phage to sheep that were experimentally inoculated with *E. coli* O157:H7 resulted in a decrease in intestinal populations within 2 days (Raya *et al.*, 2006). Following up on this study, oral addition of a phage cocktail to sheep reduced experimentally inoculated *E. coli* O157:H7 populations numerically in the rumen and significantly in the caecum and rectum (Callaway *et al.*, 2008). Interestingly, in this study, the effect of different phage:target ratios was examined (see Abedon, Chapter 17, this volume, for additional discussion of such ratios), and a ratio of 1:1 (phage:target) reduced experimentally inoculated *E. coli* O157:H7 populations to a greater extent than a 10:1 or 100:1 ratio (Callaway *et al.*, 2008). In later studies, some of the early phages isolated from ruminants (Kudva *et al.*, 1999) were effective in reducing *E. coli* O157:H7 populations when added at a multiplicity of infection of 100 directly to the recto-anal junction (Sheng *et al.*, 2006). This reduction was significant, but again pathogen shedding was not eliminated (Sheng *et al.*, 2006).

Further studies by the Lethbridge Research Group led by T.A. McAllister and K. Stanford indicated that phage treatment reduced experimentally inoculated *E. coli* O157:H7, but total *E. coli* populations were not affected. The need for phages to bypass the abomasum, however, was determined to be important by these researchers (Bach *et al.*, 2009). In a follow-up study, these researchers found that orally phage-treated steers shed less *E. coli* O157:H7 cells than rectally or oral and rectally phage-treated steers, but there was no elimination of shedding by any treatment. The beneficial effect of phages on shedding was possibly due to the ability of phages to propagate extensively in the rumen and constantly be washed to the lower intestinal tract where they can affect EHEC populations (Rozema *et al.*, 2009). Interestingly, in this study the spread of phages from treated to non-treated animals was noted, indicating that the concept of spreading a biocontrol agent among cattle may be simply accomplished through the use of a few 'seeder' animals rather than treating a herd or pen as a whole (Rozema *et al.*, 2009). When a polymer-encapsulated bacteriophage was used in experimentally inoculated *E. coli* O157:H7 animals, pathogen shedding was decreased, again suggesting that protection of phages from hydrolysis may improve phage treatment efficacy (Stanford *et al.*, 2010).

In another study, researchers found that sheep naturally colonized with an O157:H7 infecting phage were more resistant to experimentally inoculated *E. coli* O157:H7 colonization than non-phage-colonized or experimentally phage-colonized sheep, suggesting that natural phage populations may play a role in the transient nature of *E. coli* O157:H7 colonization in cattle (Raya *et al.*, 2011). In this study, phage addition was also found to reduce experimentally inoculated *E. coli* O157:H7 populations (Raya *et al.*, 2011). Furthermore, several studies have demonstrated that the use of a phage cocktail is preferable to the use of a single phage in controlling *E. coli* O157:H7 populations (Raya *et al.*, 2011; Viazis *et al.*, 2011).

Collectively, these results indicate that real-world use of phage in food animals requires further development and optimization. Elucidating factors such as appropriate phage selection and preparation (including from which environment), and protection of phages from acidic and enzymatic hydrolysis, as well as administration strategy (i.e. dose, frequency and delivery system) is critical for successful utilization of phage as a method to reduce foodborne pathogenic bacteria within the gut of all food animals (Johnson *et al.*, 2008; Rozema *et al.*, 2009; Stanford *et al.*, 2010). For additional discussion of *E. coli* O157:H7, see Kuhl *et al.*, Christie *et al.*, Williams and LeJeune, and Goodridge and Steiner (Chapters 3, 4, 6 and 11 this volume)

Phages on hides

While *E. coli* O157:H7 is a member of the microbial ecosystem of cattle, it, along with *Salmonella* spp., can also be found on the hide of cattle as they enter the abattoir. The hide-based pathogens can then be transferred to the carcass during processing; as a result, several interventions have been proposed to reduce this pathogen transfer including acid and chemical rinses (Bosilevac *et al.*, 2005; Arthur *et al.*, 2007; Loretz *et al.*, 2011). Interestingly, in spite of what was thought to be a high regulatory barrier to the use of bacteriophages, a hide spray has been approved for use on cattle in the US (FDA, 2006). Finalyse^{TM} is a hide spray product that has begun use in field studies to specifically reduce *E. coli* O157 on cattle prior to processing and thus far has exhibited promising results in field trials. Other potential hide reduction sprays are in development around the world and have shown promise in initial studies (Coffey *et al.*, 2011).

Phages and environmental contamination

Other issues that involve live animals as a source of human foodborne illnesses involve direct faecal contact (Chapman *et al.*, 2000; Pritchard *et al.*, 2000) as well as runoff from farms that can penetrate water supplies used for drinking or crop irrigation (Anon., 2000; Manshadi *et al.*, 2001; Ferens and Hovde, 2011). Further vectors of pathogens from the farm to other crops and humans include flies (Talley *et al.*, 2009) and other animal vectors (e.g. wild pigs) that have been responsible for large-scale human outbreaks (Jay *et al.*, 2007). In an effort to combat this potential source of foodborne illness, phages isolated from sewage effluent have been used as a cocktail to reduce *Salmonella* Typhimurium more than 100-fold within 4 h within composted manure, a potential source of pathogens (Heringa *et al.*, 2010). This phage cocktail offers a potential on-farm treatment to reduce the environmental impact and spread of pathogen shedding.

Phages in foods and in processing plants

One of the most direct and logical ways to utilize phages to improve food safety comes in the realm of use on foods and surfaces in processing plants (Greer, 2005; Hudson *et al.*, 2005; Karmali *et al.*, 2010). Phages have been approved by the FDA for use in food processing plants for cleaning of surfaces, and amendments to the rules have been applied for, in order to allow use in ready-to-eat products as of mid-2011 by Intralytix (FDA, 2006). Thus, we have entered a new era for the development of phage products for use in food processing.

Research studies have shown that phages can reduce populations of *Salmonella* and Vero toxin-producing *E. coli* (VTEC)/STEC in fruits and vegetables (Pao *et al.*, 2004; Abuladze *et al.*, 2008; Patel *et al.*, 2011). Sprouts have been a source of human foodborne illnesses several times, and these have been linked to contaminated seeds. Among many strategies aimed at reducing this contamination (Nandiwada *et al.*, 2004; Singla *et al.*, 2011), phages have been examined and were found to effectively reduce *Salmonella* populations (Pao *et al.*, 2004). Spinach and other green leafy vegetables have also been sources of several outbreaks (Ilic *et al.*, 2008). Phages have been sprayed on to spinach to reduce *E. coli* O157:H7 populations (Abuladze *et al.*, 2008), and recent studies have found that phage treatment of harvester blades could reduce *E. coli* O157:H7 populations (Patel *et al.*, 2011). Other researchers have found that phage treatment at high densities could reduce experimentally inoculated *E. coli* O157:H7 populations on broccoli, spinach and tomatoes (Abuladze *et al.*, 2008).

Phages have also been used experimentally to reduce pathogens on retail and ground beef (both raw and cooked) products (Abuladze *et al.*, 2008; Bigwood *et al.*, 2008). Other research has indicated that the inclusion of phages on chicken carcasses or retail cuts can effectively reduce *Salmonella* and *Campylobacter* contamination (Atterbury *et al.*, 2003b; Goode *et al.*, 2003; Higgins *et al.*, 2005). In one of the first real-world type of proof-of-concept studies, I.F. Connerton's research group demonstrated that phage application to chicken skin inoculated with *Campylobacter jejuni* resulted in a 1 log_{10} CFU reduction within 2 days (Atterbury *et al.*, 2003b). Other researchers found that the addition of a low ratio (1:1) of phage to experimentally inoculated *Campylobacter* and *Salmonella* resulted in a modest decrease in pathogen populations (Goode *et al.*, 2003). When the phage:target ratio was increased, the populations decreased by more than 2 log_{10} CFU, and when sprayed on carcasses with a very low pathogen presence with a multiplicity of infection of 1000, *Salmonella* was eliminated entirely (Goode *et al.*, 2003; see also Abedon (Chapter 17, this volume) for additional discussion of phage:bacterial ratios). Raw and cooked beef inoculated with *Campylobacter* and *Salmonella* had reduced pathogen counts in the order of 2–5.9 log_{10} CFU cm^{-2} dependent on incubation temperature and food type (Bigwood *et al.*, 2008). Other studies have found that the use of sprays and carcass rinses using a cocktail of up to 72 phage varieties could reduce carcass populations of experimentally inoculated *Salmonella* Enteritidis (Higgins *et al.*, 2005). Finally, the addition of a relatively high dose of phage (10^8–10^{10} plaque-forming units) reduced experimentally inoculated *E. coli* O157:H7 populations from 1 to 4 log_{10} CFU in ground beef (Abuladze *et al.*, 2008).

Dairy products have long been known to contain phages due to the common use of *Lactococcus* and *Streptococcus* in cheese and yogurt production (Coetzee *et al.*, 1960; Zhang *et al.*, 2006). Raw milk cheeses are known to harbour pathogenic bacteria and have caused outbreaks of human illness, and studies have shown that phage addition can reduce *Salmonella* populations in raw milk and pasteurized cheeses (Modi *et al.*, 2001). Further exciting studies have demonstrated that *Listeria* populations can be significantly reduced in raw cheeses by the addition of phage (Carlton *et al.*, 2005; Schellekens *et al.*, 2007; Guenther *et al.*, 2009). Collectively, these results indicate that phages have definite potential in reducing foodborne pathogen carriage in dairy products.

Cross-contamination of foods within the plant is a significant route of human foodborne illnesses. The use of phages as cleaning agents within processing plants therefore offers some significant opportunities to reduce the pathogen load entering the food chain, especially with regard to the creation of biofilms (see below). From use as a method to clean machinery used in harvesting/processing (Patel *et al.*, 2011) to the cleaning of food preparation surfaces (Abuladze *et al.*, 2008; Brooks, 2009), phage cleaning products have a significant role to play in the disinfection of processing facilities. The use of anti-listerial phage sprays is especially pertinent to facilities that prepare ready-to-eat foods, which are not further cooked by consumers and thus are a direct route of contamination (Guenther *et al.*, 2009).

Role of Biofilms in Food Contamination and Prospects of Phage Treatment

Biofilms are a concern in food production (Brooks, 2009; Abedon, 2011; Myszka and Czaczyk, 2011), and have been addressed primarily in the postharvest or processing type of environment as a method to install another hurdle to pathogen entry to the continuum of food production (e.g. acid washes, UV light and salt treatments) (Sofos and Smith, 1998; Koohmaraie *et al.*, 2005; Fox *et al.*, 2008). One of the most significant pathogens associated with biofilms is *Listeria*, which can form or take part in other bacterial biofilms on food-processing surfaces and equipment (Bernbom *et al.*, 2011). Other common foodborne pathogenic bacteria such as *Salmonella*, *Campylobacter* and *E. coli* O157:H7 can be found participating in biofilms in various environments (Dourou *et al.*, 2011; Hasegawa *et al.*, 2011). Contact between foods or equipment and these biofilms can transfer the bacteria to the foodstuff, thus ensuring the presence of spoilage organisms or, more significantly, foodborne pathogenic bacteria. Furthermore, the presence of a biofilm can isolate bacteria from these kinds of physical treatments, preventing pH or osmotic shocks or even UV light penetration to targeted pathogens (Bernbom *et al.*, 2011; Dourou *et al.*, 2011), thus ensuring their survival and passage into the food supply on raw, processed or ready-to-eat products.

Phages are an alternative treatment that has the potential to obviate some of the protective benefits of biofilms (Abedon, 2011). Of note, phages can produce polysaccharide depolymerases to degrade the extracellular polysaccharide matrix of the biofilm (Hughes *et al.*, 1998; see also Shen *et al.*, Chapter 15, this volume). Accordingly, phages were used to eliminate *Pseudomonas fluorescens* biofilms in the first 5 days of development (up to 80% of biofilm removal) in a model system (Sillankorva *et al.*, 2004). An *in vitro* study showed that an engineered T7 phage, with the capability of producing a biofilm-degrading enzyme, completely reduced *E. coli* counts in biofilms and improved biofilm removal compared with a wild-type phage that lacked this activity (Lu and Collins, 2007). Further studies have indicated that phage could disperse biofilms made by *C. jejuni* (Siringan *et al.*, 2011). Thus, it appears that phage sprays can be used against pathogens that would otherwise be impervious to conventional anti-pathogen treatments due to the presence of a biofilm on the work surface or processing machinery. Further research is under way into this exciting area of phage treatment.

Concluding Remarks

Because foodborne pathogenic bacteria can penetrate the food supply chain at many points from the farm to the fork (known as the hazard analysis critical control point, or HACCP, approach), the ability to counteract these pathogens directly is crucial. Because of the nature of phage activity, they have been proposed for a variety of uses to improve human public health through a variety of mechanisms. Phage therapy offers a natural, safe and effective means to treat and in some cases prevent bacterial disease, as pathogens can be specifically targeted without affecting commensal flora in the environment. As the problem of antibiotic resistance has grown, researchers have striven to develop phage-based strategies to treat or prevent bacterial diseases, which may be the only means of controlling pathogens that exhibit resistance to multiple antibiotics. By using phage as treatments in the field, on animals and crops as they enter the processing facility, as a cleaning spray and as a preparation on ready-to-eat foods, we can use phage as an adjunct to existing pathogen-reduction strategies. By erecting multiple hurdles to prevent pathogens reaching human consumers, we can significantly reduce human illnesses. As can be seen by the wide variety of approaches of phage treatment and usage in the food industry, we are well on the way to implementing these predators of pathogens to further ensure the safety of our food supply.

References

Abedon, S.T. (2011) *Bacteriophages and Biofilms: Ecology, Phage Therapy, Plaques.* Nova Science Publishers, Hauppauge, NT.

Abuladze, T., Li, M., Menetrez, M.Y., Dean, T., Senecal, A. and Sulakvelidze, A. (2008) Bacteriophages reduce experimental contamination of hard surfaces, tomato, spinach, broccoli and ground beef by *Escherichia coli* O157:H7. *Applied and Environmental Microbiology* 74, 6230–6238.

Adams, J.C., Gazaway, J.A., Brailsford, M.D., Hartman, P.A. and Jacobson, N.L. (1966) Isolation of bacteriophages from the bovine rumen. *Experientia* 22, 717–718.

Alali, W.Q., Thakur, S., Berghaus, R.D., Martin, M.P. and Gebreyes, W.A. (2010) Prevalence and distribution of *Salmonella* in organic and conventional broiler poultry farms. *Foodborne Pathogens and Disease* 7, 1363–1371.

Anon. (2000) Waterborne outbreak of gastroenteritis associated with a contaminated municipal water supply, Walkerton, Ontario, May–June 2000.

Canadian Communicable Disease Report 26, 170–173.

Arthur, T.M., Bosilevac, J.M., Brichta-Harhay, D.M., Kalchayanand, N., Shackelford, S.D., Wheeler, T.L. and Koohmaraie, M. (2007) Effects of a minimal hide wash cabinet on the levels and prevalence of *Escherichia coli* O157:H7 and *Salmonella* on the hides of beef cattle at slaughter. *Journal of Food Protection* 70, 1076–1079.

Arthur, T.M., Bosilevac, J.M., Brichta-Harhay, D.M., Kalchayanand, N., King, D.A., Shackelford, S.D., Wheeler, T.L. and Koohmaraie, M. (2008) Source tracking of *Escherichia coli* O157:H7 and *Salmonella* contamination in the lairage environment at commercial U.S. beef processing plants and identification of an effective intervention. *Journal of Food Protection* 71, 1752–1760.

Atterbury, R.J., Connerton, P.L., Dodd, C.E.R., Rees, C.E.D. and Connerton, I.F. (2003a) Isolation and characterization of *Campylobacter* bacteriophages from retail poultry. *Applied and Environmental Microbiology* 69, 4511–4518.

Atterbury, R.J., Connerton, P.L., Dodd, C.E.R., Rees, C.E.D. and Connerton, I.F. (2003b) Application of host-specific bacteriophages to the surface of chicken skin leads to a reduction in recovery of *Campylobacter jejuni*. *Food Science* 69, 6302–6306.

Bach, S.J., McAllister, T.A., Veira, D.M., Gannon, V.P. and Holley, R.A. (2002) Evaluation of bacteriophage DC22 for control of *Escherichia coli* O157:H7. *Journal of Animal Science* 80 (Suppl. 1), 263.

Bach, S.J., McAllister, T.A., Veira, D.M., Gannon, V.P.J. and Holley, R.A. (2003) Effect of bacteriophage DC22 on *Escherichia coli* O157:H7 in an artificial rumen system (RUSITEC) and inoculated sheep. *Animal Research* 52, 89–101.

Bach, S.J., Johnson, R.P., Stanford, K. and McAllister, T.A. (2009) Bacteriophages reduce *Escherichia coli* O157:H7 levels in experimentally inoculated sheep. *Canadian Journal of Animal Science* 89, 285–293.

Barrow, P.A. (2001) The use of bacteriophages for treatment and prevention of bacterial disease in animals and animal models of human infection. *Journal of Chemical Technology and Biotechnology* 76, 677–682.

Bernbom, N., Vogel, B.F. and Gram, L. (2011) *Listeria monocytogenes* survival of UV-C radiation is enhanced by presence of sodium chloride, organic food material and by bacterial biofilm formation. *International Journal of Food Microbiology* 147, 69–73.

Bicalho, R.C., Santos, T.M., Gilbert, R.O., Caixeta, L.S., Teixeira, L.M., Bicalho, M.L. and Machado, V.S. (2010) Susceptibility of *Escherichia coli* isolated from uteri of postpartum dairy cows to antibiotic and environmental bacteriophages. Part I. Isolation and lytic activity estimation of bacteriophages. *Journal of Dairy Science* 93, 93–104.

Bigwood, T., Hudson, J.A., Billington, C., Carey-Smith, G.V. and Heinemann, J.A. (2008) Phage inactivation of foodborne pathogens on cooked and raw meat. *Food Microbiology* 25, 400–406.

Borland, E.D. (1975) *Salmonella* infection in poultry. *Veterinary Record* 97, 406–408.

Bosilevac, J.M., Shackelford, S.D., Brichta, D.M. and Koohmaraie, M. (2005) Efficacy of ozonated and electrolyzed oxidative waters to decontaminate hides of cattle before slaughter. *Journal of Food Protection* 68, 1393–1398.

Brooks, J.D. (2009) Biofilms in the food industry: problems and potential solutions. *Food Science and Technology* 23, 30–32.

Callaway, T.R., Anderson, R.C., Edrington, T.S., Genovese, K.J., Harvey, R.B., Poole, T.L. and Nisbet, D.J. (2004) Recent pre-harvest supplementation strategies to reduce carriage and shedding of zoonotic enteric bacterial pathogens in food animals. *Animal Health Research Reviews* 5, 35–47.

Callaway, T.R., Edrington, T.S., Brabban, A.D., Keen, J.E., Anderson, R.C., Rossman, M.L., Engler, M.J., Genovese, K.J., Gwartney, B.L., Reagan, J.O., Poole, T.L., Harvey, R.B., Kutter, E.M. and Nisbet, D.J. (2006a) Fecal prevalence of *Escherichia coli* O157, *Salmonella*, *Listeria*, and bacteriophage infecting *E. coli* O157:H7 in feedlot cattle in the southern plains region of the United States. *Foodborne Pathogens and Disease* 3, 234–244.

Callaway, T.R., Harvey, R.B. and Nisbet, D.J. (2006b) The hygiene hypothesis and foodborne illnesses: too much of a good thing, or is our food supply too clean? *Foodborne Pathogens and Disease* 3, 217–219.

Callaway, T.R., Edrington, T.S., Brabban, A.D., Anderson, R.C., Rossman, M.L., Engler, M.J., Carr, M.A., Genovese, K.J., Keen, J.E., Looper, M.L., Kutter, E.M. and Nisbet, D.J. (2008) Bacteriophage isolated from feedlot cattle can reduce *Escherichia coli* O157:H7 populations in ruminant gastrointestinal tracts. *Foodborne Pathogens and Disease* 5, 183–192.

Callaway, T.R., Edrington, T.S., Brabban, A.D., Kutter, E., Karriker, L., Stahl, C., Wagstrom, E., Anderson, R., Genovese, K., McReynolds, J., Harvey, R. and Nisbet, D.J. (2010) Occurrence of *Salmonella*-specific bacteriophages in swine

feces collected from commercial farms. *Foodborne Pathogens and Disease* 7, 851–856.
Callaway, T.R., Edrington, T.S., Brabban, A.D., Kutter, E.M., Karriker, L., Stahl, C., Wagstrom, E.A., Anderson, R.C., Poole, T.L., Genovese, K.J., Krueger, N., Harvey, R. and Nisbet, D.J. (2011) Evaluation of phage treatment as a strategy to reduce *Salmonella* populations in growing swine. *Foodborne Pathogens and Disease* 8, 261–266.
Carlton, R.M., Noordman, W.H., Biswas, B., de Meester, E.D. and Loessner, M.J. (2005) Bacteriophage P100 for control of *Listeria monocytogenes* in foods: genome sequence, bioinformatic analyses, oral toxicity study, and application. *Regulatory Toxicology and Pharmacology* 43, 301–312.
Chapman, P.A., Cornell, J. and Green, C. (2000) Infection with verocytotoxin-producing *Escherichia coli* O157 during a visit to an inner city open farm. *Epidemiology and Infection* 125, 531–536.
Christou, L. (2011) The global burden of bacterial and viral zoonotic infections. *Clinical Microbiology and Infection* 17, 326–330.
Cleaveland, S., Laurenson, M.K. and Taylor, L.H. (2001) Diseases of humans and their domestic mammals: pathogen characteristics, host range and the risk of emergence. *Philosophical Transactions of the Royal Society of London B: Biological Sciences* 356, 991–999.
Cody, S.H., Glynn, K., Farrar, J.A., Cairns, K.L., Griffin, P.M., Kobayashi, J., Fyfe, M., Hoffman, R., King, A.S., Lewis, J.H., Swaminathan, B., Bryant, R.G. and Vugia, D.J. (1999) An outbreak of *Escherichia coli* O157:H7 infection from unpasteurized commercial apple juice. *Annals of Internal Medicine* 130, 202–209.
Coetzee, J.N., de Klerk, H.C. and Sacks, T.G. (1960) Host-range of *Lactobacillus* bacteriophages. *Nature* 187, 348–349.
Coffey, B., Rivas, L., Duffy, G., Coffey, A., Ross, R.P. and McAuliffe, O. (2011) Assessment of *Escherichia coli* O157:H7-specific bacteriophages e11/2 and e4/1c in model broth and hide environments. *International Journal of Food Microbiology* 147, 188–194.
Connerton, P.L., Loc Carrillo, C.M., Dillon, E., Scott, A., Rees, C.E.D., Dodd, C.E.R., Connerton, I.F., Swift, C. and Frost, J. (2004) Longitudinal study of *Campylobacter jejuni* bacteriophages and their hosts from broiler chickens. *Applied and Environmental Microbiology* 70, 3877–3883.
Davies, P.R. (2011) Intensive swine production and pork safety. *Foodborne Pathogens and Disease* 8, 189–201.
Doane, C.A., Pangloli, P., Richards, H.A., Mount, J.R., Golden, D.A. and Draughon, F.A. (2007) Occurrence of *Escherichia coli* O157:H7 in diverse farm environments. *Journal of Food Protection* 70, 6–10.
Dourou, D., Beauchamp, C.S., Yoon, Y., Geornaras, I., Belk, K.E., Smith, G.C., Nychas, G.J.E. and Sofos, J.N. (2011) Attachment and biofilm formation by *Escherichia coli* O157:H7 at different temperatures, on various food-contact surfaces encountered in beef processing. *International Journal of Food Microbiology* 149, 262–268.
Durso, L.M., Reynolds, K., Bauer, N. and Keen, J.E. (2005) Shiga-toxigenic *Escherichia coli* O157:H7 infections among livestock exhibitors and visitors at a Texas county fair. *Vector-Borne Zoonotic Disease* 5, 193–201.
ERS/USDA (2001) ERS estimates foodborne disease costs at $6.9 billion per year. <http://www.ers.usda.gov/Emphases/SafeFood/overview.htm>.
FDA (2006) Food additives permitted for direct addition to food for human consumption; bacteriophage preparation; final rule. Part 172 of Title 21 (Food and Drugs) of the United States Code of Federal Regulations. Federal Register, Washington, DC, pp. 47729–47732.
Ferens, W.A. and Hovde, C.J. (2011) *Escherichia coli* O157:H7: animal reservoir and sources of human infection. *Foodborne Pathogens and Disease* 8, 465–487.
Fox, J.T., Renter, D.G., Sanderson, M.W., Nutsch, A.L., Shi, X. and Nagaraja, T.G. (2008) Associations between the presence and magnitude of *Escherichia coli* O157 in feces at harvest and contamination of preintervention beef carcasses. *Journal of Food Protection* 71, 1761–1767.
Gill, J.J., Pacan, J.C., Carson, M.E., Leslie, K.E., Griffiths, M.W. and Sabour, P.M. (2006) Efficacy and pharmacokinetics of bacteriophage therapy in treatment of subclinical *Staphylococcus aureus* mastitis in lactating dairy cattle. *Animicrobial Agents and Chemotherapy* 50, 2912–2918.
Goode, D., Allen, V.M. and Barrow, P.A. (2003) Reduction of experimental *Salmonella* and *Campylobacter* contamination of chicken skin by application of lytic bacteriophages. *Applied and Environmental Microbiology* 69, 5032–5036.
Goodridge, L.D. (2008) Phages, bacteria, and food. In: Abedon, S.T. (ed.) *Bacteriophage Ecology*. Cambridge University Press, Cambridge, UK, pp. 302–331.

Goodridge, L.D. (2010) Designing phage therapeutics. *Current Pharmaceutical Biotechnology* 11, 15–27.

Greer, G.G. (2005) Bacteriophage control of foodborne bacteria. *Journal of Food Protection* 68, 1102–1111.

Guenther, S., Huwyler, D., Richard, S. and Loessner, M.J. (2009) Virulent bacteriophage for efficient biocontrol of *Listeria monocytogenes* in ready-to-eat foods. *Applied and Environmental Microbiology* 75, 93–100.

Guo, C., Hoekstra, R.M., Schroeder, C.M., Pires, S.M., Ong, K.L., Hartnett, E., Naugle, A., Harman, J., Bennett, P., Cieslak, P., Scallan, E., Rose, B., Holt, K.G., Kissler, B., Mbandi, E., Roodsari, R., Angulo, F.J. and Cole, D. (2011) Application of Bayesian techniques to model the burden of human salmonellosis attributable to U.S. food commodities at the point of processing: adaptation of a Danish model. *Foodborne Pathogens and Disease* 8, 509–516.

Hagens, S. and Loessner, M.J. (2010) Bacteriophage for biocontrol of foodborne pathogens: calculations and considerations. *Current Pharmaceutical Biotechnology* 11, 58–68.

Hasegawa, A., Hara-Kudo, Y. and Kumagai, S. (2011) Survival of *Salmonella* strains differing in their biofilm-formation capability upon exposure to hydrochloric and acetic acid and to high salt. *Journal of Veterinary Medical Science* 73, 1163–1168.

Heringa, S.D., Kim, J., Jiang, X., Doyle, M.P. and Erickson, M.C. (2010) Use of a mixture of bacteriophages for biological control of salmonella enterica strains in compost. *Applied and Environmental Microbiology* 76, 5327–5332.

Higgins, J.P., Higgins, K.L., Huff, H.W., Donoghue, A.M., Donoghue, D.J. and Hargis, B.M. (2005) Use of a specific bacteriophage treatment to reduce *Salmonella* in poultry products. *Poultry Science* 84, 1141–1145.

Hill, V.R. and Sobsey, M.D. (2003) Performance of swine waste lagoons for removing *Salmonella* and enteric microbial indicators. *Transactions of the ASABE* 46, 781–788.

Hudson, J.A., Billington, C., Carey-Smith, G. and Greening, G. (2005) Bacteriophages as biocontrol agents in food. *Journal of Food Protection* 68, 426–437.

Huff, W.E., Huff, G.R., Rath, N.C., Balog, J.M., Xie, H., Moore, P.A. and Donoghue, A.M. (2002) Prevention of *Escherichia coli* respiratory infection in broiler chickens with bacteriophage (SPR02). *Journal of Poultry Science* 81, 437–441.

Huff, W.E., Huff, G.R., Rath, N.C., Balog, J.M. and Donoghue, A.M. (2005) Alternatives to antibiotics: utilization of bacteriophage to treat colibacillosis and prevent foodborne pathogens. *Poultry Science* 84, 655–659.

Hughes, K.A., Sutherland, I.W. and Jones, M.V. (1998) Biofilm susceptibility to bacteriophage attack: the role of phage-borne polysaccharide depolymerase. *Microbiology* 144, 3039–3047.

Hutchison, M.L., Thomas, D.J.I., Walters, L.D. and Avery, S.M. (2006) Shiga toxin-producing *Escherichia coli*, faecal coliforms and coliphage in animal feeds. *Letters in Applied Microbiology* 43, 205–210.

Hynes, N.A. and Wachsmuth, I.K. (2000) *Escherichia coli* O157:H7 risk assessment in ground beef: a public health tool. In: *Proceedings of the 4th International Symposium and Workshop on Shiga toxin (Verocytotoxin)-producing Escherichia coli infections*, 29 October–2 November, Kyoto, Japan, p. 46.

Ilic, S., Odomeru, J. and LeJeune, J.T. (2008) Coliforms and prevalence of *Escherichia coli* and foodborne pathogens on minimally processed spinach in two packing plants. *Journal of Food Protection* 71, 2398–2403.

Jamalludeen, N., Johnson, R.P., Friendship, R., Kropinski, A.M., Lingohr, E.J. and Gyles, C.L. (2007) Isolation and characterization of nine bacteriophages that lyse O149 enterotoxigenic *Escherichia coli*. *Veterinary Microbiology* 124, 47–57.

Jamalludeen, N., Johnson, R.P., Shewen, P.E. and Gyles, C.L. (2009) Evaluation of bacteriophages for prevention and treatment of diarrhea due to experimental enterotoxigenic *Escherichia coli* O149 infection of pigs. *Veterinary Microbiology* 136, 135–141.

Jay, M.T., Cooley, M., Carychao, D., Wiscomb, G.W., Sweitzer, R.A., Crawford-Miksza, L., Farrar, J.A., Lau, D.K., O'Connell, J., Millington, A., *et al.* (2007) *Escherichia coli* O157:H7 in feral swine near spinach fields and cattle, central California coast. *Emerging Infectious Diseases* 13, 1908–1911.

Johnson, R.P., Gyles, C.L., Huff, W.E., Ojha, S., Huff, G.R., Rath, N.C. and Donoghue, A.M. (2008) Bacteriophages for prophylaxis and therapy in cattle, poultry and pigs. *Animal Health Research Reviews* 9, 201–215.

Jorgensen, F., Bailey, R., Williams, S., Henderson, P., Wareing, D.R.A., Bolton, F.J., Frost, J.A., Ward, L. and Humphrey, T.J. (2002) Prevalence and numbers of *Salmonella* and *Campylobacter* spp. on raw, whole chickens in relation to sampling methods. *International Journal of Food Microbiology* 76, 151–164.

Karmali, M.A., Gannon, V. and Sargeant, J.M.

(2010) Verocytotoxin-producing *Escherichia coli* (VTEC). *Veterinary Microbiology* 140, 360–370.

Kay S. (2003) *E. coli* O157:H7: the costs during the past 10 years. *Meat and Poultry News* 3, 26–34.

Kim, J.W., Siletzky, R.M. and Kathariou, S. (2008) Host ranges of *Listeria*-specific bacteriophages from the turkey processing plant environment in the United States. *Applied and Environmental Microbiology* 74, 6623–6630.

Klieve, A.V., Gregg, K. and Bauchop, T. (1991) Isolation and characterization of lytic phages from *Bacteroides ruminicola* ss *brevis*. *Current Microbiology* 23, 183–187.

Koohmaraie, M., Arthur, T.M., Bosilevac, J.M., Guerini, M., Shackelford, S.D. and Wheeler, T.L. (2005) Post-harvest interventions to reduce/eliminate pathogens in beef. *Meat Science* 71, 79–91.

Kudva, I.T., Jelacic, S., Tarr, P.I., Youderian, P. and Hovde, C.J. (1999) Biocontrol of *Escherichia coli* O157 with O157-specific bacteriophages. *Applied and Environmental Microbiology* 65, 3767–3773.

LeJeune, J. and Kersting, A. (2010) Zoonoses: an occupational hazard for livestock workers and a public health concern for rural communities. *Journal of Agricultural Safety and Health* 16, 161–179.

LeJeune, J.T. and Wetzel, A.N. (2007) Preharvest control of *Escherichia coli* O157 in cattle. *Journal of Animal Science* 85 (Suppl.), E73–E80.

LeJeune, J.T. and Rajala-Schultz, P.J. (2009) Unpasteurized milk: a continued public health threat. *Clinical Infectious Diseases* 48, 93–100.

Loc Carrillo, C.M., Atterbury, R.J., El-Shibiny, A., Connerton, P.L., Dillon, E., Scott, A. and Connerton, I.F. (2005) Bacteriophage therapy to reduce *Campylobacter jejuni* colonization of broiler chickens. *Applied and Environmental Microbiology* 71, 6554–6563.

Loneragan, G.H. and Brashears, M.M. (2005) Pre-harvest interventions to reduce carriage of *E. coli* O157 by harvest-ready feedlot cattle. *Meat Science* 71, 72.

Loretz, M., Stephan, R. and Zweifel, C. (2011) Antibacterial activity of decontamination treatments for cattle hides and beef carcasses. *Food Control* 22, 347–359.

Lu, T.K. and Collins, J.J. (2007) Dispersing biofilms with engineered enzymatic bacteriophage. *Proceedings of the National Academy of Sciences USA* 104, 11197–11202.

Mackey, B.M. and Derrick, C.M. (1979) Contamination of the deep tissues of carcasses by bacteria present on the slaughter instruments or in the gut. *Journal of Applied Bacteriology* 46, 355–366.

Manshadi, F.D., Gortares, P., Gerba, C.P., Karpiscak, M. and Frentas, R.J. (2001) Role of irrigation water in contamination of domestic fresh vegetables. In: *101st General Meeting of the American Society for Microbiology*, 20–24 May, Orlando, FL, p. 561.

McLaughlin, M.R. and King, R.A. (2008) Characterization of *Salmonella* bacteriophages isolated from swine lagoon effluent. *Current Microbiology* 56, 208–213.

McLaughlin, M.R., Balaa, M.F., Sims, J. and King, R. (2006) Isolation of *Salmonella* bacteriophages from swine effluent lagoons. *Journal of Environmental Quality* 35, 522–528.

Miller, R.W., Skinner, E.J., Sulakvelidze, A., Mathis, G.F. and Hofacre, C.L. (2010) Bacteriophage therapy for control of necrotic enteritis of broiler chickens experimentally infected with *Clostridium perfringens*. *Avian Diseases* 54, 33–40.

Modi, R., Hirvi, Y., Hill, A. and Griffiths, M.W. (2001) Effect of phage on survival of *Salmonella enteritidis* during manufacture and storage of cheddar cheese made from raw and pasteurized milk. *Journal of Food Protection* 64, 927–933.

Myszka, K. and Czaczyk, K. (2011) Bacterial biofilms on food contact surfaces – a review. *Polish Journal of Food and Nutrition Sciences* 61, 173–180.

Nandiwada, L.S., Schamberger, G.P., Schafer, H.W. and Diez-Gonzalez, F. (2004) Characterization of an E2-type colicin and its application to treat alfalfa seeds to reduce *Escherichia coli* O157:H7. *International Journal of Food Microbiology* 93, 267–279.

Natvig, E.E., Ingham, S.C., Ingham, B.H., Cooperband, L.R. and Roper, T.R. (2002) *Salmonella enterica* serovar Typhimurium and *Escherichia coli* contamination of root and leaf vegetables grown in soils with incorporated bovine manure. *Applied and Environmental Microbiology* 68, 2737–2744.

Niu, Y.D., Xu, Y., McAllister, T.A., Rozema, E.A., Stephens, T.P., Bach, S.J., Johnson, R.P. and Stanford, K. (2008) Comparison of fecal versus rectoanal mucosal swab sampling for detecting *Escherichia coli* O157:H7 in experimentally inoculated cattle used in assessing bacteriophage as a mitigation strategy. *Journal of Food Protection* 71, 691–698.

Niu, Y.D., Johnson, R.P., Xu, Y., McAllister, T.A., Sharma, R., Louie, M. and Stanford, K. (2009a) Host range and lytic capability of four bacteriophages against bovine and clinical human isolates of Shiga toxin-producing

Escherichia coli O157:H7. *Journal of Applied Microbiology* 107, 646–656.

Niu, Y.D., McAllister, T.A., Xu, Y., Johnson, R.P., Stephens, T.P. and Stanford, K. (2009b) Prevalence and impact of bacteriophages on the presence of *Escherichia coli* O157:H7 in feedlot cattle and their environment. *Applied and Environmental Microbiology* 75, 1271–1278.

Niu, Y.D., McAllister, T.A., Johnson, R.P., Kropinski, A.M., Xu, Y. and Stanford, K. (2010) A newly isolated lytic bacteriophage AKFV33 is highly virulent against shiga toxin-producing *Escherichia coli* O157:H7. In: *Canadian Society of Microbiologists 60th Annual Conference*, 14–17 June, Hamilton, Canada, pp. 317–323.

O'Flaherty, S., Coffey, A., Meaney, W., Fitzgerald, G.F. and Ross, R.P. (2005) The recombinant phage lysin LysK has a broad spectrum of lytic activity against clinically relevant staphylococci, including methicillin-resistant *Staphylococcus aureus*. *Journal of Bacteriology* 187, 7161–7164.

Oliveira, A., Sereno, R. and Azeredo, J. (2010) *In vivo* efficiency evaluation of a phage cocktail in controlling severe colibacillosis in confined conditions and experimental poultry houses. *Veterinary Microbiology* 146, 303–308.

Oliver, S.P., Jayarao, B.M. and Almeida, R.A. (2005) Foodborne pathogens in milk and the dairy farm environment: food safety and public health implications. *Foodborne Pathogens and Disease* 2, 115–129.

Oliver, S.P., Patel, D.A., Callaway, T.R. and Torrence, M.E. (2008) ASAS Centennial Paper: developments and future outlook for preharvest food safety. *Journal of Animal Science* 87, 419–437.

Orpin, C.G. and Munn, E.A. (1973) The occurrence of bacteriophages in the rumen and their influence on rumen bacterial populations. *Experientia* 30, 1018–1020.

Pao, S., Randolph, S.P., Westbrook, E.W. and Shen, H. (2004) Use of bacteriophages to control *Salmonella* in experimentally contaminated sprout seeds. *Journal of Food Science* 69, M127–M130.

Patel, J., Sharma, M., Millner, P., Callaway, T.R. and Singh, M. (2011) Inactivation of *E. coli* O157:H7 attached to spinach harvester blade using bacteriophage. *Foodborne Pathogens and Disease* 8, 541–546.

Popoff, M.Y., Bockemühl, J. and Gheesling, L.L. (2004) Supplement 2002 (no. 46) to the Kauffmann–White scheme. *Research in Microbiology* 155, 568–570.

Porter, J., Mobbs, K., Hart, C.A., Saunders, J.R., Pickup, R.W. and Edwards, C. (1997) Detection, distribution, and probable fate of *Escherichia coli* O157 from asymptomatic cattle on a dairy farm. *Journal of Applied Microbiology* 83, 297–306.

Prazak, A.M., Murano, E.A., Mercado, I. and Acuff, G.R. (2002) Prevalence of *Listeria monocytogenes* during production and postharvest processing of cabbage. *Journal of Food Protection* 65, 1728–1734.

Pritchard, G.C., Willshaw, G.A., Bailey, J.R., Carson, T. and Cheasty, T. (2000) Verocytotoxin-producing *Escherichia coli* O157 on a farm open to the public: outbreak investigation and longitudinal bacteriological study. *The Veterinary Record* 147, 259–264.

Raya, R.R., Varey, P., Oot, R.A., Dyen, M.R., Callaway, T.R., Edrington, T.S., Kutter, E.M. and Brabban, A.D. (2006) Isolation and characterization of a new T-even bacteriophage, CEV1, and determination of its potential to reduce *Escherichia coli* O157:H7 levels in sheep. *Applied and Environmental Microbiology* 72, 6405–6410.

Raya, R.R., Oot, R., Maley, M., Dyen, M., Wieland, J., Callaway, T.R., Kutter, E. and Brabban, A.D. (2011) Naturally resident and exogenously applied bacteriophages can reduce *Escherichia coli* O157:H7 levels in ruminant guts. *Bacteriophage* 1, 15–24.

Ronner, A.B. and Cliver, D.O. (1990) Isolation and characterization of a coliphage specific for *Escherichia coli* O157:H7. *Journal of Food Protection* 53, 944–947.

Rostagno, M.H. and Callaway, T.R. (2012) Pre-harvest risk factors for *Salmonella enterica* in pork production. *Food Research International* 45, 634–640.

Rozema, E.A., Stephens, T.P., Bach, S.J., Okine, E.K., Johnson, R.P., Stanford, K. and McAllister, T.A. (2009) Oral and rectal administration of bacteriophages for control of *Escherichia coli* O157:H7 in feedlot cattle. *Journal of Food Protection* 72, 241–250.

Santos, T.M.A., Gilbert, R.O., Caixeta, L.S., Machado, V.S., Teixeira, L.M. and Bicalho, R.C. (2010) Susceptibility of *Escherichia coli* isolated from uteri of postpartum dairy cows to antibiotic and environmental bacteriophages. Part II: in vitro antimicrobial activity evaluation of a bacteriophage cocktail and several antibiotics. *Journal of Dairy Science* 93, 105–114.

Sargeant, J.M., Amezcua, M.R., Rajic, A. and Waddell, L. (2007) Pre-harvest interventions to reduce the shedding of *E. coli* O157 in the faeces of weaned domestic ruminants: a systematic review. *Zoonoses and Public Health* 54, 260–277.

Scallan, E., Hoekstra, R.M., Angulo, F.J., Tauxe, R.V., Widdowson, M.-A., Roy, S.L., Jones, J.L. and Griffin, P.L. (2011) Foodborne illness acquired in the United States – major pathogens. *Emerging Infectious Diseases* 17, 7–15.

Scharff, R.L. (2010) Health-related costs from foodborne illness in the United States. Report prepared under the Produce Safety Project at Georgetown University. Produce Safety Project <http://www.producesafetyproject.org/admin/assets/files/Health-Related-Foodborne-Illness-Costs-Report.pdf-1.pdf>.

Schellekens, M.M., Woutersi, J., Hagens, S. and Hugenholtz, J. (2007) Bacteriophage P100 application to control *Listeria monocytogenes* on smeared cheese. *Milchwissenschaft* 62, 284–287.

Sheng, H., Kneccht, H.J., Kudva, I.T. and Hovde, C.J. (2006) Application of bacteriophages to control intestinal *Escherichia coli* O157:H7 levels in ruminants. *Applied and Environmental Microbiology* 72, 5359–5366.

Sillankorva, S., Oliveira, R., Vieira, M.J., Sutherland, I.W. and Azeredo, J. (2004) Bacteriophage S1 infection of *Pseudomonas fluorescens* planktonic cells versus biofilms. *Biofouling* 20, 133–138.

Sillankorva, S., Pleteneva, E., Shaburova, O., Santos, S., Carvalho, C., Azeredom, J. and Krylov, V. (2010) *Salmonella enteritidis* bacteriophage candidates for phage therapy of poultry. *Journal of Applied Microbiology* 108, 1175–1186.

Singla, R., Ganguli, A. and Ghosh, M. (2011) An effective combined treatment using malic acid and ozone inhibits *Shigella* spp. on sprouts. *Food Control* 22, 1032–1039.

Siringan, P., Connerton, P.L., Payne, R.J.H. and Connerton, I.F. (2011) Bacteriophage-mediated dispersal of *Campylobacter jejuni* biofilms. *Applied and Environmental Microbiology* 77, 3320–3326.

Sklar, I.B. and Joerger, R.D. (2001) Attempts to utilize bacteriophage to combat *Salmonella enterica* serovar Enteritidis infection in chickens. *Food Safety* 21, 15–29.

Sofos, J.N. and Smith, G.C. (1998) Nonacid meat decontamination technologies: model studies and commercial applications. *International Journal of Food Microbiology* 44, 171–188.

Stanford, K., McAllister, T.A., Niu, Y.D., Stephens, T.P., Mazzocco, A., Waddell, T.E. and Johnson, R.P. (2010) Oral delivery systems for encapsulated bacteriophages targeted *Escherichia coli* O157:H7 in feedlot cattle. *Journal of Food Protection* 73, 1304–1312.

Steinmuller, N., Demma, L., Bender, J.B., Eidson, M. and Angulo, F.J. (2006) Outbreaks of enteric disease associated with animal contact: not just a foodborne problem anymore. *Clinical and Infectious Diseases* 43, 1596–1602.

Talley, J.L., Wayadande, A.C., Wasala, L.P., Gerry, A.C., Fletcher, J., DeSilva, U. and Gilliland, S.E. (2009) Association of *Escherichia coli* O157:H7 with filth flies (Muscidae and Calliphoridae) captured in leafy greens fields and experimental transmission of *E. coli* O157:H7 to spinach leaves by house flies (Diptera: Muscidae). *Journal of Food Protection* 72, 1547–1552.

Taylor, L.H., Latham, S.M. and Woolhouse, M.E.J. (2001) Risk factors for human disease emergence. *Philosophical Transactions of the Royal Society of London B: Biological Sciences* 356, 983–989.

Toro, H., Price, S.B., Hoerr, F.J., Krehling, J., Perdue, M., McKee, S. and Bauermeister, L. (2005) Use of bacteriophages in combination with competitive exclusion to reduce *Salmonella* from infected chickens. *Avian Diseases* 49, 118–124.

Tsuei, A.C., Carey-Smith, G.V., Hudson, J.A., Billington, C. and Heinemann, J.A. (2007) Prevalence and numbers of coliphages and *Campylobacter jejuni* bacteriophages in New Zealand foods. *International Journal of Food Microbiology* 116, 121–125.

USDA/FSIS (2006) Serotypes profile of *Salmonella* isolates from meat and poultry products January 1998 through December 2005. No. 2007. USDA/FSIS, Washington, DC.

Vandeplas, S., Dubois Dauphin, R., Beckers, Y., Thonart, P. and Thewis, A. (2010) *Salmonella* in chicken: current and developing strategies to reduce contamination at farm level. *Journal of Food Protection* 73, 774–785.

Viazis, S., M. Akhtar, Feirtag, J., Brabban, A.D. and Diez-Gonzalez, F. (2011) Isolation and characterization of lytic bacteriophages against enterohaemorrhagic *Escherichia coli*. *Journal of Applied Microbiology* 110, 1323–1331.

Wagenaar, J.A., Bergen, M.A.P.V., Mueller, M.A., Carlton, R.M. and Wassenaar, T.M. (2005) Phage therapy reduces *Campylobacter jejuni* colonization in broilers. *Veterinary Microbiology* 109, 275–283.

Wall, S.K., Zhang, J., Rostagno, M.H. and Ebner, P.D. (2010) Phage therapy to reduce preprocessing *Salmonella* infections in market-weight swine. *Applied and Environmental Microbiology* 76, 48–53.

Wetzel, A.N. and LeJeune, J.T. (2007) Isolation of *Escherichia coli* O157:H7 strains that do not produce Shiga toxin from bovine, avian and environmental sources. *Letters in Applied Microbiology* 45, 504–507.

Zhang, J., Kraft, B.L., Pan, Y., Wall, S.K., Saez, A.C. and Ebner, P.D. (2010) Development of an anti-*Salmonella* phage cocktail with increased host range. *Foodborne Pathogens and Disease* 7, 1415–1419.

Zhang, X., Kong, J. and Qu, Y. (2006) Isolation and characterization of a *Lactobacillus fermentum* temperate bacteriophage from Chinese yogurt. *Journal of Applied Microbiology* 101, 857–863.

17 Phage-therapy Best Practices

Stephen T. Abedon[1]
[1]*Department of Microbiology, The Ohio State University.*

Phage therapy, or phage-based bacterial biocontrol (Abedon, 2009c), is the use of bacteriophages to combat nuisance or pathogenic bacteria. This is done to augment the use of chemical antibacterials or, especially, to serve as alternatives to antibiotics. Phage therapy is useful, particularly: (i) in light of bacterial evolution to antibiotic resistance; (ii) given the limitations to chemical antibacterial efficacy such as against bacterial biofilms; (iii) in terms of the potential for certain antibiotics to cause relatively severe side effects; and (iv) due to concerns over the negative impact of chemical antibacterials on environments. The latter can be a consequence, for example, of antibiotic release into municipal sewage systems following disposal or elimination 'down the drain', or due to the use of chemical antibacterials outdoors to control bacterial pathogens of plants (Balogh *et al.*, 2010).

Phage therapy can be impressively safe, highly efficacious and even relatively economical; see Olszowska-Zaremba *et al.* (Chapter 12, this volume) for a discussion of phage safety and Olszowska-Zaremba *et al.*, Loc-Carrillo *et al.* and Burrowes and Harper (Chapters 12–14, this volume), as well as the review by Abedon *et al.* (2011), for a discussion of phage-therapy efficacy. It is also important, nevertheless, for phage formulations and protocols to be developed knowledgeably. Specifically, there exist a number of approaches to phage choice (Gill and Hyman, 2010), application and experimentation that should only be deviated from for explicit reasons rather than avoided due solely to a lack of appreciation of their importance or utility. Here, I provide a guide to such phage-therapy 'best practices', principles that have been formulated in the course of reviewing the modern phage-therapy and phage-based bacterial biocontrol literature (Abedon and Thomas-Abedon, 2010; Kutter *et al.*, 2010; Abedon, 2010a,b, 2011b; Abedon *et al.*, 2011) in combination with extensive consideration of related issues of phage ecology (Abedon, 2006, 2008a,b, 2009a,d, 2011b,c; Abedon *et al.*, 2009; Chan and Abedon, 2012a). These ideas I present as answers to a series of questions, twelve in all, which I believe should be generally considered over the course of phage-therapy research and development. Four more issues are also presented at the end of the chapter – those of host range, unusual results, the dependence of phage-mediated bacterial killing on phage titres, and a discussion of the meaning or lack thereof of positive results. To avoid misinterpretation of the third point, which seems to occur all too often – that is, dependence of phage-mediated bacterial killing on phage titres – I will be redundant in noting that it is phage densities in particular that determine rates of bacterial killing during phage therapy and not bacterial densities or ratios of added phages to bacteria.

In addition to helping guide the design of more effective phage-therapy experimentation, the ideas presented can also be used towards rational debugging of phage-therapy protocols that have proven to be insufficiently efficacious. The hope is to improve on the overall rigour as well as the effectiveness of phage therapy and to do so particularly by providing consideration of how phage therapy should *not* be practised. This undertaking can also be viewed as an extension of ongoing efforts to tackle issues of phage-therapy pharmacology (Abedon, 2009b,c, 2010a 2011a,b,d, 2012; Abedon and Thomas-Abedon, 2010; Curtright and Abedon, 2011; Ryan *et al.*, 2011; Chan and Abedon, 2012b).

Phage-therapy Pharmacology

In phage therapy, phages are applied to bodies specifically to impact on the constitution of a bacterial community. As bodies exist, pharmacologically, as a combination of both body cells and associated flora (see Letarov, Chapter 2, this volume), this use of phages as antibacterial agents can be viewed as a form of body modification intentionally undertaken for the sake of combating disease. In other words, phages can be productively considered as variations on the concept of an antibacterial *drug* (Abedon, 2012). As the primary goal of pharmacology is to enhance drug utility, so too should the perspective of this chapter be viewed as basically pharmacological. Accordingly, in this section, I provide an introduction to the ideas of pharmacology in general and phage-therapy pharmacology in particular.

Pharmacology is the study of drug–body interactions. Relevant to phage-therapy pharmacology, symbiont–host interactions and organism–environment interactions can be studied from ecological perspectives. Just also as ecological interactions can be considered as occurring between environments and organisms (Abedon, 2011c), so too can drugs be viewed in terms of impacting on bodies or, instead, in terms of bodies impacting on drugs. In pharmacology, these distinct views are described respectively in terms of pharmacodynamics and pharmacokinetics. Pharmacodynamics in particular can be differentiated into positive versus negative consequences of drug action on the body, such as control of bacterial infections versus causing adverse side effects.

Pharmacodynamic effects occur as a function of drug densities as found in specific, target tissues within the body (with the body defined as a combination of both eukaryotic, i.e. human cells and the associated microbiome). Thus, for example, we can consider the impact of antibiotic densities found in the vicinity of bacteria-infected tissues on those bacteria (anti-bacterial effects) as well as side effects affecting human cells and tissues (which have typically negative consequences). Drug density in turn is a consequence of a combination of dosing and pharmacokinetic considerations. Specifically, most drugs following dosing will access the systemic circulation of the blood (absorption), penetrate into non-blood tissues (distribution), be subject to inactivation (metabolism) and may also be eliminated from the body (excretion). For certain drugs, including phages as well as pro-drugs, metabolism – that is, chemical reactions as they occur in for example human cells, human tissues or associated microorganisms – can also lead to increases in active pharmaceutical densities *in situ*.

As with most antibacterials, successful bacterial control can require substantially more phages than there are bacteria present. The generation of these phage densities via dosing alone has been described as a 'passive' treatment, whereas the harnessing of phage population growth to achieve phage densities that are sufficient to result in bacterial control instead has been termed 'active' treatment (Payne *et al.*, 2000; Payne and Jansen, 2001, 2003; Abedon and Thomas-Abedon, 2010; Abedon, 2012). In either case, antibacterial activity with phage therapy is accomplished via a bactericidal action and, other than in terms of phage properties, is entirely dependent on the establishment of sufficient phage densities within the vicinity of the target bacteria, such as the achievement of densities of approximately 10^8 free phages ml^{-1}.

Phage-therapy Best Practices I: Basic Considerations

It is my premise that phage-therapy experimentation could benefit from a more rigorous pharmacological approach. Although in part increases in pharmacological rigour may be addressed in terms of better modelling of phage impact or greater consideration of phage pharmacokinetics, in fact I am referring to much simpler as well as more achievable practices, such as improved consideration of basic microbiology, controls and experimental design. In this section, I consider phage-therapy best practice numbers 1–4, which address the issues of phage isolation, phage choice, phage characterization and disease models. I then provide a discussion within this section before presenting, in subsequent sections, phage-therapy best practice numbers 5–8 (controls) and then 9–12 (dosing and enumeration).

1. Has a good strategy been used to isolate highly efficacious phages?

As with chemical antibiotics, the first step in phage therapy development is the isolation of appropriate phages (Balogh *et al.*, 2010; Gill and Hyman, 2010). Appropriate phages are those that are capable of killing bacteria under the expected *in situ* conditions. This criterion should be viewed as basic to phage choice for phage-therapy use, with phage-isolation protocols designed with this goal in mind. Methods for such isolation ideally will involve enrichment-culture techniques that use bacterial hosts and conditions that are well matched to the expected hosts and environments during treatment, perhaps especially the former (Abedon, in press). Isolation without enrichment, such as via direct plating of environmental or tissue samples, will often provide numerous phages that can kill bacteria but not necessarily those phages that are the most effective at doing so. A red flag is the use of phages for phage-therapy studies that are obtained using mechanisms biased primarily towards convenience, such as through the American Type Culture Collection, unless the goal is to characterize specific phage isolates.

Phage-therapy publications should provide sufficient detail on approaches employed for phage isolation including strengths, shortcomings and rationales justifying their use; enriching for phages at 37°C for use at 15°C, for example, should represent an obvious error in experimental design. Alternatively, it is important for researchers to keep in mind that *in situ* phage growth characteristics are relevant to phage choice only if phage-therapy protocols are to be reliant on active rather than passive treatment. This is because passive treatment is dependent solely on phage bactericidal effects rather than also depending on phage *in situ* amplification. Of course, to a large degree, a demonstration of efficacy (positive results) in conjunction with phage-therapy protocols will negate phage isolation concerns. As will be discussed subsequently, however, not all phage therapy positive results are equivalently efficacious. An additional issue concerning phage choice and isolation is that of host range, which I address towards the end of the chapter.

2. Have the phages been adequately characterized *ex situ*?

From a pharmacodynamic perspective, phages should be chosen for phage therapy based not only on their ability to kill bacteria but also in terms of their relative safety (Abedon and Thomas-Abedon, 2010; Abedon, 2012; see also Olszowska-Zaremba *et al.*, Chapter 12, this volume). With both of these criteria in mind, there are generally three characteristics that should be avoided in phage therapeutics, the first two perhaps more importantly avoided than the third. These include an inability to display lysogenic cycles (i.e. non-temperate phages), a failure to encode bacterial virulence factors (see Christie *et al.*, Chapter 4, this volume) and an inability to display generalized transduction of bacterial genes (see Abedon, Chapter 1, this volume).

Fortunately, *lack* of these three properties can positively correlate with phage ability to kill bacteria. A phage's potential to display lysogenic cycles, however, may not always be obvious or necessarily easy to establish. Notwithstanding the latter caveat, (i) an inability to display lysogenic cycles means that no bacteria will become phage immune in the course of successful phage infection (Fogg *et al.*, 2010); (ii) phage encoding of bacterial virulence factors seems to be associated particularly with temperate phages or their immediate descendants (Hyman and Abedon, 2008); and (iii) non-temperate phages often disrupt bacterial chromosomes in a manner that makes generalized transduction less likely (Wilson *et al.*, 1979; Kutter *et al.*, 1994).

All phage-therapy studies if at all possible should employ phages that have been shown, as explicitly indicated in publications, to form clear plaques when plated on potential target bacteria (and, for a higher level of stringency, demonstration that phage-resistant bacteria isolated from the plaques in fact are not lysogens). This is rather than forming plaques with turbid centres, which serves as a marker for phage temperance, although not necessarily with 100% accuracy (Stent, 1963). Furthermore, not all temperate phages form lysogens on all bacteria they are capable of infecting productively. None the less, at a minimum, a failure to produce clear plaques should be taken as a red flag with regard to the use of a given phage isolate for phage-therapy purposes.

Under certain circumstances, the use of temperate phages may be necessary, such as when there is a lack of availability of non-temperate phages. The choice of using temperate phages, however, must be justified in publications, and even defended. Given the association of bacterial toxin genes in particular with temperate phages, it is also reasonable to expect greater levels of phage characterization by investigators who are using either temperate phages or phages that are derived from temperate phages, that is, when employing phages that are *not* professionally lytic (Curtright and Abedon, 2011). This is additional characterization, particularly in terms of virulence factor encoding.

It is my view that this more stringent characterization should not be an absolute requirement for publication, at least in terms of the early-stage development of phage-therapy protocols. For later-stage development of phage-therapy protocols, especially ones using human subjects, phage encoding of bacterial toxins and virulence factors should be ruled out prior to phage use, even for professionally lytic phages. By contrast, study of the phage potential to inadvertently transduce bacterial virulence factor genes via generalized or specialized transduction has not been a priority among phage-therapy researchers. Presumably, the development of standardized as well as relatively inexpensive protocols for such characterization would be useful, however. In addition, where possible, the propagation of phages *in vitro* for subsequent use *in vivo* should be conducted using bacterial hosts that lack virulence factor genes.

3. Has reasonable *ex situ* phage anti-bacterial virulence been observed?

In vivo or *in situ* experimentation can be expensive. In addition, *in vivo* studies can be subject to ethical considerations, and this is so even if such experiments have a reasonable likelihood of successful outcome. For human testing, of course, these latter considerations are explicit, but issues of expense and researcher time also suggest obvious benefits to developing techniques using simplified experimental or theoretical models prior to more involved testing.

Although various authors have presented *in silico* approaches to understanding phage therapy processes, these methods are neither well developed nor well tested for predictive accuracy (Gill, 2008). Less complicated approaches exist that can allow researchers to calculate the minimum number of supplied phages necessary to achieve phage-therapy success (Abedon and Thomas-Abedon, 2010; Abedon, 2011a,b). These too, however, are not always adequately robust means of phage characterization with regard to bacterial killing ability, as they may be performed prior to ramping up to more complex procedures such as animal testing. What then constitutes reasonably robust *in vitro* characterization of phage-mediated bacterial killing?

First, if at all possible, *in vitro* (i.e. *ex situ*) characterization should at least attempt to simulate *in situ* conditions (Nasser *et al.*, 2002; Gill *et al.*, 2006b; Bull *et al.*, 2010), perhaps particularly in terms of bacterial physiology and density. Thus, if bacteria *in situ* are presumed to be in stationary phase and present at densities of 10^4 CFU ml^{-1}, then *in vitro* testing should employ stationary-phase bacteria and bacterial densities of 10^4 CFU ml^{-1}. Similarly, if biofilms are explicitly serving as antibacterial targets, then a phage's anti-biofilm capabilities might also be explored *in vitro* before moving on to animal testing (Abedon, 2010a; 2011b).

Secondly, measuring bacterial killing in terms of bacterial viable counts can provide more information than optical density or total bacterial cell count determinations. In part, this is because the latter readily determines only a few log reduction in bacterial density. Instead, a good rule of thumb is that one should be able to attain at least 4 log reductions, *in vitro*, over reasonable spans of time (Kasman *et al.*, 2002; Abedon and Thomas-Abedon, 2010; Abedon, 2011b, 2012) when using an approximation of those phage densities that are actually obtainable *in vivo* or *in situ*. Clearly, it is hard to defend moving up to *in vivo* testing using phages for which this substantial amount of bacterial killing has not been or cannot first be demonstrated *in vitro* using presumed *in situ* phage densities.

Thirdly, during *in vitro* testing, phage–bacterial community interactions should be followed over relatively short time intervals so that bacterial growth, giving rise to substantial increases in bacterial densities, does not occur over the course of experimentation. This means, in particular, that overnight incubations of phages with bacteria do not represent an optimal means of characterizing phage impact on bacteria (Abedon, 2011a). Kinetic analyses also are typically preferable to end point determinations when studying phage–bacterial interactions, whether in terms of phage adsorption, phage growth characteristics or phage potential to clear cultures (Hyman and Abedon, 2009).

A standard assay for phage anti-bacterial virulence consists of adding phages at different initial titres to a standardized bacterial culture. Those phages that can clear and otherwise reduce culture viable counts from the lowest starting titre are deemed the most virulent. From Smith and Huggins (1983, p. 2660): 'Ten ml amounts of nutrient broth were inoculated with 3×10^8 viable *E. coli* organisms and three-fold falling numbers of phage particles and incubated for 5 h. The lowest inoculum of phage particles that produced complete clearing of the bacterial cultures was then recorded.' Such assays are a measure of multiple phage parameters that can contribute to phage population growth rates as well as anti-bacterial activity, although not all of these parameters are necessarily relevant to all phage-therapy protocols (Abedon, 2009b).

Alternative approaches to measuring a phage's bacterial killing ability have been described by Abedon and Thomas-Abedon (2010) and Abedon (2011a,b, 2012). Furthermore, unless actively designed to do so, *in vitro* assays are not necessarily a measure of a phage's ability to penetrate to target bacteria *in vivo*, such as into biofilms or in terms of phage movement between body compartments. Still, before moving on to animal or otherwise more costly *in situ* experimentation, phages should at least display *in vitro* therapeutic abilities, such as 4 log reductions in bacterial viable counts at phage densities that are readily attainable *in situ*. In any case, it is not so much that robust and effective *in vitro* phage characterization is crucial to successful phage therapy development but instead that such efforts are cheaper, less time-consuming and frankly more humane than relying entirely on animal testing to characterize phage isolates for bacterial killing ability.

4. Has bacterial colonization been established prior to phage application?

While *ex situ* testing is important for practical as well as economic reasons, it is crucial to make sure that subsequent testing employs animal or other models that reasonably represent the actual circumstances under which phage therapy is envisaged. What constitutes a good model for *in vivo* or *in situ* phage therapy efficacy? At a minimum, effort should be made to make sure that bacterial colonization along with tissue invasiveness and other infection details occur to an extent that is similar to that seen with typical infection presentation (see, for example, Loc-Carrillo *et al.*, Chapter 13, this volume). For instance, if biofilms are normally present, and have developed for days, weeks or months prior to the onset of antibacterial treatment, then such biofilms should also be present in the disease model employed to explore the potential for phage-therapy efficacy (Ramage *et al.*, 2010; Abedon, 2010a; 2011b).

The ultimate indication of poor technique in terms of failing to allow establishment of bacterial colonization prior to phage application is the mixing of phages with bacteria *prior* to bacterial challenge, although simultaneous or even just short delays between bacteria and phage addition should also be viewed as suspect. Indeed, often the result of such practices is what appear to be bacterial reductions only to below minimal lethal densities rather than a 'curing' of existing disease. At a minimum, therefore, effort should be made during phage-therapy experimentation to avoid mixing bacterial challenges with phages in a manner that mimics phage–bacterial interactions as they can occur in broth cultures. Usually such avoidance will be the case given substantial delays between bacterial challenge and phage addition.

Researchers should consider treatment difficulties that may be encountered because of such delays as an indication of a need for further development of phage-therapy protocols and/or a need for identification of more effective phages. Such considerations would contrast with observation of reductions in efficacy – given substantial delays between bacterial challenge and phage application – serving instead as study end points, as too often is the case in the phage-therapy literature. Such a reduction in efficacy rather should serve as the *starting* point in terms of *in situ* protocol development. In particular, unless protocols have been designed specifically to test for the prevention of initiation of bacterial infections (prophylaxis), then at a minimum multiple hours should separate bacterial challenge and subsequent phage addition (and if testing of prophylaxis is envisioned, then phage addition should *precede* bacterial challenge by a substantial length of time rather than being simultaneous or near simultaneous to bacterial application). An alternative approach involves applying phages and bacteria to different body compartments, although determining whether such approaches are truly good models for determining phage-therapy efficacy should be the subject of rigorous pharmacokinetic testing rather than simply assumed.

Discussion

Given the above considerations, I envisage a logical four-step process of *initial* phage-therapy development: (i) rational phage isolation including in terms of avoidance of temperate phages as well as the carriage of virulence-factor genes; (ii) *in vitro* phage characterization for anti-bacterial virulence; (iii) *in vivo* or *in situ* proof-of-principle efforts using models in which efficacy is highly expected while at the same time limitations of technique may be identified (particularly in terms of delays between bacterial challenge and phage application); and then (iv) use of more realistic disease models to develop clinically useful therapeutic techniques, that is, that improve on efficacy at or beyond the limits of proof-of-principle success (again, particularly following substantial delays between bacterial challenge and the initiation of phage treatment). In short, the field of phage therapy, even in Western literature, has moved beyond the point of proof of principle (e.g. see Abedon *et al.*, 2011). Thus, if animal experimentation is indicated, then that experimentation really ought to be properly done. This means, perhaps more than anything else, that effort should be made to improve the accuracy of disease models (see Loc-Carrillo *et al.*, Chapter 13, this volume).

Ultimately phage-therapy protocols need to be tested against naturally occurring infections, although in my opinion this represents a later step in development. Thus, in terms of publication, I envisage possibly a four-step process: Step (iv), as described above, might be the first publication in the series. Next, a step (v) could seek to expand the step (iv) type efforts to a greater range of bacterial target strains or treatment conditions. Step (vi) might then incorporate treatment of a small number of naturally occurring infections, representing essentially a limited trial. Step (vii) would then entail a fairly substantial trial, perhaps under more real-world conditions. For human application, there will additionally be phase I safety testing, prior to limited efficacy testing (phase II), and then full-blown clinical trial (phase III), all of which will follow what testing can be accomplished using animal models. Animal testing of the treatment of human diseases, however, may not always be extendable beyond step (v).

By way of example, consider the development of phage-therapy treatment against *Pseudomonas* ear infections (see Burrowes and Harper, Chapter 14, this volume, for references). Here, the first publication addressed phage treatment of a naturally acquired infection in a single dog, the 'second' publication considered the treatment of naturally acquired infections in ten additional dogs, the 'third' publication was a safety and limited efficacy trial in humans, and currently a phase III clinical trial is either planned or under way (note that the quotes are because the 'third' publication came out prior to the 'second' publication).

Phage-therapy Best Practices II: Controls

The subject of controls during phage-therapy experimentation is complicated by the issue of double blinding of well-designed clinical trials. Thus, in addition to researcher blinding, it is important not to overlook the more basic experimental controls – negative controls, positive controls and controlled variables. In this section, I consider best practice numbers 5–8, all of which involve questions of how to properly control as well as reduce biases during phage-therapy experimentation.

5. Has a positive phage-killing control been used?

In this section, I will concentrate on positive controls; that is, can a phage-therapy protocol, under some approximation of the best of all possible circumstances, achieve phage-therapy efficacy even if it fails to do so under other, perhaps more clinically realistic, conditions? Generally, the best of all possible circumstances will involve procedures that assure that sufficient phage densities are present to allow substantial killing of target bacteria such as may be equivalent to that observed during *in vitro* testing (above). This would particularly be the use of an approximation of passive treatment, meaning supplying via dosing enough phages to achieve bacterial clearing, including by means of repeated phage dosing (below). A good rule-of-thumb definition of 'enough', whether provided by active or passive means, may be the achievement of a density of at least 10^9 phages ml^{-1} to the actual physical location of target bacteria, or at least 10^8 given somewhat ideal conditions (Abedon and Thomas-Abedon, 2010; Abedon, 2011a,b, 2012; Curtright and Abedon, 2011). Keep in mind, furthermore, that the dosing of animals may need to be higher than that of broth cultures to account for phage loss (excretion), dilution (associated with absorption, distribution and inefficiencies in movement between body compartments), decay (as due to 'metabolism') or inefficiencies of phage penetration to bacteria (inefficiencies in distribution as described above but also inefficiencies in penetration into biofilms).

It should be strongly stressed that positive controls are most relevant, and arguably perhaps only relevant, if phage-therapy protocols fail to otherwise achieve reasonable levels of efficacy (i.e. sufficiently positive results). In other words, negative experimental results generally are more meaningful to the extent that positive experimental results – phage-mediated bacterial clearing positive controls – are demonstrably possible, such as might be achieved by adding more phages, employing more effective phages, preparing infections better for phage treatment and/or dosing more frequently (Abedon and Thomas-Abedon, 2010; Abedon, 2010a, 2011a,b). To improve on experimental phage therapy outcomes, it therefore may be necessary for issues of convenience, or of economy of use of materials, to not be allowed to interfere with the achievement of greater efficacy, particularly should results otherwise prove disappointing. Studies should not be published if only poor or negative results are obtained unless a robust effort has been made to improve on these results.

Another way of making the latter point is that publication of negative results is unquestionably important but only if those negative results are assuredly *not* simply a consequence of poor or otherwise insufficient experimental technique. Before publishing negative results, there thus should always be concern that sufficient effort has been made to achieve desired outcomes. Above all, one should always be worried that the phages employed were in some manner inadequate, that the doses used were in some manner insufficient or that the rates or duration of dosing were to some degree in need of further increase. Thus, try to establish especially whether supplying more phages will result in greater overall levels of bacterial killing.

Note that a phage-killing positive control is not the only kind of positive control that is possible in phage experiments. For example, if one is comparing phage performance with a normal standard of care, then an antibiotic-positive control may be employed. The use of such a control, however, does not substitute for the use of a phage-killing positive control. This is because the question that is being addressed with phage-killing positive controls is not that of phage therapy efficacy in comparison with other treatment options but instead whether, in the *absence* of desired levels of bacterial-treatment success, insufficiencies in dosing might be to blame. In other words, there is little information that may be obtained should phage application result in poor outcomes under circumstances in which antibiotics nevertheless are effective, except that more effort may need to be put into phage-therapy protocol development.

6. Has a reasonable negative-treatment control been used?

The best negative control to employ during phage-therapy protocols is not necessarily either obvious or straightforward. None the less, the use of a reasonable negative control is crucial to determining phage-therapy efficacy. To simplify the discussion, it can be easier to view negative controls not in terms of negative efficacy but rather in terms of positive infection outcomes, such as continued subject morbidity or resulting subject death. Thus, the negative control is an approximation of the normal phage-therapy protocol but one that allows a normal infection outcome, as seen without phage treatment. Clearly, the easiest means of attaining this outcome is to avoid applying phage formulations altogether. This approach, however, is complicated by the need to differentiate phage-associated therapy efficacy from efficacy that is associated instead with phage carriage material or other aspects of treatment protocols that might not be employed in the absence of phage application. Particularly, non-phage components of phage formulations can contain complex bacterial-lysis products (Gill and Hyman, 2010). Therefore, as phage-therapy negative controls, it is better practice to use some kind of mock treatment that is a good approximation of the phage formulation, only lacking in active phages.

Phages can be removed from formulations via filtration or destroyed via heating. Formulations otherwise can be created in a phage-free manner that approximates the phage-generated material. Alternatively, phages can be sufficiently purified prior to use such that a simple buffer, as equivalent to that in which the phages have been suspended, may be employed as the negative control. Unfortunately, not much effort has been put into determining whether these various substitutions provide similar or even reasonable performance as true phage-negative controls. None the less, it is crucial to employ negative controls during phage-therapy testing that allow a definitive distinction between phage-mediated efficacy and efficacy that is due instead to non-phage components of the treatment.

7. Have experiments been blinded?

It is the gold standard of clinical trials to employ double blinding, such that nobody knows who has received the treatment versus the placebo (negative control) until the data has been collected from the trials. These methods are important for the sake of removing biases from what can be subjective observations, such as how sick the subject is. In addition, blinding is also important to remove biases in the treatments themselves, particularly if the treatments are relatively complicated or, instead, if there is any potential that subjects may be chosen or taken care of differently depending on the treatment type.

It is my opinion that if clinical trials should be expected to display double blinding, then at a minimum some effort should be made to remove biases from *in vivo* or *in situ* testing as well, such as by blinding researchers who are directly involved in a protocol from the treatment being employed. This approach can be seen, for example, in the mastitis study of Gill *et al.* (2006a). More generally, subtle differences between treatment results, even if shown to be statistically significantly different, should not be considered to be definitive indicators of efficacy unless, at a minimum, some form of bias reduction such as blinding has been employed.

8. Has testing for phage formulation toxicity been explicitly performed?

An ongoing concern with regard to phage therapy is the potential for introduction of bacterial toxins into patients in the course of phage application. This occurs because bacterial lysis is involved in phage stock preparation, and lysis has the potential to release a number of bacterial toxins, in particular endotoxin from Gram-negative bacteria (Gill and Hyman, 2010). To guard against the introduction of these toxins into patients, phage preparations can be purified. Even given such purification, in the course of experimentation an important control is to show that phage application in and of itself will not lead to toxicity. Indeed, it can be important to ascertain what level of phage purification is actually necessary to protect subjects from side effects.

Fortunately, toxicity testing is easily accomplished through the application of phage doses to animals. The importance of such controls is not just to show that formulations as well as the phages themselves that are employed in a given study are non-toxic but also to build up in the literature a record of phage toxicity testing. Should formulation toxicity in fact turn out to be the case, then production or delivery approaches may be modified. For example, greater lysate purification or less invasive application – the latter being, for example, topical use rather than systemic use – may be employed. Note that *in vitro* testing of phage formulations, especially for endotoxin presence, is also routinely available (Boratynski *et al.*, 2004). See Olszowska-Zaremba *et al.* (Chapter 12, this volume) for further discussion on issues of the potential for phages to display toxicity.

Phage Therapy Best Practices III: Dosing and Enumeration

Certain additional details can be relevant to successful phage therapy experimentation. In this section, I consider issues of dosing number, dosing level and dosing description, as well as bacterial enumeration. These constitute best practice numbers 9–12.

9. Has multiple or continuous dosing been attempted?

A conceit too often found in the phage-therapy literature is the notion that phages, as potentially self-amplifying antibacterials, ought to be efficacious even if applied only once. By contrast, it is atypical among pharmaceuticals to demand efficacy following only a single application, although certainly that can be desirable from a treatment perspective. Notwithstanding such issues, it seems unreasonable to expect a priori that phages should consistently work despite such intentional hobbling in terms of dosing. For that reason, phage-therapy protocols, particularly where human treatment is the goal, should not be intentionally designed to employ only a single dosing unless benefits gained from such an approach are substantial – such as in terms of economics, convenience or reduced invasiveness – while effectiveness is only minimally lost.

The alternative perspective is that, to the extent that a phage-therapy protocol has failed to provide sufficiently efficacious results, strategies ought to be explored that are aimed at improving outcomes. Such strategies might include providing phages multiple times over the course of treatment, providing phages with increasing frequency (e.g. such as multiple times per day rather than only once per day) or even providing phages continuously (Abedon, 2011a, 2012). Of course, if single dosing can be shown to be efficacious and to provide sufficient benefits in terms of convenience and/or economy, then it would not be unreasonable to limit phage application to just one dose per treatment. It is dangerous, however, to assume that single dosing represents a default means by which phage therapy is implemented. In particular, where phage therapy is actually practised clinically, such as in Georgia and Poland, single dosing does not represent the standard of care (Kutter *et al.*, 2010; Abedon *et al.*, 2011). See Loc-Carrillo *et al.* (Chapter 13, this volume) for additional discussion of this issue.

10. Has there been inappropriate reliance on active treatment?

Similar to the idea of a utility associated with multiple rather than single dosing, so too there exists a usefulness to supplying adequate phage numbers per dose. Supplying sufficient phage numbers is an obvious dosing criterion given consideration of phage therapy from a pharmacological perspective. Even so, the prospect of phages multiplying while in the presence of target bacteria, potentially resulting in active treatment, can lead to assumptions that such multiplication might always occur, plus occur to a degree that phages will achieve sufficient densities to eradicate bacteria (Abedon and Thomas-Abedon, 2010; Abedon, 2010a, 2011a,b, 2012).

Notwithstanding this potential, and unless there is strong reason to believe otherwise, in the course of debugging phage-therapy protocols – that is, experimentally addressing the question of why results may not be as efficacious as anticipated or desired – one should always question whether sufficient phages were applied per dose. In addition, one should question whether a sufficient frequency or number of doses was applied, as well as whether the phages themselves have sufficient antibacterial activity. Thus, for example, if one observes a single log killing with an application of 10^7 phages per dose, where such minimal killing should not be a desired end point of any phage-therapy research and development programme, then it would not be unrealistic to consider whether adding 10^8 phages per dose might improve outcomes or supplying, for instance, ten rather than a single phage dose to increase levels of bacterial eradication.

Further considerations include the following: (i) when employing phages that have been engineered to not reproduce, such as to avoid *in situ* bacterial lysis, which will release endotoxin (Bull and Regoes, 2006; Goodridge, 2010; Abedon and Thomas-Abedon, 2010), then only those phages supplied will be present and passive treatment occurs by default; (ii) the rate at which phages find, adsorb and then kill bacteria is a function of phage density – that is, efficient bacterial adsorption by phages generally is not a function of either bacterial densities or ratios of phages added to bacteria but instead is almost entirely dependent on the phage titres that are present within an environment. Consequently, even if an excess of phages over bacteria is employed, this does not mean that phages will rapidly adsorb and then kill the bacteria. Indeed, at low phage *and* low bacterial densities it can take hours, days, weeks or even longer for a reasonable fraction of bacteria to become phage adsorbed (Goodridge, 2008; Hagens and Loessner, 2010; Abedon, 2011a,d, 2012).

To reiterate these first two points, if insufficient phage densities are present within the environment, then insufficient bacterial killing will occur and such insufficiencies in bacterial killing are possible even given a substantial excess of phages over bacteria. The latter is the case because generally the only absolute determinant of the likelihood or rate at which a given bacterium becomes phage adsorbed, other than phage properties, is the *density* of phage particles that are present in the environment of the bacterium. Furthermore, this statement is different from the question of the overall rate of phage-to-bacterium adsorption that will occur within an environment, which will increase based on phage densities, bacterial densities or indeed ratios of phages present to bacteria. Thus, if more phages and bacteria are present within an environment, then of course there will be more phage-to-bacterium encounters. The important issue in phage therapy, however, is the per capita rate at which bacteria become adsorbed by phages, and that rate is entirely a function of phage densities, phage adsorption properties and the ability of phages to reach target bacteria, that is, the basic pharmacological parameters of drug dosing, drug functionality given access to targets and drug ability to reach targets. Similarly, we expect rates at which bactericidal antibiotics kill bacteria to be a function particularly of antibiotic densities.

Alternatively, (iii) interference between phages making up phage cocktails may be greater when phages are supplied at higher versus lower densities, particularly given co-infection of individual bacteria by heterogeneous phages (Abedon, 1994). The result could be a utility associated with supplying phages at lower densities per dose, perhaps in combination with repeated or continuous dosing. See Chan and Abedon (2012b) for further discussion of this admittedly somewhat hypothetical issue, as well as Abedon (2011e) for a related issue – that of lysis from without. Lastly, (iv) if contamination of formulations with bacterial lysis products such as endotoxin is a problem, then using more phages per treatment will result in patient exposure to more of these toxins. Thus, while using more phages per dose can be an important consideration in optimizing phage-therapy protocols, there are potential tradeoffs associated with this strategy, although these are all tradeoffs that may be readily explored experimentally.

11. Has the concept of 'multiplicity of infection' been used correctly?

To achieve bacterial killing, phages must be present in sufficient densities such that at best only a small fraction of susceptible bacteria remain uninfected. These phage densities can be supplied as a direct consequence of dosing (passive treatment) or can be generated *in situ* as a result of phage multiplication on target or other bacteria (active treatment). In either case, it is the titre of phages that are present in the vicinity of target bacteria that determines the degree to which bacterial densities are reduced, rather than necessarily the *ratio* of phages supplied to bacteria present (Abedon and Thomas-Abedon, 2010). The latter has been described as an *input* multiplicity of infection (Kasman *et al.*, 2002) and is a flawed approach to dosing considerations (Abedon, 2011a,b).

The reason to avoid framing phage dosing in terms of input multiplicities is that what matters in terms of bacterial killing is the multiplicity of phages that actually succeed in adsorbing (*actual* multiplicity of infection). This multiplicity, however, will vary not only with phage densities, phage properties and *in situ* conditions but also with bacterial densities. Thus, if bacterial densities happen to be low, then the input multiplicity of a given dosage of phages (titre) will be high, but if bacterial densities happen to be high, then the input multiplicity associated with that same phage dose will be low. Furthermore, in a treatment situation, actual bacterial densities often will not be known and therefore dosing expressed as an input multiplicity of infection will not be possible.

Phage dosing thus should be described in terms of phage titres along with formulation volumes if phages are added directly to bacteria. Alternatively, absolute virion numbers supplied may be more relevant if phage application to bacteria is less direct, such as oral dosing. The bottom line is that not only is describing phage dosing solely in terms of multiplicities of infection usually not useful, it can also be misleading, and in most or all cases should be avoided.

12. Have the bacteria been adequately protected during enumeration?

There are a number of means of assessing phage impact during treatment of bacterial infections. These include reductions in subject mortality, improvement in subject health and various measures of infection severity. In terms of infection severity, the most objective measure can be of *in situ* bacterial density, and this is particularly so since the phage impact on bacteria during phage therapy is to directly reduce bacterial viable counts.

Although it often is desirable to assess phage-therapy efficacy in terms of determining bacterial viable counts, actually determining those counts, unfortunately, is complicated by the potential for phages to infect bacteria during the enumeration process itself. The result of such infection can be perception of a reduction in bacterial densities that exceeds actual *in situ* reductions. Thus, enumeration procedures if nothing else should be designed as well as assessed in such a manner that little doubt exists that the observed bacterial killing occurred *in situ* rather than *ex situ* during the enumeration process. Note that the primary reason that this is a concern is when the enumeration step involves disruption of spatial structure associated with bacterial infections such that phage penetration to bacteria is more efficient during the enumeration procedure than it had been during the actual treatment process. Such disruption might occur following tissue or biofilm suspension and/or homogenization into liquid solutions.

There are three ways to address this concern of phage *ex situ* killing of bacteria: (i) inactivate phages prior to the enumeration step (particularly prior to any bacterial resuspension in buffer step); (ii) dramatically dilute phage densities in the course of enumeration (e.g. 100-fold or more depending on phage concentrations *in situ* since, as discussed above as well as in the following section, the phage killing potential is a function of phage densities rather than of ratios of phages to bacteria); and (iii) employ enumeration controls that assure the integrity of enumeration results. Such controls involve determining phage densities during enumeration and/or the degree to which bacterial densities decline in the course of prolonged *ex situ* but pre-plating incubation. In short, a carefully designed phage therapy experiment will include vigilant consideration of the shortcomings associated with various means of outcome determination, and an obvious concern is the potential for phages to adsorb bacteria following homogenization of phage- and bacteria-containing samples within buffer. In some cases, protection of phages during their enumeration can also be an issue, as discussed by Loc-Carrillo *et al.* (Chapter 13, this volume).

Additional Issues

In this section, I discuss additional issues that can be relevant to phage-therapy experimentation. These are the relevance of phage host range, particularly to laboratory phage-therapy experimentation as typically practised, and the need to follow up unexpected results. In addition, I provide a quick overview of the logic of phage adsorption theory, which results in phage titres being more relevant to rates of bacterial decline due to phage adsorption than bacterial densities, and, finally, a reiteration about the relevance, or lack thereof, of positive results.

Host range and laboratory experimentation

In the presented discussions, I have mostly ignored the topic of phage host range (Hyman and Abedon, 2010; Abedon, 2011a; Chan and Abedon, 2012a). Host range is typically measured in terms of phage impact on bacteria in solid or semi-solid media, that is, in terms of plaque or spot formation. Measured in this way, host range is often of primary concern within phage-therapy publications, and sometimes is seemingly emphasized more than issues of actual phage therapy efficacy. Notwithstanding this emphasis, host range is relevant mostly in terms of modifying the spectrum of activity of phage therapeutics, including in terms of the use of fewer versus greater numbers of phage types per formulation (Chan and Abedon, 2012b).

By contrast, the range of bacterial strains that phages can 'spot' on should *not* directly impact the approach that is taken to most phage-therapy experiments, particularly as practised in the laboratory (versus in the clinic or on the farm). This is because such experiments typically involve the application of specific phage isolates to infections caused by specific target bacterial strains, that is, bacteria against which phages have already been characterized. In any case, it is phage anti-bacterial virulence, *in situ* phage-infection performance and the various details of phage-formulation application that are salient to phage-therapy success. Thus, while phage host range is relevant to phage choice for subsequent experimentation, the range of bacteria a phage can plaque or spot on itself does not *directly* impact that experimentation, at least as usually practised in the laboratory, unless treatments specifically are of infections by bacteria that a priori display unknown phage susceptibilities. See Williams and LeJeune as well as Cox (Chapters 6 and 10, this volume) for further consideration of phage spotting and plaque formation.

Follow-up of unexpected results

Experimentation can be both expensive and time-consuming. Good scientific method none the less involves not only implementing all necessary controls but also employing a valid theoretical framework within which experiments are conducted. This framework includes knowledgeable hypothesis making and testing, as well as prediction of experimental outcomes. More than occasionally, results will be unexpected in some manner. To the extent that such results are germane to the study being conducted, it is then imperative that additional experiments and/or controls be designed and conducted to achieve a greater understanding of one's experimental system.

If adding phages to bacteria does not result in the degree of bacterial killing that one expects (see Abedon (2011a,b, 2012) for tips on how to make such predictions), then clearly factors not taken into account may be operating within the system under study. More importantly, if one observes unexpectedly poor efficacy, then it can be important for these results to be followed up with additional experimentation, if possible with these additional experiments published at the same time as the original unexpected observation. An important example of such a result occurs when dose-response curves (phage density added relative to treatment outcome) are non-linear, and particularly so where adding more phages results in substantial reductions in efficacy (Abedon and Thomas-Abedon, 2010; Abedon, 2012) or, hypothetically, should multiple dosing prove inferior to single-dosing protocols. In such circumstances it is necessary – for both the development of the field as a whole and the development of the specific system under study – to show that such results can be replicated *ex situ*, explained mechanistically and/or overcome via better treatment design. For example, *in vitro* exploration might be attempted when lysis from without is suspected (Abedon, 2011e). In short, publication of unexpected results can be confusing to the field and particularly so when such results are not subject to reasonable, subsequent exploration.

Importance of phage titres to phage-therapy efficacy

A basic premise in pharmacology, only rarely violated, is that the administration of more drug results in greater physiological impact (although not always generating *better* results). The same is true for phage application, at least in terms of rates of phage adsorption by bacteria. In particular, basic phage adsorption theory dictates that the rates of phage-bacterial encounters will be a function of both phage and bacterial densities. Crucially, however, the rate at which a given bacterium becomes phage adsorbed is a function of phage density and *not* of bacterial density, except to the extent that bacterial densities can impact on phage densities.

Specifically, the rate of phage–bacterial adsorptions in a given environment is equal to PNk where P is phage density, N is bacterial density and k is the likelihood of encounter, and subsequent adsorption of one phage to one bacterium. The number of phage adsorptions to a specific bacterium, per unit time, thus is equal to Pk, as N has been defined, in this scenario, as equal to 1 (that is, a 'specific' bacterium). Indeed, no matter how many bacteria are present, the likelihood of phage adsorption to any one bacterium remains Pk because this view is from the bacterium's perspective rather than that of the phage. Key to this calculation, however, is that P is being held constant, meaning in effect that phages are present in excess.

By analogy, in war the more ammunition (phage equivalent) that one throws at one's enemy (bacterial culture or infection equivalent), then the greater the likelihood that any given enemy soldier (bacterium equivalent) will die. The more enemy soldiers that are present, the more enemy soldiers that will die, but unless soldiers are shielding other soldiers, the per capita rate of enemy soldier loss will be a function of how much ammunition you throw at them rather than how many enemy soldiers happen to be present. Keep in mind, however, that greater impact in the short term (e.g. winning battles) does not always translate into greater impact in the longer term (e.g. winning wars), just as using more phages per dose does not always translate into consistently better treatment outcomes (above). For additional consideration of these ideas, see, for example, Abedon (1990, 1999), Kasman *et al.* (2002), Goodridge (2008), Abedon (2009c), Hagens and Loessner (2010), Abedon and Thomas-Abedon (2010) and Abedon (2011a,b,d, 2012).

Positive results

The goal of phage-therapy research and development should be an improvement in treatment results. Well-documented studies that achieve positive results therefore can be inherently more valuable than studies that fail to demonstrate efficacy, although good documentation in the latter case may provide clues as to what strategies *not* to employ, or indeed what may have gone wrong. Notwithstanding these considerations, the concept of a positive treatment result in fact has some ambiguity. One concern is that there is a difference between statistically significant positive results and biologically or clinically meaningful outcomes: very small but none the less reproducible results may not be worth exploring further, although they might hint instead at a potential for greater real-world significance, for example were more or different phages employed.

An additional ambiguity concerns the suitability of experimental models. Again, although positive results are preferred to negative experimental results, positive results are valuable only to the extent that they are meaningful. Thus, infection models that do not provide realistic inhibitions on phage penetration to bacteria, or which otherwise fail to allow bacterial colonization prior to the initiation of treatment, may bias results unrealistically towards efficacy. Authors should strive to recognize as well as to fully explain flaws in the systems they are using and, of course, endeavour to improve upon those systems such that major or substantive problems are reduced or eliminated. Although such advice should be obvious when experimental results are disappointing, self-criticism is important even given positive experimental outcomes.

Conclusion

The development of any field requires advancements in techniques, not just in terms of technology but also such that poor practices are discouraged or eliminated. Phage therapy has benefited from ongoing improvements in techniques over the last century since the discovery of phages, especially in terms of phage molecular characterization. This increasing technical mastery does not always translate into an implementation of best practices, however. In addition, techniques, no matter how powerful, always have limitations. The quality of actual scientific output thus will depend on a substantial appreciation of which practices are most relevant towards achieving meaningful phage-therapy efficacy along with a reasonable appreciation of the limitations of these techniques.

The greatest strength of phages as antibacterial agents is that, in many instances, they seem to give rise to bacterial-control efficacy even without a robust pharmacological understanding (Abedon and Thomas-Abedon, 2010). A weakness, however, is that phages are relatively easy to work with, thereby lowering barriers of entry into the field. The result can be a lack of subtle understanding of phage biology, phage therapy, experimental design and even of the phage literature in the course of designing phage-therapy approaches. In this chapter, I have endeavoured to point out technical flaws that seem to be commonly associated with modern phage-therapy studies. It is my hope that this discussion will help to provide a roadmap for researchers towards improvement in real-world phage-therapy efficacy. More generally, phages possess vast potential in terms of combating bacterial disease and otherwise improving health, but effective realization of that potential will require both innovative and well-considered research and development.

Acknowledgements

Thank you to Catherine Loc-Carrillo, Jason Gill and, of course, Paul Hyman, for their help in editing this chapter.

References

Abedon, S.T. (1990) Selection for lysis inhibition in bacteriophage. *Journal of Theoretical Biology* 146, 501–511.

Abedon, S.T. (1994) Lysis and the interaction between free phages and infected cells. In: Karam, J.D. (ed.) *The Molecular Biology of Bacteriophage T4*. ASM Press, Washington, DC, pp. 397–405.

Abedon, S.T. (1999) Bacteriophage T4 resistance to lysis-inhibition collapse. *Genetical Research* 74, 1–11.

Abedon, S.T. (2006) Phage ecology. In: Calendar, R. and Abedon, S.T. (eds) *The Bacteriophages*, 2nd edn. Oxford University Press, Oxford, UK, pp. 37–46.

Abedon, S.T. (2008a) *Bacteriophage Ecology: Population Growth, Evolution, and Impact of Bacterial Viruses*. Cambridge University Press, Cambridge, UK.

Abedon, S.T. (2008b) Ecology of viruses infecting bacteria. In: Mahy, B.W.J. and van Regenmortel, M.H.V. (eds) *Encyclopedia of Virology*, 3rd edn. Elsevier, Oxford, UK, pp. 71–77.

Abedon, S.T. (2009a) Bacteriophage intraspecific cooperation and defection. In: Adams, H.T. (ed.) *Contemporary Trends in Bacteriophage Research*. Nova Science Publishers, Hauppauge, NY, pp. 191–215.

Abedon, S.T. (2009b) Impact of phage properties on bacterial survival. In: Adams, H.T. (ed.) *Contemporary Trends in Bacteriophage Research*. Nova Science Publishers, Hauppauge, NY, pp. 217–235.

Abedon, S.T. (2009c) Kinetics of phage-mediated biocontrol of bacteria. *Foodborne Pathogens and Disease* 6, 807–815.

Abedon, S.T. (2009d) Phage evolution and ecology. *Advances in Applied Microbiology* 67, 1–45.

Abedon, S.T. (2010a) Bacteriophages and biofilms, In: Bailey, W.C. (ed.) *Biofilms: Formation, Development and Properties*. Nova Science Publishers, Hauppauge, NY, pp. 1–58.

Abedon, S.T. (2010b) The 'nuts and bolts' of phage therapy. *Current Pharmaceutical Biotechnology* 11, 1.

Abedon, S. (2011a) Phage therapy pharmacology: calculating phage dosing. *Advances in Applied Microbiology* 77, 1–40.

Abedon, S.T. (2011b) *Bacteriophages and Biofilms: Ecology, Phage Therapy, Plaques*. Nova Science Publishers, Hauppauge, NY.

Abedon, S.T. (2011c) Communication among phages, bacteria, and soil environments. In:

Witzany, G. (ed.) *Biocommunication of Soil Microorganisms*. Springer, New York, NY, pp. 37–65.

Abedon, S.T. (2011d) Envisaging bacteria as phage targets. *Bacteriophage* 1, 228–230.

Abedon, S.T. (2011e) Lysis from without. *Bacteriophage* 1, 46–49.

Abedon, S.T. (in press) Bacteriophages as drugs: the pharmacology of phage therapy. In: Borysowski, J., Miêdzybrodzki, R. and Górski, A. (eds) *Phage Therapy: Current Research and Applications*. Caister Academic Press, Norfolk, UK.

Abedon, S.T. and Thomas-Abedon, C. (2010) Phage therapy pharmacology. *Current Pharmaceutical Biotechnology* 11, 28–47.

Abedon, S.T., Duffy, S. and Turner, P.E. (2009) Bacteriophage ecology. In: Schaecter, M. (ed.) *Encyclopedia of Microbiology*. Elsevier, Oxford, UK, pp. 42–57.

Abedon, S.T., Kuhl, S.J., Blasdel, B.G. and Kutter, E.M. (2011) Phage treatment of human infections. *Bacteriophage* 1, 66–85.

Balogh, B., Jones, J.B., Iriarte, F.B. and Momol, M.T. (2010) Phage therapy for plant disease control. *Current Pharmaceutical Biotechnology* 11, 48–57.

Boratynski, J., Syper, D., Weber-Dabrowska, B., Lusiak-Szelachowska, M., Pozniak, G. and Górski, A. (2004) Preparation of endotoxin-free bacteriophages. *Cellular and Molecular Biology Letters* 9, 253–259.

Bull, J.J. and Regoes, R.R. (2006) Pharmacodynamics of non-replicating viruses, bacteriocins and lysins. *Proceedings of the Royal Society of London Series B Biological Sciences* 273, 2703–2712.

Bull, J.J., Vimr, E.R. and Molineux, I.J. (2010) A tale of tails: sialidase is key to success in a model of phage therapy against K1-capsulated *Escherichia coli*. *Virology* 398, 79–86.

Chan, B.K. and Abedon, S.T. (2012a) Bacteriophage adaptation, with particular attention to issues of phage host range. In: Quiberoni, A. and Reinheimer, J. (eds) *Bacteriophages in Dairy Processing*. Nova Science Publishers, Hauppauge, NY, pp. 25–52.

Chan, B.K. and Abedon, S.T. (2012b) Phage therapy pharmacology: phage cocktails. *Advances in Applied Microbiology* 78, 1–23.

Curtright, A.J. and Abedon, S.T. (2011) Phage therapy: emergent properrty pharmacology. *Journal of Bioanalysis and Biomedicine*, S6.

Fogg, P.C., Allison, H.E., Saunders, J.R. and McCarthy, A.J. (2010) Bacteriophage lambda: a paradigm revisited. *Journal of Virology* 84, 6876–6879.

Gill, J.J. (2008) Modeling of bacteriophage therapy. In: Abedon, S.T. (ed.) *Bacteriophage Ecology*. Cambridge University Press, Cambridge, UK, pp. 439–464.

Gill, J.J. and Hyman, P. (2010) Phage choice, isolation and preparation for phage therapy. *Current Pharmaceutical Biotechnology* 11, 2–14.

Gill, J.J., Pacan, J.C., Carson, M.E., Leslie, K.E., Griffiths, M.W. and Sabour, P.M. (2006a) Efficacy and pharmacokinetics of bacteriophage therapy in treatment of subclinical *Staphylococcus aureus* mastitis in lactating dairy cattle. *Antimicrobial Agents and Chemotherapy* 50, 2912–2918.

Gill, J.J., Sabour, P.M., Leslie, K.E. and Griffiths, M.W. (2006b) Bovine whey proteins inhibit the interaction of *Staphylococcus aureus* and bacteriophage K. *Journal of Applied Microbiology* 101, 377–386.

Goodridge, L.D. (2008) Phages, bacteria, and food. In: Abedon, S.T. (ed.) *Bacteriophage Ecology*. Cambridge University Press, Cambridge, UK, pp. 302–331.

Goodridge, L.D. (2010) Designing phage therapeutics. *Current Pharmaceutical Biotechnology* 11, 15–27.

Hagens, S. and Loessner, M.J. (2010) Bacteriophage for biocontrol of foodborne pathogens: calculations and considerations. *Current Pharmaceutical Biotechnology* 11, 58–68.

Hyman, P. and Abedon, S.T. (2008) Phage ecology of bacterial pathogenesis. In: Abedon, S.T. (ed.) *Bacteriophage Ecology*. Cambridge University Press, Cambridge, UK, pp. 353–385.

Hyman, P. and Abedon, S.T. (2009) Practical methods for determining phage growth parameters. *Methods in Molecular Biology* 501, 175–202.

Hyman, P. and Abedon, S.T. (2010) Bacteriophage host range and bacterial resistance. *Advances in Applied Microbiology* 70, 217–248.

Kasman, L.M., Kasman, A., Westwater, C., Dolan, J., Schmidt, M.G., and Norris, J.S. (2002) Overcoming the phage replication threshold: a mathematical model with implications for phage therapy. *Journal of Virology* 76, 5557–5564.

Kutter, E., White, T., Kashlev, M., Uzan, M., McKinney, J., and Guttman, B. (1994) Effects on host genome structure and expression. In: Karam, J.D. (ed.) *Molecular Biology of Bacteriophage T4*. ASM Press, Washington, DC, pp. 357–368.

Kutter, E., de Vos, D., Gvasalia, G., Alavidze, Z., Gogokhia, L., Kuhl, S., and Abedon, S.T. (2010) Phage therapy in clinical practice: treatment of human infections. *Current Pharmaceutical Biotechnology* 11, 69–86.

Nasser, A.M., Glozman, R. and Nitzan, Y. (2002) Contribution of microbial activity to virus reduction in saturated soil. *Water Research* 36, 2589–2595.

Payne, R.J.H. and Jansen, V.A.A. (2001) Understanding bacteriophage therapy as a density-dependent kinetic process. *Journal of Theoretical Biology* 208, 37–48.

Payne, R.J.H. and Jansen, V.A.A. (2003) Pharmacokinetic principles of bacteriophage therapy. *Clinical Pharmacokinetics* 42, 315–325.

Payne, R.J.H., Phil, D. and Jansen, V.A.A. (2000) Phage therapy: the peculiar kinetics of self-replicating pharmaceuticals. *Clinical Pharmacology and Therapeutics* 68, 225–230.

Ramage, G., Culshaw, S., Jones, B. and Williams, C. (2010) Are we any closer to beating the biofilm: novel methods of biofilm control. *Current Opinion in Infectious Diseases* 23, 560–566.

Ryan, E.M., Gorman, S.P, Donnelly, R.F. and Gilmore, B.F. (2011) Recent advances in bacteriophage therapy: how delivery routes, formulation, concentration and timing influence the success of phage therapy. *Journal of Pharmacy and Pharmacology* 63, 1253–1264.

Smith, H.W. and Huggins, M.B. (1983) Effectiveness of phages in treating experimental *Escherichia coli* diarrhoea in calves, piglets and lambs. *Journal of General Microbiology* 129, 2659–2675.

Stent, G.S. (1963) *Molecular Biology of Bacterial Viruses*, W.H. Freeman and Co., San Francisco, CA.

Wilson, G.G., Young, K.K.Y., Edlin, G.J. and Konigsberg, W. (1979) High-frequency generalized transduction by bacteriophage T4. *Nature* 280, 80–82.

Index